Hormonal Carcinogenesis II

Springer
New York
Berlin
Heidelberg
Barcelona
Budapest
Hong Kong
London
Milan
Paris
Santa Clara
Singapore
Tokyo

Jonathan J. Li Sara Antonia Li
Jan-Åke Gustafsson Satyabrata Nandi
Lea I. Sekely
Editors

Hormonal Carcinogenesis II

Proceedings of the
Second International Symposium

With 118 Illustrations in 131 Parts

Springer

Jonathan J. Li, Ph.D.
Professor of Pharmacology, Toxicology, and Therapeutics
Director, Division of Etiology and Prevention of Hormonal Cancers
Hormonal Carcinogenesis Laboratory
University of Kansas Cancer Center
University of Kansas Medical Center
Kansas City, Kansas
USA

Sara Antonia Li, Ph.D.
Associate Professor of Pharmacology, Toxicology, and Therapeutics
Division of Etiology and Prevention of Hormonal Cancers
Hormonal Carcinogenesis Laboratory
University of Kansas Cancer Center
University of Kansas Medical Center
Kansas City, Kansas
USA

Jan-Åke Gustafsson, M.D., Ph.D.
Professor and Chairman, Department of Medical Nutrition
Director, Center for Biotechnology
Huddinge University Hospital
NOVUM
Huddinge
Sweden

Satyabrata Nandi, Ph.D.
Professor of Cell and Development Biology
Department of Molecular and Cell Biology
Cancer Research Center
University of California
Berkeley, California
USA

Lea I. Sekely, Ph.D.
Program Director, Division of Cancer Etiology, Chemical and Physical Carcinogenesis Program
National Cancer Institute
National Institutes of Health
Rockville, Maryland
USA

Library of Congress Cataloging-in-Publication Data applied for.

Printed on acid-free paper.

Production managed by Laura Carlson; manufacturing supervised by Jeffrey Taub.
Camera-ready copy prepared by the editors.

9 8 7 6 5 4 3 2 1

ISBN-13: 978-1-4612-7506-0 e-ISBN-13: 978-1-4612-2332-0
DOI: 10.1007/978-1-4612-2332-0

This volume is dedicated to two pioneers in estrogen action in developing the concept of estrogen receptor and hormone interaction at the cellular and molecular levels. Their findings have not only laid the foundation for elucidating mechanisms for all of steroid hormone action, but contributed profoundly to our understanding of the underlying mechanism(s) involved in estrogen carcinogenesis and hormone dependency of endocrine-associated malignancies.

Elwood V. Jensen

Institute Fur Hormon-und
Fortpflanzungsforshung
Hamburg, Germany

Jack Gorski

College of Agriculture & Life Sciences
University of Wisconsin-Madison
Madison, Wisconsin

Acknowledgments

Patrons

National Institute of Environmental Health Sciences
National Cancer Institute
National Institute of Child Health & Human Development
Office of Research on Women's Health
Karolinska Institute, NOVUM Research Park
Center for Nutrition and Toxicology
Medical Research Council of Sweden
Stockholm Cancer Society

Sponsors

Wyeth-Ayerst Research
R.W. Johnson Pharmaceutical Research Institute
Schering AG
International Life Sciences Institute, Risk Science Institute
Phytosyn, Inc.
Zenayako Kogyo Co., Ltd.

Contributors

Bayer
Berlex Laboratories
American Tobacco Institute
National Institute of Diabetes and Digestive and Kidney Diseases
The Upjohn Company
Nitta Gelatin, Inc.

Executive Board and N.I.H. Associates

Top Row, L to R: M. Metzler, R. Vihko, J.-Å. Gustafsson, G. Lucier, D. Longfellow; *Middle Row, L to R:* P. Siiteri, J.S. Norris, J. Yager, S. Nandi, S. Liao; *Bottom Row, L to R:* N. Terakawa, V.W. Pinn, J.J. Li, J. Daling, L.I. Sekely, S.A. Li (Missing R. Schulte-Hermann)

Participants of the Second International Hormonal Carcinogenesis Symposium

Preface

For those of us who have labored in the field of hormonal carcinogenesis, it has been most gratifying to see its rapid growth and increasing relevance in recent years. Although many factors and forces have contributed to this phenomenon, a few appear particularly significant. Perhaps foremost is the realization that two of the most prevalent cancers which afflict women and men; that is, breast and prostate, have essential hormonal component(s) to their etiologies. This should not surprise us since the high frequency of these cancers in human populations has to date not been attributed to any exogenous physical, environmental, or dietary factor(s). A similar argument may be applied to other less prevalent but equally important cancers including ovarian, endometrial, testicular, cervico-vaginal, pituitary, thyroid, and sex hormone-associated hepatic neoplasms. The Office of Research on Women's Health and many women's interest groups have been instrumental in fostering research and public awareness on women's cancers. Similar concern is beginning to emerge for solely male cancers by other groups.

To illustrate an example of the potential pervasive role of hormones in the human, particularly sex hormones, Figure 1 depicts the endogenous and known exogenous exposures to estrogens and progestins during a woman's lifetime, which may contribute to increased risk for hormonally-associated cancers.

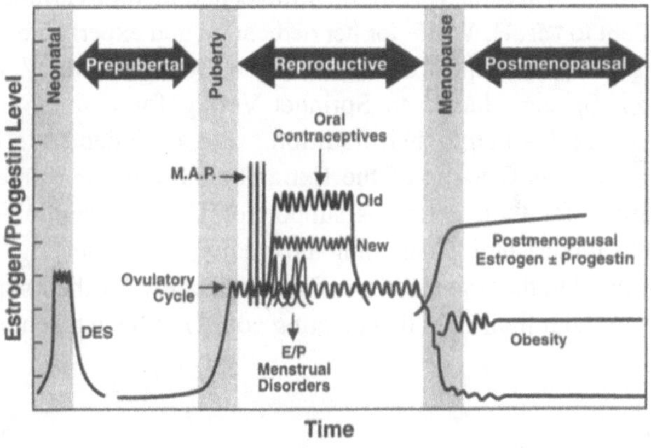

Figure 1. Exposures to estrogens and progestins during a woman's lifetime.

The neonatal exposure of the fetus to diethylstilbestrol (DES) in women who have ingested this estrogen to prevent miscarriages is well documented. Early menarche and late menopause increase the total number of ovulatory cycles and hence estrogen and progesterone exposure, thus increasing the risk for breast cancer. The frequent use of the "morning-after pill" (MAP) among many college-age women further increases exposure to estrogen, representing a relatively high dose at an early age. The ingestion of oral contraceptives, whether the early high (old) dose or the late low (new) dose of estrogen plus progestin elevates women's exposure to female sex hormones. The use of similar hormonal agents to correct menstrual disorders further increases the hormonal burden. Postmenopausal obese women are able to convert androgens to estrogens via the aromatase enzyme residing in adipose tissue. Finally, the widespread use of hormone replacement therapy by peri- and postmenopausal women, that is, estrogen in the presence or absence of progestin, may often span the remaining 25-30 years in a woman's lifetime.

In regard to acknowledgments, we would like to express our deep gratitude to members of the Executive and Scientific Advisory Boards for their time and contributions in the development of an outstanding scientific program. The editors are grateful to Drs. J. Daling, S. Liao, G.A. Lucier, D. Medina, J.S. Norris, J. Yager, and A. Parkinson (KUMC) for their assistance in critically reviewing the manuscripts presented herein. We are especially indebted to Dr. J.H. Pickar for arranging and supporting the Clinical Forum, a new addition to these Symposia. We thank the staff of the Department of Medical Nutrition, Novum, the Local Organizing Committee headed by Drs. I. Porsch-Hällstrom and A. Blanck, as well as Dr. D. Liao for special assistance. We also thank International Tours, Mission, KS, for their travel and facilities arrangements. We owe a great debt to Ms. H. Wolfe for her dedication and expert care in formatting and preparing the manuscripts for publication, and we thank Ms. V. Hahn for her editorial help. Special thanks to Springer-Verlag for maintaining the high publishing standard which is their tradition. We are indebted to Dr. Kumiko Iwamoto and Dr. Iris Obrams of the Extramural Programs Branch, National Cancer Institute, for their generous support of U. S. epidemiologists to our Symposia, thus immeasurably enriching the quality of this meeting. Finally, we are deeply grateful to the organizations listed separately for their support of this Symposium; without their help this meeting could not have taken place.

Kansas City
1995

Jonathan J. Li
Sara Antonia Li

Contents

Participants

ERIKO AIZU-YOKOTA Kyoritsu College of Pharmacy, Minato-ku, Tokyo, Japan

BORGE S. ANDERSSON MD Anderson Cancer Center, Houston, TX, USA

YOHANNES ASSEFAW-REDDA Department of Medical Nutrition, Huddinge University Hospital, Karolinska Institute, Huddinge, Sweden

ETIENNE-EMILE BAULIEU Institut National de La Santé et de la Recherche Médicale, Collège de France, Le Kremlin-Bicêtre, France

RALF BENNEMERSCHULT Schering AG, Berlin, Germany

VALERIE BERAL ICRF Cancer Epidemiology Unit, The Radcliffe Infirmary, Oxford, United Kingdom

LEIF BERGKVISt Department of Surgery, Central Hospital, Västerås, Sweden

AGNETA BLANCK Department of Medical Nutrition, Novum, Huddinge University Hospital, Huddinge, Sweden

ROBERT W. BRUEGGEMEIER College of Pharmacy, The Ohio State University, Columbus, OH, USA

ANTONY M. CARR MRC Cell Mutation Unit, Sussex University, Falmer, United Kingdom

R. JEFFREY CHANG UC Davis School of Medicine, Sacramento, CA, USA

DE HUI CHEN Institute of Basic Medical Sciences, Beijing, P.R. China

W. R. CHRISTENSON Miles, Inc., Stillwell, KS, USA

LELAND CHUNG University of Texas, M.D. Anderson Cancer Center, Houston, TX, USA

JANET R. DALING Division of Public Health Sciences, Fred Hutchinson Cancer Research Center, Seattle, WA, USA

JAMES O. DRIFE Academic Unit of Obstetrics and Gynecology, Leeds, United Kingdom

JOHN A. EDEN Royal Hospital for Women, Frank Rundle House, Paddington, Australia

J. CHARLES ELDRIDGE Department of Physiology-Pharmacology, Bowman Gray School of Medicine of Wake Forest University, Winston-Salem, NC, USA

TOMMY FERNANDER Department of General Oncology, Stockholm Suder Hospital, Stockholm, Sweden

MICHAEL A. GALLO The Cancer Institute of New Jersey, Piscataway, NJ, USA

FERNANDO U. GARCIA Department of Pathology, Hahnemann University, Philadelphia, PA, USA

KAJ GRANDIEN Department of Medical Nutrition, Huddinge University Hospital, Karolinska Institute, Huddinge, Sweden

DAVID A. GRIMES Department of Obstretrics, Cynecology, and Reproductive Sciences, San Francisco General Hospital, University of California, San Francisco, CA, USA

JAN-ÅKE GUSTAFSSON Department of Medical Nutrition, Huddinge University Hospital, NOVUM, Huddinge, Sweden

BARRY A. GUSTERSON Haddow Laboratories, Sutton, Surrey, United Kingdom

JOHAN HÄGGBLAD KaroBio Ab, Huddinge, Sweden

ROBERT W. HAILE Department of Epidemiology, School of Public Health, University of California - Los Angeles, Los Angeles, CA, USA

INGER PORSCH HÄLLSTRÖM Department of Medical Nutrition, Novum, Huddinge University Hospital, Huddinge, Sweden

LYNN C. HARTMANN Division of Medicial Oncology, Mayo Clinic, Rochester, MN, USA

WILLIAM HELFERICH Food Science and Human Nutrition, Michigan State University, East Lansing, MI, USA

SHUK-MEI HO Department of Biology, Tufts University, Medford, MA, USA

SUE INGLES Department of Epidemiology, School of Public Health, University of California - Los Angeles, Los Angeles, CA, USA

HELENA C.B. JERNSTRÖM Department of Oncology, University Hospital of Lund, Lund, Sweden

ALEXANDER C. KAMB Myriad Genetics, Inc., Salt Lake City, UT, USA

CORNELIUS KNABBE Department of Clinical Chemistry, University of Krankenhaus Eppendorf, Hamburg, Germany

ROLF KRATTENMACHER Schering AG, Fertility Control/Hormone Therapy Research, Berlin, Germany

JONATHAN J. LI Division of Etiology and Prevention of Hormonal Cancers, University of Kansas Cancer Center, KUMC, Kansas City, KS, USA

SARA ANTONIA LI Division of Etiology and Prevention of Hormonal Cancers, University of Kansas Cancer Center, KUMC, Kansas City, KS, USA

CHUN-HAI LI Institute of Basic Medical Science, Beijing, P.R. China

JOSHUA DEZHONG LIAO Department of Medical Nutrition, Novum, Huddinge University Hospital, Huddinge, Sweden

SHUTSUNG LIAO The Ben May Institute, Chicago, IL, USA

ANTONIO LLOMBART-BOSCH Department of Pathology, Facultad de Medicina, Valencia, Spain

DAVID LONGFELLOW National Institutes of Health, National Cancer Institute, Bethesda, MD, USA

KATHY E. MALONE Division of Public Health Sciences, Fred Hutchinson Cancer Research Center, Seattle, WA, USA

EERO T.E. MÄNTYLÄ Orion-Farmos Pharmaceuticals, Farmos Research, Department of Pharmacology and Toxicology, Turku, Finland

CHARLES P. MARTUCCI Strang-Cornell Cancer Research Laboratory, New York, NY, USA

JOHN A. MCLACHLAN Tulane/Xavier Center for Bioenvironmental Research, Tulane University, New Orleans, LA,

DANIEL MEDINA Department of Cell Biology, Baylor College of Medicine, Houston, TX, USA

MANFRED METZLER Laboratory of Environmental Toxicology, Department of Chemistry, University of Kaiserslautern, Germany

ALFREDO A. MOLINOLO Department of Experimental Medicine, Buenos Aires, Argentina

SURESH MOHLA National Institutes of Health, National Cancer Institute, Rockville, MD, USA

PETER MUHN Schering AG, Fertility Control/Hormon Thereapy Research, Berlin, Germany

SATYABRATA NANDI Cancer Research Laboratory, Department of Molecular and Cell Biology, University of California, Berkely, CA, USA

JAMES S. NORRIS Department of Medicine, Medical University of South Carolina, Charleston, SC, USA

LENA OHLSON Department of Medical Nutrition, Novum, Huddinge University Hospital, Huddinge, Sweden

LARS H. OLSSON Department of Oncology, University Hospital of Lund, Lund, Sweden

MONIQUE OP TEN BERG NV Organon, The Netherlands

JULIE R. PALMER Slone Epidemiology Unit, Boston University, Brookline, MA, USA

KARSTEN PAREZYLL Schering AG, Experimental Oncology, Berlin, Germany

MALCOLM G. PARKER Molecular Endocrinology Laboratory, Imperial Cancer Research Fund, London, United Kingdom

VITO PATELLA Institute of Medical Oncology, Bologna, Italy

JEFFREY A. PERLMAN National Cancer Institute, Bethesda, MD, USA

INGEMAR PERSSON University Hospital, Uppsala, Sweden

JAMES H. PICKAR Wyeth-Ayerst Research, Philadelphia, PA, USA

VIVIAN W. PINN Associate Director for Research on Women's Health, Department of Health and Human Services, National Instituës of Health, Bethesda, MD, USA

GENEVIERE PLU-BUREAU INSERM U351 Gustave Roussy Institute, Villejuif, France, and Department of Endocrinology, Hospital Necker, Paris, France

DEODUTTA ROY Department of Environmental Health Sciences, University of Alabama, Birmingham, AL, USA

JOSE RUSSO Department of Pathology, Fox Chase Cancer Center, Philadelphia, PA, USA

GÖRAN SAMSIOE Departments of Obstetrics and Gynecology, Lund University Hospital, Lund, Sweden

ROLF SCHULTE-HERMANN Institute of Tumore Biologie, Krebforschung, University of Wein, Vienna, Austria

BEATE SEIBERT Schering AG, Institute of Experimental Toxicology, Berlin, Germany

LEA I. SEKELY National Cancer Institute, Bethesda, MD, USA

CLAUS SMED-SORENSEN Department of Gynecology and Obstetrics, Orebro Medical Centre, Orebro, Sweden

LEWIS L. SMITH Toxicology Unit, University of Leicester, Leicester, United Kingdom

FRANK Z. STANCZYK Departments of Obstetrics and Gynecology, Urology and Preventive Medicine, University of Southern California School of Medicine, Los Angeles, CA, USA

NAOKI TERAKAWA Department of Obstetrics and Gynecology, Totori University School of Medicine, Nishimachi, Yonago, Japan

DONALD J. TINDALL Mayo Clinic, Rochester, MN, USA

PHILIPPE TOURAINE Department of Endocrinology, Hopital Necker, Paris, France

LAUFEY TRYGGVADOTTIR Icelandic Cancer Society, Reykjavik, Iceland

KARIN VAN DER KOOY Netherlands Cancer Institute, Amsterdam, Netherlands

REIJO K. VIHKO Department of Clinical Chemistry, University of Oulu, Oulu, Finland

CORNELIUS KNABBE Universität-Krankenhaus Eppendorf, Medical University Clinic, Department of Clinical Chemistry, Hamburg, Germany

FREDERIC M. WALDMAN Department of Laboratory Medicine, University of California - San Francisco, San Francisco, CA, USA

ANNI M. WÄRRI Orion Corporation, Cancer Research Laboratory, BioCity, Turku, Finland

COLIN K. W. WATTS Cancer Biology Division, Garvan Institute of Medical Research, St. Vincent's Hospital, Sydney, Australia

NOEL S. WEISS Division of Public Health Sciences, Fred Hutchinson Cancer Research Center, Seattle, WA, USA

PHYLLIS A. WINGO American Cancer Socity, Atlanta, GA, USA

JAMES YAGER Department of Environmental Health, Division of Toxicological Sciences, Johns Hopkins University, Baltimore, MD, USA

JOANNE ZURLO Department of Environmental Health, Division of Toxicological Sciences, Johns Hopkins University, Baltimore, MD, USA

Contributors

ERIKO AIZU-YOKOTA Kyoritsu College of Pharmacy, Minato-ku, Tokyo, Japan

RAMASWAMY ANBAZHAGAN Haddow Laboratories, Sutton, Surrey, United Kingdom

MARK S. AUSTENFELD Department of Surgery, University of Kansas Medical Center, Kansas City, KS, USA

HARLAND AUSTIN Emory University, Atlanta, GA, USA

CAROL D. BANNER Department of Medical Nutrition, Karolinska Institute, Huddinge Hospital, Huddinge, Sweden

NANDITA BARNABAS Department of Pathology, Fox Chase Cancer Center, Philadelphia, PA, USA

KEVIN L. BASHMAN Departments of Pathology and Urology, University of Kansas Medical Center, Kansas City, KS, USA

ETIENNE-EMILE BAULIEU Institut National de La Santé, Et De La Recherche Médicale, Collége De France, Le Kremlin-Bicêtre, France

VALERIE BERAL ICRF Cancer Epidemiology Unit, The Radcliffe Infirmary, Oxford, United Kingdom

SHIRLEY A.A. BERESFORD Division of Public Health Sciences, Fred Hutchinson Cancer Research Center, Seattle, WA, USA

LEIF BERGKVISt Department of Surgery, Central Hospital, Västerås, Sweden

AGNETA BLANCK Department of Medical Nutrition, Novum, Huddinge University Hospital, Huddinge, Sweden

LEEN J. BLOK Mayo Clinic, Rochester, MN, USA

NANCY BOWER Environmental and Occupational Health Sciences Institute, Piscataway, NJ, USA

H. LEON BRADLOW Strang-Cornell Cancer Research Laboratory, New York, NY, USA

ROBERT W. BRUEGGEMEIER College of Pharmacy, The Ohio State University, Columbus, OH, USA

DOMINIQUE BUREAU Ministere des Finances, Paris, France

W. BURSCH Institute of Tumore Biologie, Krebforschung, University of Wein, Vienna, Austria

GLORIA CALAF Department of Pathology, Fox Chase Cancer Center, Philadelphia, PA, USA

BO CARLSSON KaroBio Ab, Huddinge, Sweden

ANTONY M. CARR MRC Cell Mutation Unit, Sussex University, Falmer, United Kingdom

I. CARTIER Department of Endocrinology, Hopital Necker, Paris, France

WILLIAM CHANG College of Veterinary Medicine, The Ohio State University, Columbus, OH, USA

CHING-JEY G. CHANG College of Veterinary Medicine, The Ohio State University, Columbus, OH, USA

EDUARDO H. CHARREAU Department of Experimental Medicine, Buenos Aires, Argentina

DE HUI CHEN Institute of Basic Medical Sciences, Beijing, P.R. China

CHIAO-WEN CHEN Department of Environmental Health Sciences, University of Alabama, Birmingham, AL, USA

LELAND W.K. CHUNG University of Texas, M.D. Anderson Cancer Center, Houston, TX, USA

VIRGINIA CORTES Department of Pathology, Facultad de Medicina, Valencia, Spain

VICTORIA K. CORTESSIS Department of Epidemiology, School of Public Health, University of California - Los Angeles, Los Angeles, CA, USA

ANTONIO CREMADES Department of Pathology, Facultad de Medicina, Valencia, Spain

JANET R. DALING Division of Public Health Sciences, Fred Hutchinson Cancer Research Center, Seattle, WA, USA

SOPHIE DAUVOIS Molecular Endocrinology Laboratory, Imperial Cancer Research Fund, London, United Kingdom

ANH DIEP Department of Epidemiology, School of Public Health, University of California - Los Angeles, Los Angeles, CA, USA

GRACIELA DRAN Department of Experimental Medicine, Buenos Aires, Argentina

JAMES O. DRIFE Academic Unit of Obstetrics and Gynecology, Leeds, United Kingdom

PIERRE A. DRIGUEZ Department of Endocrinology, Hopital Necker, Paris, France

JOHN A. EDEN Royal Hospital for Women, Frank Rundle House, Paddington, Australia

J. CHARLES ELDRIDGE Department of Physiology-Pharmacology, Bowman Gray School of Medicine of Wake Forest University, Winston-Salem, NC, USA

WILLIAM ELEY Emory University, Atlanta, GA, USA

PATRICIA ELIZALDE Department of Experimental Medicine, Buenos Aires, Argentina

KATARINA ENGLUND Departments of Obstetrics and Gynecology, Huddinge University Hospital, Huddinge, Sweden

WILLIAM B. FARRAR College of Medicine, The Ohio State University, Columbus, OH, USA

PER FLODBY Center for Biotechnology, Novum, Huddinge University Hospital, Huddinge, Sweden

MICHAEL A. GALLO The Cancer Institute of New Jersey, Piscataway, NJ, USA

CHUAN GAO University of Texas, M.D. Anderson Cancer Center, Houston, TX, USA

FERNANDO U. GARCIA Department of Pathology, Hahnemann University, Philadelphia, PA, USA

NANCY E. GILBERT College of Veterinary Medicine, The Ohio State University, Columbus, OH, USA

B. GRASL-KRAUPP Institute of Tumore Biologie, Krebforschung, University of Wein, Vienna, Austria

MARIA AURELIA GREGORI Department of Pathology, Facultad de Medicina, Valencia, Spain

DAVID A. GRIMES San Francisco General Hospital, University of California, San Francisco, CA, USA

FABIANA GUERRA Department of Experimental Medicine, Buenos Aires, Argentina

SHAN-CHUN GUO Institute of Basic Medical Science, Beijing, P.R. China

JAN-ÅKE GUSTAFSSON Department of Medical Nutrition, Karolinska Institute, Huddinge Hospital, Huddinge, Sweden

INGER GUSTAFSSON Departments of Obstetrics and Gynecology, Huddinge University Hospital, Huddinge, Sweden

BARRY A. GUSTERSON Haddow Laboratories, Sutton, Surrey, United Kingdom

RAPHAEL C. GUZMAN Cancer Research Laboratory, Department opf Molecular and Cell Biology, University of California, Berkely, CA, USA

JOHAN HÄGGBLAD KaroBio Ab, Huddinge, Sweden

ROBERT W. HAILE Department of Epidemiology, School of Public Health, University of California - Los Angeles, Los Angeles, CA, USA

PIRKKO L. HÄRKÖNEN University of Turku, Department of Anatomy and MediCity Research Laboratory, Turku, Finland

STUART HARTZ Medical Research International, Boston, MA, USA

NISSIM HAY The Ben May Institute, Chicago, IL, USA

CAROL HERMON ICRF Cancer Epidemiology Unit, The Radcliffe Infirmary, Oxford, United Kingdom

RICHARD A. HIIPAKKA The Ben May Institute, Chicago, IL, USA

SHUK-MEI HO Department of Biology, Tufts University, Medford, MA, USA

MICHAEL B. HOLLAND Department of Environmental Health Sciences, University of Alabama, Birmingham, AL, USA

YUNFU HU College of Veterinary Medicine, The Ohio State University, Columbus, OH, USA

KATRIN HUBER University of Kaiserslautern, Germany

SUE INGLES Department of Epidemiology, School of Public Health, University of California - Los Angeles, Los Angeles, CA, USA

JORMA J. ISOLA Department of Laboratory Medicine and Pathology, University of California - San Francisco, School of Medicine, San Francisco, CA, USA

VELI V. ISOMAA Department of Clinical Chemistry, University of Oulu, Oulu, Finland

HELENA C.B. JERNSTRÖM Department of Oncology, University Hospital of Lund, Lund, Sweden

OLLI-PEKKA KALLIONIEMI Department of Laboratory Medicine and Pathology, University of California - San Francisco, School of Medicine, San Francisco, CA, USA

ANNE KALLIONIEMI Department of Laboratory Medicine and Pathology, University of California - San Francisco, School of Medicine, San Francisco, CA, USA

ALEXANDER C. KAMB Myriad Genetics, Inc., Salt Lake City, UT, USA

STEFAN H. KARLSSON Orion-Farmos Pharmaceuticals, Farmos Research, Department of Pharmacology and Toxicology, Turku, Finland

TIMOTHY KEY ICRF Cancer Epidemiology Unit, The Radcliffe Infirmary, Oxford, United Kingdom

JOHN M. KOKONTIS The Ben May Institute, Chicago, IL, USA

EDITH C. KORDON Department of Experimental Medicine, Buenos Aires, Argentina

ANDRZEJ KOSINSKI Emory University, Atlanta, GA, USA

SABINE KULLING Laboratory of Environmental Toxicology, Department of Chemistry, University of Kaiserslautern, Germany

SAMUEL KULP College of Veterinary Medicine, The Ohio State University, Columbus, OH, USA

M. VIJAY KUMAR Mayo Clinic, Rochester, MN, USA

F. KUTTENN Department of Endocrinology, Hopital Necker, Paris, France

CLAUDIA LANARI Department of Experimental Medicine, Buenos Aires, Argentina

MONIQUE G. LÊ INSERM U351 Gustave Roussy Institute, Villejuif, France

JONATHAN J. LI Director of the Division of Etiology and Prevention of Hormonal Cancers, University of Kansas Cancer Center, KUMC, Kansas City, KS, USA

SARA ANTONIA LI Division of Etiology and Prevention of Hormonal Cancers, University of Kansas Cancer Center, KUMC, Kansas City, KS, USA

CHUN-HAI LI Institute of Basic Medical Science, Beijing, P.R. China

DEZHONG LIAO Department of Medical Nutrition, Novum, Huddinge University Hospital, Huddinge, Sweden

SHUTSUNG LIAO The Ben May Institute, Chicago, IL, USA

YOUNG C. LIN College of Veterinary Medicine, The Ohio State University, Columbus, OH, USA

BO LINDBLOM Uppsala University, Uppsala, Sweden

ANTONIO LLOMBART-BOSCH Department of Pathology, Facultad de Medicina, Valencia, Spain

ROGERIO A. LOBO Departments of Obstetrics and Gynecology, Urology and Preventive Medicine, University of Southern California School of Medicine, Los Angeles, CA, USA

CARL J. LOVELY College of Veterinary Medicine, The Ohio State University, Columbus, OH, USA

ISABEL LUTHY Department of Experimental Medicine, Buenos Aires, Argentina

KATHLEEN E. MALONE Division of Public Health Sciences, Fred Hutchinson Cancer Research Center, Seattle, WA, USA

EERO T.E. MÄNTYLÄ Orion-Farmos Pharmaceuticals, Farmos Research, Department of Pharmacology and Toxicology, Turku, Finland

CHARLES P. MARTUCCI Strang-Cornell Cancer Research Laboratory, New York, NY, USA

PIERRE MAUVAIS-JARVIS Department of Endocrinology, Hopital Necker, Paris, France

DANIEL MEDINA Department of Cell Biology, Baylor College of Medicine, Houston, TX, USA

SUSAN MERTES Departments of Obstetrics and Gynecology, Urology and Preventive Medicine, University of Southern California School of Medicine, Los Angeles, CA, USA

MANFRED METZLER Laboratory of Environmental Toxicology, Department of Chemistry, University of Kaiserslautern, Germany

DANIEL G. MILLER Strang-Cornell Cancer Research Laboratory, New York, NY, USA

ROBERT C. MILLIKAN Department of Epidemiology, School of Public Health, University of California - Los Angeles, Los Angeles, CA, USA

ALFREDO A. MOLINOLO Department of Experimental Medicine, Buenos Aires, Argentina

FERNANDA MONTECCHIA Department of Experimental Medicine, Buenos Aires, Argentina

L. MÜLLAUER Institute of Tumore Biologie, Krebforschung, University of Wein, Vienna, Austria

E. A. MUSGROVE Cancer Biology Division, Garvan Institute of Medical Research, St. Vincent's Hospital, Sydney, Australia

LOUIS MYLECRAINE Environmental and Occupational Health Sciences Institute, Piscataway, NJ, USA

SATYABRATA NANDI Cancer Research Laboratory, Department of Molecular and Cell Biology, University of California, Berkely, CA, USA

LAURI S. NIEMINEN Orion-Farmos Pharmaceuticals, Farmos Research, Department of Pharmacology and Toxicology, Turku, Finland

JAMES S. NORRIS Department of Medicine, Medical University of South Carolina, Charleston, SC, USA

H. OCHS Institute of Tumore Biologie, Krebforschung, University of Wein, Vienna, Austria

LENA C.E. OHLSON Department of Medical Nutrition, Novum, Huddinge University Hospital, Huddinge, Sweden

LARS HÅKAN OLSSON Department of Oncology, University Hospital of Lund, Lund, Sweden

HOWARD ORY Emory University, Atlanta, GA, USA

JULIE R. PALMER Slone Epidemiology Unit, Boston University, Brookline, MA, USA

MALCOLM G. PARKER Molecular Endocrinology Laboratory, Imperial Cancer Research Fund, London, United Kingdom

W. PARZEFALL Institute of Tumore Biologie, Krebforschung, University of Wein, Vienna, Austria

CHRISTIANE DOSNE PASQUALINI Department of Experimental Medicine, Buenos Aires, Argentina

PATRICIA PAZOS Department of Experimental Medicine, Buenos Aires, Argentina

E. HELLEVI PELTOKETO Department of Clinical Chemistry, University of Oulu, Oulu, Finland

JEFFREY A. PERLMAN National Cancer Institute, Bethesda, MD, USA

INGEMAR PERSSON University Hospital, Uppsala, Sweden

HANS L. PETERSE Netherlands Cancer Institute, 1066 CX Amsterdam, Netherlands

BERT PETERSON Centers for Disease Control and Prevention, Atlanta, GA, USA

AMANDO PEYDRO-OLAYA Department of Pathology, Facultad de Medicina, Valencia, Spain

ERIKA PFEIFFER Cancer Biology Division, Garvan Institute of Medical Research, St. Vincent's Hospital, Sydney, Australia

VIVIAN W. PINN Associate Director for Research on Women's Health, Department of Health and Human Services, National Institues of Health, Bethesda, MD, USA

GENEVIÈVE PLU-BUREAU INSERM U351 Gustave Roussy Institute, Villejuif, France, and Department of Endocrinology, Hospital Necker, Paris, France

INGER PORSCH-HÄLLSTRÖM Department of Medical Nutrition, Novum, Huddinge University Hospital, Huddinge, Sweden

JANE PORTER Medical Research International, Boston, MA, USA

MATTI H. POUTANEN Department of Clinical Chemistry, University of Oulu, Oulu, Finland

JOSEPH J. RAFTER Department of Medical Nutrition, Karolinska Institute, Huddinge Hospital, Huddinge, Sweden

JEAN-PIERRE RAYNAUD ARIBIO, Paris, France

GILLIAN REEVES ICRF Cancer Epidemiology Unit, The Radcliffe Infirmary, Oxford, United Kingdom

KENNETH R. REUHL Environmental and Occupational Health Sciences Institute, Piscataway, NJ, USA

MATTI A. ROOKUS Netherlands Cancer Institute, 1066 CX Amsterdam, Netherlands

BRIGITTE ROSENBERG Laboratory of Environmental Toxicology, Department of Chemistry, University of Kaiserslautern, Germany

LYNN ROSENBERG Slone Epidemiology Unit, Boston University, Brookline, MA, USA

RONALD R. ROSS Departments of Obstetrics and Gynecology, Urology and Preventive Medicine, University of Southern California School of Medicine, Los Angeles, CA, USA

DEODUTTA T.E. ROY Department of Environmental Health Sciences, University of Alabama, Birmingham, AL, USA

JOSE RUSSO Department of Pathology, Fox Chase Cancer Center, Philadelphia, PA, USA

IRMA H. RUSSO Department of Pathology, Fox Chase Cancer Center, Philadelphia, PA, USA

B. RUTTKAY-NEDECKY Institute of Tumore Biologie, Krebforschung, University of Wein, Vienna, Austria

THENAA K. SAID Department of Cell Biology, Baylor College of Medicine, Houston, TX, USA

GÖRAN SAMSIOE Departments of Obstetrics and Gynecology, Lund University Hospital, Lund, Sweden

YOSHIHIRO SATO Kyoritsu College of Pharmacy, Minato-ku, Tokyo, Japan

GUIDO SAUTER Department of Laboratory Medicine and Pathology, University of California - San Francisco, School of Medicine, San Francisco, CA, USA

JAMES J. SCHLESSELMAN University of Pittsburgh, Pittsburgh, PA, USA

MAIK SCHULER University of Kaiserslautern, Germany

ROLF SCHULTE-HERMANN Institute of Tumore Biologie, Krebforschung, University of Wein, Vienna, Austria

LEA I. SEKELY National Cancer Institute, National Institutes of Health, Bethesda, MD, USA

DANIEL W. SEPKOVIC Strang-Cornell Cancer Research Laboratory, New York, NY, USA

SAMUEL SHAPIRO Slone Epidemiology Unit, Boston University, Brookline, MA, USA

RÉGINE SITRUK-WARE Department of Endocrinology, Hospital Necker, Paris, France

PETER SJOBLOM Departments of Obstetrics and Gynecology, Huddinge University Hospital, Huddinge, Sweden

EILA C. SKINNER Departments of Obstetrics and Gynecology, Urology and Preventive Medicine, University of Southern California School of Medicine, Los Angeles, CA, USA

MARYANN F. SPAHN Departments of Obstetrics and Gynecology, Urology and Preventive Medicine, University of Southern California School of Medicine, Los Angeles, CA, USA

ROBERT S. SPARKES Department of Epidemiology, School of Public Health, University of California - Los Angeles, Los Angeles, CA, USA

LEWIS L. SMITH Toxicology Unit, University of Leicester, Leicester, United Kingdom

FRANK Z. STANCZYK Departments of Obstetrics and Gynecology, Urology and Preventive Medicine, University of Southern California School of Medicine, Los Angeles, CA, USA

JANET L. STANFORD Division of Public Health Sciences, Fred Hutchinson Cancer Research Center, Seattle, WA, USA

ADAM P. STEIN Environmental and Occupational Health Sciences Institute, Piscataway, NJ, USA

JAMES T. STEVENS Department of Toxicology, CIBA Plant Protection, Greensboro, NC, USA

MICHAEL STRATTON Haddow Laboratories, Sutton, Surrey, United Kingdom

YASURO SUGIMOTO College of Veterinary Medicine, The Ohio State University, Columbus, OH, USA

R. L. SUTHERLAND Cancer Biology Division, Garvan Institute of Medical Research, St. Vincent's Hospital, Sydney, Australia

NAOKI TERAKAWA Department of Obstetrics and Gynecology, Totori University School of Medicine, Nishimachi, Yonago, Japan

JEAN-CHRISTOPHE THALABARD Universite Claude Bernard, Faculte Lyon Sud, Lyon, France

DONALD J. TINDALL Mayo Clinic, Rochester, MN, USA

G. STEPHEN TINT Department of Veterans' Affairs Medical Center, East Orange, NJ, USA

MERRILL O. TISDEL Department of Toxicology, CIBA Plant Protection, Greensboro, NC, USA

PHILIPPE TOURAINE Department of Endocrinology, Hopital Necker, Paris, France

KARIN VAN DER KOOY Netherlands Cancer Institute, 1066 CX Amsterdam, Netherlands

FLORA E. VAN LEEUWEN Netherlands Cancer Institute, 1066 CX Amsterdam, Netherlands

REIJO K. VIHKO Department of Clinical Chemistry, University of Oulu, Oulu, Finland

LYNDA F. VOIGT Division of Public Health Sciences, Fred Hutchinson Cancer Research Center, Seattle, WA, USA

FREDERIC M. WALDMAN Department of Laboratory Medicine and Pathology, University of California - San Francisco, School of Medicine, San Francisco, CA, USA

A. WARLTERS Cancer Biology Division, Garvan Institute of Medical Research, St. Vincent's Hospital, Sydney, Australia

ANNI M. WÄRRI Orion Corporation, Cancer Research Laboratory, BioCity, Turku, Finland

COLIN K.W. WATTS Cancer Biology Division, Garvan Institute of Medical Research, St. Vincent's Hospital, Sydney, Australia

NOEL S. WEISS Division of Public Health Sciences, Fred Hutchinson Cancer Research Center, Seattle, WA, USA

LAWRENCE T. WETZEL Department of Physiology-Pharmacology, Bowman Gray School of Medicine of Wake Forest University, Winston-Salem, NC, USA

ANNE WHITE Department of Cell Biology, Baylor College of Medicine, Houston, TX, USA

EMILY WHITE Division of Public Health Sciences, Fred Hutchinson Cancer Research Center, Seattle, WA, USA

IAN N.H. WHITE Toxicology Unit, University of Leicester, Leicester, United Kingdom

N. WILCKEN Cancer Biology Division, Garvan Institute of Medical Research, St. Vincent's Hospital, Sydney, Australia

PHYLLIS A. WINGO American Cancer Society, Atlanta, GA USA

BARRY G. WREN Royal Hospital for Women, Frank Rundle House, Paddington, Australia

KLEANTHIS G. XANTHOPOULOS Department of Medical Nutrition, Novum, Huddinge University Hospital, Huddinge, Sweden

YUAN-JI XU Institute of Basic Medical Science, Beijing, P.R. China

JAMES D. YAGER Department of Environmental Health, Division of Toxicological Sciences, Johns Hopkins University, Baltimore, MD, USA

ZHI-JIE YAN Department of Environmental Health Sciences, University of Alabama, Birmingham, AL, USA

H. YANEVA Department of Endocrinology, Hopital Necker, Paris, France

JASON YANG Cancer Research Laboratory, Department of Molecular and Cell Biology, University of California, Berkely, CA, USA

HEINRICH ZANKL University of Kaiserslautern, Germany

PEI-LI ZHANG Department of Pathology, Fox Chase Cancer Center, Philadelphia, PA, USA

XUE-MIN ZHANG Institute of Basic Medical Sciences, Beijing, P.R. China

YUE-TANG ZHAO Institute of Basic Medical Science, Beijing, P.R. China

HAIYEN E. ZHAU University of Texas M.D. Anderson Cancer Center, Houston, TX, USA

JIANG ZHOU Institute of Basic Medical Science, Beijing, P.R. China

JOANNE ZURLO Department of Environmental Health, Division of Toxicological Sciences, Johns Hopkins University, Baltimore, MD, USA

Introductory Remarks

Lea I. Sekely

It is a privilege and a delight to address the participants of the Second International Symposium on Hormonal Carcinogenesis in this beautiful setting. I am a Program Director in the Chemical and Physical Carcinogenesis Branch, which is part of the Division of Cancer Etiology, National Cancer Institute of the United States National Institutes of Health, the NIH. As most of you are aware, the NIH supports intramural, multidisciplinary research programs; but its much more extensive commitment in terms of funding is its extramural support of research at universities and other institutions all over the world through peer-reviewed grants. It also provides support to a number of national and international meetings deemed of superior caliber as well as contemporary significance. The current conference was peer-reviewed and earned the support of four institutes in addition to that provided by the Karolinska Institute and private sources.

Chemically and physically induced carcinogenesis has been a major focus in cancer etiology for many years in our institute, but it has been less than a decade that hormonal carcinogenesis has been appreciated as a major contributor to some types of common cancers, including breast, prostate and endometrium. In recognition of this heightened awareness, the First International Symposium on Hormonal Carcinogenesis was held in Mexico in 1991. It elicited wide interest and was very successful in bringing together leading researchers in the field who recognized the need for continued emphasis on and support for these programs.

It is with immense pleasure that I address this conference at Novum Research Park, Stockholm, the beautiful southern campus of the Karolinska Institute. I should like to take this occasion to express the gratitude of all the participants to Professor Jan-Åke Gustafsson, Director of the Department of Medical Nutrition and Toxicology, Karolinska Institute, his staff and Professor Jonathan J. Li, Director of the Division of Etiology and Prevention of Hormonal Cancers, University of Kansas Cancer Center for providing the site and making all arrangements; and to the city of Stockholm for its extraordinary hospitality in providing banquette at the City (Nobel) Hall, a memorable event. Our thanks too to Her Majesty, Queen of Sweden, who, though unable to attend, sent personal greetings. I don't think any of us have been privileged to meet in a more enchanting setting than Stockholm, with its myriad waterways and bridges, breathtaking gardens and illustrious history. It is truly a fairy-tale city.

Opening Address

Important Questions in Women's Health Research: Determining the Influence of Hormones on Health

Vivian W. Pinn

In this Second International Symposium on Hormonal Carcinogenesis, we shall all benefit from the sharing of stimulating concepts and investigative results in this vital and expanding field of inquiry.

The Constitution of the World Health Organization states that, "The enjoyment of the highest attainable standard of health is one of the fundamental rights of every human being without distinction of race, religion, political belief, economic or social condition" (1).

At the National Institutes of Health (NIH), through the Office of Research on Women's Health (ORWH), we are working to expand medical knowledge through scientific inquiry to ensure that every human being enjoys the highest attainable standard of health without regard to race, religion, political belief, economic or social condition, or gender.

Through remarkable advances in biomedical, epidemiological, and behavioral research, the average life expectancy for both men and women in developed countries has increased by some 30 years since 1900. The increase in life expectancy has been particularly marked for women, whose lives, throughout history, have been shortened by childbirth and complications stemming from maternity. And, when women survived childhood, adolescence and childbirth, they faced other causes of premature death, including, importantly, cancer.

Until recently, health services around the world have viewed women's health as synonymous with their reproductive health. Scant attention has been given to women's health issues beyond their reproductive capacity. This has been particularly true of medical research of the past, which has used the male body as the model for clinical research (2).

Four years ago, in June 1990, in response to a directive from the United States Congress, the United States General Accounting Office (GAO) reported that women were routinely excluded from medical research studies (3). The report noted that scientists considered male biology to be the standard or norm for medical research, even when conditions affecting both men and women were studied.

Just three months after the release of the GAO report, in September 1990, the ORWH was established at the NIH (4). The threefold mandate of the office addresses the major barriers that hinder women's full participation in medical research, both as research subjects and as investigators, and the gaps in knowledge that preclude women from receiving optimum health care.

The threefold mandate of the office is:

> First, to strengthen, develop, and increase research into diseases, disorders, and conditions that affect women, determining gaps in knowledge about such conditions and diseases, and establish a research agenda for the NIH for future directions in women's health research.
>
> Second, to ensure that women are appropriately represented in biomedical and biobehavioral research studies, especially clinical trials, that are supported by the NIH.
>
> Third, to create direct initiatives to increase the number of women in biomedical careers and to facilitate their advancement and promotion.

Our basic mandate has now evolved to encompass a myriad of responsibilities and activities. For example, the ORWH has the responsibility for establishing and monitoring the NIH's agenda for biomedical and behavioral research into the causes, treatment, and prevention of diseases and conditions that affect women. In our research agenda, the parameters of women's health have been redefined, and research has been directed toward providing better information on gender differences between women and men in health and in disease.

Our research agenda is based upon an expanded concept of women's health; one that addresses women's health over the entire life span, including studies to better define normal development, physiology, and aging in women. Our agenda addresses health conditions that are unique to women as well as those that affect both men and women. We direct many of our resources, both human and financial, toward supporting basic, clinical, and epidemiological research designed to discover gender differences in the etiology, prevention, progression, and treatment of diseases and conditions that affect both men and women, as well as supporting research on diseases and conditions unique to women.

We maintain that studies of women's health must cut across scientific disciplines and medical specialties. We stress that biomedical studies must reach across the life span of women, from birth, through adolescence and the adult years, to the menopausal and advanced years of life. In addition, we emphasize that considerations of ethnicity and socioeconomic status must be considered in research on women's health, along with innovative methods for recruiting populations of women who have been underrepresented as research subjects.

The NIH women's health research agenda entails not just clinical research, but investigations across the spectrum from genetic and molecular to epidemiological studies. We direct many of our resources, both human and financial, toward supporting basic, clinical, and epidemiological research designed to determine if gender differences exist in the etiology, prevention, progression, and treatment of diseases and conditions that affect both men and women, as well as supporting research on diseases and conditions unique to women. Our office fosters and facilitates research on issues in women's health carried out by the NIH's 24 constituent institutes, centers, and divisions through supplemental grants, co-funding, and collaboration on special initiatives.

Central to our efforts is a focus on gender differences caused by women's hormones and the major events in the lives of women related to hormones, such as menarche and menopause. Understanding the role of therapeutically administered exogenous hormones in women's health and disease is also of scientific importance. Environmental sources of estrogen and their bioaccumulation are assuming increasing importance as we look for contributors to diseases, especially in relationship to possible carcinogenic activity.

Cancer: A High Priority for Women's Health Research

Supporting basic and clinical research on the causes, diagnosis, treatment, and prevention of cancer is a high priority for all components of the NIH and especially for the ORWH because of cancer mortality among women. In fact, in recent years we have witnessed an increase in mortality from lung, breast, and prostate cancer for persons aged 45 years and older in Europe, the United States, East Asia, Australia, New Zealand, and Micronesia (5). Cancer is the second leading cause of death among women and men in the United States, after heart disease, and cancers remain major contributors to the mortality of women and men in the industrialized world.

Today, in the United States, breast cancer continues to be the most common form of cancer in women, although lung cancer has exceeded breast cancer as the leading cause of death for women since 1987. In addition, among men and women, we have witnessed an increase in deaths from other malignancies, such as liver, prostate, and kidney cancers.

The fact that more than 25 percent of human tumors originate in endocrine tissues or their target organs (6) has made expansion of research in hormonal carcinogenesis a top priority for the NIH. Since the discovery, in 1971, that young women who had been exposed *in utero* to diethylstilbestrol (DES) were developing vaginal cancer at higher than usual rates, the NIH has funded major, interdisciplinary studies of the hormonal determinants of various cancers and their risk factors, integrating studies of hormones and their metabolism with studies of environmental exposure and dietary factors.

Breast Cancer

From 1969 to 1986, mortality from breast cancer increased among premenopausal and postmenopausal women in most industrialized countries, except among women under the age of 45 in the United States and Scandinavia, and women over 45 in Australia, New Zealand, and Micronesia (5). Overall, the incidence of breast cancer in the United States has increased since 1973, but the rate has begun to level off since 1987 and even slightly decreased. Nevertheless, mortality rates have not dramatically changed, nor have survival rates increased markedly overall. However, the incidence and mortality rates have increased among African-American women, especially those over the age of 50 years (7). The greatest increases in breast cancer mortality, among women of all ages, have been in the Eastern European countries of the former Soviet bloc (5).

Breast cancer, which now affects one in eight American women and is the most common cause of death for American women between the ages of 40 and 44, presents exciting challenges and opportunities for researchers concerned with discovering the roles of exogenous and endogenous hormones in malignancy. The hormonal epidemiology of breast cancer is a pivotal focus of ongoing NIH-supported studies and serves as a paradigm for research aimed at uncovering the etiologic roles of specific hormones acting alone, and in concert with other potentially pathogenic nonhormonal factors, for other hormonally influenced malignancies.

In December 1993, Health and Human Services Secretary Donna Shalala convened a conference to establish a national action plan on breast cancer. More than 300 scientists, advocates, and breast cancer survivors met to develop and implement an action plan to prevent and cure breast cancer, and to address the concerns of the many women who have, or fear, breast cancer, as well as the concerns of those involved in scientific investigation or clinical care of this disease. The resulting action plan identifies opportunities for a collaborative implementation plan for health care, research, and policy priorities in conquering breast cancer (8).

Some of the conference's major recommendations for research focused on:

- Strengthening collaborative multidisciplinary research.
- Establishing comprehensive patient registries and tumor/blood (DNA tissue) banks as research tools.
- Increasing the number of qualified basic and clinical investigators in breast cancer research.
- Expanding support for investigator-initiated research on breast cancer.

Elucidating environmental influences on breast cancer.
Speeding the translation of novel therapeutic opportunities from the laboratory to the clinic.
Improving methods for breast cancer detection.
Reducing barriers to participation of patients and health care providers in breast cancer clinical trials.

Some of the therapeutic interventions to be considered in the implementation plan include hormonal modulation, chemoprevention, vaccines, gene therapy, antiangiogenesis, anti-growth factors, and anti-metastatic agents. We are hopeful that the medical community's expanding knowledge of the role of genetics in breast cancer will be helpful in developing vaccines or other prevention strategies.

The conference recommendations may benefit not only breast cancer research but also a broad spectrum of investigative efforts. Fortunately, many of the conference's recommended research initiatives are already under way at the National Cancer Institute (NCI) and other components of the NIH.

Ovarian Cancer

Ovarian cancer is the leading cause of death from gynecologic malignancies in the United States. In 1994, approximately 24,000 new cases of ovarian cancer will be diagnosed, and 13,600 women will die of the disease (9). The high mortality rate from this malignancy can be attributed, in large part, to the difficulty in detecting this cancer during its early stage, when cure rates are higher.

The NIH convened a consensus conference in April 1994 to formulate recommendations for ovarian cancer research, and has recently launched a new initiative to encourage research on ovarian cancer. Because ovarian cancer is so difficult to detect, several of the conference recommendations center on the need to develop better screening modalities. Much of the work to develop these modalities will involve the identification of genetic and biological markers through basic research. However, increasing our understanding of the role of hormones in the onset and progression of ovarian cancer could provide the key to preventive and curative interventions.

Further study is needed of the possible protective effect of taking oral contraceptives (OC) and of the role of estrogen replacement therapy as a possible protective intervention against ovarian cancer (10). Indeed, increasing our understanding of how steroids may influence the methylation of genes is a challenge that will require the united efforts of basic and clinical researchers involved in the study of hormonal carcinogenesis. Through the new ovarian cancer initiative, we also hope to encourage further research aimed at understanding the effects of environmental factors, such as synthetic hormones and pesticides, which are predominantly estrogenic, on the development and progress of ovarian and other cancers. Answering the questions related to environmental factors in the etiology of ovarian, endometrial, and other cancers could prove particularly important for improving the health of women, not only in the United States and other industrialized countries, but also in the Third World, where the majority of women work in the agricultural sector and may have high exposure to pesticides and other toxins.

Women's Longer Life Span: Issues to Address Through Research

In developed nations, women's increased life span presents the medical community with a number of challenging health issues, ranging from prevention of osteoporosis and Alzheimer's disease to prevention and treatment of breast and endometrial cancers.

With women's average life expectancy at 79 years in the United States and 81 years in Sweden (11), and the average age for menopause at approximately 52 years, women today spend about one-third of their lives in the postmenopausal state. As the use of hormone replacement therapy (HRT) becomes more and more widespread, we must determine the long-term health effects of such therapy on women who have a wide variety of medical histories and familial predispositions.

While HRT appears to reduce the risk for cardiovascular disease and osteoporosis in postmenopausal women, some studies indicate that long-term use may put some women at higher risk for some cancers. Further research must be conducted in order to provide physicians and women with hard data and analysis of the relative risks and benefits associated with long-term use of HRT, especially with regard to possible increased risks of endometrial cancer and breast cancer.

Increasing our understanding of the effects of long-term HRT is one of the goals of the NIH's Women's Health Initiative, a 15-year preventive study that will involve approximately 163,000 postmenopausal women from communities across the United States. The study is designed to answer questions about the benefits and risks of long-term HRT on prevention of coronary heart disease and osteoporotic fractures; evaluate the effect of a low-fat dietary pattern on prevention of breast and colon cancer and coronary heart disease; and evaluate the effect of calcium and vitamin D supplementation on prevention of osteoporotic fractures and colon cancer. It will also include a study of community approaches to developing healthful behaviors.

Diverse populations of women are being recruited for the Women's Health Initiative. It is anticipated that the results of this initiative, in conjunction with findings from other studies, over the next 10 to 15 years, will provide information to help us better understand the role of hormones and HRT in the lives and health of women.

We also need to learn more about the relationship between fertility treatments and cancer. To address this issue, the NCI has launched an epidemiological study of cancer risk following evaluation and treatment for infertility. The NCI is collaborating with the National Institute of Child Health and Human Development to expand an *in-vitro* fertilization registry to study women who have received fertility treatments, as well as to study hormonally based health problems in their descendants.

There is also a need to determine how changes in the onset of the menstrual cycle, combined with social changes, may affect women's health. With menses starting today at the age of 12 or 13, compared with age 17 or 18 some years ago, combined with a reduction in the number of children born to the typical woman and a reduction in time spent breast-feeding, women in the developed world are, as one writer has put it, leading a "strange new life" in terms of hormones (12). Since all these factors contribute to an increase in the average number of menstrual cycles a woman experiences over her lifetime, increasing her lifetime exposure to sex hormones, they raise issues of particular importance for women's health. Gaining a better understanding of the health effects of this increased lifetime exposure to steroids, and its influence on the relative risks for breast, endometrial or other malignancies, is essential to our understanding of hormones and carcinogenesis.

Conclusions

Research aimed at discovering the effects of sex hormones in the development of cancer is central to our efforts to improve women's health. We must gain a better understanding of hormones and cancer, including the carcinogenic potential of natural and synthetic hormones, and the role of exposure to hormones in the environment. Also, it is imperative to have scientifically confirmed documentation of the risks versus benefits of OC use in women and postmenopausal treatment with HRT, including data to compare the effects of unopposed estrogen with the effects of estrogen and progestin combined. The NIH is supporting basic and clinical research on hormonal carcinogenesis. Much of this basic research will benefit not only women, but also men who suffer from prostate and pancreatic cancer.

Studies supported by the NCI are focusing on the relationship between the level and duration of hormone exposure and malignant transformation in hormonally sensitive tissue, and the effect of diet and hormones on cancer risks. Like many of the participants at this conference who are focusing on these issues, the NCI is targeting the molecular interactive elements that determine cancer risk, including hormonal, genetic, and environmental factors. The NIH is also supporting research to investigate these interrelationships in order to address prevention and develop biomarkers of cancers. We must focus on genetic determinants, viruses, and chemoprevention. Then, appropriate clinical trials must be implemented, so that clinical applications can be rapidly translated into standards of practice for men and women.

With our focused efforts to learn more about gender in health and disease, the NIH is working to establish gender equality in health care through inclusive policies for clinical trials and investigative efforts. Governments and individuals in many other countries are also focusing on improving the health of women, recognizing that better health for women is the key to strengthening the health of men, children, families, and communities. For inherent in the effort to improve women's health through biomedical investigation is a commitment to raising the general level of health enjoyed by all people.

The German-born psychoanalyst and educator, Erik Erikson, once observed that, "True equality can only mean the right to be uniquely creative" (13). Achieving the highest attainable standard of health for all people also requires equality in the scientific community, equality that allows each and every researcher, regardless of gender, to be uniquely creative.

Over the next few days, we will enjoy the privilege of hearing presentations by some of the most uniquely creative individuals in medical research. Women and men worldwide are counting on the unique creativity of scientists, such as those assembled in this symposium, to effect new interventions against cancer and allow the highest standard of health to be truly attainable by all.

References

1. World Health Organization (WHO) (1992) Women's Health Across Age and Frontier, Geneva.
2. LaRosa JH, Pinn VW (1993) Gender bias in biomedical research. J Am Med Wom Assn 48:145-151.
3. US General Accounting Office (GAO) (1990) NIH Problems in Implementing Policy on Women in Study Populations. Washington:US Government Printing Office, Publication no GAO/T-HRD-90-38.
4. Pinn VW (1992) Commentary: Women, research and the national institutes of health. Am J Preven Med 8:324-327.
5. Hoel DG (1992) Trends in cancer mortality in 15 industrialized countries, 1969-1986. J Natl Cancer Inst 84:312-20.
6. National Cancer Institute (1993) 1995 Budget Estimate. Bethesda: National Cancer Institute, p 203.

7. US Centers for Disease Control and Prevention (1994) Deaths from breast cancer, 1991. Morbidity and Mortality Weekly Report 43:273, 279-281.
8. Proceedings: Secretary's conference to establish a national action plan on breast cancer (1993) Bethesda: National Institutes of Health.
9. NIH consensus development conference statement (1994) Ovarian cancer: screening, treatment, and follow-up. Bethesda: NIH Office of Medical Applications of Research, p 2.
10. Spirtas R, Kaufman SC, Alexander NJ (1993) Fertility drugs and ovarian cancer: Red alert or red herring? Fert Steril 59:291-93.
11. US Bureau of the Census (1989) World population profile: 1989. Washington: US Government Printing Office, Table 8.
12. Judson, OP (1993) Towards healthier infertility. Nature 365:15-16.
13. Erikson, EH (1965) Inner and outer space: Reflections on womanhood. In Lifton R (ed): The Woman in America. Boston: Houghton Mifflin Co, pp 1-26.

Symposium Presentation

Hormones and the Cellular Origin of Mammary Cancer: A Unifying Hypothesis

Satyabrata Nandi, Jason Yang, and Raphael C. Guzman

Mammogenic hormones regulate the proliferation of mammary epithelial cells, and this hormonally induced proliferation is also prerequisite for breast carcinogenesis. Based on our extensive *in vivo* and *in vitro* studies, and on analysis of the literature, a unifying hypothesis that offers an explanation of the role that hormones may play in the genesis of mammary cancers in mice, rats, and humans is presented. It is our hypothesis that the hormonal environment present during critical periods of early development of the mammary gland determines proportion of mammary epithelial cells which proliferate as a direct response to hormones. During carcinogenesis of the mammary gland, the hormone-responsive cells give rise to mammary cancers that are hormone-dependent, and the mammary cells that are not directly hormone-responsive give rise to cancers that are hormone- independent. Our hypothesis attempts to explain 1) how hormones regulate mammary epithelial cell proliferation; 2) why hormones are required for the genesis of mammary cancers of heterogeneous phenotypes and genotypes, including those classified as hormone dependent or hormone independent; and 3) why mice, rats, and humans have consistently different ratios of hormone- dependent/hormone-independent mammary tumor types. This hypothesis deals with the roles of hormones in mammary carcinogenesis up to the time of their earliest detection/diagnosis, and does not deal with later events involved in the progression and metastasis of these cancers.

General Features of the Mammary Gland and Mammary Cancers

This section is intended to provide general information pertinent to the analysis of the roles of hormones in mammary carcinogenesis (1-2). The mammary gland is a compound tubuloalveolar gland which is divided into two major compartments: an epithelial compartment and a stromal compartment consisting of connective tissue, adipose tissue, and a vascular network. The epithelial compartment consists of a nipple connected to ducts and alveolar units, and is composed of two kinds of epithelial cells: luminal epithelial cells and myoepithelial cells. Luminal mammary epithelial cells (LMEC) cover the inner lining of ducts and alveoli and are involved in the synthesis and passage of milk. The myoepithelial cells, with their contractile elements, are involved in the expulsion of milk. Hormones from different endocrine glands regulate proliferation and milk product synthesis by the LMECs.

The majority of mammary cancers originate in the LMECs, and are chiefly but not exclusively confined to the female sex. Hormone-regulated proliferation of LMECs is a prerequisite to the genesis of mammary cancers of all phenotypes and genotypes.

Mammary cancers at their earliest detection/diagnosis are classified into two major varieties: hormone-dependent (HDT) and hormone-independent (HIT) tumors (3, 4). HDTs are dependent on the host mammogenic hormones for their proliferation, whereas HITs continue to proliferate in the absence of such hormones. Both varieties of tumors can be heterogeneous, and are subdivided into various categories according to their phenotypes.

Mammary Cancers in Mice, Rats and Human: Comparison and Questions Raised

There are similarities and differences in the pathogenesis of mammary neoplasia and in the phenotype of mammary cancers among mice, rats and humans. Our strategy has been to identify the main characteristics distinguishing the cancers in these three species in order to develop a hypothesis which could account for both the similarities and the differences.

The chief similarities among these species in terms of mammary carcinogenesis are as follows: First, mammary cancers occur predominantly in females. Second, hormone-regulated proliferation of LMECs, the targets for most mammary cancers, occurs during postpubertal life and primarily in females. Third, proliferating mammary epithelial cells are prerequisite to the genesis of mammary cancers of all phenotypes and genotypes in the three species.

The major differences among these species are as follows: First, mammary cancers in the three species can be either HDTs or HITs at the time of detection, although the ratios of these types vary with the species. Approximately 85-90% of the mammary cancers originating in virgin rats exposed to chemical carcinogens are of the HDT type (32). In contrast, the majority (80-100%) of the mammary cancers in most strains of mice exposed to mouse mammary (MMTV) tumor virus or chemical carcinogens are of the HIT type (5, 6). However, a high incidence of hormone-dependent tumors has been observed in a few mouse strains (GR, RIII, DD) which have a variant form of MMTV (6). In human breast cancers, the ratio HDT/HIT between can be approximated either from the estrogen receptor status in primary mammary cancer, or from the response rate of metastatic mammary cancer to hormone therapy. About 60-65% of primary mammary cancers are positive for estrogen receptor and can therefore be considered hormone responsive (4). However, about one third of all patients with advanced mammary cancer respond to hormone therapy. In either case, the proportion of HDT to HIT in humans seems to be intermediate between those of mice and rats.

Second, the range of the histology of mammary cancers in the three species is different. Pathologists have identified many different morphological types of breast cancer in humans (7), and mammary tumors originating in mice are also of many different morphological types (8). In rats, however, mammary cancers are much less morphologically heterogeneous and are classified primarily into two categories: adenocarcinomas and papillary carcinomas (9).

Third, only the human mammary tumors are highly metastatic, spreading to other organs even when the primary tumor has a relatively small mass. In contrast, rat and mouse mammary cancers can be invasive but rarely metastasize.

These similarities and differences among the three species under consideration have raised the following questions: 1) How do hormones regulate LMEC proliferation? 2) What is the role of hormone-regulated LMEC proliferation in the genesis of mammary cancers? 3) Why are hormones required for the genesis of mammary cancers of heterogeneous phenotypes, including both HDTs and HITs? 4) Why do these three species show consistently different ratios of HDT/HIT? In the following sections we summarize the current status of our knowledge related to these questions, and the answers that have led us to formulate our unifying hypotheses.

Mammogenic Hormones and the Regulation of LMEC Proliferation

Hormones from multiple endocrine organs, especially hormones from the ovary and pituitary, are known to induce LMEC proliferation. Classic studies, reviewed recently (10, 11), have shown that ablation of endocrine organs (ovary and/or pituitary) results in atrophic, non-functional mammary glands in all three species under consideration. In these studies, both *in-vivo* (10, 11) and *in-vitro* organ cultures (10-12) have demonstrated that proliferation of different compartments of the mammary tree in rats and mice require different combinations of mammogenic hormones. Cell culture studies (10, 11) have clearly shown that growth factors, most of which are produced locally, also have a profound influence on the growth of LMEC. The following is a brief summary of the known influences of hormones and growth factors on LMEC proliferation in mice, rats and humans.

In vivo, hormones from the pituitary and the ovary are the principle regulators of mammary growth in rats and mice (10). However, rats appear to be more sensitive to hormone-induced LMEC proliferation than mice. For example, estrogen treatment results in full lobulo alveolar growth in ovariectomized rats (13), but only ductal growth in ovariectomized mice (5, 14). Our knowledge of *in vivo* regulation of human LMEC proliferation is extremely limited. In humans, the endocrine control of breast development has been studied by correlating the circulating level of various hormones with changes in the growth of breast epithelial cells (11). Also, information has been derived from incidences of abnormal hormonal environment as a consequence of disease, accident, or therapy (48). Biopsy materials have shown more proliferating LMEC during the luteal phase than in the follicular phase of the human ovarian cycle, suggesting that both estrogens and progestins play a role (15, 16). Studies of human breast xenografts in nude mice have shown estrogen to be the most potent inducer of DNA synthesis (17).

Several growth factor families have now been identified as playing possible roles in the regulation of mammary growth. Insulin-like growth factors (IGFs) and epidermal growth factor (EGF) have been found in milk. Transcripts and proteins of several growth factors have been found in the mammary glands of mice, rats, and humans (10, 11). These observations provide strong evidence for the local synthesis and potential involvement of these factors in mammary growth regulation.

In-vitro organ culture studies using mouse and rat mammary glands also suggest that the rat mammary gland is sensitive to hormones (10, 11). Dubois and Elias (12) showed that different parts of the mouse mammary epithelial tree (ducts, end buds, alveoli, etc.) may respond differently to hormones and to growth factors, thus paralleling the *in-vivo* findings cited above. Human breast cultures reportedly show an increased labelling index after exposure to progesterone or ovine prolactin, but the hormones that regulate the proliferation of human LMECs have not been clearly defined (11).

Serum-free cell culture studies provide the clearest picture of the factors which stimulate LMEC proliferation. In insulin-containing serum-free collagen matrix cultures, the minimal requirements for proliferation of rat LMECs are prolactin and progesterone (18). In contrast, mouse LMECs can respond to a variety of individual hormones (prolactin, progesterone, aldosterone, corticosterone) and growth factors (EGF, basic-fibroblast growth factor, keratinocye growth factor) (10, 11). A combination of progesterone, prolactin, and insulin markedly stimulated LMEC proliferation in mouse and rat cultures (10, 18) . Proliferation of normal human breast epithelial cells has been reported under various culture conditions, including supplementation with EGF, transforming growth factor α and cholera toxin (11). However, growth stimulation of human mammary cells has not yet been demonstrated unequivocally in response to hormones considered mammotropic *in vivo* (e.g., ovarian steroid and pituitary hormones) (11).

In conclusion, *in-vivo* endocrine ablation studies have clearly demonstrated that hormones are the primary factors regulating LMEC proliferation in all three species. This well-established observation suggests that mammotropic hormones trigger a cascade of events associated with LMEC proliferation. Although the major sources of hormones are discrete endocrine glands, growth factors are produced locally by either epithelial or stromal compartments within the mammary gland itself. Both hormones and growth factors, alone or in combinations, can influence LMEC proliferation *in vivo* and *in vitro*. These two groups of chemical mediators are thus considered to be mammary mitogens. Although all three species can be shown to be hormone dependent for LMEC proliferation *in vivo*, clear species-specific differences are shown by *in-vitro* studies. The LMECs of mice respond to both hormones and growth factors *in vitro*. In rats, mammotropic hormones appear to be specifically required, as they are *in vivo*. In humans, growth factors seem to be the primary requirement for LMEC growth *in vitro*.

Our current concept of the hormonal regulation of LMEC proliferation is that mammogenic hormones stimulate both epithelial compartment cells (luminal epithelial and myoepithelial cells) and stromal compartment cells (adipocytes, fibroblasts, vascular cells) to synthesize and secrete mammogenic growth factors. Hormones and/or these locally produced growth factors then create microenvironments, depending on the physiological state of the organism, which are responsible for the proliferation of the different luminal epithelial cells ultimately responsible for the synthesis and release of milk products. As will be shown in section IV, these different mammary mitogenic microenvironments may explain the occurrence of mammary tumors which are of different phenotypes and genotypes at the time of their detection.

Hormonal Role in LMEC Proliferation, and Preneoplastic and Neoplastic Transformation

Studies performed *in-vivo* and *in-vitro* have shown that proliferating target cells serve as a prerequisite for carcinogenesis by oncogenic agents (19). Recently, there has been much discussion of the possible mechanisms by which cell proliferation may be involved (19-21). At the time of earliest detection, however, none of the many hypotheses put forward, account for the origin of cancers of heterogeneous phenotypes and genotypes in a single organ. In this section our aim is to demonstrate that the mitogenic microenvironment of mammary cells

around the time of carcinogen exposure is an important factor in determining the genotype and/or the phenotype of the resultant mammary lesions (22).
In earlier *in-vivo* studies, the incidence of mammary cancer in mice and rats resulting from chemical carcinogen administration (23, 24) is closely related to the frequency of LMEC division at the time of action of the carcinogen on the gland. Beattie and his colleagues (25), in a series of recent studies, have provided evidence to support the concept that the prevailing hormonal profile at the time of tumor initiation modulates the subsequent nature of the tumors induced.

Our own objective has been to develop an *in-vitro* transformation system in which mammary lesions could be induced with the same characteristics as those present in the animals (26, 27). Our serum-free collagen matrix culture system allows three-dimensional growth of mouse LMECs in the presence of specific hormones, growth factors, and diffusible signalling molecules (10, 11). Studies *in vivo* have shown that chemical carcinogens induce ductal hyperplasias (DH) in virgin mice, and direct tumors and/or hyperplastic alveolar nodules (HAN) in mice bearing pituitary isografts (26,28). When LMECs from adult virgin BALB/c mice were cultured in media containing any one of three different mitogenic combinations of equivalent proliferative potential [progesterone + prolactin (PPRL), EGF, or lithium] in our collagen matrix culture system, exposing them to MNU (N-methyl-N-nitrosourea) and inducing their transformation (26, 27). The design of the transformation system (26, 27) was such that all parameters were kept constant with the exception of the mitogenic environment of the cells in culture. Our results are summarized below.

1) *In vivo*-like DHs, HANs, and HITs were observed following exposure of LMEC *in vitro* to MNU and transplantation to host mice.

2) The nature of the induced lesions differed among the three groups. Lithium and PPRL induced primarily HANs and HITs, and EGF induced primarily DHs.

3) The incidence of lesions was high in the PPRL and lithium groups compared to that in the EGF group, despite the fact that the proliferative potential of the three mitogens was equivalent.

4) Upon transplantation into mammary fat pads of syngeneic hosts, both DHs and HANs gave rise to mammary cancers. HANs from the lithium and PPRL groups were morphologically indistinguishable from each other (5).

5) The morphology of tumors arising directly or from preneoplastic lesions was different in each of the three experimental groups (8). The EGF group was predominantly type A carcinomas, while the lithium group was primarily type B. Tumors from the PPRL group were cystic adenocarcinomas with squamous metaplasias.

6) Genetic analysis of the lesions from these groups indicated that the activation of protooncogenes was also dependent on the mitogen used around the

time of carcinogen treatment. Eighty percent of the HANs and carcinomas in the PPRL group had an activation of protooncogene *c-Ki-ras* by a specific G35→A35 point mutation at codon 12 (26, 27). Since this same lesion was detected in the preneoplastic HANs, activation of *c-Ki-ras* was considered to be an early event in the transformation process (26, 27). Five of eight tumors from the lithium group contained an overexpressed putative protooncogene *MAT-1*, which has been shown to effect neoplastic transformation of NIH 3T3 cells and the mouse mammary epithelial cell line TM3 (27). Neither the lithium nor the EGF group showed activation of *c-Ki-ras* in preneoplastic lesions or tumors.

These results led us to postulate (22) that the mitogenic microenvironment around the time of carcinogen exposure determines both the incidence and the phenotype of resultant mammary lesions, as well as the molecular events associated with mammary carcinogenesis. Recent *in-vivo* experiments in our laboratory provided further evidence for this. Activation of *c-Ki-ras* mutation and tumors of the same histotypes as those in the PPRL group described above have also been observed in pituitary-isografted BALB/c female mice injected with MNU (29). Pituitary isografts in mice raise the blood levels of PPRL (30). *In-vivo* experiments in rats have also shown a correlation between the frequency of G35→A35 mutated *H-ras* protooncogene with the stage of the estrous cycle at the time of MNU administration (31).

These findings have encouraged us to interpret our *in vitro* results as representative of events which can occur under physiologic conditions. We suggest the following possible mechanisms: 1) Normal mammary glands contain many subcompartments, each with its own specific mitogenic microenvironment resulting from hormonal stimulation; 2) The specific hormonal milieu within each microenvironment will induce specific sets of gene expression associated with mitogenesis within the subcompartment; 3) Following chemical carcinogen exposure, adduct formation will occur in one or more of these expressed genes. Upon further cell division, fixation of the mutation will occur within specific genes, giving rise to initiated cells; 4) Further promotion will then result in a number of specific phenotypes.

Hormones and the Cellular Origin of Mammary Cancer: A Unifying Hypothesis

We propose a model (Figure 1) which can account for known similarities and differences among mice, rats, and humans. To support this model, we will present current data from our own studies as well as pertinent information from the literature. Moreover, we will attempt to account for the roles of hormones in the origin of mammary cancers of different phenotypes and genotypes, including HDTs and HITs.

Model of Mammary Carcinogenesis

The model offers an explanation for the origin of HDTs and HITs, and for the requirement of prior exposure to mammogenic hormones for the genesis of both these types of mammary cancers. It also attempts to account for the generation

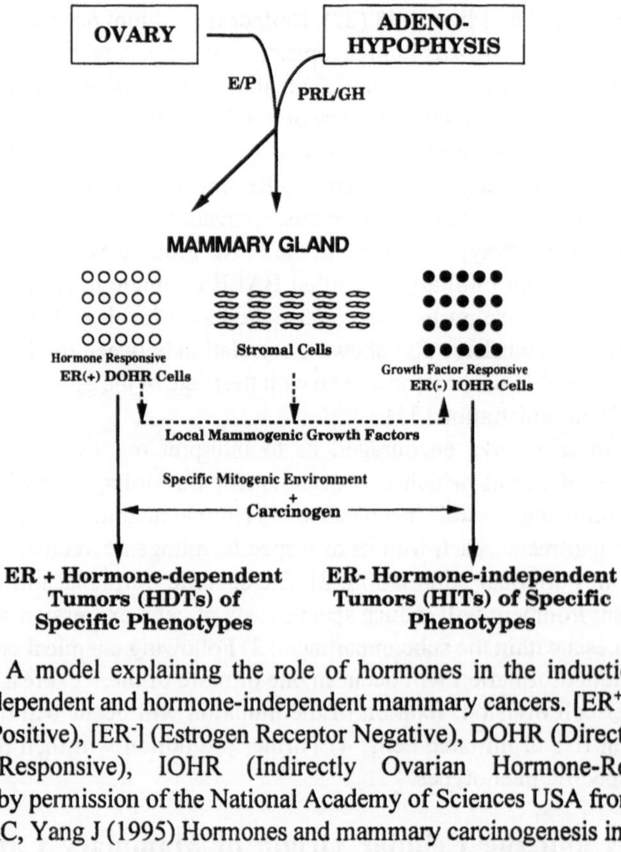

Figure 1. A model explaining the role of hormones in the induction of both hormone-dependent and hormone-independent mammary cancers. [ER^+] (Estrogen Receptor Positive), [ER^-] (Estrogen Receptor Negative), DOHR (Directly Ovarian Hormone-Responsive), IOHR (Indirectly Ovarian Hormone-Responsive). Reprinted by permission of the National Academy of Sciences USA from Nandi S, Guzman RC, Yang J (1995) Hormones and mammary carcinogenesis in mice, rats, and humans: A unifying hypothesis. PNAS USA 92:3650-3657.

of HDTs and of HITs at the time of earliest detection. The proposal suggests that there are two populations of LMECs in the mammary gland: directly ovarian hormone-responsive cells (DOHR cells with functional [ER^+]) and indirectly ovarian hormone-responsive cells (IOHR cells lacking functional ER, or [ER^-]). When exposed to hormones from the ovary and pituitary, DOHR cells proliferate, as well as synthesize and secrete local mammogenic growth factors. IOHR cells, on the other hand, are unresponsive to hormones and proliferate only in response to the hormone-induced stimulation and secretion of growth factors produced locally by DOHRs or other cell types (stromal fibroblasts, etc.). Thus the IOHRs are indirectly dependent on hormones for their initial development. DOHR and IOHR cells are considered to be the progenitors of HDTs and HITs, respectively. Particular subcompartments of DOHR cells may give rise to HDTs of specific phenotypes and/or genotypes, depending on the hormonal milieu around the time of carcinogen exposure. These HDTs may undergo further progression, giving rise to heterogeneous phenotypes which may include HITs. In similar fashion, specific IOHR subcompartments responding to particular mitogenic growth factors available at the time of carcinogen exposure will give rise to HITs of specific phenotypes and/or genotypes. These also may undergo further progression to HITs of heterogeneous phenotypes. This scenario explains why all of the tumors, HDT or HIT, initially depend on hormones. It also explains the manner in which normal LMECs can give rise to tumors with such a variety of phenotypes and genotypes. The differences between species can be explained by differences in the ratio of DOHR to IOHR undergoing DNA synthesis at the time of carcinogen exposure.

Model Validation

Validation of this proposed model requires evidence to support the following:

1) There must be two pathways for the development of HITs, one with and one without an intermediate HDT state; 2) Two populations of LMECs must be demonstrable, one of which is directly hormone-responsive (DOHR), and the other nonresponsive (IOHR) except through intermediate growth factors produced in response to hormones by DOHR or other mammary cell compartments; 3) DOHRs must give rise to HDTs, and IOHRs must be shown to give rise to HITs; 4) A corresponding proliferating DOHR/IOHR ratio should be found at the time of carcinogen exposure to match the HDT/HIT ratio at the time of tumor diagnosis/detection in the three species studied; 5) An explanation must be found for the species differences observed at any of these levels.

1) In studying pathways for genesis of HITs, the mouse and rat experimental models have provided well-controlled data, including the exact timing of carcinogen action. Mice develop primarily HITs with no intermediate HDT state, but with precancerous precursors (HANs) which are generally unresponsive to ovarian hormones. In female rats, most mammary tumors are HDTs, although there are a small number of HITs. The absolute number of the latter are unchanged under circumstances which markedly reduce the number of HDTs, namely, carcinogen treatment after parity (32). This leads one to suspect that some HITs arise directly from normal LMECs, not through a HDT precursor. Other studies have (3, 33) shown another pathway, in which HDTs do act as

precursors to HITs in rats and mice. These observations support the idea that there are two pathways from LMEC→HIT.

2) The existence of two LMEC populations (DOHR and IOHR) is supported by recent immunocytochemical studies with monoclonal antibodies to ER proteins (34, 35). Four to 26% of human and 8-21% of mouse LMECs are [ER$^+$], implying that [ER$^-$] cells are predominant in these two species (correlated with the lower incidence of HDTs). It is assumed that DOHR cells would be [ER$^+$], and IOHR cells [ER$^-$]. Much evidence exists that hormones can induce growth factor expression in various mammary compartments (11), providing a pathway for indirect hormone action on [ER$^-$] LMECs. The best evidence to date for the existence of cells with differing requirements for growth is the demonstration that mouse LMECs can proliferate *in vitro* in the presence of either hormones or growth factors. Similar evidence in rats and humans is suggestive but inconclusive at this time (11).

3) Recent studies of the transfection of the ER gene into human breast cancer cell lines support the view that HDTs and HITs arise from separate cell populations (36, 37). There is an inverse correlation between EGF receptor and ER in primary human breast cancers, with most tumors positive for only one of these receptors (38, 39). Human mammary cancers exhibit a high degree of stability of the nuclear DNA content (49, 50) and estrogen receptor status (51) during the progression of the disease. These observations suggest that mammary cancer originates as diploid and/or estrogen receptor positive tumors which then progress into aneuploid and/or estrogen receptor negative tumors associated with high malignancy.

4) Evidence cited above showing that most LMECs in mice and humans lack functional ER supports our requirement that the ratios of replicating DOHR/IOHR cells at the time of the carcinogenic insult determine the subsequent frequencies of HDTs/HITs. Quantification of [ER$^+$] cells in rat mammary glands has not yet been accomplished (one would expect a higher ratio of [ER$^+$]/[ER$^-$] cells, corresponding to the high incidence of HDT in rats). The extreme sensitivity of rat LMECs to hormones *in vivo*, while not constituting proof, suggests support for our hypothesis.

5) Can we determine the reasons for species differences in the ratios of cell (DOHR/IOHR) and tumor (HDT/HIT) types? Are these differences genetically determined? We believe the answers to these questions lie in the nature of the prenatal and early postnatal environment, and their clarification may lead us to methods of breast cancer prevention. These ideas are further elaborated below.

Effects of the Perinatal Environment on LMEC Development

Human epidemiological evidence (40) has shown that conditions associated with high levels of estrogen during pregnancy are correlated with increased breast cancer risk in daughters, whereas pre-eclampsia/eclampsia (associated with low estrogen levels) may be associated with decreased risk, suggesting an important influence of the intrauterine hormonal environment. Numerous studies in mice have shown that administration of steroids (androgens, estrogens, progestins) to pregnant females has drastic effects on nipple development, mammary morphogenesis, growth patterns, future susceptibility to mammary carcinogenesis, and also on the nature (HDT/HIT) of subsequent tumors (41-44). Thus, the fetal and/or perinatal environment during mammary development can influence the future phenotypes of mammary epithelial cells.

Mammary growth patterns in early postnatal life are very different in mice and rats (Nandi, unpublished data). Fetal endbuds persist and undergo growth and differentiation in rats, but these endbuds undergo regression during the first week of postnatal life in mice. We are currently studying the relationship of persistence or regression of fetal endbud cells to the nature of future induced tumors. Along similar lines, DES-induced persistence of Muellerian ducts has been implicated in early human vaginal cancers (45), and primitive cells in the kidneys of hamsters are reported to be the targets of estrogen-induced hormone-dependent cancers (46). Therefore, it appears that the perinatal hormonal environment profoundly influences the persistence of fetal cells postnatally, which in turn may influence the risks for later hormonal carcinogenesis. In humans, mammary growth patterns may have similarities to those observed in mice and rats. Considerable variation is observed in the development of the ductal system in the newborn human, ranging from rudimentary ducts with no branches to well developed branching ducts with terminal lobules (47).

Conclusions and Summary

Our conclusions concerning the roles of hormones in mammary carcinogenesis in mouse, rat, and human, are summarized in Figure 2. The hypothesis is intended to account for the part hormones play in the origins of DOHR and IOHR cells, in the development of microenvironments for mitogenic stimulation in various subcompartments of the mammary tree, in the risks of formation of HDTs or HITs in these species, as well as in the eventual development of tumors of different genotypes and phenotypes.

1) Future phenotypes of luminal mammary epithelial cells, such as, hormone/growth factor responsiveness, susceptibility to carcinogenesis, and the phenotypes of subsequent tumors, are determined by the fetal hormonal

environment (androgens, estrogens, and perhaps also progestins and growth factors) which exists duuring mammary development.

2) Two distinct types of LMECs occur: directly ovarian hormone responsive (DOHR) and indirectly ovarian hormone responsive (IOHR). Functional estrogen receptors are found only in DOHR cells.

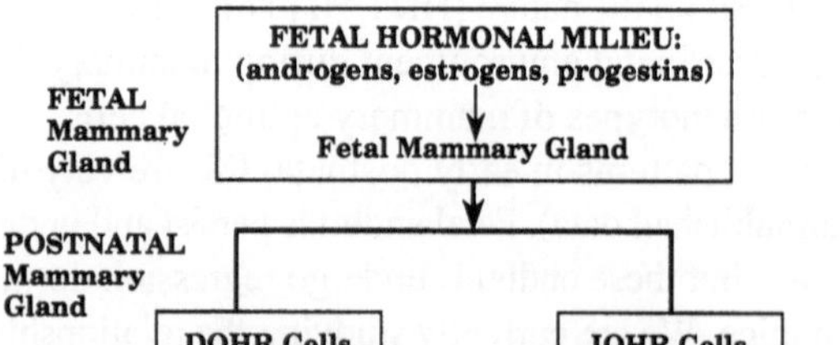

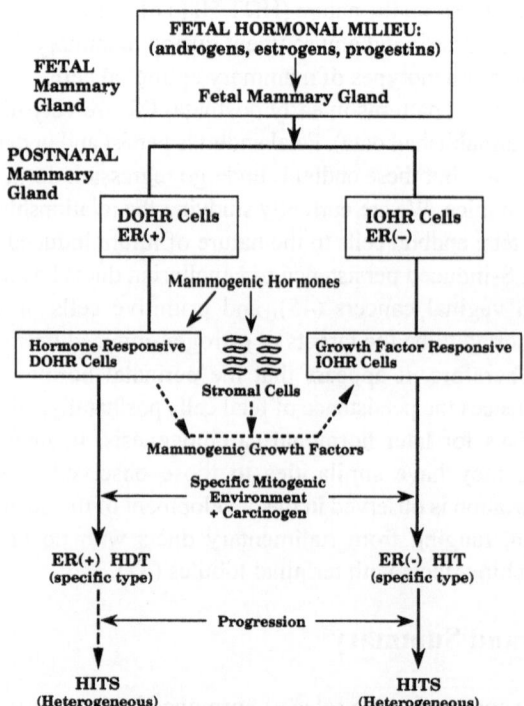

Figure 2. A model for a unifying hypothesis explaining the role of hormones in mammary carcinogenesis in mice, rats, and humans. DOHR (Directly Ovarian Hormone-Responsive), IOHR (Indirectly Ovarian Hormone-Responsive), [ER^+] (Estrogen Receptor Positive), [ER^-] (Estrogen Receptor Negative), HDT (Hormone- Dependent Tumor), HIT (Hormone-Independent Tumor). Reprinted by permission of the National Academy of Sciences USA from Nandi S, Guzman RC, Yang J (1995) Hormones and mammary carcinogenesis in mice, rats, and humans: A unifying hypothesis. PNAS USA 92:3650-3657.

3) Mammogenic hormones, especially from the ovary and adenohypophysis, directly stimulate DOHR cells to proliferate and to produce growth factors. These hormones also stimulate growth factor production by mammary stromal cells.

4) Growth factors produced, as described above, act by paracrine pathways to stimulate the proliferation of IOHR cells.

5) DOHR and IOHR cells are the direct progenitors of HDTs and HITs, respectively. Therefore the ratios of replicating DOHR/IOHR cells at the time of carcinogen exposure determine the eventual frequencies of hormone dependent and hormone independent tumors (HDTs/HITs).

6) Based on these conclusions, we suggest that the mitogenic environment (including hormones and growth factors) of a particular subcompartment of the DOHR and IOHR cell populations around the time of carcinogenic insult will determine the eventual phenotype and/or genotype of subseqent cancers.

Future Perspectives

Clearly, the two animal species that have been most thoroughly studied differ markedly in factors controlling mammary tissue growth and carcinogenesis, while human breast development and cancer seem to share some features of each species. We have known for many years that hormones play some role in both normal mammary growth and carcinogenesis, but explanations for the differences as well as similarities in the relationships between hormones and cancer in mice, rats, and humans have not been previously forthcoming.

The unifying hypothesis presented here is not intended to be a formal theory, but rather a starting point to stimulate new avenues of research. A hypothesis of such broad scope would have been impossible to formulate without the kind of information that has been obtained with experimental model systems, and indeed can only be tested and verified in such systems. We firmly believe that only by continuing to refine (or refute) these ideas by careful laboratory and epidemiological analyses will we eventually understand human breast cancer development sufficiently to prevent or to cure it.

Acknowledgments

We wish to thank our past and present students, postdoctoral fellows, and research assistants for their many contributions to the development of the ideas presented herein. We would also like to thank Carol Slatten and Yun Kit Hom for their aid in the preparation of this manuscript. We also acknowledge the generous financial support of the National Cancer Institute (CA05388, CA63369, CA09041) for the past 34 years, as well as the American Cancer Society; and the University of California.

References

1. Harris JR, Hellman S, Henderson IC, Kinne DW (eds) (1991) Breast Diseases. Philadelphia: Lippincott.
2. Harris JR, Lippman MC, Veronesi U, Willett W (1992) Breast cancer. N Engl J Med 327:319-328.
3. Furth J (1975) Hormones as etiological agents in neoplasia In Becker GF (ed): Cancer: A Comprehensive Treatise. New York: Plenum, vol 1, pp 75-120.
4. Stoll BA (1981) Breast cancer: rationale for endocrine therapy In Stoll BA (ed): Hormonal Management of Endocrine-related Cancer. London: Lloyd-Luke Medical Books Ltd, pp 77-89.
5. Shimkin MB (1945) Hormones and mammary cancer in mice. In Moulton FR (ed): A Symposium on Mammary Tumors in Mice. Washington: AAAS, pp 85-122.
6. Heston WE, Vlahakis G (1971) Mammary tumors, plaques, and hyperplastic alveolar nodules in various combinations of mouse inbred strains and the different lines of mammary tumor viruses. Int J Cancer 7:141-148.
7. McDivitt RW, Stewart FW, Berg JW (1968) Tumors of the breast, Atlas of tumor pathology. Washington: Armed Forces Institute of Pathology.
8. Dunn TB (1959) Morphology of mammary tumors in mice. In Homburger F (ed): Physiopathology of cancer. New York : Hoeber, pp 38-84.
9. Komitowski D, Saas B, Laub W (1982) Rat mammary tumor classification: Notes on comparative aspects. J Natl Cancer Inst 68:147-156.
10. Imagawa W, Bandyopadhyay GK, Nandi S (1990) Regulation of mammary epithelial cell growth in mice and rats. Endocrine Rev 11:494-523.
11. Imagawa W, Yang J, Guzman R, Nandi S (1994) Control of mammary gland development In Knobil E, Neil JD (eds): The Physiology of Reproduction. New York: Raven, pp 1033-1063.
12. DuBois M, Elias JJ (1984) Subpopulations of cells in immature mouse mammary glands as detected by proliferative response to hormones in organ culture. Dev Biol 106:70-75.
13. Neumann F (1991) Early factors for carcinogenesis in sex-hormone-sensitive organs. Mutation Res 248:341-356.
14. Bresciani F (1968) Topography of DNA synthesis in the mammary gland of C3H mouse and its control by ovarian hormones. Cell Tissue Kinet 1:51-63.
15. Masters JRW, Drife JO, Scarisbrick JJ (1977) Cyclic variation of DNA synthesis in human breast epithelium. J Natl Cancer Inst 58:1263-1265.

16. Meyer JS (1977) Cell proliferation in normal human breast ducts, fibroadenomas, and other ductal hyperplasias measured by nuclear labeling with tritiated thymidine. Human Pathol 8:67-81.
17. McManus MJ, Welsch CW (1984) The efffect of estrogen, progesterone, thyroxine, and human placental lactogen on DNA synthesis of human breast ductal epithelium maintained in nude mice. Cancer 54:1920-1927.
18. McGrath M, Palmer S., Nandi S (1985) Differential response of normal rat mammary epithelial cells to mammogenic hormones and EGF. J Cell Physiol 125: 1182-1191.
19. Cayama E, Tsuda H, Sarma DSR, Farber E (1978) Initiation of chemical carcinogenesis required cell proliferation. Nature 275:60-62.
20. Preston-Martin S, Pike MC, Ross RK, et al. (1990) Increased cell division as a cause of human cancer. Cancer Res 50:7415-7421.
21. Weinstein IB (1991) Mitogenesis is only one factor in carcinogenesis. Science 251: 387-388.
22. Nandi S, Guzman RC, Miyamoto S (1992) Hormones, Cell Proliferation and Mammary Carcinogenesis. In Li J, Nandi S, Li SA (eds): Hormonal carcinogenesis. New York: Springer-Verlag, pp 73-77.
23. Nagasawa H (1979) The cause of species differences in mammary carcinogenesis: Significance of mammary gland DNA synthesis. Medical Hypotheses 5:499-510.
24. Russo J, Russo IH (1978) DNA labelling index and structure of the rat mammary gland as determinants of its susceptibity to carcinogenesis. J Natl Cancer Inst 61: 1451-1459.
25. Anderson CH, Hussain RA, Han MC, Beattie CW (1991) Estrous cycle dependence of nitrosomethylurea (NMU)-induced preneoplastic lesions in rat mammary gland. Cancer Lett 56:77-84.
26. Miyamoto S, Guzman RC, Osborn RC, Nandi S (1988) Neoplastic transformation of mouse mammary epithelial cells by in vitro exposure to N-methyl-N-nitrosourea. Proc Natl Acad Sci USA 85:477-481.
27. Bera TK, Guzman RC, Miyamoto S, et al. (1994) Identification of a mammary transforming gene (MAT 1) associated with mouse mammary carcinogenesis. Proc Natl Acad Sci USA 91:9789-9793.
28. Medina D (1981) The preneoplastic state in mouse mammary tumorigenesis. Carcinogenesis 9:1113-1119.
29. Guzman RC, Osborn RC, Swanson SM, et al. (1992) Incidence of c-Ki-ras activation in N-methyl- N-nitrosourea-induced mammary carcinomas in pituitary isografted mice. Cancer Res 52:5732-5737.
30. Christov L, Swanson SM, Guzman RC, Thordarson G, Jin E, Talamantes F, Nandi S (1993) Kinetics of mammary epithelial cell proliferation in pituitary isografted mice. Carcinogenesis 14:2019-2025.

31. Pascual RV, Hwang S-i, Swanson SM, et al. (1994) c-H-*ras* activation in MNU induced rat mammary cancers is regulated by ovarian hormones. Proc Am Assoc Cancer Res 35:262.
32. Moon RC (1981) Influence of pregnancy and lactation on experimental mammary carcinogenesis In Pike MC, Siiteri PK, Welsch CW (eds): Hormones and Breast Cancer. Cold Spring Harbor: Banbury Rep 8, pp 353-361.
33. Foulds L (1956) The histological analysis of mammary tumors in mice. II. The histology of responsiveness and progression: The origins of tumors. J Natl Cancer Inst 17:713-753.
34. Jacquemier JD, Hassoun J, Torrente M, Martin P-M (1990) Distribution of estrogen and progesterone receptors in healthy tissues adjacent to breast lesions at various stages-immunohistochemical study of 107 cases. Breast Cancer Res Treat 15:109-117.
35. Haslam SZ, Nummy KA (1992) The ontogeny and cellular distribution of estrogen receptors in normal mouse mammary gland. J Steroid Biochem Molec Biol 42: 589-595.
36. Jiang S-Y, Jordan CV (1992) Growth regulation of estrogen receptor-negative breast cancer cells transfected with complementary DNA for estrogen receptor. J Natl Cancer Inst 84:580-591.
37. Zajchowski DA, Sager R, Webster L (1993) Estrogen inhibits the growth of estrogen receptor-positive human mammary epithelial cells expressing recombinant estrogen receptor. Cancer Res 53:5004-5011.
38. Barker S, Panahy C, Puddefoot JR, et al. (1989) Epidermal growth factor receptor and estrogen receptors in the non-malignant part of the cancerous breast. Brit J Cancer 60:673-677.
39. Dittadi R, Donisi PM, Brazzale A, et al (1993) Epidermal growth factor receptor in breast cancer. Comparison with the non-malignant breast tissue. Brit J Cancer 67:7-9.
40. Anbazhagan R, Gusterson BA (1994) Prenatal factors may influence predisposition to breast cancer. Europ J Cancer 30A:1-3.
41. Raynaud A (1961) Morphogenesis of the mammary gland. In Kon S K, Cowie AT (eds): Milk: The Mammary Gland and its Secretion, New York: Academic Press, pp 3-46.
42. Bern HA, Talamantes FJ Jr (1981) Neonatal mouse models and their relation to disease in the human female.In Herbst AL, Bern HA (eds): Developmental Effects of Diethylstilbestrol (DES) in Pregnancy, New York: Thieme-Stratton, pp 129-147.
43. Mori T, Nagasawa H, Bern HA (1980) Long-term effects of exposure to hormones on normal and neoplastic mammary growth in rodents: A review. J Environ Pathol Toxicol 3:191-205.

44. Boylan ES, Calhoon RE (1981) Prenatal exposure to diethylstilbestrol: Ovarian-independent growth of mammary tumors induced by 7,12-dimethylbenz(a)anthracene. J Natl Cancer Inst 66:649-652.
45. Scully RE, Welch WR, (1981) Pathology of the female genital tract after prenatal exposure to diethylstilbestrol. In Herbst AL, Bern HA (eds): Developmental Effects of Diethylstilbestrol (DES) in Pregnancy, New York: Thieme-Stratton, pp 26-45.
46. Oberley TD, Gonzalez A, Lauchner LJ, Oberley LW, Li JJ (1991) Characterization of early kidney lesions in estrogen-induced tumors in the syrian hamster. Cancer Res 51:1922-1929.
47. Anbazhagan R, Bartek J, Monaghan P, Gusterson BA (1991) Growth and development of the human infant breast. Am J Anat 192:407-417.
48. Laurence DJ, Monaghan P, Gusterson BA (1991) The development of the normal human breast. Oxford Rev Reprod Biol 13:149-174.
49. Pallis L, Skoog L, Falkmer U, Wilking N, Rutquist LE, Auer G, Cedermark B (1992) The DNA profile of breast cancer in situ. Eur J Surg Oncology 18:108-111.
50. Aue, GU, Fallenius AG, Erhardt KY, Sundelin B (1984) Progression of mammary adenocarcinomas as reflected by nuclear DNA content. Cytometry 5: 420-425.
51. Habel LA, Stanford JL (1993) Hormone receptors and breast cancer. Epidemiol Rev 15:209-219.

Symposium Presentation

Regulation of Cytochrome P-450 and Carcinogenesis: Peroxisome Proliferator-Activated Receptor and Dioxin Receptor

Carol D. Banner, Joseph J. Rafter, and Jan-Åke Gustafsson

Introduction

Chemical carcinogenesis often involves metabolic activation of precursor compounds to reactive species which may initiate the carcinogenic process by interacting with the genetic material, leading to formation of DNA-adducts and other chemical insults. A key enzyme in this scenario is cytochrome P-450, which can carry out a multitude of chemical reactions whereby both xenobiotics and sometimes endogenous compounds may be transformed into highly reactive metabolites. Alternatively, the reactions may lead to the generation of highly reactive side products, e.g., free radicals. Accordingly, the levels of different isoforms of cytochrome P-450 in various tissues are enormously important in determining the extent of these activation processes in the organism. Therefore, regulation of cytochrome P-450 is a most significant issue in the elucidation of chemical carcinogenesis. Often, hormones are important regulators of cytochrome P-450; thus, hormonal carcinogenesis may involve hormonal regulation of cytochrome P-450. We have chosen to illustrate regulation of cytochrome P-450 with two examples. The first is the peroxisome proliferator-activated receptor (PPAR), which regulates P-450 of the 4A type and for which involvement in liver carcinogenesis has been claimed. We and others have discovered that PPAR-regulated genes may be expressed by dehydroepiandrosterone sulphate; this paper addresses possible mechanisms involved. Secondly, we describe certain aspects of Ah (dioxin)-receptor activation, an important transcription factor for development of the immune system and liver (1). It has been claimed that the dioxin receptor is involved in chemical carcinogenesis; we will show that common air pollutants may be important in activating the receptor and thus causing an increased transcription of cytochrome P-450.

Dehydroepiandrosterone (DHEA) is a C_{19} steroid synthesized in the human adrenal cortex and serving as a precursor for production of testosterone and 17β-estradiol (2). DHEA circulates *in vivo* principally as its 3β sulphate conjugate (DHEAS) and, as such, is the most abundant steroid hormone in circulation (3). Despite this fact, the physiological role of DHEA in humans remains unclear. However, in experimental animals that make little or no DHEA, long-term administration of this steroid inhibits many pathological processes, including the development of tumors, diabetes, obesity, and atherosclerosis (3, 4). It has been suggested that some of DHEA's effects on obesity, when administered in pharmacological doses, are related to its ability to serve as a peroxisome proliferator. Like exogenous chemicals (which compose the majority of known peroxisome proliferators), DHEA administered to rats and mice induces a pleiotropic response characterized by hepatomegaly and hepatic peroxisomal proliferation, as well as by induction of peroxisomal β-oxidation enzymes, microsomal fatty acid ω-hydroxylase, and several other enzymes associated with lipid metabolism (5-8). Recent studies have revealed that PPAR, a member of the nuclear receptor superfamily, mediates the action of peroxisome proliferators with different chemical structures (9). However, this receptor, which is transcriptionally active via peroxisome-proliferator response elements (PPRE) in the regulatory region of the genes encoding the lipid metabolism enzymes indicated above (10-16), appears responsive to neither DHEA nor DHEAS (9, 17, 18), and the mechanism of action of DHEA remains unknown.

In the present studies, we report the induction of P-450IVA1 mRNA levels in rat liver and hepatocytes in primary culture as a marker for the pleiotropic effects of DHEA and DHEAS as peroxisome proliferators. In an attempt to discover a role of PPAR in these events, proteins in nuclear extracts from DHEA-treated livers were tested for their ability to associate specifically with a radiolabeled PPRE oligonucleotide probe in a gel mobility shift assay. Despite the clearly demonstrated peroxisomal proliferative effects of DHEA and DHEAS, no evidence was observed to suggest an increase in PPRE binding. Such data, however, do not exclude a possible role for PPAR in the induction of fatty-acid metabolism by DHEA. Alternatively, DHEA may activate an as yet unidentified isoform of PPAR or a PPAR-related factor.

Polycyclic aromatic compounds (PACs) are generated in large quantities in our environment during combustion processes (19, 20), and many are potentially very toxic. Metabolic activation of these compounds in mammalian tissues leads to more chemically reactive species and constitutes a key step in the events leading to genotoxicity and carcinogenicity. Cytochromes P-450IA1 and IA2 preferentially catalyze this activation and are known generally as aryl hydrocarbon hydroxylase (AHH) (21). PACs themselves can induce cytochrome P-450IA1 activity (21, 22), thus increasing their own metabolism to possibly highly toxic end products. Studies have shown that this induction occurs via binding of these compounds to a cytosolic receptor protein that specifically binds 2,3,7,8-tetrachlorodibenzo-p-dioxin (TCDD), the dioxin or Ah receptor (23, 24). Having bound a ligand, the dioxin receptor undergoes an "activation" process resulting in its heterodimerization to a protein known as Arnt (25). This heterodimer then translocates to the cell nucleus, where it interacts with specific DNA sequences to alter the transcription rates of target genes (26, 27).

The ability of compounds to induce AHH has been shown to correlate with their affinity for the dioxin receptor (23, 28, 29). For example, vehicle exhaust particulate extracts have been shown to induce AHH in the cultured cell line H4-II-E, and the extent of this induction correlates with their dioxin-receptor affinity (30), indicating that these compounds have the potential to produce effects in living cells. Dioxin-receptor binding has been associated with the initiation of a variety of toxic effects, including teratogenesis, tumor promotion, and cellular hyperplasia and differentiation (23). Thus, binding to the dioxin receptor produces a measure of the toxic potential of these compounds.

It is generally accepted that human cancer rates are higher in urban areas than those in more rural areas. In recent years, we have been interested in studying the presence of dioxin-receptor ligands in urban air in an effort to understand the mechanisms underlying this phenomenon. We have shown that extracts of urban air and vehicle exhaust particulates contain significant amounts of dioxin-receptor binding activity and could provoke the induction of cytochrome P-450IA1 in cultured rat hepatoma cells (31). In addition, studies with a variety of diesel fuels show that the amounts of dioxin-receptor ligands present in exhaust emissions is fuel-dependent, and that substantial amounts of these ligands were also present in the semi-volatile phase of exhaust emissions (31). Therefore, it is of considerable interest to attempt to characterize more thoroughly the nature of the compounds present in vehicle exhaust emissions that are responsible for the observed dioxin-receptor binding activity. To this end, we fractionated a model emission particulate material from a diesel fuel, analyzed which fractions elicited dioxin-receptor binding, and carried out a chemical analysis on the fractions. In addition, we used multivariate data analysis to determine which compounds present in the emission samples contributed most to the observed dioxin-receptor binding activity.

Results and Discussion

DHEA Induces Liver P-4504A1 *In Vivo*

Studies carried out in intact rats have demonstrated that peroxisome proliferator compounds such as clofibric acid induce expression of several cytochromes

P-450 belonging to family 4A (P-4504A). This induction may correspond to an early, perhaps obligatory step in the peroxisome-proliferator pathway (32-35). In order to characterize the peroxisome-proliferator activity of DHEA, the liver P-4504A1 mRNA levels of DHEA-treated rats (gavage in corn oil vehicle 225 mg DHEA/100g body weight) were compared with those of control animals (corn oil gavage only) (Figure 1). This dose is considered pharmacological, as there is little evidence that DHEA is made in appreciable amounts in rats. Low levels of P-4504A1 mRNA found in the livers of control animals at 1 h and 24 h post dose were comparable with that described previously (36). Of interest is the lack of induction of P-4504A1 by the corn oil vehicle, as peroxisome proliferation has been associated with a high fat diet (37). One hour after dosing, a slight elevation (approximately 2.5-fold) in P-4504A1 mRNA levels was observed in the DHEA-treated animals compared with the controls. A major increase in mRNA encoding P-4504A1 was observed 24 h post dose, with approximately 20.0-fold higher levels in the treated animals than in the controls. Similar data have been observed by Prough, et al., using Northern analysis (38). The increase in P-4504A1 expression by DHEA has also been demonstrated by measurement of the cytochrome P-4504A1 protein levels and lauric hydroxylase activity (39). The magnitude of DHEA induction of P-4504A1 mRNA levels is comparable with that by xenobiotic peroxisome proliferators, e.g., a single dose of methylclofenapate induced liver P-4504A1 mRNA levels 15.0-fold above those of controls, 24 h after dosing (40). However, the high doses of DHEA required to elicit an inductive effect observed by ourselves and others (6, 38, 39) make it unlikely that this steroid is a direct activator of the peroxisome-proliferator pathway.

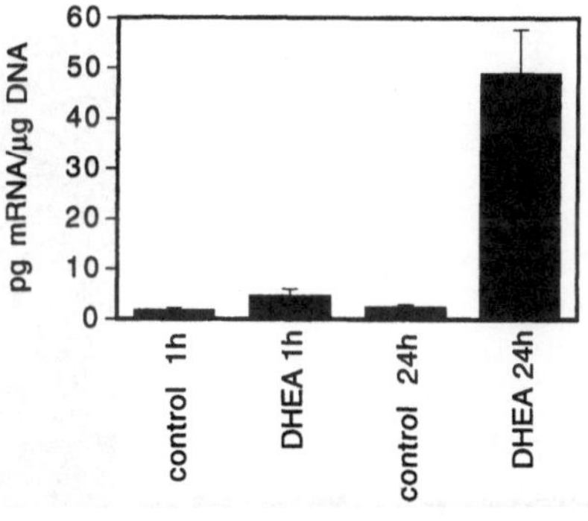

Figure 1. Induction of hepatic microsomal P-4504A1 mRNA by DHEA *in vivo*. DHEA was administered to male rats (150 g) in corn oil by a single gavage dose of 225 mg/100 g body weight. P-4504A1 mRNA levels were quantified by solution

hybridization. Each bar represents the mean P-4504A1 mRNA concentration of 5 animals.

DHEAS but not DHEA Induces P-4504A1 mRNA in Primary Cultures of Rat Hepatocytes

Primary rat hepatocytes are known to be responsive to clofibric acid and other peroxisome proliferators (41, 42). We therefore utilized this cell system to study whether the *in-vivo* peroxisome-proliferator response of DHEA could be achieved by direct administration of the steroid to hepatocytes. In light of the well-documented *in-vivo* rat metabolism of DHEA to DHEAS (43), this sulphate conjugate of DHEA was also studied *in vitro.* These studies (Figure 2) revealed that 100 μM DHEAS, but not 100 μM DHEA, substantially induced mRNA levels of the P-4504A1 gene. This finding is consistent with those of Ram and Waxman (44), who observed increases in both P-4504A protein and mRNA levels after exposure of primary hepatocytes to the sulphated, but not the unconjugated, DHEA. However, Yamada, et al. (45), did observe DHEA induction of peroxisomal β-oxidation activity in cultured rat hepatocytes, albeit substantially weaker than that of DHEAS. This low or absent peroxisome-proliferator activity of DHEA in cellular systems contrasts with its high peroxisome-proliferator activity observed *in vivo.*

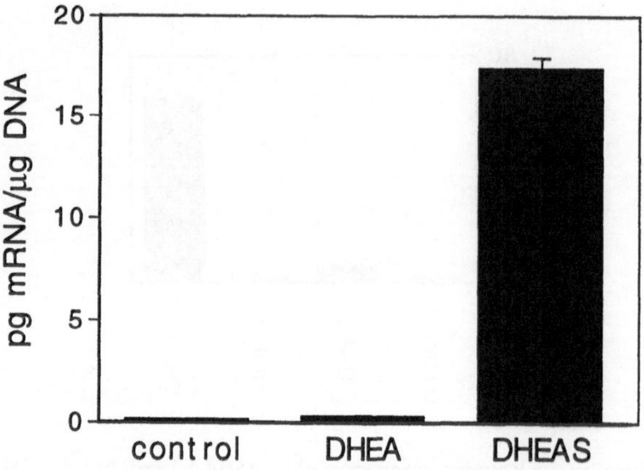

Figure 2. P-4504A1 mRNA levels were determined in TNA samples from hepatocytes treated for 3 days with 0.1% DMSO (control), 100 μM DHEA and 100

μM DHEAS using solution hybridization. Data represent mean ± SEM of 3 samples.

Taken together, these data suggest that DHEAS is an activated form of DHEA and that peroxisome-proliferator effects observed after DHEA administration *in vivo* are mediated by its 3-sulphate conjugate. Therefore, sulphation of DHEA, which occurs both in the liver and at the site of steroid synthesis in the adrenal gland, not only facilitates systemic transport of the steroid, but also activates the steroid to a form that stimulates the peroxisome-proliferator pathway. The 3-sulphate conjugate is a hydrophobic anion, a structure that fulfills minimum requirements for the peroxisome-proliferator activity of a wide range of chemicals. A variation in sulphotransferase activity between different cell culture systems may explain, therefore, the observed inconsistent peroxisome-proliferator responses by DHEA. *In-vitro* studies in which inhibitors of sulphate conjugation, such as acetaminophen and chlorate, suppress DHEA induction of peroxisomal β-oxidation offer further evidence for the DHEAS mediation of DHEA peroxisome-proliferator response (46). That DHEAS is an activated form of DHEA is consistent with DHEAS inducing P-4504A1 mRNA *in vivo*, and its effects correlate well with those of DHEA (44). Whether DHEAS itself is the ultimate activator of P-4504A1 expression, or requires further metabolism into an active form, cannot yet be determined.

Time Course and Dose Dependence of DHEAS Induction of P-4504A1 *In Vitro*

To characterize further the induction of P-4504A1 by DHEAS in cultured rat hepatocytes, time course and dose studies were performed. Treatment of cultured rat hepatocytes with 100μM DHEAS for up to 5 days resulted in a progressive increase in P-4504A1 mRNA levels (Figure 3). The basal expression of P-4504A1 mRNA (day 0) is reduced in hepatocytes compared with intact liver, possibly because of the absence of endogenous inducers; however, responsiveness to peroxisome proliferators, such as DHEAS, is maintained. The progressive increase in peroxisome-proliferator response in cultured rat hepatocytes during extended exposure to DHEAS was also observed by Yamada, et al. (45), who studied peroxisomal β-oxidation over a five-day period. In contrast to the extended time course of DHEAS-induced peroxisomal proliferation response, Wy 14,643 and fatty acids induce maximal P-4504A1 mRNA levels in primary rat hepatocytes after only two days of treatment (47). The mRNA levels induced by 10 μM Wy 14,643 at two days were approximately half those induced by 100 μM DHEAS at five days. The reason for this discrepancy in the time course of responses *in vitro* between various peroxisome proliferator compounds is not clearly understood. Rates of metabolism to the direct activating compound and/or the solubility of such compounds in the cell media and therefore their availability to the cells may contribute to this phenomenon.

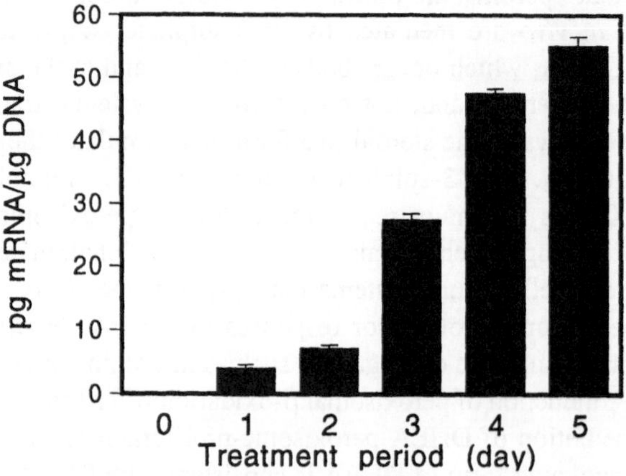

Figure 3. Time-related increase in mRNA levels of P-4504A1 in cultured rat hepatocytes treated with DHEAS 100 µM. Each bar represents the mean ± SD of 3 samples.

Treatment with DHEAS at 50- to 200-µM concentrations induced P-4504A1 mRNA levels in a dose-dependent manner (data not shown). However, 200 µM DHEA was incapable of inducing a response greater than that of 100 µM of the unconjugated steroid. Ram and Waxman observed induction of P-4504A1 mRNA levels in cultured rat hepatocytes at DHEAS concentrations as low as 10 µM (44), which is within the range of physiological circulatory levels of this adrenal steroid in man. The activity of peroxisomal β-oxidation was maximal in the presence of 50-100 µM DHEAS as reported by Yamada, et al. (45). However, no change in cell condition, such as detachment from culture dishes, was observed in the 100-250 µM dose range.

Specificity of DHEAS Induction of P-4504A1 *In Vitro*

In order to determine the specificity of DHEAS as a peroxisome proliferator in cultured rat hepatocytes, several related compounds (DHEA-3-acetate, cholesterol, cholesterol-3-sulphate, pregnenolone-3-sulphate, androstenediol, androstenediol- 3-acetate, cholic acid, deoxycholic acid, ursodeoxycholic acid, and chenodeoxycholic acid) were tested for their ability to induce P-4504A1 mRNA levels. Exposure to 100 µM of any of these compounds for three days did not increase gene expression in cultured rat hepatocytes above basal levels. DHEAS was the only tested steroid that could induce P-4504A1 mRNA levels *in vitro,* as neither precursors nor metabolites of DHEAS could elicit the same response. The ineffectiveness of DHEA acetate, a conjugate that is more hydrophilic than DHEA, suggests the response from DHEAS is not associated with its greater solubility in cell medium compared with the unconjugated steroid.

Other groups have indicated that androstenediol and its 3-sulphate have as much ability as DHEA and its 3-sulphate (DHEAS) to induce peroxisome proliferation *in vivo* (38) and *in vitro*, respectively (48). These two steroids, which at pharmacological doses lead to transcriptional activation of P-4504A1 in rats, appear to be the only natural intermediates of the human steroid biosynthetic pathway. However, the active forms of these steroids which induce peroxisomal proliferation are not known.

The peroxisome-proliferator activity of DHEA and DHEAS, as demonstrated by their induction of P-4504A1 mRNA in this study, is inconsistent with their reported lack of PPAR activity in transactivation studies (8, 17, 18). Therefore, PPAR activity may be mediated by a metabolite of DHEAS formed in hepatocytes but not in the heterologous cell systems used in the receptor reporter transfection experiments. We have yet to determine whether PPAR itself is activated by DHEAS, or whether the effects of this steroid sulphate on P-4504A1 gene transcription reported here are mediated by a PPAR-related receptor protein or a so far unidentified isoform of PPAR.

Role of PPAR in DHEA-Induced Peroxisome Proliferation

In an attempt to obtain evidence for a role for PPAR (or other novel receptor that interacts with its target gene at a peroxisome-proliferator response element [PPRE]) in DHEA-induced peroxisome proliferation, gel shift analysis was employed. A [^{32}P]-labeled PPRE from the acyl CoA oxidase gene (ACOX) was used as a probe to identify PPRE-interacting proteins in liver nuclear extracts prepared 24 h after oral treatment of control rats and those dosed with Wy 14,643 and DHEAS (Figure 4). It has been previously demonstrated that PPAR forms a heterodimer with the retinoid X receptor (RXR) and that they can cooperatively stimulate expression of peroxisome proliferation target genes (13, 49). The specific binding of this heterodimer to the ACOX PPRE is illustrated in Lanes 3 and 4 (marked with an arrow). Liver nuclear extracts from treated rats form complexes with the ACOX PPRE of similar mobility whatever the type of treatment, i.e., control (Lanes 5 and 6), DHEA (Lanes 7-9), and Wy 14,643 (Lanes 10 and 11). Therefore, no peroxisome proliferator-induced PPRE binding proteins are apparent. Three complexes appear specific to the PPRE, as they are strong competitors at 80.0-fold excess of unlabeled PPRE oligonucleotide. One of these specific complexes has a similar mobility to that of the standard heterodimer complex (Lane 3 marked with an arrow). As both PPAR and RXR are highly expressed in the liver (9, 50), it is possible that this specific band is that of the PPAR RXR heterodimer bound to the PPRE.

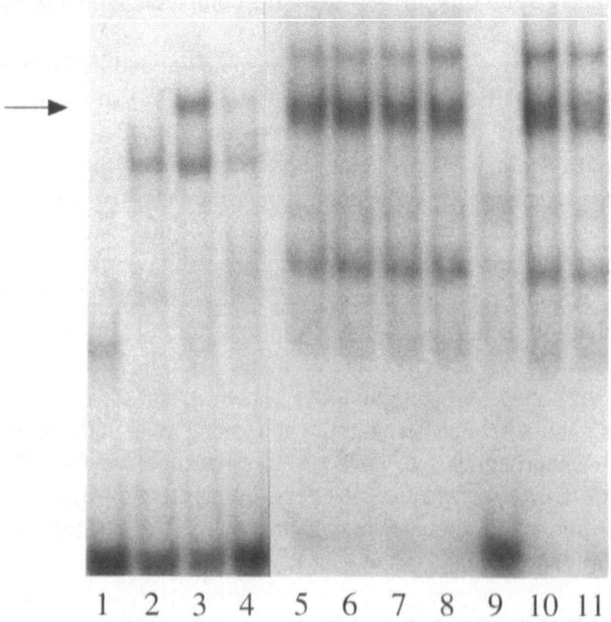

Figure 4. Gel shift analysis of rat liver nuclear extracts prepared 24 h post oral treatment with peroxisome proliferators. Lane 1 rRXRα *in vitro* translated; Lane2 rPPARα baculovirus expressed as described by Gearing, et al. (41); Lane 3 rRXRα and rPPARα; Lane 4 same as Lane 3 but with the addition of 80.0-fold excess unlabeled PPRE oligonucleotide; Lanes 5 and 6 control liver from two animals treated with carboxymethyl cellulose (gavage vehicle only); Lanes 7 and 8 liver from two animals treated with 225 mg DHEA/100 g body weight; Lane 9 same as Lane 8 but with the addition of 80.0-fold excess unlabeled PPRE oligonucleotide; Lanes 10 and 11 liver from two animals treated with 50 mg Wy 14,643/kg body weight. The specific PPAR-RXR-PPRE complex is indicated (arrow).

Of particular interest, however, is that the intensity of the specific bands in the control rat liver nuclear extract is not influenced by treatment with peroxisome proliferators, e.g., DHEA and Wy 14,643. Similar observations were made with nuclear extracts taken from similarly treated animals at one and three hours after treatment. Taken together, these observations indicate that such peroxisome proliferators do not influence the binding of PPAR RXR heterodimer or any other nuclear protein to ACOX PPRE.

Therefore, the mechanism by which peroxisome-proliferator compounds activate transcription of their target genes remains to be established. It can be speculated that binding of PPAR RXR heterodimer to the PPRE may occur in the absence of peroxisome-proliferator compounds. Target gene expression activated in the presence of these compounds may occur via *in-situ* post-translational modification of the function of those or other transcription factors, as demonstrated in the case of the progesterone receptor (51).

Dioxin-Receptor Binding Activity of Diesel Exhaust Particulate Material

To obtain the large amounts of particulate material from light-duty diesel vehicles required for the present study, particulate material was collected using a cyclone particulate separator, as described previously (52). The particulate material collected is used as a model containing compounds present in urban air originating from vehicular traffic. The cyclone particulate sample was Soxhlet extracted with dichloromethane for 24 hours. The DCM extract was evaporated under reduced pressure to near dryness and diluted to a known volume with acetone; an aliquot of this crude extract was stored at -4°C until dioxin-receptor binding assay was performed. The remaining crude extract was fractionated, based on polarity, into five fractions I to V, on a silica gel column, as described previously (52). Fractions I, II, and III were further subfractionated on the silica column as described previously (52). PAC analysis of fraction II and its subfractions was carried out using gas chromatography-mass spectrometry as described previously (20). 1-nitropyrene analysis of fraction III and its subfractions was performed by HPLC with fluorescence detection as previously described (20). The crude extract, recombined sample (fraction I to V), fractions I to V, and subfractions I-1, I-2, II-1 to II-7, and III-1 to III-6 were assayed for relative dioxin-receptor binding affinity. The dioxin-receptor binding affinities of samples were measured using a HAP adsorption assay developed in our laboratory (53, 54). Relative binding affinity was calculated as IC^{50}, the concentration of sample (μg particulate matter/ml cytosol) required to compete for 50% of the $[^3H]$-TCDD binding sites. Interpretation of the data was simplified by use of the inverted IC_{50} value (1 IC^{50-1}), which correlates positively with both receptor binding affinity and potential toxicity. Principal components analysis was performed as described above.

The distribution of dioxin-receptor binding activity over the fractions and subfractions from the diesel exhaust particulate material are shown in Figure 5. Most of this activity elutes in fractions II, III and IV. Fraction II contained all of the 33 PAC analyses and elicited high receptor binding with a 1 IC_{50}^{-1} value of 8.33^{-3}. However, fraction III contained the somewhat more polar mono-nitro PAC, as was evident by our 1-nitropyrene analysis, and was a much better competitor of [^{3}H]-TCDD binding, with a 1 IC_{50}-1 value of 2.00^2 . Fraction IV contains more polar dinitro-PAC and elicited high dioxin-receptor binding activity, with a 1 IC_{50}-1 value of 1.54^{-2}.

On subfractionation of fraction II, more than 90% of the PACs detected were in subfractions II-3 to II-5. Subfraction II-3 contained the largest proportion of PAC (64%) and elicited poor receptor binding. However, subfraction II-4, which contained the second largest amount of PAC (35%), displayed the highest dioxin-receptor binding activity of all the subfractions (Figure 5). Subfraction II-5 contained <1% PAC and showed the second highest receptor binding of

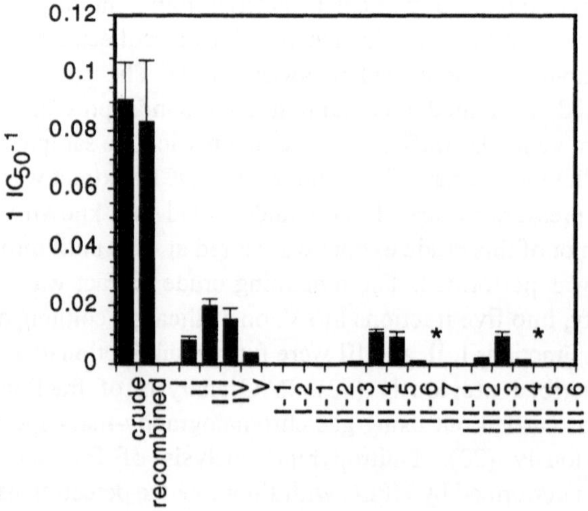

Figure 5. Dioxin-receptor binding affinities (mean 1 IC_{50}^{-1} with standard deviations, n ≥3) of crude extract, recombined sample, fractions and subfractions of diesel exhaust particulate extracts. Subfractions annotated with (*) displayed very low receptor binding affinity with 1 IC_{50}^{-1} values in the range 4.08^{-5}-1.08^{-4}. Other fractions and subfractions without data bars displayed unquantifiable dioxin-receptor binding affinity, i.e., 1 IC_{50}^{-1} < 4.08^{-5}.

fraction II subfractions. Although subfraction II-3 contains the majority of the PACs (64%), most of the compounds present in subfraction II-3 are mainly three-ringed PACs, e.g., anthracene, phenanthrene, their methyl derivatives, and fluoranthenes. Such compounds have been reported to be poor ligands for the dioxin receptor (55, 56). We have previously shown that PAC retention on silica gel depends on the number of rings, i.e., PACs elute in order of increasing number of rings, from 2 to 6 (57). Subfractions II-4 and II-5 elicited the highest dioxin-receptor binding activity but contain only a small proportion of the PACs measured. However, the PACs in subfractions II-4 and II-5 are larger molecular weight compounds such as the four-ringed benzo(a)anthracene, chrysene, benzo(b and k)fluoranthene, and the five-ringed benzo(a and e)pyrene, all of which, with the exception of benzo(e)pyrene, have been shown to bind well to the dioxin receptor (55, 56).

Thus, the distribution of the dioxin-receptor binding activities determined over these subfractions agrees with that predicted by PAC content. However, it is not known whether the measured PACs account for all of the observed binding activity or whether other unidentified compounds contribute to the binding activity. Principal components analysis indicated that dioxin-receptor binding activity observed in the samples is positively correlated with certain PACs. These included benzo(a)anthracene, chrysene, benzo(b)fluoranthene, benzo(a)pyrene, and indeno(1,2,3-cd)pyrene.

In the present study, we have used 1-nitropyrene as a marker for mono-nitro PAC and examined its distribution over the subfractions of fraction III. More than 80% of the 1-nitropyrene detected eluted in subfractions III-3 and III-4. Subfraction III-3 contained the largest proportion of 1-nitropyrene (68%) and elicited the highest receptor binding among subfractions of fraction III. Subfraction III-4, which contained the second largest amount of 1-nitropyrene (12%), displayed a lower dioxin-receptor binding activity. Thus, the major dioxin-receptor binding activity of fraction III co-elutes with most of the marker nitro-PACs. In light of the low binding activity of 1-nitropyrene, these data indicate that compounds as yet unidentified in fraction III are responsible for the high binding activity observed. The high dioxin-receptor binding activity in fraction IV may be associated with di-nitro-PACs. Further investigation into the chemical composition and dioxin-receptor binding affinity of fraction IV is needed.

In conclusion, upon fractionation of diesel vehicle particulate emission, the major dioxin-receptor binding activity is recovered in those fractions containing nitro-PACs. There is also a considerable amount of receptor binding activity in that fraction, which contained most of the PAC analyses. Interestingly, this binding activity did not subfractionate with the majority of the PACs in this fraction. This divergence may be explained by the dependence of dioxin-receptor binding activities on PAC molecular weight and structure-activity properties. In addition, compounds other than PACs may play a role in the observed dioxin-receptor binding. However, principal components analysis did indicate that dioxin-receptor binding is correlated with PAC content, most closely with benzo(a)anthracene, chrysene, and benzo(a)pyrene. Thus, it is likely that nitro-PACs and the PACs originating from vehicle exhaust emissions (specified above), are present in urban air. It can be speculated that such compounds may contribute to the higher cancer risk associated with urban areas, although the magnitude of this risk requires further study.

In summary, we have shown that DHEAS, a major steroid hormone metabolite in human blood induces cytochrome P-450 of the 4A1 type. This induction is seen in primary hepatocytes when cultured *in vitro* and thus represents a direct action on the liver cell. In a sense, cytochrome P-450 4A1 may be regarded as a marker for the peroxisome proliferator-like activity of DHEAS, since induction of this cytochrome and peroxisomal proliferation are often coupled. The biological function of DHEAS is still unclear, although the high production rates of this particular steroid metabolite in the human may indicate an important physiological role. Anti-tumorigenic characteristics have been claimed for DHEA, and further mechanistic studies on the various biochemical effects of this interesting steroid are clearly warranted. Furthermore, we have shown in this paper that common air pollutants are ligands to the dioxin receptor and may thus play a role in activating this transcription factor, leading to an increased transcription of cytochrome P-450. It is well known that the particular cytochrome P-450 isozymes induced are 1A1 and 1A2, which have been shown to play important roles in the activation of polycyclic aromatic hydrocarbons and heterocyclic amines, respectively. Furthermore, studies from several laboratories indicate that ligands to the dioxin receptor may have quite potent anti-estrogenic activity and that this mechanism may be of significance in explaining the alleged protective effect of certain foodstuffs against breast carcinogenesis. The description in the present paper of relatively potent ligands to the dioxin receptor in car exhaust is yet another example of the potential impact of the environment on carcinogenesis.

Acknowledgments

Roger Westerholm is acknowledged for carrying out the chemical analyses on the diesel vehicle particulate emission. This investigation was financed by the Swedish Cancer Society and the Swedish Environmental Protection Agency.

References

1. Fernandez-Salguero P, Pineau T, et al. (1995) Science, in press.
2. Zumoff B, Rosenfeld R, Strain GW, et al. (1980) Sex differences in the twenty-four-hour mean plasma concentrations of dehydroisoandrosterone (DHA) and dehydroisoandrosterone sulfate (DHAS). J Clin Endocrinol Metab 51:330-333.
3. Gordon GB, Shantz LM, Talalay P (1987) Modulation of growth, differentiation and carcinogenesis by dehydroepiandrosterone. Adv Enzyme Regul 26:355-382.

4. Schwartz AG, Whitcomb JM, Nyce JW, et al. (1988) Dehydroepiandrosterone and structural analogs: A new class of cancer chemopreventive agents. Adv Cancer Res 51:391-423.
5. Frenkel RA, Slaughter CA, Orth K, et al. (1990) Peroxisome proliferation and induction of peroxisomal enzymes in mouse and rat liver by dehydroepiandrosterone feeding. J Steroid Biochem 35:333-342.
6. Yamada J, Sakuma M, Ikeda T, et al. (1991) Characteristics of dehydroepiandrosterone as a peroxisome proliferator. Biochim Biophys Acta 1092:233-243.
7. Sakuma M, Yamada J, Suga T (1992) Comparison of the inducing effect of dehydroepiandrosterone on hepatic peroxisome proliferation-associated enzymes in several rodent species. A short term administration study. Biochem Pharmacol 43:1269-73.
8. Rao MS, Musunuri S, Reddy JK (1992) Dehydroepiandrosterone-induced peroxisome proliferation in the rat liver. Pathobiology 60:82-86.
9. Issemann I, Green S (1990) Activation of a member of the steroid hormone superfamily by peroxisome proliferators. Nature 347:645-650.
10. Dreyer C, Krey G, Keller H, et al. (1992) Control of the peroxisomal beta-oxidation pathway by a novel family of nuclear hormone receptors. Cell 68:879-887.
11. Tugwood JD, Issemann I, Anderson RG, et al. (1992) The mouse peroxisome proliferator activated receptor recognizes a response element in the 5' flanking sequence of the rat acyl CoA oxidase gene. EMBO J 11:433-439.
12. Zhang B, Marcus SL, Sajjadi FG, et al. (1992) Identification of a peroxisome proliferator-response element upstream of the gene encoding rat peroxisomal enoyl-CoA hydratase/3-hydroxyacyl-CoA dehydrogenase. Proc Natl Acad Sci USA 89:7541-7545.
13. Kliewer SA, Umesono K, Noonan DJ, et al. (1992) Convergence of 9-cis retinoic acid and peroxisome proliferator signaling pathways through heterodimer formation of their receptors. Nature 358:771-774.
14. Muerhoff AS, Griffin KJ, Johnson EF (1992) The peroxisome proliferator-activated receptor mediates the induction of CYP4A6, a cytochrome P-450 fatty acid omega-hydroxylase. J Biol Chem 267:19051-19053.
15. Rodriguez JC, Gil-Gomez G, Hegardt FG, Haro D (1994) Peroxisome proliferator-activated receptor mediates induction of the mitochondrial 3-hydroxy-3-methylglutaryl-CoA synthase gene by fatty acids. J Biol Chem 269:18767-18772.
16. Castelein H, Gulick T, Declercq PE, et al. (1994) The peroxisome proliferator activated receptor regulates malic enzyme gene expression. J Biol Chem 269:26754-26758.

17. Göttlicher M, Widmark E, Li Q, Gustafsson J-Å (1992) Fatty acids activate a chimera of the clofibric acid-activated receptor and the glucocorticoid receptor. Proc Natl Acad Sci USA 89:4653-4657.
18. Issemann I, Prince RA, Tugwood JD, Green S (1993) The peroxisome proliferator-activator receptor: Retinoid X receptor heterodimer is activated by fatty acids and fibrate hydrolipidaemic drugs. J Mol Endocrinol 11:37-47.
19. Alsberg T, Stenberg U, Westerholm R, et al. (1985) Chemical and biological characterization of organic material from gasoline exhaust particles. Environ Sci Technol 19:43-50.
20. Westerholm RN, Almén J, Li H, et al. (1991) Chemical and biological characterization of particulate-, semivolatile-, and gas-phase-associated compounds in diluted heavy duty diesel exhausts: A comparison of three different semivolatile-phase samplers. Environ Sci Technol 25:332-338.
21. Conney AH (1982) Induction of microsomal enzymes by foreign chemicals and carcinogenesis by polyaromatic hydrocarbons. Cancer Res 42:4875-4917.
22. Nebert DW, Eisen HJ, Negishi M, et al. (1981) Genetic mechanisms controlling the induction of polysubstrate monooxygenase (P-450) activities. Ann Rev Toxicol Pharmacol 21:431-462.
23. Poland A, Knutson JC (1982) 2,3,7,8-tetrachlorodibenzo-p-dioxin and related halogenated aromatic hydrocarbons: Examination of the mechanism of toxicity. Ann Rev Toxicol Pharmacol 22:517-554.
24. Whitlock JP Jr (1986) The regulation of cytochrome P-450 gene expression. Ann Rev Toxicol Pharmacol 26:333-369.
25. Hoffman EC, Reyes H, Chu FF, et al. (1991) Cloning of a factor required for activity of the Ah (dioxin) receptor. Science 252:954-958.
26. Denison MS, Fisher JM, Whitlock JP Jr (1988) Inducible, receptor-dependent protein-DNA interactions at a dioxin-responsive transcriptional enhancer. Proc Natl Acad Sci USA 85:2528-2532.
27. Fujisawa-Sehara A, Yamane M, Fujii-Kuriyama Y (1988) A DNA binding factor specific for xenobiotic responsive elements of P-450c gene exists as a cryptic form in cytoplasm: Its possible translocation to nucleus. Proc Natl Acad Sci USA 85:5859- 5863.
28. Mason GF, Sawyert T, Keys B, et al. (1985) Polychlorinated dibenzofurans (PCDFs): Correlation between *in-vivo* and *in-vitro* structure-activity relationships. Toxicol 37:1-12.
29. Mason GF, Farrell K, Keys B, et al. (1986) Polychlorinated dibenzo-p-dioxins: Quantitative *in-vitro* and *in-vivo* structure-activity relationships. Toxicol 41:21-31.

30. Franzen B, Haaparanta T, Gustafsson J-Å, Toftgård R (1988) TCDD receptor ligands present in extracts of urban air particulate matter induce aryl hydrocarbon hydroxylase activity and cytochrome P-450c gene expression in rat hepatoma cells. Carcinogenesis 9:111-115.
31. Mason GF (1994) Dioxin-receptor ligands in urban air and vehicle exhaust. Environ Health Perspect 102:111-116.
32. Kimura S, Hanioka N, Matsunaga E, Gonzalez FJ (1989) The rat clofibrate-inducible CYP4A gene subfamily. I. Complete intron and exon sequence of the CYP4A1 and CYP4A2 genes, unique exon organization, and identification of conserved 19-bp upstream element. DNA 8:503-516.
33. Sundseth SS, Waxman DJ (1992) Sex-dependent expression and clofibrate inducibility of cytochrome P-4504A fatty acid omega hydroxylases. Male specificity of liver and kidney CYP4A2 mRNA and tissue-specific regulation by growth hormone and testosterone. J Biol Chem 267:3915-3921.
34. Sharma R, Lake BG, Foster J, Gibson GG (1988) Co-induction of microsomal cytochrome P-452 and the peroxisomal fatty acid beta-oxidation pathway in the rat by clofibrate and di-(2-ethylhexyl)phthalate. Dose response studies. Biochem Pharmacol 37:1193-1201.
35. Kaikaus RM, Chan WK, Lysenko N, et al. (1993) Induction of peroxisomal fatty acid beta-oxidation and liver fatty acid-binding protein by peroxisome proliferators. Mediation via the cytochrome P-450IVA1 omega-hydroxylase pathway. J Biol Chem 268:9593-9603.
36. Tamburini PP, Masson HA, Bains SK, et al. (1984) Multiple forms of hepatic cytochrome P-450. Purification, characterization, and comparison of a novel clofibrate-induced isozyme with other major forms of cytochrome P-450. Eur J Biochem 139:235-246.
37. Ishii H, Fukumori N, Horie S, Suga T (1980) Effects of fat content in the diet on hepatic peroxisomes of the rat. Biochim Biophys Acta 617:1-11.
38. Prough RA, Johnson Webb S, Wu H-Q, et al. (1994) Induction of microsomal and peroxisomal enzymes by dehydroepiandrosterone and its reduced metabolite in rats. Cancer Res 54:2878-2886.
39. Wu H-Q, Masset-Brown J, Tweedie DJ, et al. (1989) Induction of microsomal NADPH-cytochrome P-450 reductase and cytochrome P-450IVA1 (P-450LA omega) by dehydroepiandrosterone in rats: A possible peroxisomal proliferator. Cancer Res 49:2337-2343.
40. Bell DR, Bars RG, Gibson GG, Elcombe CR (1991) Localization and differential induction of cytochrome P-450IVA and acyl-CoA oxidase in rat liver. Biochem J 275:247-252.
41. Gray TJB, Lake BG, Beamand JA, et al. (1983) Peroxisome proliferation in primary cultures of rat hepatocytes. Toxicol Appl Pharmacol 67:15-25.

42. Bell DR, Elcombe CR (1991) Induction of acyl-CoA oxidase and cytochrome P-450IVA1 RNA in rat primary hepatocyte culture by peroxisome proliferators. Biochem. J 280:249-253.
43. Hamilton SR, Gordon GB, Floyd J, Golightly S (1991) Evaluation of dietary dehydroepiandrosterone for chemoprotection against tumorigenesis in premalignant colonic epithelium of male F344 rats. Cancer Res 51:476-480.
44. Ram PA, Waxman DJ (1994) Dehydroepiandrosterone 3 beta sulfate is an endogenous activator of the peroxisome-proliferation pathway: Induction of cytochrome P-450 4A and acyl-CoA oxidase mRNAs in primary rat hepatocyte culture and inhibitory effects of Ca(2+)-channel blockers. Biochem J 301:753-758.
45. Yamada J, Sakuma M, Suga T (1992) Induction of peroxisomal beta-oxidation enzymes by dehydroepiandrosterone and its sulfate in primary cultures of rat hepatocytes. Biochim Biophys Acta 1137:231-236.
46. Yamada J, Sakuma M, Ikeda T, Suga T (1994) Activation of dehydroepiandrosterone as a peroxisome proliferator by sulfate conjugation Arch Biochem Biophys 313:379-381.
47. Tollet P, Strömstedt M, Fröyland L, et al. (1994) Pretranslational regulation of cytochrome P-450A1 by free fatty acids in primary cultures of rat hepatocytes. J Lipid Res 35:248-254.
48. Sakuma M, Yamada J, Suga T (1993) Induction of peroxisomal beta-oxidation by structural analogues of dehydroepiandrosterone in cultured rat hepatocytes: Structure-activity relationships. Biochim Biophys Acta 1169:66-72.
49. Keller H, Dreyer C, Medin J, et al. (1993) Fatty acids and retinoids control lipid metabolism through activation of peroxisome proliferator-activated receptor-retinoid X receptor heterodimers. Proc Natl Acad Sci USA 90:2160-2164.
50. Mangelsdorf DJ, Ong ES, Dyck JA, Evans RM (1990) Nuclear receptor that identifies a novel retinoic acid response pathway. Nature 345:224-229.
51. Denner LA, Weigel NL, Maxwell BL, et al. (1990) Regulation of progesterone receptor-mediated transcription by phosphorylation. Science 25:1740-1743.
52. Li H, Banner CD, Mason GF, et al. (1995) Determination of polycyclic aromatic compounds and dioxin receptor ligands in diesel exhaust particulate extracts. Atmospheric Environment, in press.
53. Toftgård R, Lofroth G, Caristedt-Duke J, et al. (1983) Compounds in urban air compete with 2,3,7,8-tetrachlorodibenzo-p-dioxin for binding to the receptor protein. Chem Biol Interactions 46:335-346.

54. Poellinger L, Lund J, Dahlberg E, Gustafsson J-Å (1985) A hydroxyapatite microassay for receptor binding of 2,3,7,8-tetrachlorodibenzo-p-dioxin and 3-methylcholanthrene in various target tissues. Anal Biochem 144:371-384.
55. Piskorska-Pliszczynska J, Keys B, Safe S, Newman MS (1986) The cytosolic receptor binding affinities and AHH induction potencies of 29 polynuclear aromatic hydrocarbons. Toxicol Lett 34:67-74.
56. Toftgård R, Franzen B, Gustafsson J-Å (1985) Characterization of TCDD-receptor ligands present in extracts of urban particulate matter. Environm Int 11:369-374.
57. Mason GF, Gustafsson J-Å, Westerholm RN, Li H (1992) Chemical fractionation of particulate extracts from diesel vehicle exhaust: Distribution of ligands for the dioxin receptor. Environ Sci Technol 26:1635-1638.

STATE OF THE ART LECTURES

1

Cell Cycle Control

Antony M. Carr

Introduction

All organisms must control their cell division. Unicellular organisms have to coordinate nuclear division, cytokinesis (cell separation) and DNA synthesis so that the correct order of events is maintained (1, 2). In addition, the cell cycle must be coordinated with nutrient availability and differentiation into the meiotic, or sexual, cycle. Multicellular organisms, such as humans, also have to maintain the correct order of events within the cell cycle, and must, in addition, regulate the growth and division of different tissues so that uncontrolled proliferation does not lead to tumorigenesis (3, 4). This complex task of controlling the timing of cell proliferation in response to both external stimuli and internal status is not yet fully understood. The study of cell cycle controls in a number of experimental systems has led to the discovery that much of the basic machinery underlying control of the cell cycle has been conserved in all eukaryotic organisms (5). In this chapter, I will attempt to describe these fundamental mechanisms that control the cell cycle, and to relate them to the etiology of cancer development.

The Universal Conservation of Cell Cycle Controls

Early observational studies on the cell cycle identified two major events, DNA synthesis (S phase) and mitosis (M phase). Two "gap" periods (G_1 and G_2) separate these two events. Commitment to the cell cycle (i.e., DNA duplication and subsequent mitosis and cell division) is controlled at a point termed the restriction point or "start" (6). Passage through this point is commensurate with the transition from G_1 into S phase (Figure 1). Cells which are not in the mitotic cycle generally rest in a state referred to as G_0 and do not enter the cell cycle unless stimulated to divide. When cells are stimulated to divide, they pass through "start," DNA synthesis is initiated, and the cell will complete an entire cycle. Genetic analysis of the cell cycle in unicellular yeasts has led to the identification of a similar control point in the yeast cell cycle (5). Based on such factors as cell size, nutrient availability, and the presence or absence of mating pheromone, yeast cells ascertain whether to pass through "start" and commit themselves to a whole cell cycle. If such cells do not pass through "start," they remain capable of mating and sexual differentiation or of entering a G_0 stationary phase. Once cells have passed through the "start" decision point, they can be thought of as unable to enter these alternative states until the completion of the entire cell cycle.

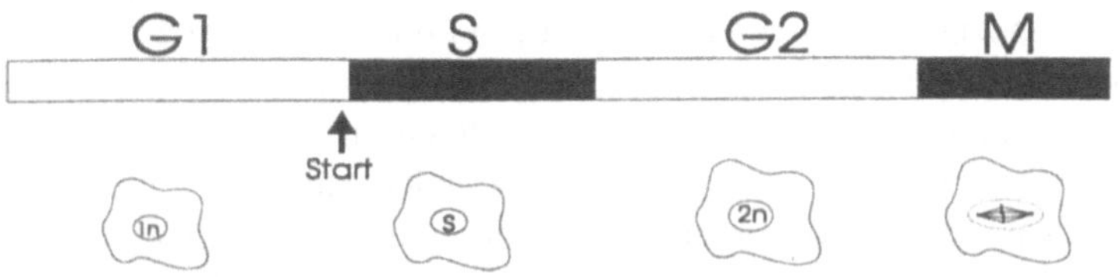

Figure 1. The cell cycle is divided into four sections. After passage through "start," cells progress into S phase where the DNA is duplicated. During G_2 phase, cell growth continues and preparations for mitosis are made. At M phase, the spindle forms and the chromosomes are segregated.

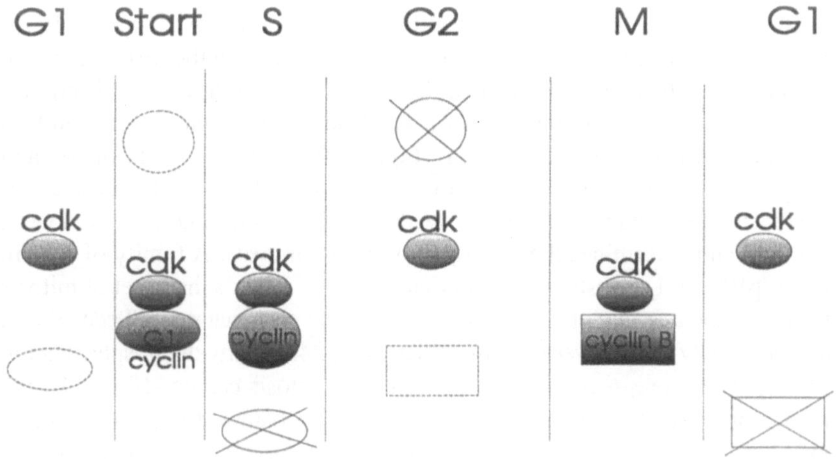

Figure 2. Different cdk molecules associate with distinct cyclin subunits during different stages of the cell cycle. The availability of the cyclin subunits is controlled by cell cycle-specific transcription (dotted lines) and proteolysis (crossed figures).

Work on the control of the cell cycle in several distinct experimental systems has recently converged, and much is now known about the proteins composing the machinery that drives the cell cycle in all eukaryotic cells. Key regulators of cell cycle progression are the cdc2 protein kinase family (p34cdc2). This family of proteins is highly conserved in all eukaryotic cells (5, 7). While yeast cells contain a single p34cdc2 protein, mammalian cells have at least four closely related proteins.

Because the kinase activity of all p34cdc2-like proteins is dependent on their association with another highly-conserved class of proteins called cyclins, the p34cdc2-related proteins are often referred to as cdk's (cyclin-dependent kinases). All cells contain a large number of cyclin proteins which act during different stages of the cell cycle (8). Cyclin concentrations fluctuate during cell cycle progression, individual cyclins peak at the different stages of the cell cycle where they are required, and they are then destroyed once that point is over. For example, the B-type mitotic cyclins accumulate before mitosis and are destroyed during mitosis (Figure 2). The destruction of B-type cyclins is thought to be required for proper exit from mitosis, suggesting that these cyclins have a further role in ensuring the order of events within the cell cycle (9).

Different combinations of p34cdk and cyclin subunits form kinase complexes at different points in the cell cycle. For example, the association of some human p34cdk's with mitotic B-type cyclins forms the kinase complex(es) responsible for initiation and control of mitosis, while the association of different p34cdk's with the G_1 cyclins forms the kinase complex(es) responsible for passage through "start." In addition to the association with the different cyclin subunits at different stages of the cell cycle, p34cdk proteins are regulated by a number of phosphorylation events (10, 11). The best-characterized example is inhibitory tyrosine phosphorylation (Figure 3). A balance of kinase and phosphatase activities contributes to the tyrosine phosphorylation state of p34cdk's (at the residue coincident with the tyrosine 15 residue of fission yeast p34cdc2) and determines the timing of entry into mitosis. A family of protein kinase (p107wee1) tyrosine phosphorylate p34cdk's inhibits the onset of mitosis (12, 13), while a family of p34cdk-specific tyrosine phosphatase (p80cdc25) acts antagonistically to p107wee1 kinase to activate p34cdk's by dephosphorylation (14, 15). Therefore, regulation of the timing of mitosis can be affected by the balance of activity of competing kinase and phosphatase. Both these protein families are themselves regulated by multiple phosphorylation events (Figure 3) and, in the case of p80cdc25, by cyclical accumulation prior to mitosis (16). Therefore, one possible mechanism which can control the timing of mitosis would be the activity of p107wee1 and p80cdc25 in response to internal cell status and possibly external stimuli. It is apparent from work performed in a number of experimental systems (17, 18) that tyrosine phosphorylation of p34 is not the only mechanism regulating its activity at mitosis.

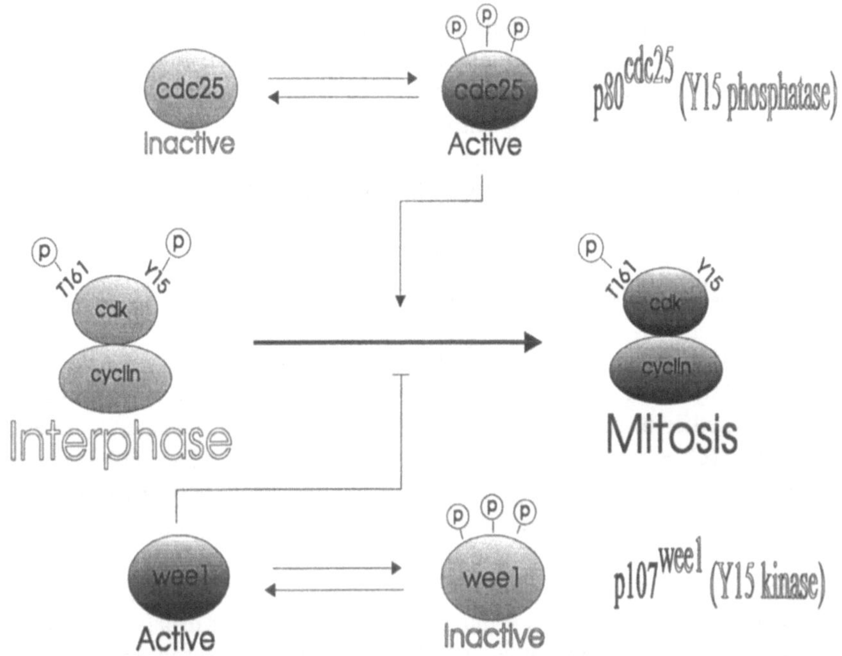

Figure 3. The Activity of p34cdk proteins at mitosis is controlled by inhibitory tyrosine phosphorylation of residue 15 by a balance of activity between the p107wee1 kinase and the p80cdc25 phosphatase. These two proteins are themselves regulated by phosphorylation (p). Active proteins are shown in darker shade.

In addition to their involvement in regulating the timing of mitosis, the p107wee1 and p80cdc25 proteins are also involved in an auto-activation loop for mitotic p34cdk activity. The activation of p34cdk at mitosis is a sudden event. This "all or nothing" activation is achieved by an auto-activation feedback loop whereby, once p34cdk reaches a crucial threshold value, the p34cdk kinase complex serves to direct its own activation (19). This is achieved, at least in part, by controlling (either directly or indirectly) the activity of the p34cdk-specific tyrosine phosphatase and kinase (p80cdc25 and p107wee1). This leads to the rapid dephosphorylation of the tyrosine 15 residue, as well as the rapid and full activation of the kinase complex.

A second phosphorylation event can also regulate p34cdk function. A protein kinase (CAK, cdc-activating kinase) has been biochemically identified which phosphorylates p34cdk's on the serine 161 residue (20). This phosphorylation event is required for p34cdk kinase activity (11) and is thus another potential mechanism for regulating the activity of cdk kinase complexes.

It is not yet apparent what role phosphorylation of the different p34cdk kinases plays in regulating passage through "start" and commitment to the cell cycle. In the yeast model systems, the p34cdk activity at "start" appears to be regulated by association with G_1-specific cyclins rather than by phosphorylation. In mammalian cells, a number of specific inhibitors of cyclin/cdk complexes have recently been identified (21-25), again suggesting that multiple mechanisms exist which serve to regulate the activity of the cyclin-dependent kinase at specific stages in the cell cycle and in response to specific signals.

Extracellular Signal Transduction

In mammalian cells, the major decision point in the cell cycle is at the G_1/S boundary ("start") and it is before passage through "start" that cells make the decision whether or not to commit themselves to division, whether to enter a nondividing (G_0) state, or whether to differentiate. These decisions will generally be made on the basis of extracellular signals, such as growth factors. Extracellular signal molecules bind to receptor proteins on the cell surface, many of which either have tyrosine kinase activity or become associated with tyrosine kinase when activated. While there are numerous specific signaling molecules and receptors, the binding of a signal molecule to a receptor usually activates a cascade of kinase events which will ultimately determine the proliferative status of the cell. Many, but not all, signal molecules activate a protein kinase known as mitogen-activated protein (MAP) kinase (26, 27). The actions of MAP kinase include phosphorylation of a ribosomal protein (a microtubule-associated protein) and changing the transcription patterns in the nucleus. It is not fully understood how the activities of MAP kinase interact with the cdk/cyclin complexes previously discussed.

Activation of the MAP kinase entails a series of phosphorylation events involving at least four proteins (Figure 4). The basic structure of this signal-responsive kinase cascade has been conserved throughout eukaryotic evolution (28), and several MAP kinase-related signal transduction pathways operate in the yeast *S. cerevisiae*. The best-characterized of these is the mating factor response pathway. Yeast cells can release one of two mating pheromones (short polypeptides), dependent on their mating type. A cell of one mating type responds to the presence of pheromone from a cell of the opposite mating type by arresting the cell progression cycle at "start" and initiating a pattern of transcription which results in differentiation into the sexual cycle. The signal transduction pathway mediating this response acts as follows: mating factor binds to a receptor molecule, thus initiating a kinase cascade involving several kinases (each of which is structurally related to elements within the MAP kinase cascade).

This cascade results in the activation of Fus1p and Kss1p, both of which are homologues of MAP kinase. The activation of these kinases promotes both cell cycle arrest in $G_1$by production of a specific inhibitor of cdk/cyclin complexes (FAR1) and transcription pattern changes which promote sexual differentiation. The yeast pheromone response pathway is a direct paradigm of the MAP kinase response in mammalian cells (growth factor binds to a receptor, stimulating tyrosine kinase activity, which results in MAP kinase activation and changes in cell cycle progression and transcription). Interestingly, several structurally related kinase cascades which respond to different stimuli are found in yeast. For example, an osmotically sensitive cascade has been identified (29), elements of which have structural homology to the equivalent element in the MAP and kinase/mating pheromone cascades. This suggests that evolution has utilized the same basic pathway to transmit information concerning the extracellular environment, and that the mitogen-activated kinase cascade stimulated by growth factors is only one of a number of related pathways in mammalian cells (Figure 4).

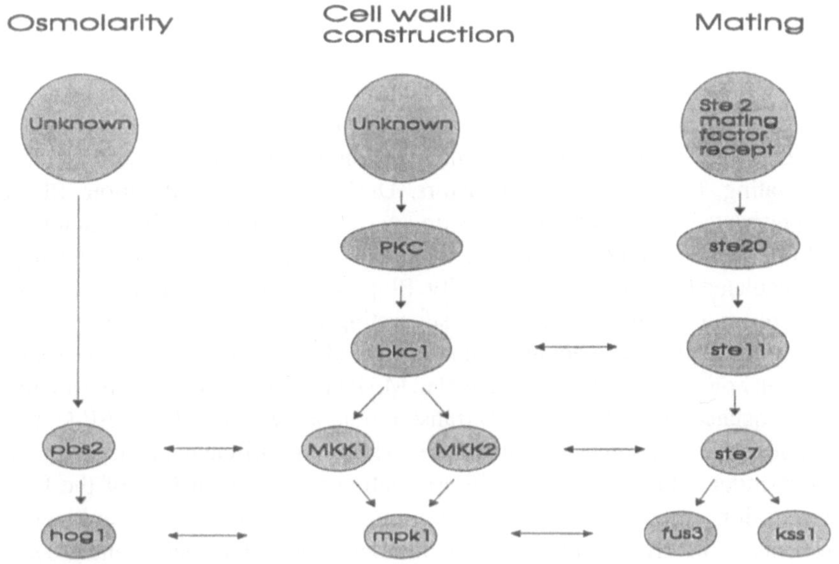

Figure 4. Multiple MAP kinase-related pathways in *S. cerevisiae*. Horizontal arrows represent the structurally-conserved elements in the pathways.

Tumor Suppressors

The importance of the MAP kinase signal transduction cascade is supported by the observation that many growth factors, cytokines, and tumor promoters cause activation of MAP kinase. Some tumors are associated with activating mutations in receptor-associated protein kinase which activate the MAP kinase pathway. This inappropriate activation is associated with inappropriate cell proliferation, as the cell responds to a growth factor which is not actually present. However, the activation of the MAP kinase pathway alone is not sufficient to cause malignant cell growth. Most tumors are a result of a number of mutations, one of which often inactivates a tumor suppressor gene. Several loci have been identified as tumor suppressors, the two major ones being the Retinoblastoma (Rb) gene and the p53 gene. The exact roles of these tumor suppressor genes in controlling inappropriate proliferation are not fully understood, but both provide insights as to how the cell cycle is controlled and how the MAP kinase pathway and the cell cycle machinery might be coordinated.

The Rb protein is of particular interest, as it provides a link among the cdk/cyclins, cell-cycle machinery, and transcription (30). Rb appears to act as a negative regulator of a subclass of the E2F transcription factors. These transcription factors are active during the transition from G_1 into S phase of the cell cycle and are required (and possibly regulatory) for entry into the cell cycle. Rb seems to prevent cell cycle progression by binding to, and thereby inactivating, these transcription factors. During normal proliferation, Rb is phosphorylated, possibly by a cdk/cyclin kinase, resulting in the dissociation of Rb from the transcription factor. In some tumors, Rb protein is inactive, leading to unregulated E2F transcription activity. Rb protein can be found in association with cdk and cyclin proteins (31), suggesting cross talk occurs among the cdk/cyclins, cell-cycle machinery, and control of transcription. While it is still unknown how the growth factor-regulated MAP kinase cascade interacts with the E2F-associated (and Rb-regulated) transcription, activation of the MAP kinase cascade by an activated Ras oncogene, while not sufficient on its own for transforming cultured cells, can, in association with deregulation of the E2F transcription complex, cause foci to form. This suggests that overlapping mechanisms (involving cyclin/cdk proteins, transcriptional regulation and growth factor receptors) are interacting in a complex manner to regulate cell growth.

The p53 tumor suppressor is an inessential gene involved in a number of pathways influencing cell cycle progression. p53 is often mutated in human tumors and is itself a transcription factor. For example, in response to DNA damage, p53 appears to direct the transcription of the CIP1 gene (32). Cip1p is a member of the newly emerging class of proteins which specifically inhibit cdk/cyclin kinase complexes; the obvious interpretation of this is that p53-dependent CIP1 induction is responsible for arresting the cell cycle in G_1

following DNA damage. At least some members of this new class of cdk/cyclin inhibitors are found to be frequently mutated in human tumors (33), suggesting a potential role as a tumor suppressor protein. The fact that these proteins interact directly with cdk/cyclin complexes would provide an enticing model whereby this class of tumor suppressors might prevent inappropriate cell proliferation, although how this activity is regulated, and what signals indicate that growth is inappropriate, as opposed to appropriate, is largely unknown.

Summary

Four elements of the cell cycle control mechanisms have been briefly covered (Figure 5).

1. The cdk/cyclin complexes which appear to regulate and coordinate the cell cycle and which are themselves regulated by several phosphorylation events and by subunit availability.
2. The MAP kinase cascade which is one of the signal transduction pathways stimulated by the presence of external signaling molecules.
3. The cell cycle-coordinated regulation of transcription, part of which is controlled by the Rb tumor suppressor protein.
4. The p53 tumor suppressor gene plays a role in monitoring the internal integrity of the genome. It also controls the production of at least one of a family of proteins that specifically inhibit the activity of cdk/cyclin complexes, which apparently prevents cell cycle progression.

How do these control systems interact? In multicellular organisms, cell growth is highly controlled, and only cells stimulated to divide by the correct growth factor in the correct context actually proliferate. For example, peripheral blood lymphocytes require two sequential mitogenic signals in order to proliferate. One, stimulation of T-cell antigen receptor, stimulates cyclin synthesis, while the other, provided by interleukin 2, leads to inactivation of a specific inhibitor of cyclin/cdk complexes (34). Presumably, the cell integrates a number of distinct positive and negative growth signals which together converge on common targets to influence progression into and through the cell cycle by controlling cdk/cyclin activity through a number of potential mechanisms (i.e., phosphorylation, subunit availability and cdk-specific inhibitor molecules). The cross talk between these different mechanisms and the pathways to which they respond must be considerable. The inappropriate stimulation of the growth factor-responsive MAP kinase pathway is one aspect of a considerable number of human cancers. Understanding how this activation bears on cell cycle regulation and cdk activity should provide insight into the nature of carcinogenesis and potential treatment of cancers.

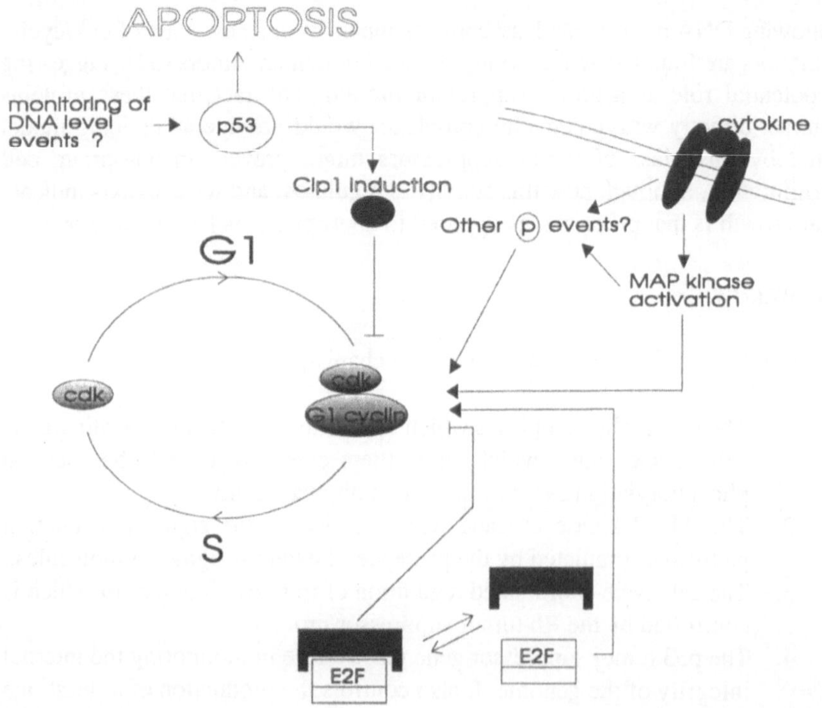

Figure 5. The different pathways involved in cell cycle control converge at the "Start" transition point to determine cell proliferation. Notice that p53 is also involved in an apoptotic pathway. It is possible (but not indicated here) that p53 could feed into the Rb/E2F control pathway, since Rb is phosphorylated by cdk/cyclin complexes, and since p53 acts to induce CIP1, a potent inhibitor of cdk/cyclin activity.

References

1. Hartwell LH, Weinert TA (1989) Checkpoints: Controls which ensure the order of cell cycle events. Science 246:629-634.
2. Enoch T, Nurse P (1991) Coupling M phase and S phase: Control maintaining the dependence of mitosis on chromosome replication. Cell 65:921-923.
3. Murray AW (1992) Creative blocks: Cell cycle checkpoints and feedback controls. Nature 359:599-604.
4. Hartwell LH (1992) Defects in cell cycle checkpoints may be responsible for the genomic instability of cancer cells. Cell 71:543-546.
5. Nurse P (1990) Universal control mechanism regulating onset of M-phase. Nature 344:503-507.

6. Nurse P (1975) Genetic control of cell size and cell division in yeast. Nature 256 547:551.
7. Lee MG, Nurse P(1987) Complementation used to clone a human homologue of the fission yeast cell cycle control gene cdc2. Nature 327:31-35.
8. Pines J (1993) Cyclins and cyclin dependent kinases: Take your partners. Trends in Bioch Sci 6:195-197.
9. Holloway SL, Glotzer M, King RW, Murray AW (1993) Anaphase is inactivated by proteolysis rather than by the inactivation of maturation promoting factor. Cell 73: 1393-1402.
10. Gould KL, Nurse P (1989) Tyrosine phosphorylation of the fission yeast cdc2+ protein kinase regulates entry into mitosis. Nature 342:39-45.
11. Gould KL, Moreno S, Owen DJ, et al. (1991) Phosphorylation at Thr167 is required for Schizosaccharomyces pombe p34cdc2 function. EMBO J 11:3297-3309.
12. Russel P, Nurse P (1987) Negative regulation of mitosis by weel+, a gene encoding a protein kinase homolog. Cell 49:559-567.
13. Parker LL, Atherton-Fessler S, Piwnica-Worms H (1992) p107weel is a dual-specificity kinase that phosphorylates p34cdc2 on tyrosine 15. Proc Nat Acad Sci USA 89:2971-2921.
14. Russel P, Nurse P (1996) cdc25+ functions as an inducer in the mitotic control of fission yeast. Cell 45:145-153.
15. Millar JBA, Russell P (1992) The cdc25 M-Phase inducer: An unconventional protein phosphatase. Cell 68:407-410.
16. Moreno S, Nurse P, Russell P (1990) Regulation of mitosis by cyclic accumulation of p80cdc25 mitotic inducer in fission yeast. Nature 344:549-552.
17. Dunphy WG (1994) The decision to enter mitosis. TICB 4:202-207.
18. Sheldrick KS, Carr AM (1993) Feedback controls and G2 checkpoints: Fission yeast as a model system. BioEssays 15:775-782.
19. Novak B, Tyson JJ (1993) Numerical analysis of a comprehensive model of M-phase control in xenopus oocytes and intact embryos. J Cell Sci 4:1153-1168.
20. Solomon MJ, Lee T, Kirschner MW (1992) Role of phosphorylation in p34cdc2 activation: Identification of an activating kinase. Mol Biol Cell 3:13-27.
21. Harper JW, Adami GR, Wei N, et al. (1993) The p21 cdk-interacting protein cip1 is a potent inhibitor of G_1 cyclin-dependent kinases. Cell 75:805-816.
22. El-Deiry WS, Tokino T, Velculescu VE, et al. (1993) WAF1, a potential mediator of p53 tumor suppression. Cell 75:817-825.

23. Serrano M, Hannon GL, Beach D (1993) A new regulatory motif in cell-cycle control causing specific inhibition of cyclin D/CDK4. Nature 366:704-707.
24. Xiong Y, Hannon GL, Zhang H, et al. (1993) p21 is a universal inhibitor of cyclin kinases. Nature 366: 701-704.
25. Xiong Y, Hannon GL, Zhang H, et al. (1993) Inhibition of CDK2 activity *in vivo* by associated 20K regulatory subunit. Nature 366:707-710.
26. Pelech SL, Charest DL, Mordret GP, et al. (1993) Networking with mitogen-activated protein kinases. Molec Cell Biochem 127/128:157-169.
27. Ahn NG (1993) The MAP kinase cascade. Discovery of a new signal transduction pathway. Molec Cell Biochem 127/128:201-209.
28. Elion EA, Brill JA, Fink GR (1991) FUS3 represses CLN1 and CLN2 and in concert with KSS1 promotes signal transduction. Proc Nat Acad Sci U S A 88:9392-9396.
29. Brewster JL, de Valoir T, Dwyer ND, et al. (1993) An osmosensing signal transduction pathway in yeast. Science 259:1760-1763.
30. La Thangue N (1993) Cell cycle--Transcriptional complexity. Current Biology 8:554-557
31. Hall FL, Williams RT, Wu F, et al. (1993) Two potentially oncogenic cyclins, cyclin A and cyclin D1, share common properties of subunit configuration, tyrosine phosphorylation and physical association with the Rb protein. Oncogene 8:1377-1384.
32. Dulic V, Kaufmann WK, Wilson SJ, et al. (1994) p53 dependent inhibition of cyclin dependent kinase activated in human fibroblasts during radiation induced G_1 arrest. Cell 76:1013-1023.
33. Kamb A, Gruis NA, Weaver-Feldhaus J, et al. (1994) A cell cycle regulator potentially involved in genesis of many tumour types. Science 264: 436-440.
34. Firpo EJ, Koff A, Solomon M, Roberts JM (1994) Inactivation of a cdk2 inhibitor during interleukin 2-induced proliferation of human T lymphocytes. Molec and Cell Biol 14: 4889-4901.

2

A Cell Cycle Regulator and Cancer

Alexander C. Kamb

Introduction

During cell division, eukaryotic cells execute a basic set of steps referred to as the cell cycle (1). The cell cycle consists of 4 general time periods, or phases: G_1, during which the cell grows and prepares for DNA synthesis; S, during which DNA replication occurs; G_2, during which the cell continues to grow and prepares for mitosis; and M, during which cytokinesis occurs. The vast majority of cells in most tissue compartments are not continuously dividing, but rather in a resting state known as G_0. When the appropriate stimulus is received, G_0-arrested cells enter the cell cycle.

Regulation of the cell cycle involves integration of a variety of signals, many of which originate outside the cell. External factors influence the duration of cell cycle phases at specific points in the cycle. These points of control are known as checkpoints, and the primary ones are located between G_1 and S phases, and between G_2 and M phases. For example, brief irradiation with ultraviolet light induces a state of temporary arrest in cycling cells at the G_1/S checkpoint, during which time DNA damage is repaired. Many of the signaling molecules, which influence the frequency with which cells enter the cell cycle as well as the rate of the process itself, are known. These signals include peptide growth factors, inhibitory factors, and steroid hormones. The signaling molecules bind to receptors either on the cell surface or within the cytoplasm, transducing the signal from an external to an internal one. The internal signals are in many cases amplified by a cascade of protein interactions that ultimately converge on the cell cycle apparatus itself. This apparatus consists of enzymes called cyclin-dependent kinases (cdk's) plus a large set of ancillary factors that together act as the ultimate determinants of decisions to proceed through the various checkpoints in the cell cycle.

The cdk's are a family of protein kinases that, when activated, phosphorylate a variety of substrates (2). The best studied cdk's include cdk1 (or cdc2), which appears to control the G_2/M transition; cdk2, which participates in the G_1/S transition; and cdk4, whose precise role in the cell cycle is still obscure. The spectrum of substrates varies among the different cdk's. One of the most significant substrates appears to be the retinoblastoma protein, Rb. cdk4, for example, readily phosphorylates Rb *in vitro*, while it does not phosphorylate one of the preferred *in-vitro* substrates of cdk2, histone H1. The relevance of histone H1 as a substrate for cdk's *in vitro* is not certain. However, the role of Rb as a substrate has great potential significance. Rb in its hypophosphorylated form binds a variety of transcription factors such as E2F, thereby blocking transcriptional activation functions. Phosphorylation of Rb by cdk's dissociates Rb from these factors, leading to expression of several genes involved in progression through the cell cycle.

The immediate regulators of cdk activity include the cyclins, a family of molecules whose pattern of expression varies through the cell cycle, as well as an emerging family of cdk-specific phosphatases and kinases that modify cdk activity (2, 3). A set of negative regulators whose names derive from their apparent mobility on denaturing gels, p16, p21, and p27, bind various cdk's *in vivo* and *in vitro* and inhibit kinase activity *in vitro* (Figure 1) (2). In the following sections, genetic evidence will be provided to demonstrate that one of these negative regulators, p16, may be an important tumor suppressor gene which is frequently mutated in human cancers.

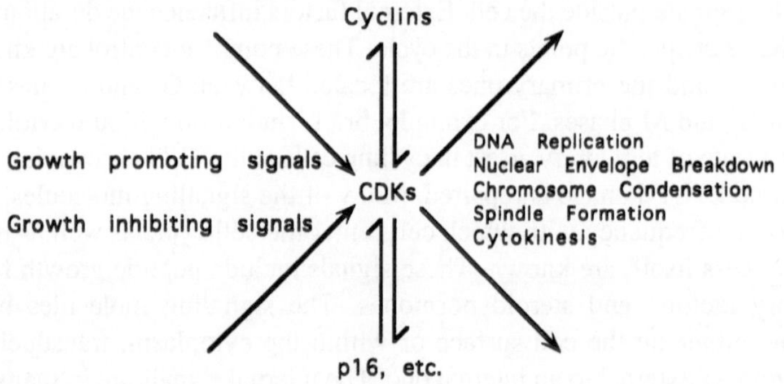

Figure 1. Role of cdk's in the cell cycle.

Melanoma Predisposition and 9p21

The 9p21 region of human chromosome 9 was known to contain chromosome aberrations at high frequency in cell lines and tumors of various origins, including leukemias, gliomas and melanomas (4). The region was shown to contain homozygous deletions in some cell lines. In the past few years, studies of primary tumors have demonstrated that 9p21 is deleted heterozygously at high frequency in several cancer types. For example, loss of heterozygosity (LOH) occurs in 80% of primary bladder carcinomas and 90% of melanomas (5, 6).

These findings imply the presence of a tumor suppressor gene located in 9p21 that is involved in a variety of cancers.

In 1992, an independent group of investigators demonstrated that 9p21 is also the site of a gene named MLM that predisposes to melanoma. In a set of 7 Utah families and one Texas family, Cannon-Albright, et al., mapped MLM to 9p21 using a group of closely-linked markers with a combined Lod score of 12.4, indicating statistical confidence of more than one trillion to one in favor of linkage to the 9p21 markers (7). The MLM locus includes predisposing alleles, some of which increase the probability of developing melanoma by a factor of 20 (8). These predisposing alleles are believed to act as dominant mutations in hereditary cancer predisposition, while acting as recessive mutations in somatic cells. Thus, susceptibility to melanoma results from inheritance of a single predisposing MLM allele, while malignancy requires loss or inactivation of both alleles.

The 9p21 region is suspected, based on analysis of somatic chromosomal aberrations, of harboring a tumor suppressor gene. MLM's genetic linkage to markers within 9p21 suggests that MLM may be not only a melanoma-predisposing gene but also a tumor suppressor gene involved in different forms of cancer. This hypothesis jibes with previous observations of familial cancer genes. In all cases so far examined, the familial gene is also frequently inactivated in sporadic cancers (9). Thus, a strategy for identifying MLM utilizes cell lines that contain 9p21 deletions to localize the gene targeted by those deletions. This strategy rests on the assumption that the gene targeted by deletion is the same as the melanoma susceptibility gene MLM.

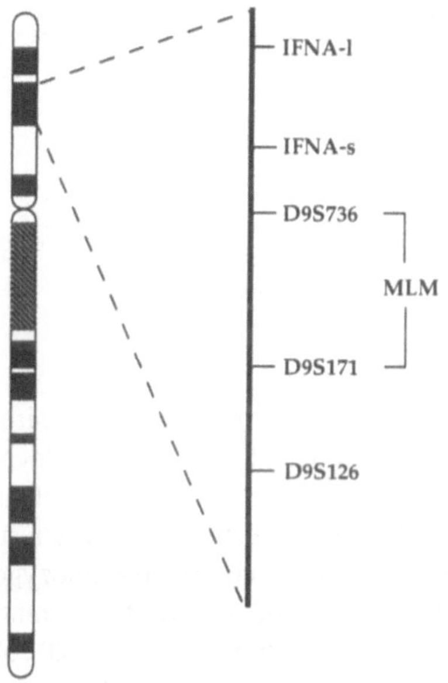

Figure 2. Site of the MLM gene.

Identification of CDKN2 (MTS1)

To begin the process of localizing the 9p21 tumor suppressor gene, several markers linked to melanoma predisposition were used to test for homozygous deletions in a small set of cell lines known from the literature to contain homozygous deletions in 9p21. This preliminary deletion analysis identified the region between markers D9S171 and IFNA as the probable location for the gene (Figure 2). These two markers were then used as the departure point to construct a physical map of the area with YAC, P1 and cosmid clones (10). The physical map permitted generation of a set of 54 new markers between D9S171 and IFNA. These new markers were tested against genomic DNA from nearly 100 melanoma cell lines. Roughly 58% of the lines displayed homozygous deletions of at least some sequences within the region. The most frequently deleted marker fell within a cosmid, c5, suggesting that a tumor suppressor gene lay at least partly within c5. Figure 3 depicts the DNA sequence analysis of cosmid c5 revealed 2 related sequences, CDKN2 (also called MTS1, cdk4I, and p16^{INK4}) and MTS2 (11). CDKN2 proved to be the gene for p16, discovered previously based on its inhibitory activity against cdk's (12). MTS2, a close relative of CDKN2, may encode a molecule with properties similar to those of p16. CDKN2 consists of 3 coding exons (E1, E2, and E3) and encodes a protein composed of 4 tandem ankyrin repeats. Ankyrin repeats are present in cytoskeletal proteins but have also been found in at least one other negative cell cycle regulator: pho81 from yeast (13).

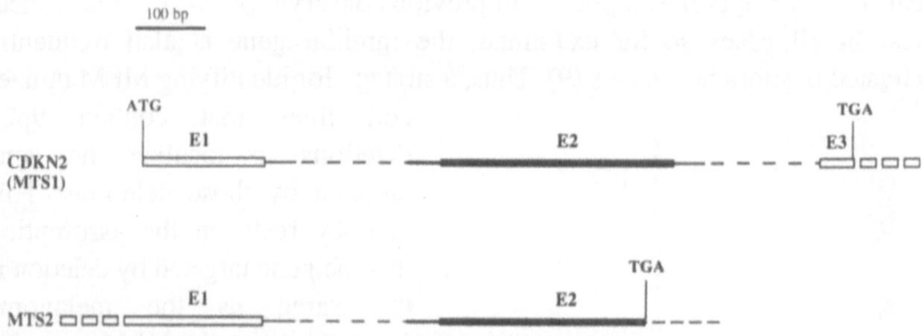

Figure 3. DNA sequence analysis of cosmid c5.

Role of CDKN2 in Cancer

The identification of CDKN2 as a potential target of 9p21 deletion inspired a large-scale deletion analysis of tumor cell lines. A set of 9 markers, located either within CDKN2 or flanking the gene, was used to test 290 cell lines representing 14 different tumor types for homozygous deletions (11). The results were astonishing; nearly 50% of this set of cell lines contained homozygous deletions of CDKN2. The deletion frequencies varied widely. For example, 14/17 (82%) of astrocytoma lines had homozygous deletions, whereas 0/20 colon cancer lines were deleted. The deletions centered around a marker located immediately 3' to E2 of CDKN2. No deletions were found that unequivocally removed MTS2 sequences alone. Thus, based on analysis of homozygous deletions, CDKN2 appeared to be a tumor suppressor gene targeted by the homozygous deletions.

To help exclude the possibility that the deletions were removing another unidentified tumor suppressor gene tightly linked to CDKN2, the subset of cell lines that did not contain CDKN2 homozygous deletions was examined for point mutations in CDKN2 (14). Of this subset, 18% contained detectable sequence differences that were shown not to be common polymorphisms. These mutations included nonsense, missense, frameshift, and splicing mutations. However, in each type of cell line tested, the frequency of point mutations was lower than the frequency of homozygous deletion. For example, in the case of melanoma, 57 homozygous deletions were detected, roughly 4 times the number of point mutations detected. In some cell line types, no point mutations were found, though the lines exhibited homozygous deletions. This result suggests that homozygous deletion may be the predominant mechanism for inactivating CDKN2 in many cell types. Alternatively, the homozygous deletions may remove other tumor suppressor genes that are important in tumorigenesis. Notably, no mutations in E2 from MTS2 have been detected so far, though E1 has yet to be screened.

The question of when during tumor formation CDKN2 is inactivated has not been resolved. It is clear, however, that a significant fraction, perhaps all, of CDKN2 mutations occur in primary tumors while in the body. Analysis of primary melanomas and bladder carcinomas has demonstrated that point mutations are detected in these tumors at about the same frequencies as in the corresponding cell lines (15). In addition, a recent report suggests that nearly half of all primary esophageal carcinomas contain point or frameshift mutations in CDKN2 (16). Though one study suggests that homozygous deletions of CDKN2 may occur at a lower frequency *in vivo* than in cultured cells, a careful analysis of homozygous deletions in primary tumors has not been published. The issue of the overall frequency of CDKN2 mutation frequency in primary tumors thus awaits thorough analysis of a large number of tumor samples of different types (17).

Surprisingly, the role of CDKN2 in genetic susceptibility to melanoma has not been simple to determine. An extensive analysis of kindreds that segregate 9p21-linked predisposing MLM alleles has uncovered only 2 potential predisposing mutations (17). Both are nonconservative missense substitutions (gly->trp and val->asp), rare in the normal population, and linked to the MLM carrier chromosome in their respective pedigrees. However, without biochemical information, it is difficult to exclude the possibility that these mutations are neutral. No clearly disruptive genetic lesions such as frameshift or nonsense mutations have been found. In addition, the overall frequency of mutations in the coding sequence of CDKN2 is low: 2/13 pedigrees contain mutations. This result indicates that either: 1) the coding sequence of CDKN2 rarely, if ever, contains

MLM-predisposing mutations; or 2) CDKN2 is not MLM, and the original assumption about the identity of MLM and the 9p21 tumor suppressor gene was wrong. To distinguish between these 2 alternatives will require an extensive search for mutations in other parts of the CDKN2 gene and, possibly, in other 9p21 genes. It is again worth noting that the pedigree samples used to screen for CDKN2 mutations do not contain mutations in E2 from MTS2.

Pathways of Tumor Suppression Involving CDKN2

The precise role of CDKN2 in the cell cycle is currently unknown. The protein it encodes, p16, inhibits cdk4 but not certain other cdk's *in vitro* (12). Thus, p16 may be a specific inhibitor of cdk4 in the cell. This finding in turn suggests that cdk4 may be an exceptionally important regulator of the cell cycle, at least as far as tumorigenesis is concerned. Unfortunately, there is to date no direct evidence that cdk4 is involved in regulating the cell cycle. An experiment designed to test its role failed to provide any proof of its function in the cell cycle (19). Indeed, in transformed cells cdk4 is frequently found in tight complex with p16, an observation that seems to contradict the role of cdk4 as a presumptive activator of the cell cycle (12). Nevertheless, by analogy with other cdks and based on its affinity for cyclin D1, a primary activator of the cell cycle, cdk4 has been proposed to drive the cell from G_1 into S phase (20). In addition, a substrate for cdk4 *in vitro* is Rb, also thought to regulate genes involved in the G_1 to S transition. Thus, it is attractive to consider cdk4 as at least one determinant of the G_1 to S transition and, therefore, to propose p16 as a regulator of checkpoint control at the G_1/S boundary. On the other hand, a role for p16 in other parts of the cell cycle should be considered seriously. For example, p16 and perhaps cdk4 may be involved in modulating the duration of G_0 arrest (Figure 4).

The upstream regulators of p16 are even less well understood. One possibility is that p53 controls p16 expression in the same manner as it controls the expression of p21, another cdk inhibitor (12, 21). If true, tumor cell lines and primary tumors should not contain mutations in both genes. Mutation of either one would be sufficient. Analysis of CDKN2 and TP53 mutations in melanoma and bladder primary tumors and cell lines suggests that p53 does not regulate p16. Though many tumors and cell lines contain mutations in either TP53 or CDKN2 alone, some contain mutations in both genes (15). The frequency with which such double mutant cells are observed is roughly the product of the mutation frequencies for the individual genes alone. This result suggests that

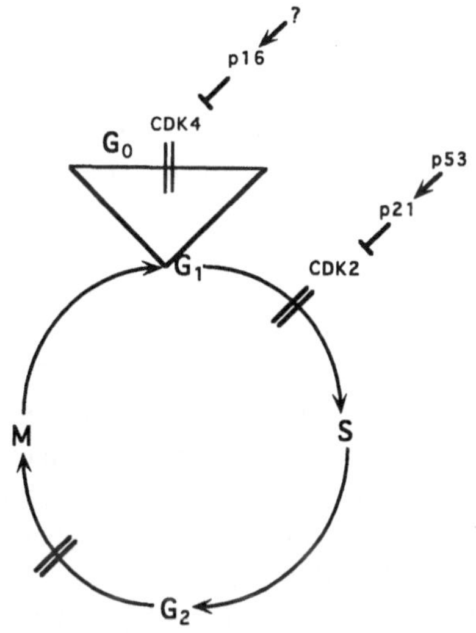

Figure 4. Possible determinants of the G_1 to S transition.

mutations occur independently in the two genes. It follows, therefore, that p53 and p16 act in different pathways of tumor suppression. Inactivation of both pathways by mutation of one of the elements TP53 or CDKN2 may be required for frank malignancy (Figure 5).

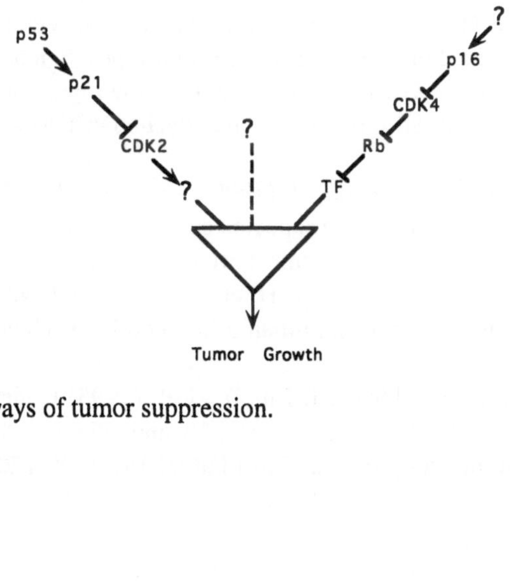

Figure 5. Pathways of tumor suppression.

Conclusion

Clearly, much is still unknown about CDKN2 and its role in tumor formation. The cell cycle promises to continue as an exciting area of research for those interested in elucidating the basic mechanisms of carcinogenesis.

References

1. Jacobs T (1992) Control of the cell cycle. Dev Biol 153:1-15.
2. Heichman KA, Roberts JM (1994) Rules to replicate by. Cell 79:557-562.
3. Sherr CJ (1993) Mammalian G_1 cyclins. Cell 73:1059-1065.
4. Kamb A (1995) Role of a cell cycle regulator in hereditary and sporadic cancer. In: Cold Spring Harbor Symposia on Quantitative Biology. New York: Cold Spring Harbor Laboratory Press, pp 39-47.
5. Fountain JW, Bale SJ, Housman DE, Dracopoli NC (1990) Genetics of melanoma. Cancer Sur 9: 645-650.
6. Dalbagni G, Presti J, Reuter V, et al. (1993) Genetic alterations in bladder cancer. Lancet 342:469-471.
7. Cannon-Albright LA (1992) Assignment of a locus for familial melanoma MLM, to chromosome 9p13-22. Science 258:1148-1152.
8. Cannon-Albright LA, Meyer LJ, Goldgar DE, et al. (1995) Penetrance and expressivity of the chromosome 9p melanoma susceptibility locus (MLM). Cancer Res 54:6041-6044.
9. Knudson AG (1993) All in the (cancer) family. Nature Genet 5:103-104.
10. Weaver-Feldhaus J, Gruis NA, Neuhausen S, et al. (1995) Localization of a putative tumor suppressor gene by using homozygous deletions in melanomas. Proc Natl Acad Sci USA 91:7563-7567.
11. Kamb A, Gruis NA, Weaver-Feldhaus J, et al. (1994) A cell cycle regulator potentially involved in genesis of many tumor types. Science 264:436-440.
12. Serrano M, Hannon GJ, Beach D (1993) A new regulatory motif in cell-cycle control causing specific inhibition of cyclin D/CDK4. Nature 366:704-707.
13. Lux SE, John KM, Bennett V (1990) Analysis of cDNA for human erythrocyte ankyrin indicates a repeated structure with homology to tissue-differentiation and cell-cycle control proteins. Nature 344:36-42.
14. Liu Q, Neuhausen S, Skolnick MH, et al. (1995) CDKN2 (MTS1) tumor suppressor gene mutations in human tumor cell lines. Oncogene 10:1061-1067.
15. Gruis NA, Weaver-Feldhaus J, Liu Q, et al. (1995) Genetic evidence in melanoma and bladder cancers that p16 and p53 function in separate pathways of tumor suppression. Am J Pathol 146:1199-1206.

16. Mori T, Aoki T, Nishihira T, et al. (1994) Frequent somatic mutation of the MTS1/CDK4I (Multiple Tumor Suppressor/Cyclin-dependent Kinase 4 Inhibitor) gene in esophageal squamous cell carcinoma. Cancer Res 54:3396-3397.
17. Kamb A, Liu Q, Harshman K, et al. (1995) Rates of p16 (MTS1) mutation in primary tumors with 9p loss (response). Science 265:416-417.
18. Kamb A, Shattuck-Eidens D, Eeles R, et al. (1995) Analysis of CDKN2 (MTS1) as a candidate for the chromosome 9p melanoma susceptibility locus (MLM). Nature Genet 8:22-30.
19. Van den Heuvel S, Harlow E(1993) Distinct roles for cyclin-dependent kinases in cell cycle control. Science 262:2050-2054.
20. Matsushime H, Ewan ME, Strom DK, et al. (1992) Identification and properties of an atypical catalytic subunit for mammalian D type cyclins. Cell 71:323-334.
21. El-Diery WS, Tokino T, Velculescu VE, et al. (1993) WAF-1, a potential mediator of p53 tumor suppression. Cell 75:817-825.

3

Molecular Cytogenetics of Solid Tumor Progression

Frederic M. Waldman, Guido Sauter, Jorma Isola,
Olli Kallioniemi, and Anne Kallioniemi

Introduction

Solid tumors are thought to develop and progress by a multi-step process, involving numerous genetic events during tumor evolution (1). Even in cases of familial inheritance, when an individual may already be predisposed to tumorigenesis by carrying a mutation in one allele of a tumor suppressor gene (retinoblastoma, p53, BRCA1, etc.), it is likely that several additional genetic events are required during the progression of a tumor clone from normal epithelium through premalignant, invasive, and metastatic stages. Characterization of the mechanisms underlying tumor progression, and of the genetic events specific to individual tumor types, will lead to improved clinical management and possibly to new therapeutic approaches.

Previous studies have suggested that genetic instability is a characteristic of solid tumors (2-6). Classical cytogenetic analysis of solid tumors has been limited due to the difficulty in preparing high-quality metaphase spreads of adequate number from tumor biopsies. Even so, a number of studies have shown that there is significant heterogeneity among metaphases from individual tumors (7). Flow cytometric analysis of DNA content frequently shows multiple tumor clones (8). Immunohistochemical analyses for overexpression of *erb*B-2, *egf*-r, and other oncogenes or growth factors commonly show heterogeneity within a single section (9-12). Tumor proliferation is frequently characterized by "hot spots" of DNA synthesis, usually at the periphery of the tumor (13, 14). Similarly, neovascularity is highly variable within a tumor (15, 16). These studies show that tumor cells, although presumably derived from a single cell at an early point of tumorigenesis, are phenotypically and genotypically highly variable. We hypothesize that the degree of genetic instability of individual tumors is reflected both by the total number of genetic aberrations and by the heterogeneity of aberrations present within a tumor. Further, the degree of genetic instability present in a tumor may predict future clinical behavior since tumors with increased instability may be most likely to acquire further aberrations leading to progression (higher proliferative rate, greater invasiveness, metastatic potential) or drug resistance.

Fluorescence *In-Situ* Hybridization (FISH)

Techniques of molecular cytogenetics are especially useful in defining both clonal alterations (common to all cells) and nonclonal alterations (variable among cells). We have focused primarily on structural genomic alterations as examples of genetic instability since these alterations are especially amenable to molecular cytogenetic approaches. FISH uses labeled DNA probes to define the copy number of a target DNA sequence (17). Probe and target cell DNA are denatured, and the probe is allowed to hybridize to the intact interphase or metaphase nucleus. Visualization is by directly conjugated fluorescent dyes, or by indirect means using biotin-avidin or similar detection systems. Absorptive stains can also be used for visualization of hybridization domains, although the number of probes (colors) which can be used simultaneously is limited, and the detectability of small targets is difficult.

Three types of probes are available for tumor analysis (Figure 1). 1. "Painting" probes are libraries of multiple clones generally derived from sorted chromosomes and maintained in bacterial or phage vectors. Repetitive elements that are not chromosome specific are eliminated from the "paint" reaction by addition of unlabeled repetitive (Cot1) DNA. Paint probes are very useful for analysis of structural alterations in metaphase spreads, but are less useful for interphase analysis since the domain of the entire chromosome is large, diffuse, and thus not useful for enumerating target sequences.

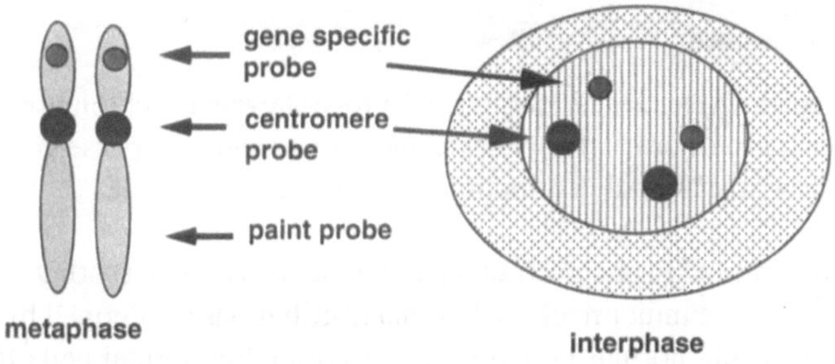

Figure 1. Probes for FISH. Painting probes detect entire chromosomes. Centromere probes detect repetitive elements at or near the centromere. Locus-specific probes detect copy number of individual genes.

2. "Centromeric" or repetitive probes are small (~1 kb) chromosome-specific sequences which are repeated 100's to 1000's of times at the centromeric region of individual chromosomes. They are part of a family of alpha-satellite or heterochromatin sequences. They are very useful for enumeration of their targets since the hybridization signal is generally large, bright, and focused. It should be remembered, however, that the centromeric target is only one locus within the chromosome, so numerical or structural abnormalities of the remainder of the chromosome are not being measured. 3. Locus-specific probes refer to a class of unique-sequence probes which are specific to known genes or are mapped to specific chromosome loci. Although small cDNA and phage probes (1-20 kb) can be detected under optimum conditions in some target nuclei (and can be used for probe mapping), they are not generally useful for tumor analyses since a high hybridization efficiency is required for analysis of a tumor cell population. Tumor cell analyses are generally limited to cosmid size (35 kb) and greater. Hybridizations using individual P1 probes (80-120 kb), which have recently become available after selection from large genomic libraries, yield large bright signals that can be counted easily. Larger YAC probes have also been used; however, these sometimes yield signals which are spread and difficult to score, and have sometimes lost or gained unwanted DNA.

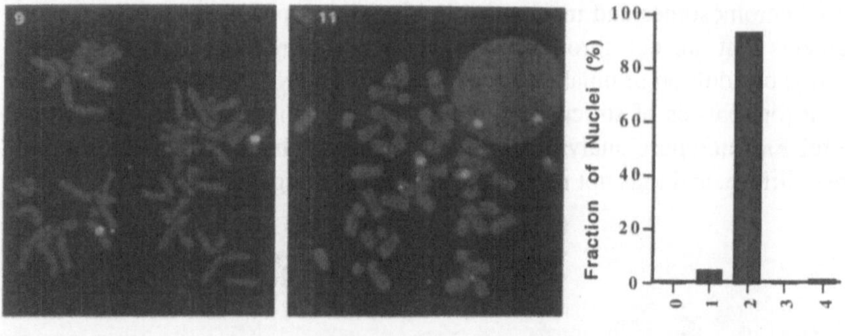

Figure 2. Repetitive element, "centromeric," probes targeted to chromosome 9 and chromosome 11 on normal lymphocyte metaphase and interphase nuclei. The histogram shows the normal distribution of signals in control cells.

FISH analysis of genetic aberrations in tumors is based on scoring of signals within the interphase tumor nucleus. Normal distributions are defined by control hybridizations using normal lymphocytes or preferably normal cells from the tissue of tumor origin (Figure 2). Normal cells show two signals for each probe in more than 90% of cells in satisfactory hybridizations. A few cells will show more signals, due to artifactual signal "splitting" or perhaps due to a low level of aneusomic (abnormal chromosome copy number) cells in the normal population, and 5-10% of the cells will usually show only one FISH signal due to overlap of signals (3-dimensional cells are flattened onto slides and then looked at in two dimensions) or due to less than perfect hybridization efficiency (signals may be too weak to appear over background fluorescence) (12, 18).

Chromosome Aneusomy

A number of conclusions can be drawn from studies of abnormal copy number using centromere probes in breast and bladder cancers. First, it is extremely common. At least one chromosome shows an abnormal copy number distribution in almost every tumor examined when at least three or four chromosomes are tested (19). Interestingly, this holds true even for tumors showing a diploid DNA content by flow cytometry. Secondly, different chromosomes within a tumor frequently show different patterns of centromeric copy number (Figure 3).

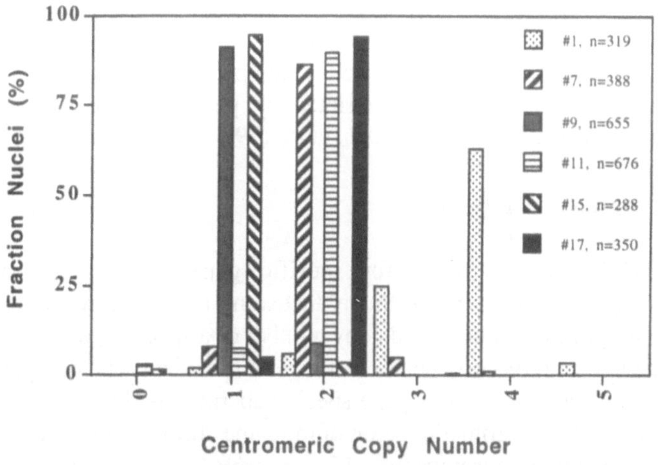

Figure 3. Distribution of centromeric copy number for a single bladder tumor. Note monosomy for chromosomes 9 and 15, disomy for 7, 11, and 17, and a heterogeneous (3/4) polysomy for chromosome 1.

Third, heterogeneity is commonly seen within a tumor for copy number of some chromosomes. This is interpreted as due to the presence of multiple individual clones, or else as instability in the number of copies per cell of a specific chromosome within an individual tumor. Fourth, heterogeneity of chromosome copy number may correlate with phenotypic characteristics, such as proliferative rate (Balasz, et al., personal communication). This is reflected by one clone within a tumor defined by chromosome copy number having a different S phase labeling index than a second clone. This pattern has been seen in a minority of breast cancer specimens labeled *in vivo* with 5-bromo-deoxyuridine. Last, chromosome copy number is a marker for tumor progression, since chromosome copy number is generally correlated with both grade and stage in bladder cancer (20) (Figure 4).

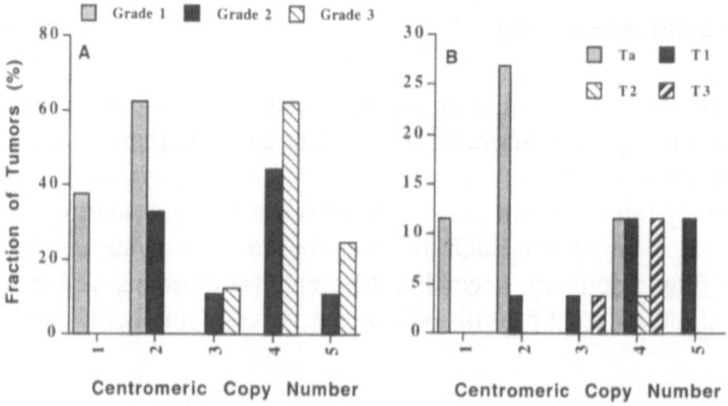

Figure 4. Chromosome 7 centromeric copy number is associated with bladder tumor grade (A) and stage (B). Reprinted by permission of the American Association for Cancer Research and the author (19).

Gene-Specific Aberrations

Interphase FISH using probes for specific genes may show different hybridization patterns (Figure 5). Since aneusomy of the entire chromosome should be distinguished from altered copy number of a specific gene, it is useful to determine the ratio of copy number of a gene-specific probe compared to a centromeric probe as a measure of gene-specific aberrations. This is usually done using a probe for the centromeric region of the same chromosome as the gene of interest (for example, chromosome 17 for *erb*B-2 or p53) (12, 18, 21). Amplifications are thus defined as more gene-specific signals than centromeres, and deletions as fewer gene-specific signals than centromeres. It should be remembered, however, that structural alterations such as translocations that lead to a dissociation of the gene from its centromere may lead to misinterpretation of the gene/centromere ratio.

We have used a probe specific for the *erb*B-2 (*her2/neu*) gene to define amplifications of this growth factor receptor whose expression has been linked to prognosis in breast cancer (21). FISH allows determination of *erb*B-2 copy number on a cell by cell basis. There was a tight correlation of amplification as determined by FISH with amplification as determined by Southern analysis, and with immunohistochemical detection of overexpression (Figure 6). Significant cell-to-cell heterogeneity was observed for *erb*B-2 copy number in this study.

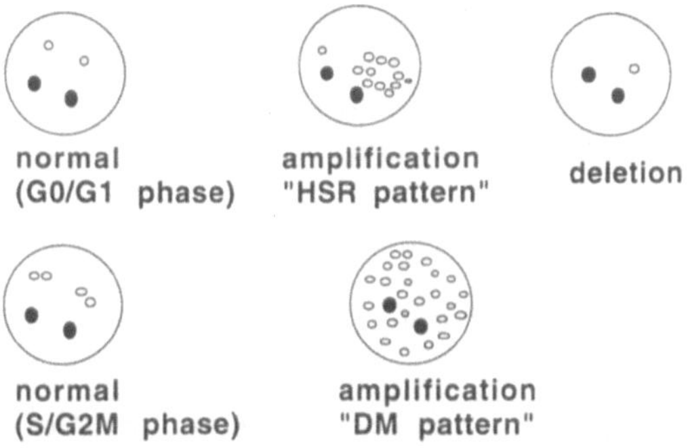

Figure 5. Patterns of FISH signals seen with centromeric (•) and gene-specific (○) probes. Note doublets due to sister chromatid signals in normal cells in S or G_2/M phase. Gene-specific signals are clustered in the Homogeneously Staining Region pattern, and diffuse in the Double Minute pattern of amplification. Gene-specific deletion is defined as fewer gene-specific signals than centromere signals, seen here as 2 centromeres and 1 gene-specific signal.

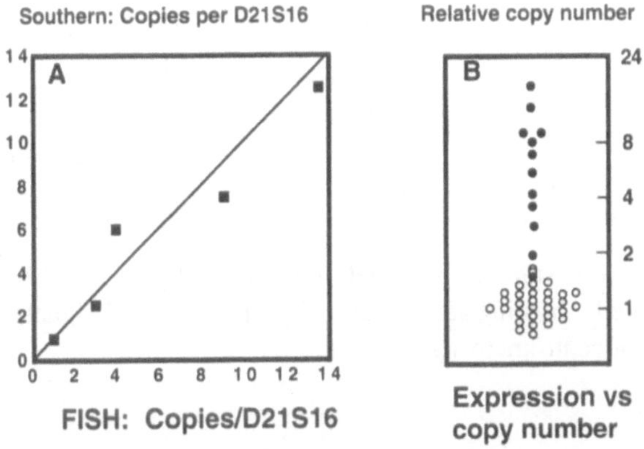

Figure 6. (A) Association between copy number of *erb*B-2 gene by Southern analysis (Y axis) and by FISH (X axis). Probe D21S16 was used as a control for both Southern and FISH to normalize the analyses. (B) Relationship between *erb*B-2 FISH copy number and immunohistochemical staining; •Positive staining, ○Negative staining. Reprinted by permission of the National Academy of Sciences USA (21).

Although each amplified tumor showed some cells with 25 copies of the *erb*B-2 gene, the actual range was extreme, varying between 5 and 100 copies in some breast tumors. A different sort of heterogeneity was present in bladder tumors.

In contrast to breast tumors, bladder tumors frequently showed *erb*B-2 overexpression immunohistochemically in the absence of gene amplification (12). Even in individual cases showing gene amplification, there were tumors with overexpression in most cells, yet the fraction of tumor cells with gene amplification was significantly less (Table 1). This was interpreted as due to coexistence of both mechanisms of *erb*B-2 overexpression in the same tumor.

Table 1. Fraction of cells showing amplification by FISH of *erb*B-2 IHC.

Tumor[a]	*erb*B-2 IHC[a] (%)	Amplified cells[b] (%)
1	40	49
2	100	80
3	100	88
4	100	94
5	80	29
6	100	18
7	100	28
8	90	41
9	100	38
10	nd[c]	89

[a] Fractions of cells staining positively for *erb*B-2 by immunohistochemistry.
[b] Fraction of cells showing amplification by FISH (*erb*B-2 signals >2 times the number of centromere signals).
[c] Not done.

Comparative Genomic Hybridization (CGH).

The total number of numeric alterations present in the entire tumor genome is a useful marker for genetic instability since instability should lead to an increased number of such clonal aberrations. CGH defines the relative copy number of all regions in the tumor genome simultaneously by a competitive hybridization reaction between tumor DNA and normal DNA to a normal metaphase spread (Figure 7 and Color Plate 1) (22-24). Tumor DNA from chromosomal regions which are present at increased copy number (duplications and amplifications) will bind relatively more than normal DNA at these regions, while DNA from regions which are present at lower copy number (deletions) will bind relatively less. If tumor DNA is labeled with one color (e.g., green, FITC-dUTP) and normal DNA is labeled with a second color (Texas red-dUTP), then the ratio of tumor to normal can be determined at each chromosomal locus throughout the genome as a green-to-red ratio. Specific probes or tumor metaphases are not required. The normal DNA does not have to come from the patient, since large genomic changes rather than single base-pair polymorphisms are being detected. Since normal metaphases are being used as a target of the hybridization, abnormalities can be mapped to their normal chromosomal locus. Although many changes can be detected from direct visualization of the CGH hybridization, greater sensitivity is achieved by averaging multiple metaphases, requiring digital microscopy and customized software.

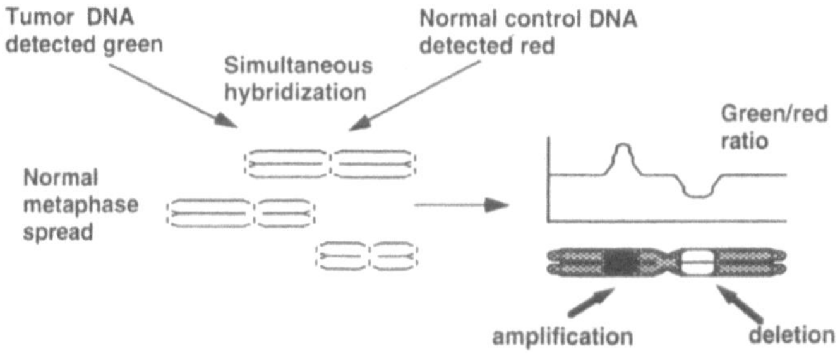

Figure 7. Comparative Genomic Hybridization uses a simultaneous hybridization of tumor DNA and a normal reference DNA to define the relative copy number of DNA sequences along a normal metaphase chromosome. Amplifications and deletions are defined by an increase or decrease in the ratio of tumor to normal fluorescent signals.

Breast Cancer CGH

We have used CGH to identify regions of the genome that are preferentially aberrant in breast tumors (25). Figure 8 shows a summary of 33 breast tumors of all stages which were analyzed by CGH. At least one DNA copy number gain was seen by CGH in over 85% of the primary tumors examined. The most common sites of aberrations included 1q+, 8q+, 17p-, 16q-, 8p-, and 20q+. The number of aberrations per tumor varied from none to 16. We have hypothesized that the number of aberrations as defined by CGH represents the degree of genetic instability which is characteristic of an individual tumor. Thus, the number of CGH abnormalities should predict the clinical behavior of an individual tumor. A recent pilot study comparing node-negative breast cancer with "good" outcome (having no recurrence with at least 5 years follow-up) and node-negative tumors having a distant recurrence within 5 years, showed that there was a significant association of copy number losses with clinical outcome (10). There was also an association with the total number of gains plus losses, although no statistical association was seen for gains alone.

Summary

Genetic aberrations can be characterized in solid tumors by molecular cytogenetic approaches of fluorescence *in-situ* hybridization and comparative genomic hybridization. A significant degree of intratumor heterogeneity is apparent from FISH analyses, using cell-by-cell measurements. CGH has shown the wide diversity of chromosomal changes between tumors. The combination of approaches has identified regions of the tumor genome with genetic aberrations which may prove prognostically useful.

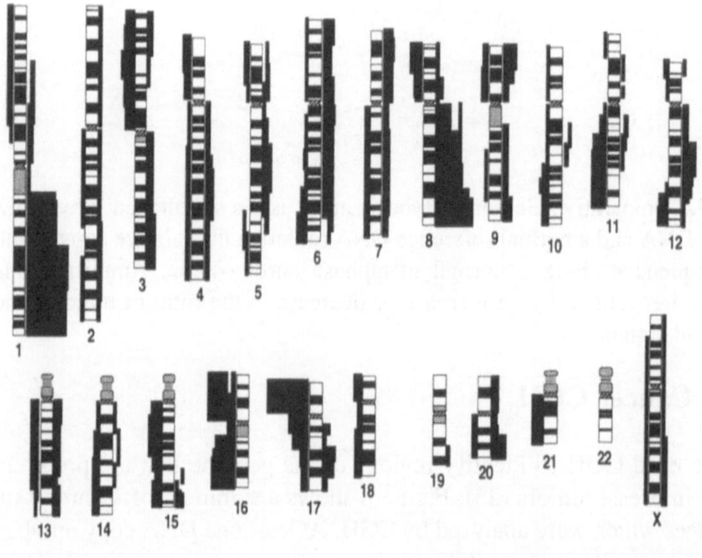

Figure 8. Breast cancer CGH. Losses are on the left, gains on the right of each chromosome ideogram. Each vertical line represents one case. The extent of each aberration is represented by the length of each line (n=33). Reprinted by permission of the author (25).

References

1. Fearon ER and Vogelstein B (1990) A genetic model for colorectal tumorigenesis. Cell 61:759-767.

2. Hartwell L (1992) Defects in a cell cycle checkpoint may be responsible for the genomic instability of cancer cells. Cell 71:543-546.
3. Livingstone LR, White A, Sprouse J, et al. (1992) Altered cell cycle arrest and gene amplification potential accompany loss of wild-type p53. Cell 70:923-935.
4. Bonsing BA, Devilee P, Cleton-Jansen AM, et al. (1993) Evidence for limited molecular genetic heterogeneity as defined by allelotyping and clonal analysis in nine metastatic breast carcinomas. Cancer Res 53:3804-3811.
5. Bronner CE, Baker SM, Morrison PT, et al. (1994) Mutation in the DNA mismatch repair gene homologue hMLH1 is associated with hereditary non-polyposis colon cancer. Nature 368:258-261.
6. Fishel R, Lescoe MK, Rao MR, et al. (1994) The human mutator gene homolog MSH2 and its association with hereditary nonpolyposis colon cancer. Cell 77:167.
7. Saint-Ruf C, Gerbault-Seureau M, Viegas-Pequignot E, et al. (1990) Proto-oncogene amplification and homogeneously staining regions in human breast carcinomas. Genes Chromosom Cancer 2:18-26.
8. Kallioniemi OP, Visakorpi T, Holli K, et al. (1991) Improved prognostic impact of S-phase values from paraffin-embedded breast and prostate carcinomas after correcting for nuclear slicing. Cytometry 12:413-421.
9. Kallioniemi OP, Holli K, Visakorpi T, et al. (1991) Association of c-*erb*B-2 protein over-expression with high rate of cell proliferation, increased risk of visceral metastasis and poor long-term survival in breast cancer. Int J Cancer 49:650-655.
10. Isola J, Visakorpi T, Holli K, Kallioniemi, OP (1992) Association of overexpression of tumor suppressor protein p53 with rapid cell proliferation and poor prognosis in node-negative breast cancer patients. J Natl Cancer Inst 84:1109-1114.
11. Thor AD, Moore DI, Edgerton SM, et al. (1992) Accumulation of p53 tumor suppressor gene protein: An independent marker of prognosis in breast cancers. J Natl Cancer Inst 84:845-855.
12. Sauter G, Moch H, Moore D, et al. (1993) Heterogeneity of *erb*B-2 gene amplification in bladder cancer. Cancer Res 53:2199-2203.
13. Goodson W3, Ljung BM, Waldman F, et al. (1991) *In vivo* measurement of breast cancer growth rate. Arch Surg 126:1220-1223.
14. Waldman FM, Carroll PR, Cohen MB, et al. (1993) 5-Bromodeoxyuridine incorporation and PCNA expression as measures of cell proliferation in transitional cell carcinoma of the urinary bladder. Mod Pathol 6:20-24.
15. Weidner N, Semple JP, Welch WR, Folkman J (1991) Tumor angiogenesis and metastasis--correlation in invasive breast carcinoma. N Engl J Med 324:1-8.

16. Weidner N, Folkman J, Pozza F, et al. (1992) Tumor angiogenesis: A new significant and independent prognostic indicator in early-stage breast carcinoma [see comments]. J Natl Cancer Inst 84:1875-1887.
17. Pinkel D, Straume T, Gray, JW (1986) Cytogenetic analysis using quantitative, high-sensitivity, fluorescence hybridization. Proc Natl Acad Sci U S A 83:2934-2938.
18. Matsumura K, Kallioniemi A, Kallioniemi O, et al. (1992) Deletion of chromosome 17p loci in breast cancer cells detected by fluorescence *in situ* hybridization. Cancer Res 52:3474-3477.
19. Waldman FM, Carroll PR, Kerschmann R, et al. (1991) Centromeric copy number of chromosome 7 is strongly correlated with tumor grade and labeling index in human bladder cancer. Cancer Res 51:3807-3813.
20. Sauter G, Haley J, Chew K, et al. (1994) Epidermal-growth-factor-receptor expression is associated with rapid tumor proliferation in bladder cancer. Int J Cancer 57:508-514.
21. Kallioniemi OP, Kallioniemi A, Kurisu W, et al. (1992) ERBB2 amplification in breast cancer analyzed by fluorescence *in situ* hybridization. Proc Natl Acad Sci U S A 89:5321-5325.
22. Kallioniemi A, Kallioniemi OP, Sudar D, et al. (1992) Comparative genomic hybridization for molecular cytogenetic analysis of solid tumors. Science 258:818-821.
23. Kallioniemi OP, Kallioniemi A, Sudar D, et al. (1993) Comparative genomic hybridization: A rapid new method for detecting and mapping DNA amplification in tumors. Semin Cancer Biol 4:41-46.
24. Speicher MR, du Manoir S, Schrock E, et al. (1993) Molecular cytogenetic analysis of formalin-fixed, paraffin-embedded solid tumors by comparative genomic hybridization after universal DNA-amplification. Hum Mol Genet 2:1907-1914.
25. Kallioniemi A, Kallioniemi OP, Piper J, et al. (1994) Detection and mapping of amplified DNA sequences in breast cancer by comparative genomic hybridization. Proc Natl Acad Sci U S A 91:2156-2160.

4

Nucleocytoplasmic Shuttling of Estrogen Receptors is Blocked by "Pure Anti-Estrogens"

Sophie Dauvois and Malcolm G. Parker

Introduction

The estrogen receptor belongs to a large family of nuclear receptors whose activity as a transcription factor depends on the binding of a hormonal ligand (1). As with other steroid receptors, in the absence of ligand the estrogen receptor is found in an oligomeric complex that is unable to bind to DNA (2-4). Following hormone binding, the complex is dissociated to allow receptor dimerization, high-affinity DNA binding, and transcriptional activation (5, 6). In the estrogen receptor, there are at least two regions called activation functions, or AFs, which are required to stimulate transcription (7, 8). The relative activities of AF-1, located in the N-terminal domain, and AF-2, in the C-terminal hormone binding domain, vary according to the responsive promoter and target cells.

Since estrogens are mitogens in approximately two thirds of breast tumors, specific estrogen antagonists have been developed for the treatment of hormone-dependent breast cancer. The most widely used anti-estrogen to date is the nonsteroidal compound tamoxifen (9, 10), which promotes high-affinity DNA binding but inhibits transcriptional activation by the receptor (11). This anti-estrogen, however, possesses partial agonist activity (10); consequently, compounds devoid of agonist activity have been developed. The best known of these "pure" anti-estrogens are ICI 164384 and ICI 182780, which are derivatives of estradiol containing an alkylamide side-chain extension in the 7α position (12, 13). In addition to their effect on transcription, the pure anti-estrogens reduce steady-state levels of the estrogen receptor by increasing the turnover of the protein (14-16).

Immunohistochemical studies have shown that estrogen receptor is located predominantly in the cell nucleus even in the absence of hormone (17, 18). Nuclear uptake of steroid receptors was initially thought to be mediated by a short cluster of basic residues conserved near the DNA binding domain, similar to that identified in SV40 large T antigen (19-21). However, as with many other nuclear targeting sequences (22), a number of basic clusters in the estrogen receptor appear to target the protein into the cell nucleus in the absence of hormone (23). Recent work indicates that, although nuclear receptors appear to be located predominantly in the nucleus, they actually shuttle continuously between the nucleus and the cytoplasm (24-27). In view of this, we investigated whether the estrogen receptor undergoes nucleocytoplasmic shuttling and analyzed the effect of anti-estrogens on this process (28).

Nucleocytoplasmic Shuttling of Estrogen Receptors Occurs in the Presence of Estradiol but Not Pure Anti-Estrogens

We have analyzed nucleocytoplasmic shuttling of the estrogen receptor by generating heterokaryons between receptor-expressing and non-expressing cells and then examining the transfer of receptors between their nuclei. For these experiments, we fused transiently transfected COS-1 cells, of which 30-40% were expressing receptor, with mouse NIH-3T3 cells that were devoid of receptor. Their nuclei were distinguished with Hoechst dye, the COS-1 nuclei being uniformly stained whereas NIH 3T3 nuclei were smaller and showed punctuate staining (Color Plate 2). After fusion, the cell hybrids were treated with cycloheximide to inhibit *de-novo* protein synthesis; the distribution of receptors with no hormone, estradiol, or ICI 182780 was determined after three hours.

In the absence of hormone or in the presence of 10 nM estradiol, receptors accumulated in NIH 3T3 nuclei showing that they were able to shuttle from COS-1 cell nuclei. In the presence of 100 nM ICI 182780, however, estrogen receptors were not detected in NIH 3T3 nuclei (Color Plate 2), suggesting that the anti-estrogen interferes with the process of nucleocytoplasmic shuttling in heterokaryons. Thus, as previously reported (28), ICI 182780 inhibits nuclear uptake of estrogen receptors.

Pure Anti-Estrogens Reduce Nuclear Levels of Estrogen Receptors in Breast Cancer Cells

Previous work demonstrated that steady-state levels of estrogen receptor, overexpressed in COS cells, were reduced when cells were treated with the pure anti-estrogen ICI 182780 (14). We have extended this analysis to MCF-7 breast cancer cells by comparing the effects of ICI 182780, 17β-estradiol and 4-hydroxytamoxifen on estrogen receptor expression. Cells were grown in the absence of estrogen for five days and then treated with 10^{-8} M 17β -estradiol, 10^{-7}M ICI 182780 or 10^{-7}M 4-hydroxytamoxifen for four hours. The receptor was then analyzed by Western blotting analysis using the monoclonal antibody H222 raised against the human estrogen receptor. Figure 1 indicates that the amount of receptor was markedly reduced by ICI 182780 but was unaffected by 4-hydroxytamoxifen treatment and only modestly reduced (by approximately 50%) in the presence of 17β-estradiol.

Indirect immunofluorescence showed that the receptor was located predominantly in the nuclear compartment when cells were grown in the absence of hormone or in the presence of 17β-estradiol or 4-OH-tamoxifen, but was barely detectable in the presence of ICI 182780 (Figure 2). As judged by fluorescence intensity, the levels of receptor detected were broadly in agreement with those determined by Western blotting. The main difference between these results and those reported in COS cells (28) was the failure to detect receptor in the cytoplasm of MCF-7 cells in the presence of ICI 182780.

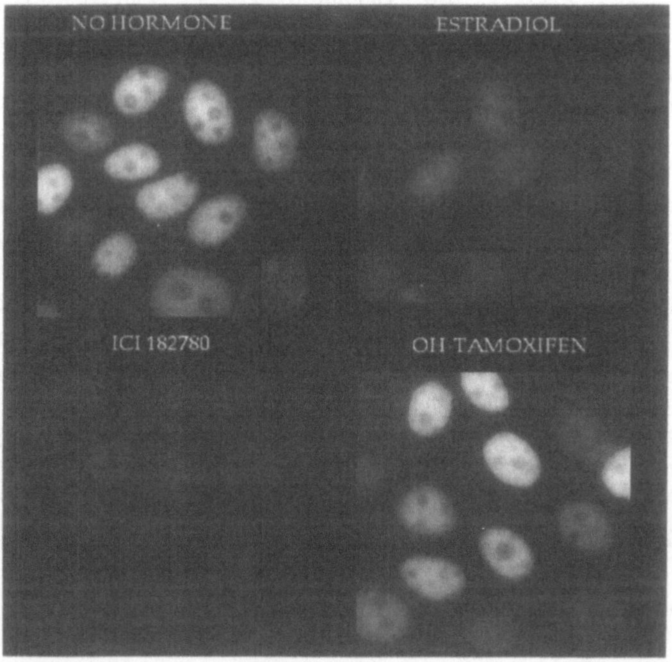

Figure 1. Effect of ligands and chloroquine on estrogen receptor levels in MCF-7 cells. Cells were incubated in the presence of no hormone (NH), 10^{-8} M 17β-estradiol (E_2), 10^{-7} M ICI 182780, or 4-hydroxytamoxifen with or without 50 mM chloroquine. Equal amounts of cell extract were separated by electrophoresis and the estrogen receptor quantitated by Western blotting using H222.

This might reflect differences in expression levels since we estimate that the concentration of receptor in COS cells exceeds that in MCF-7 cells by approximately 100.0-fold. It is conceivable that the receptor accumulates in the cytoplasm of MCF-7 cells prior to its degradation but at a level below that detectable with the H222 antibody. Thus, ICI 182780 treatment of MCF-7 breast cancer cells results in a marked reduction in total estrogen receptor content, probably brought about following a block in nuclear reentry during the nucleocytoplasmic shuttling of the receptor.

In view of the increased turnover of receptor and its speckled immunostaining appearance in COS cells treated with ICI 182780, we investigated whether the receptor was accumulating in lysosomes prior to its degradation. We tested this in two ways. First, we examined the effect of chloroquine, which is reported to inhibit lysosome function. The levels of estrogen receptor were detected by Western blotting in MCF-7 breast cancer cells in the presence and absence of ICI 182780. We found that 50 mM chloroquine prevented the reduction of receptor brought about by the anti-estrogen (Figure 1), suggesting that the receptor was being degraded in lysosomes.

Secondly, we investigated whether receptors co-localized with a known lysosomal protease, cathepsin D, in the presence of ICI 182780. The subcellular localization of estrogen receptors and cathepsin D was examined in COS cells using indirect immunofluorescence and confocal microscopy. Estrogen receptors were detected with H222 and Texas red-conjugated antibodies while cathepsin D antibodies (29) were detected with fluorescein conjugated antibodies. It can be seen in Color Plate 3 that the estrogen receptor and cathepsin D did not co-localize. Since it was possible that the receptor had undergone degradation within the lysosome, we repeated the experiment in the presence of chloroquine to inhibit lysosomal function but still failed to show co-localization of the receptor and cathepsin D. We tentatively conclude that the receptor is degraded in lysosomes in the presence of ICI 182780, but, when overexpressed in COS-1 cells, the receptor initially forms protein aggregates which produce a speckled staining appearance.

Conclusions

We have demonstrated that estrogen receptors are constantly shuttling between the nucleus and the cytoplasm in the presence of estradiol or the estrogen antagonist 4-hydroxytamoxifen. In contrast, the pure anti-estrogen ICI 182780 disrupts shuttling and increases the turnover rate of the receptor.

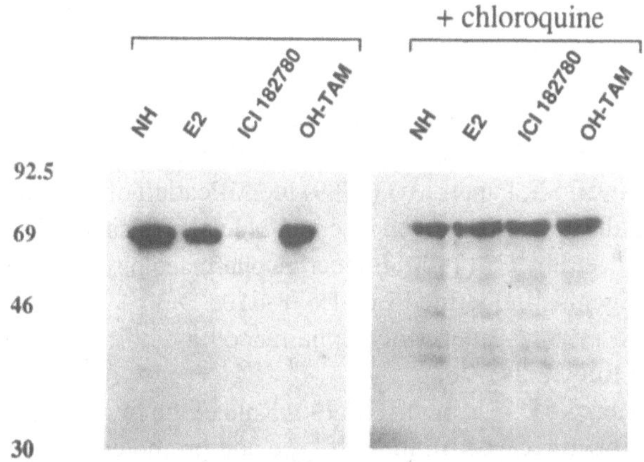

Figure 2. Effect of ligands on the subcellular localization of the estrogen receptor in MCF-7 human breast cancer cells. After four hours treatment with no hormone, 10^{-8} M 17β-estradiol, 10^{-7} M ICI 182780, or 10^{7} M 4-OH-tamoxifen, cells were fixed and receptors were detected by indirect immunofluorescence using the estrogen receptor monoclonal antibody H222.

We conclude that the resulting nuclear exclusion and reduction in the level of estrogen receptor must contribute to the action of ICI 182780 and ICI 164384 as estrogen antagonists.

References

1. Parker MG (1993) Steroid and related receptors. Curr Opinion in Cell Biol 5:499-504.
2. Catelli MG, Binart N, Jung TI, et al. (1985) The common 90-kd protein component of non-transformed '8S' steroid receptors is a heat-shock protein. EMBO J 4:3131-3135.
3. Chambraud B, Berry M, Redeuilh G, et al. (1990) Several regions of the human estrogen receptor are involved in the formation of receptor-heat shock protein 90 complexes. J Biol Chem 265:20686-20691.
4. Schlatter LK, Howard JK, Parker MG, Distelhorst CW (1992) Comparison of the 90-kilodalton heat shock protein interaction with *in vitro* translated glucocorticoid and estrogen receptors. Mol Endocrinol 6:132-140.
5. Kumar V, Chambon P (1988) The estrogen receptor binds tightly to its responsive element as a ligand-induced homodimer. Cell 55:145-156.

6. Fawell SE, Lees JA, White R, Parker MG (1990) Characterization and colocalization of steroid binding and dimerization activities in the mouse estrogen receptor. Cell 60:953-962.
7. Tora L, White J, Brou C, et al. (1989) The human estrogen receptor has two independent nonacidic transcriptional activation functions. Cell 59:477-487.
8. Lees JA, Fawell SE, Parker MG (1989) Identification of two transactivation domains in the mouse oestrogen receptor. Nucleic Acids Res 17:5477-5488.
9. Jordan VC, Murphy CS (1990) Endocrine pharmacology of antiestrogens as antitumor agents. Endocrine Rev 11:578-610.
10. Jordan VC (1984) Biochemical pharmacology of antiestrogen action. Pharmacol Rev 36:245-76.
11. Berry M, Metzger D, Chambon P (1990) Role of the two activating domains of the oestrogen receptor in the cell-type and promoter-context dependent agonistic activity of the anti-oestrogen 4-hydroxytamoxifen. EMBO J 9:2811-2818.
12. Bowler J, Lilley TJ, Pittam JD, Wakeling AE (1989) Novel steroidal pure antiestrogens. Steroids 54:71-99.
13. Wakeling AE, Dukes M, Bowler J (1991) A potent specific pure antiestrogen with clinical potential. Cancer Res 51:3867-3873.
14. Dauvois S, Danielian PS, White R, Parker MG (1992) Antiestrogen ICI 164,384 reduces cellular estrogen receptor content by increasing its turnover. Proc Natl Acad Sci USA 89:4037-4041.
15. Gibson MK, Nemmers LA, Beckman Jr WC, et al. (1991) The mechanism of ICI 164,384 antiestrogenicity involves rapid loss of estrogen receptor in uterine tissue. Endocrinology 129:2000-2010.
16. Reese JC, Katzenellenbogen BS (1992) Examination of the DNA-binding ability of estrogen receptor in whole cells: Implications for hormone-dependent transactivation and the actions of antiestrogens. Mol Cell Biol 12:4531-4538.
17. King WJ, Greene GL (1984) Monoclonal antibodies localize oestrogen receptor in the nuclei of target cells. Nature 307:745-747.
18. Welshons WV, Krummel BM, Gorski J (1985) Nuclear localization of unoccupied receptors for glucocorticoids, estrogens, and progesterone in GH3 cells. Endocrinology 117:2140-2147.
19. Kalderon D, Richardson WD, Markham AT, Smith AE (1984) Sequence requirements for nuclear location of simian virus 40 large-T antigen. Nature 311:33-38.
20. Kalderon D, Roberts BL, Richardson WD, Smith AE (1984) A short amino acid sequence able to specify nuclear location. Cell 39:499-509.
21. Lanford RE, Butel JS (1984) Construction and characterization of an SV40 mutant defective in nuclear transport of T antigen. Cell 37:801-813.

22. Dingwall C, Laskey RA (1991) Nuclear targeting sequences: A consensus? TIBS 16:478-481.
23. Ylikomi T, Bocquel MT, Berry M, et al. (1992) Cooperation of proto-signals for nuclear accumulation of estrogen and progesterone receptors. EMBO J 11:1-14.
24. Guiochon-Mantel A, Loosfelt H, Lescop P, et al. (1989) Mechanisms of nuclear localization of the progesterone receptor: evidence for interaction between monomers. Cell 57:1147-1154.
25. Guiochon-Mantel A, Lescop P, Christin-Maitre S, et al. (1991) Nucleo-cytoplasmic shuttling of the progesterone receptor. EMBO J 10:3851-3859.
26. Chandran UR, De Franco DB (1992) Internuclear migration of chicken progesterone receptor, but not simian virus-40 large tumor antigen, in transient heterokaryons. Mol Endocrinol 6:837-844.
27. De Franco DB, Qi M, Borror KC, et al. (1991) Protein phosphatase types 1 and/or 2A regulate nucleocytoplasmic shuttling of glucocorticoid receptors. Mol Endocrinol 5:1215-1228.
28. Dauvois S, White R, Parker MG (1993) The antiestrogen ICI 182780 disrupts estrogen receptor nucleocytoplasmic shuttling. J Cell Sci 106:1377-1388.
29. Henry JA, McCarthy AL, Angus B, et al. (1990) Prognostic significance of the estrogen-regulated protein, cathepsin D, in breast cancer. Cancer 65:265-271.

PART 1. EPIDEMIOLOGY: HORMONAL CANCERS

Introduction

Causality of Hormone-Associated Human Cancers

Janet R. Daling

Early epidemiologic studies linked reproductive and hormonal factors to breast cancer risk (1). These factors included early age at menarche, natural age at menopause, menstrual characteristics, nulliparity, age at first full-term pregnancy, number of live births, and lactation. A woman's family history of breast cancer has also been known for quite some time to affect breast cancer risk strongly (2).

In the 1980's, attention focused on diet (3), use of oral contraceptives (OC) (4), and estrogen replacement therapy (ERT) (5). The studies on diet have been equivocal, and little, if any, excess risk was found with the use of hormones. However, in the last few years, reports have shown increased risks associated with early OC use among women diagnosed with breast cancer at an early age (6), and concerns about the effect of combined hormone therapy have led to new studies on the incidence of postmenopausal breast cancer (7).

Recently, other exposures have received attention. Positive associations have been observed between alcohol use and breast cancer (8). However, the associations have not been consistent, and the lack of a credible biologic mechanism does not allow a causal inference at this time. The effects of body size and exercise have also been topics of research (9-13). Lean body mass has been associated with an increased risk of breast cancer in premenopausal women, whereas postmenopausal women have a higher risk if they are obese. The possible protective effect of exercise on breast cancer risk and the increased risk observed among obese women are thought to be hormonally related. Increased levels of physical activity have been associated with lower concentrations of circulating estrone. Exercise also decreases body fat tissue, which in turn is thought to produce estrone.

A recent study finding a 50% increase in risk of breast cancer among women who have had an induced abortion has led to a number of studies addressing this issue (14-18). In general, these studies have found an increase in overall risk among some subgroups of women based on the time of the abortion or their parity status. However, the inconsistencies among studies related to the magnitude of the risk and the subgroups of women affected indicate that more research is needed in this area. These findings partially accord with animal studies and are biologically plausible (19, 20). During the first trimester of pregnancy, the breast is characterized by high mitotic ability and proliferation; only in mid- to late pregnancy does cellular differentiation predominate. If the pregnancy is interrupted, some areas of the mammary gland are left with immature, undifferentiated cells that may be more susceptible to carcinogens.

The cloning of the locus of the BRCA1 gene (21) and discovery of the locus of the BRCA2 gene (22) have caused great excitement among molecular geneticists and epidemiologists. Determining the proportion of the disease that can be attributed to these genes and/or to other related genes, yet to be identified, will soon be possible. Population-based studies will be needed to further this effort, as will determination of each putative breast cancer gene's frequency in the general population. In addition, as tests become available, it will be feasible to screen large numbers of samples, and epidemiologists will be able to assess interactions between genes and environment.

References

1. Kelsey JL, Gammon MD, John EM (1993) Reproductive factors and breast cancer. Epidemiol Rev 15:36-47.
2. Macklin MT (1959) Comparison of the number of breast cancer deaths observed in relatives of breast cancer patients, and the number expected on the basis of mortality rates. J Natl Cancer Inst 22:927-951.
3. Hunter DJ, Willett WC (1993) Diet, body size, and breast cancer. Epidemiol Reviews 15:110-132.
4. Malone KE, Daling JR, Weiss NS (1993) Oral contraceptives and breast cancer risk. Epidemiol Rev 15:80-97.
5. Brinton LA, Schairer C (1993) Estrogen replacement therapy and breast cancer risk. Epidemiol Rev 15:66-79.
6. White E, Malone KE, Weiss NS, Daling JR (1994) Breast cancer among young U.S. women in relation to oral contraceptive use. J Natl Cancer Inst 86:505-514.
7. Stanford JL, Thomas DT (1993) Exogenous progestins and breast cancer. Epidemiol Rev 15:98-107.
8. Rosenberg L, Metzger LS, Palmer JR (1993) Alcohol consumption and risk of breast cancer: A review of the epidemiologic evidence. Epidemiol Rev 15:133-144.
9. Gammon MD, John EM (1993) Recent etiologic hypotheses concerning breast cancer. Epidemiol Rev 15:163-168.

10. Frisch RE, Wyshak G, Albright NL et al (1985) Lower prevalence of breast cancer and cancers of the reproductive system among former college athletes compared to non-athletes. Br J Cancer 52:885-891.
11. Vihko JV, Apter DL, Pukkala EI et al (1992) Risk of breast cancer among female teachers of physical education and languages. Acta Oncol 31:201-204.
12. Bernstein L, Henderson BE, Hanisch R et al (1994) Physical exercise and reduced risk of breast cancer in young women. J Natl Cancer Inst 86:1403-1408.
13. Dorgan JF, Brown C, Barrett M et al (1994) Physical activity and risk of breast cancer in the Framingham Heart Study. Am J Epidemiol 139:662-669.
14. Daling JR, Malone KE, Voigt LF et al (1994) Risk of breast cancer among young women: Relationship to induced abortion. J Natl Cancer Inst 86:1584-1592.
15. Lipworth L, Katsouyanni K, Ekbom A et al (1995) Abortion and the risk of breast cancer. A case-control study in Greece. Int J Epidemiol, in press.
16. Rookus MA, van Leeuwen FE (1995) Breast cancer risk after induced abortion, a Dutch case-control study. Am J Epidemiol 141:S54.
17. Bu L, Boigt LF, Yu Z et al (1995) Risk of breast cancer associated with induced abortion in a population at low risk of breast cancer. Am J Epidemiol 141:S85.
18. Newcomb PA, Storer BE, Longnecker MP et al (1995) Pregnancy termination in relation to risk of breast cancer. Am J Epidemiol 141:S54.
19. Russo J, Russo IH (1987) Biology of disease. Biological and molecular bases of mammary carcinogenesis. Lab Invest 57:112-137.
20. Russo J, Wilgus G, Russo IH (1979) Susceptibility of the mammary gland to carcinogenesis. I. Differentiation of the mammary gland as determinant of tumor incidence and type of lesion. Am J Pathol 96:721-736.
21. Miki Y, Swensen J, Shattuck-Eidens D et al (1994) A strong candidate for the breast and ovarian cancer susceptibility gene BRCA1. Science 266:66-71.
22. Wooster R, Neuhausen SL, Mangion J et al (1994) Localization of a breast cancer susceptibility gene, BRCA2, to chromosome 13q12-13. Science 265:2088-2090.

5

Breast Cancer Incidence in Women Exposed to Estrogen and Estrogen-Progestin Replacement Therapy

Ingemar Persson

Importance of the Issue

Hormone replacement therapy (HRT) is widely used, mainly to relief vasomotor symptoms, but also for long-term protection against osteoporosis and cardiovascular disease. However, the alleged relationship between HRT and increased risk of postmenopausal breast cancer is of major concern. Since breast cancer is a frequent and serious disease, an association between HRT and an enhanced risk of breast cancer could cause a substantial number of new cases of or deaths from breast cancer. The aim of this presentation is to summarize briefly the available literature.

Etiological Role of Endogenous Hormones

Established risk factors for breast cancer in aggregate indicate that the production of estrogens and/or progesterone by the ovaries is important in the development of breast cancer (1). Thus, after premature oophorectomy (2) or early menopause (3), a lowered risk is noted. On the other hand, factors reflecting a prolonged period of hormone production, such as early menarche (4), late menopause (5), a large number of menstrual cycles (6), or obesity (7) seem to enhance the risk. These associations might fit either with the "estrogen alone hypothesis" or the "estrogen plus progesterone hypothesis" (8).

Theoretically, sex hormones may affect the likelihood of transformation of normal breast epithelial cells to cancer cells through their effect on cell division. *In-vivo* studies using hyperplastic breast or benign breast disease samples have revealed higher cell turnover rates during the luteal phase than during the follicular phase of the normal menstrual cycle. These findings indicate that progesterone might exert an adverse effect on the breast by increasing its rate of cell division (9). Conversely, *in-vitro* experiments using “normal” breast epithelial cells in culture have demonstrated that progestins can counteract the growth stimulatory effects of estrogens (10).

Thus, these and similar data regarding the hormonal effects of sex hormones on normal epithelial breast cells are in conflict (11). The possibility that hormonal effects can be exerted not only through direct receptor-mediated mechanisms, but also through autocrine and paracrine factors, indicates that the regulation of normal breast tissue is complex (11).

Incidence of Breast Cancer After HRT

Ever since HRT was introduced, there has been concern regarding a possible increase in breast cancer risk. Numerous epidemiological studies exploring the relationship between exogenous hormones and breast cancer risk had been published as of 1989. Those studies were summarized in two major review articles (12, 13), and five additional studies have been published subsequently (14-18). Thus, our current knowledge is based on the results of over thirty studies, most of which were retrospective case-control, with seven prospective cohort studies.

Salient Results of Some Recent Epidemiological Studies

Table 1 depicts the results of nine retrospective case-controls studies (14-17, 19-23); Table 2 is a summary of six prospective cohort studies (18, 24-28). Reporting mainly on conjugated estrogens, the studies from North America provided evidence indicating an increased risk. Regarding years of intake, the studies by Brinton and Yang demonstrated a significant association with increased risk; their relative risk (RR) estimate values were 1.5-1.6 (Table 1). Mills and Colditz's prospective studies provided evidence of a positive risk relationship, either with long duration or current intake (Table 2). Studies from Italy, Denmark, the United Kingdom, and Sweden comprise data mainly from estradiol compounds, although three studied (23, 25, 27) estrogens combined with cyclically-added progestins. All of these European studies demonstrated an elevated RR, chiefly in association with long duration of intake, as well as with combined estrogen-progestogen regimes (Tables 1 and 2).

Detailed Risk Relationships

Overall Risk

In most studies, the overall risk (OR), i.e., the association with ever-use of replacement estrogens, was not altered. Also, when combining overall risk

Table 1. Recent Retrospective Studies of HRT and Postmenopausal BC Risk.

Author/ year	Country of origin	Design/ No. cases/ controls	Risk Relationships		
Brinton, et al../ 1986 (19)	USA	Screening program 1,960/2,258 (postmenopausal)	Overall: 20+ years:		RR=1.0 (0.9, 1.2) RR=1.5 (0.9, 2.3) Significant trend by duration.
McDonald, et al../ 1986 (20)	USA	Population-based 183/537 (age 50-74)	Overall: 6+ years		RR=0.74 (0.5, 1.1) RR=0.6 No significant trend.
La Vecchia et al./ 1986 (21)	Italy	Hospital-based 1,108/1,281 (age<75)	Overall: 2+ years		RR=1.93 (1.4, 2.8) RR=2.0 (1.0, 4.1) Interaction with age at menopause.
Wingo, et al../ 1987 (22)	USA	Population-based 1,369/1,645 (age<55)	Overall: 20+ years		RR=1.0 (0.9, 1.2) RR=1.8 (0.6, 5.8) No significant trend.
Ewertz, et al../ 1988 (23)	Denmark	Population-based 1,486/1,336 (age<70)	Overall: 12+ years		RR=1.28 (0.96-1.71) RR=2.3 (1.3-4.1) Risk increased for estrogen/ progestin/ androgen combinations.
Palmer, et al../ 1991 (14)	Canada Toronto	Population-based 607/1,214 Neighborhood (<70 years of age)	Ever use: ≥15 years: Current use:	CE only	RR=0.9 (0.6-1.2) RR=1.5 (0.6-3.8) RR=0.9 (0.4-1.9)
Kaufman, et al../ 1991 (15)	USA Eastern States	Hospital-based 1,686/2,077 Interviews (40-69 years of age)	Any use: >15 years: Current use:	E only E +P CE	RR=1.2 (1.0-1.4) RR=1.7 (0.9-3.3) RR=1.3 (1.0-1.6) RR=1.1 (0.7-1.6)
Harris, et al../ 1992 (16)	USA New York City	Hospital-based 604/520 Interviews (all ages)	Summary OR (lean women): <5 years: ≥15 years:	 E only E only	OR=2.0 (1.1-3.5) OR=2.0 OR=2.2
Yang, et al../ 1992 (17)	Canada	Population-based 1,018/1,025 (<75 years of age)	Ever use: 10+ years: Current use:	E only E +P E only 1-4 years	OR=1.0 (0.8-1.3) OR=1.2 (0.6-2.2) OR=1.6 (1.1-2.5) OR=1.3 (0.8-2.2)

Table 2. Recent Prospective Studies of HRT and Postmenopausal Breast Cancer Risk.

Author/ year	Country of origin	Design/ size	Risk relationships	
Buring, et al../ 1987 (24)	USA	Nurses' Health Study 33,335, 35-55 years 221 cases	Overall: 5-9 years: 10+ years:	RR=1.1 (0.8, 1.4) RR=1.5 (1.0-2.2) RR=0.9 (0.4-1.6)
Hunt, et al../ 1987 (25)	UK	Menopause clinics 4,544 with HRT, 45-54 years, 50 cases	Overall: 10+ years since last use:	RR=1.6 (1.2, 2,1) RR=3.1 (1.5, 5.6) 43% progestin combined use.
Mills, et al../ 1989 (26)	USA	7th Day Adventists 20,341, >25 years, 215 cases	Overall: 6-10 years: 10+ years:	RR=1.7 (1.2, 2.4) RR=2.8 (1.6, 4,6) RR=1.8 (0.9-2.5) Interaction with menopause age.
Bergkvist, et al../ 1989 (27)	Sweden	Prescription-based 23,244 with HRT, ≥35 years, 253 cases	Overall: 9+ years: 6+ years of estradiol:	RR=1.1 (1.0, 1.3) RR=1.7 (1.1-2.7) RR=2.3 (1.2-4.3) Similar risk pattern for combined use.
Colditz, et al../ 1990 (28)	USA	Nurses' Health Study 118,273, 30-55 years, 722 cases	Overall: Current use:	RR=0.98 (0.81, 1.18) RR=1.35 (1.11, 1.67)
Colditz, et al../ 1992 (18)	USA	Nurses' Health Study (continued follow-up) 1,050 cases	Overall: Current use: Current use, combined:	RR=0.91 (0.78-1.07) RR=1.33 (1.12-1.57) RR=1.54 (0.99-2.39)

Duration and Timing

Subgroup analyses would be more informative, especially regarding the relationship of dose and duration of intake with level of risk. As shown in the tables, the level of RR varied from 1.5 (19) to 3.1 (25) in association with years of estrogen use. A meta-analysis, including 16 case-control studies, demonstrated a statistically verified 30% increased risk after > 15 years of use, RR = 1.3, 95% CI 1.2-1.6 (12). Another meta-analysis based on 28 studies could not find any consistent evidence of increased risk after extended use of conjugated estrogens (13). However, these meta-analyses report a substantial variability in the results among studies, indicating that there were important differences between the studies regarding the underlying treatment characteristics, the study populations, and the quality of the various study designs. Regarding latency or recency effects, no consistent patterns have been noted.

Type of Regimens and Dose

Given the magnitude of the RR estimates reported by the Scandinavia and United Kingdom studies, it has been assumed that estradiol compounds could have greater adverse effects on the breast than conjugated estrogens (Tables 1 and 2). However, a similar conclusion cannot be drawn from the available data, since studies on conjugated estrogens have also shown RR estimates of similar magnitude to those reported for estradiol use. Regarding dose level, one meta-analysis found no change in RR after 0.625 mg of conjugated estrogens, whereas a higher dose of 1.25 mg was linked with a possibly increased risk (13).

Added Progestins

As mentioned before, there is concern that progestins may produce an additional growth stimulatory effect on the breast epithelium, and therefore further enhance the risk of cancer development in association with estrogen replacement therapy (ERT). This issue is particularly important, since estrogen-progestin combined regimens are becoming more popular. The Swedish cohort study (27) and the Danish case-control study (23) reported a similar magnitude of increase in risk associated with both estrogens and combined estrogen-progestogen regimens. Furthermore, the most recent report from the Nurses' Health study (18) provided additional evidence of an increased risk in women currently using both unopposed and progestogen-opposed replacement therapy.

Thus, progestins do not seem to eliminate the increased risk of breast cancer after long-term exposure to estrogens alone. Evidence that progestins are not protective in the breast also stems from numerous studies on the relationship between combined oral contraceptives (OCs) and breast cancer risk, showing in no instance protection but, in some studies, an increased risk (29). In addition, a study in young women receiving injectable contraceptive progestins (medroxyprogesterone acetate) reported an increased risk of breast cancer after long-term intake (30).

Interactions

The effects of HRT on breast cancer risk might be modified by certain patient characteristics. For example, HRT in a patient with family history of breast cancer, especially in first-degree relatives, might result in a higher risk (12). Given the data available for the other factors such as parity, age at first full-term pregnancy, smoking, and alcohol, there is no firm evidence of an interaction.

Clinical Course in Patients with HRT-Related Breast Cancer

Long-term unopposed ERT leads to a substantially increased incidence of endometrial cancer, but virtually without an excess risk of death from the disease (31). In studies of HRT and breast cancer, there is some evidence of a limited development of small and early-stage tumors (19). Analyses of relative survival in the Swedish prospective study showed that women with breast cancer after HRT had an advantage as compared with patients who were unexposed (32). Follow-up mortality in the same cohort study did not reveal an increase in breast cancer mortality after 12 years. The findings in the Swedish cohort may reflect earlier detection for tumors among estrogen-treated women. It is also possible that HRT-related cancers have more favorable biological characteristics.

Conclusions

The following conclusions can be drawn from a critical evaluation of the results from individual studies, as well as from meta-analyses:

Overall, i.e., with exposure to any type, dose, or duration of HRT, there is no evidence of an increased risk. However, the great heterogeneity among the studies precludes a meaningful interpretation of the data.

The intake duration affects the risk of breast cancer. In several studies, the RR is increased to 1.5-2.0, after exposure to both estradiol compounds and conjugated estrogens for 10-15 years. The magnitude of the risk increase after very long exposure can be considered small, and may be explained to a large extent by methodological problems.

The addition of progestins does not seem to alter the risk relationship.

The clinical outcome for patients with HRT-related breast cancer may be better than for unexposed breast cancer patients.

Further research in this area is essential, as are studies on the mechanisms of hormonal regulation of the normal breast.

References

1. Adami HO, Adams G, Boyle P, et al. (1990) Breast cancer etiology. Report from a working party for the Nordic Cancer Union Meeting, 1989. Int J Cancer 5:22-39.
2. Kelsey JL, Hildreth NG (1983) Breast and gynecologic cancer epidemiology CRC Press, Boca Raton, FL.
3. Negri E, La Vecchia C, Bruzzi (1988) Risk factors for breast cancer: Pooled results from three Italian case-control studies. Am J Epidemiol 128:1207-1215.
4. Boyle P (1988) Epidemiology of breast cancer. Baillière's Clinical Oncology 2:1-59.
5. Brinton LA, Schairer C, Hoover RN, Fraumeni JF (1988) Menstrual factors and risk of breast cancer. Cancer Invest 6:245-254.
6. Henderson BE, Ross RK, Judd HL, et al. (1985) Do regular ovulatory cycles increase breast-cancer risk? Cancer 56:1206-1208.
7. Rose DP (1986) Dietary factors and breast cancer. Cancer Surv 5:671-688.
8. Key TJA, Pike MC (1988) The role of oestrogens and progestogens in the epidemiology and prevention of breast cancer. Eur J Cancer Clin Oncol 24:29-43.
9. Ferguson DJP, Anderson TJ (1981) Morphological evaluation of cell turnover in relation to the menstrual cycle in the 'resting' human breast. Brit J Cancer 44:177-181.
10. Mauvais-Jarvis P, Kuttenn F, Gompel A (1986) Antiestrogen action of progesterone in breast tissue. Breast Cancer Res Treat 8:179-87.
11. Clarke CL, Sutherland RL(1990) Progestin regulation of cellular proliferation. Endocrine Rev 11:260-300.
12. Steinberg KK, Thacker SB, Smith SJ, et al. (1991) A meta-analysis of the effect of estrogen replacement therapy on the risk of breast cancer. JAMA 265:1985-1990.
13. Dupont WD, Page DL (1991) Menopausal estrogen replacement therapy and breast cancer. Arch Intern Med 151:67-72.
14. Palmer JR, Rosenberg L, Clarke EA, et al. (1991) Breast cancer risk after estrogen replacement therapy: Results from the Toronto Breast Cancer Study. Am J Epidemiol 134:1386-1395.
15. Kaufman DW, Palmer JR, de Mouzon J, et al. (1991) Estrogen replacement therapy and the risk of breast cancer: Results from the Case-Control Surveillance Study. Am J Epidemiol 134:1375-1385.
16. Harris RE, Namboodiri KK, Wynder EL (1992) Breast cancer risk: Effects of estrogen replacement therapy and body mass. J Natl Cancer Inst 84:1575-1582.

17. Yang CP, Daling JR, Band PR, et al. (1992) Noncontraceptive hormone use and risk of breast cancer. Cancer Causes Cont 3:475-479.
18. Colditz GA, Stampfer MJ, Willett WC, et al. (1992) Type of postmenopausal hormone use and risk of breast cancer: 12-year follow-up from the Nurses' Study. Cancer Causes Cont 3:433-439.
19. Brinton LA, Hoover R, Fraumeni JF Jr (1986) Menopausal oestrogens and breast cancer risk: An expanded case-control study. Br J Cancer 54:825-832.
20. McDonald JA, Weiss NS, Daling JR, et al. (1986) Menopausal estrogen use and the risk of breast cancer. Breast Cancer Res Treat 7:193-199.
21. La Vecchia C, Decarli A, Parazzini F, et al. (1986) Non-contraceptive oestrogens and the risk of breast cancer in women. Int J Cancer 38:853-858.
22. Wingo PA, Layde PM, Lee NC, et al. (1987) The risk of breast cancer in postmenopausal women who have used estrogen replacement therapy. JAMA 257:209-215.
23. Ewertz M (1988) Influence of non-contraceptive exogenous and endogenous sex hormones on breast cancer risk in Denmark. Int J Cancer 42:832-838.
24. Buring JE, Hennekens CH, Lipnick RJ, et al. (1987) A prospective cohort study of postmenopausal hormone use and risk of breast cancer in US women. Am J Epidemiol 125:939-947.
25. Hunt K, Vessey M, McPherson K, Coleman M (1987) Long-term surveillance of mortality and cancer incidence in women receiving hormone replacement therapy. Brit J Obstet Gynecol 94:620-635.
26. Mills PK, Beeson WL, Phillips RL, Fraser GE (1989) Prospective study of exogenous hormone use and breast cancer in Seventh-Day Adventists. Cancer 64:591-597.
27. Bergkvist L, Adami HO, Persson I, et al. (1989) The risk of breast cancer after estrogen and estrogen-progestin replacement. N Engl J Med 321:293-297.
28. Colditz GA, Stampfer MJ, Willett, et al. (1990) Prospective study of estrogen replacement therapy and risk of breast cancer in menopausal women. JAMA 264:2648-2655.
29. UK National Case Control Study Group (1989) Oral contraceptive use and breast cancer risk in young women. Lancet 1:973-982.
30. Paul C, Skegg DCG, Spears GFS (1989) Depot medroxyprogesterone (Depo-Provera) and risk of breast cancer. Brit Med J 299:759-762.
31. Collins J, Donner A, Allen LH, Adams O (1980) Oestrogen use and survival in endometrial cancer. Lancet II:961-964.
32. Bergkvist L, Adami HO, Persson I, et al. (1989) Prognosis after breast cancer diagnosed in women exposed to estrogen and estrogen-progestogen replacement therapy. Am J Epidemiol 130:221-228.

6

Breast Cancer Trends in Women in Sweden, the UK, and the USA in Relation to Their Past Use of Oral Contraceptives

Valerie Beral, Carol Hermon, Gillian Reeves, and Timothy Key

Introduction

Since their introduction in the early 1960's, oral contraceptives (OCs) have become one of the most popular methods of contraception. They have been used by an estimated 67 million women in developed countries and 200 million women worldwide (International Planned Parenthood Federation, personal communication). Reports that OC use might increase the risk of breast cancer came first from California (1), and subsequently from Sweden and Norway (2), the UK (3), and elsewhere in the USA (4). However, the results of many other studies, particularly those that included older women, did not suggest an increase in breast cancer risk (5). Nevertheless, some have expressed concern that the use of OCs at an early age, especially before the birth of the first child, may lead to an increase in the risk of breast cancer when women reach middle-age (6).

This is not intended to be a review of the evidence on OCs and breast cancer. Numerous epidemiological studies have collected information on the subject, and it is not possible to summarize the published results in a systematic way. This is mainly because methods of analysis and presentation of results vary from one study to another, in addition to which, the results of some studies have never been published. The original data from all studies on the topic need to be brought together and re-analyzed using a systematic approach. The Collaborative Group on Hormonal Factors in Breast Cancer was set up in 1992 with that goal in mind, and the results from this collaboration, based on around 50,000 women with breast cancer and 100,000 controls from more than 50 studies, should be available in the next few years.

The purpose of this paper is to describe breast cancer trends in Sweden, the UK, and the USA, bearing in mind that OCs have been available for 30 years, and that the generations of women who began using them some 20 or more years ago are now in their 40's and 50's.

Use of OCs by Women Who Were Aged 40-59 Years Old in 1990

The pattern of OC use by women in Sweden, the UK, and the USA, who were aged 40-59 in 1990, is shown in Table 1. Estimates in the table come from results for women selected as controls from the general population in various studies of the relation between OCs and breast cancer (1-5, 7, 8).

Table 1. Estimated prevalence of OC use and characteristics of use by women in Sweden, the UK, and the USA who were aged 40-59 in 1990.

Age in 1990	40-44 years			45-49 years			50-54 years	55-59 years
Approx year of birth	1946-50			1941-45			1936-40	1931-35
Country	Sweden[a]	UK	USA[b]	Sweden	UK	USA[c]	UK	USA[d]
% who ever used OCs	0.83	83	91	69	77	78	60	48
% who used OCs for 4 years or longer	40	52	34	33	42	34	29	18
% who began using OCs before they had a child	55	52	32	27	27	18	8	0
% who began using OCs before they were 25 years old	74	64	No data	42	45	No data	8	No data

[a] Year of birth 1946 and earlier [34% born after 1950 (2, 7)]
[b] Year of birth 1946-57 (5)
[c] Year of birth 1936-47 (5)
[d] Year of birth 1926-37 (5)

Although it is not always possible to match the age groups in 1990 exactly to the required years of birth, the pattern of use is fairly similar across countries, but varies considerably by age. In each country, the prevalence of past use of OCs was around 80-90% in women aged 40-44 in 1990, around 70-80% for women aged 45-49, and around 50-60% for women in their 50's. Use at a young age or before the birth of a first child varied markedly by age, but again there was no large variation from one country to another. About half the women aged 40-44 in 1990 had taken OCs before the birth of their first child, and two thirds before they were 25 years old. The prevalence of such use was about one quarter and one half, respectively, at age 45-49 but was rare at age 50-54 and older.

Breast Cancer Trends and Their Relation to Patterns of OC Use

Age-adjusted mortality rates for breast cancer have been fairly steady in Sweden and the USA since 1955. They have increased in England and Wales up to 1985, but have fallen ever since (Figure 1).

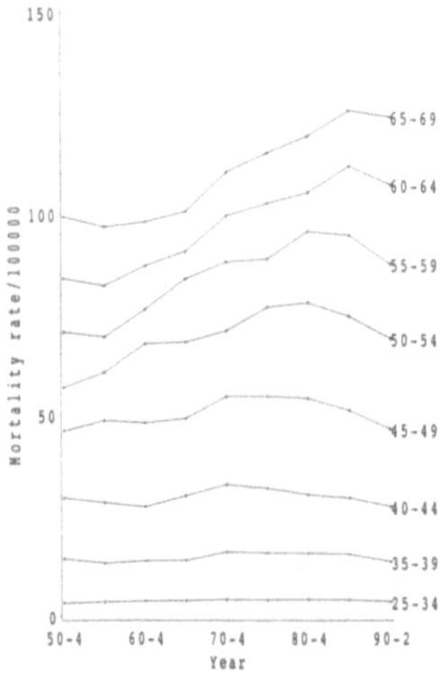

Figure 1. All-ages mortality rates for breast cancer per 100,000 women in the UK, the USA, and Sweden in 1955, 1965, 1975, 1985, and 1990 (adjusted by age to the world standard population).

However, the all-ages rates conceal trends in young women that have been declining, predominantly, in recent years. This is illustrated for England and Wales in Figure 2, which shows mortality rates by age, from 1950-54 to 1990-92. It can be seen that, against a background of stable or increasing rates during the 1950's, 60's and 70's, mortality began to decline in women < 50 years of age after 1970-74, in women aged 50-54 and 55-59 after 1980-84, and in women aged 60-64 and 65-69 after 1985-89.

In Sweden and the USA, mortality rates for breast cancer in women in their 40's and 50's have also declined, the decline tending to begin earlier than in the UK (Figure 3). The rates in Sweden fluctuate somewhat because they are based on small numbers of deaths (between 50-100 deaths each year at ages 40-44 and 45-49; and 100-150 deaths at ages 50-54 and 55-59).

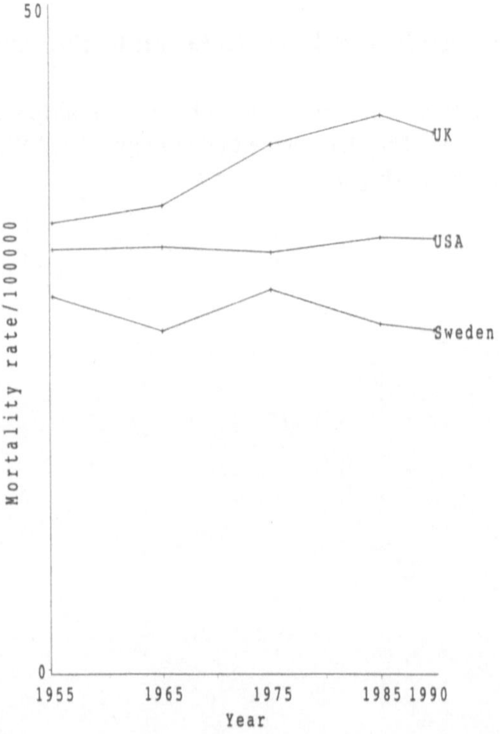

Figure 2. Age-specific mortality rates for breast cancer per 100,000 women in England and Wales, 1950-54 to 1990-92.

What is clear from Figures 2 and 3 is that, among women aged 40-59 years, breast cancer mortality has not increased substantially during the past 20 years in any of the three countries. Yet it is in these women that the pattern of OC use has changed considerably. Women aged 50-59 in 1990 were aged 20-29 in 1960, when OCs were first introduced, and about half of them have taken OCs at some time in their lives (Table 1).

Furthermore, most women who were in their 50's in 1990 began taking OCs more than 20 years ago, when preparations containing high doses of estrogens were in use. For example, women in the UK aged 50-54 in 1990 began taking

OCs in 1969 (on average 21 years before), and preparations containing more than 50 μg of estrogen were widely used until 1970 (8). Although the trends (see below) may be affected by other factors, these data provide considerable reassurance that a serious long-term hazard from OC use is unlikely, in that there appears

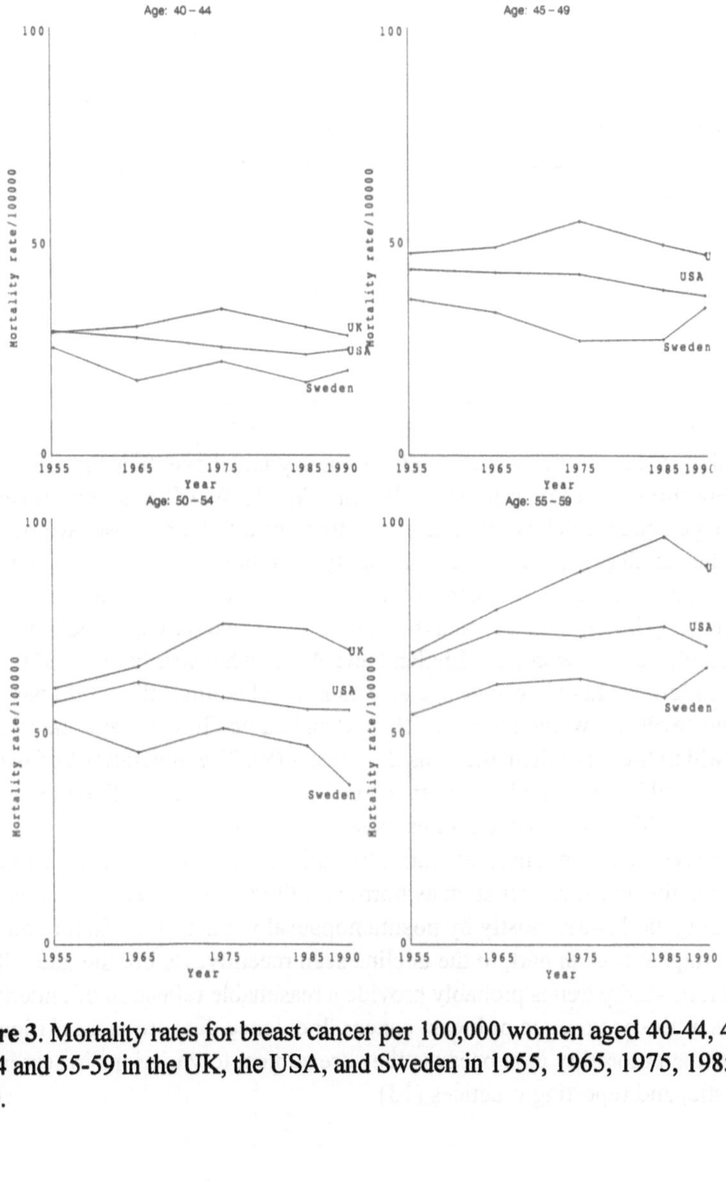

Figure 3. Mortality rates for breast cancer per 100,000 women aged 40-44, 45-49, 50-54 and 55-59 in the UK, the USA, and Sweden in 1955, 1965, 1975, 1985, and 1990.

to be no major increase in breast cancer rates in the generations of women who began taking the pill more than 20 years ago and who took preparations containing comparatively high doses of estrogen.

Very few women aged 50-59 in 1990 had used OCs before they had a child or before they were 25 years old. The trends for women aged 50-54 and 55-59 therefore have little bearing on whether early use of OCs might increase the risk of breast cancer. By contrast, about half the women aged 40-44 used OCs before they had a child, and about two thirds of those women used them before they were 25 years old (Table 1). Mortality rates at age 40-44 have fallen somewhat in all three countries since 1975. The trends in Sweden reflect the small number of deaths at that age (67 in 1955, 46 in 1965, 47 in 1975, 48 in 1985, and 63 in 1990. The fluctuations from year to year are likely to be due to chance). Thus, there is little to suggest from the trends at age 40-44 that the generations of women who first took OCs at a young age or before they had a child have substantially increased risks of breast cancer.

Other Factors That Might Influence Breast Cancer Trends

The main factors known to affect breast cancer risk are early age at menarche, childlessness, delayed childbearing, low parity, and late age at menopause. During the early years of this century, and particularly during the economic depression of the 1930's, women tended to marry late, have their first child late, have few children or no children at all. After World War II this trend reversed; the average age at which women had their first child fell, and fewer women had no children at all (9). The decline in mortality from breast cancer in women aged 40-59 in the USA has been linked to the increased fertility of women during the postwar "baby boom" (10, 11), and the same appears to have happened in the UK. For example, among women in England and Wales who were born in 1920, 21% had no children, and the average age at which they had their first child was 28.5 years; however, for women born in 1945, 10% had no children, and the average age at which they had their first was 25.9 years (9). The generations of women for whom childlessness and the average age at first birth has declined tend to be the ones in which breast cancer rates have been falling.

Improvements in survival can also affect mortality trends. Effective treatments for breast cancer such as hormonal therapy (12) began to be widely used during the 1980's, mostly by postmenopausal women. This factor could be contributing, at least in part, to the decline seen recently. Before the late 1980's, however, mortality trends probably provide a reasonable reflection of underlying changes in the occurrence of breast cancer. Incidence data are more difficult to interpret than mortality data because they are subject to variations in screening, diagnostic, and reporting practices (13).

Conclusions

Breast cancer mortality rates have been falling in generations of women who began taking OCs many years ago, including those who first used OCs more than 20 years ago, those who first used preparations containing high doses of estrogens, and those who first used them when they were young or before they had a child. The decline seems to be related to changes in childbearing patterns and recently, perhaps, to improvements in breast cancer treatment. There is no suggestion of the emergence of an epidemic of breast cancer in women who have taken OCs.

Acknowledgments

The data for Figures 1 and 3 were kindly provided by Peto, Lopez, and colleagues (14).

References

1. Pike MC, Henderson BE, Krailo MD et al (1983) Breast cancer in young women and use of oral contraceptives. Possible modifying effects of formulation and age at use. Lancet II: 926-930.
2. Meirik O, Adami H-O, Christoffersen T et al (1986) Oral contraceptive use and breast cancer in young women. Lancet II: 650-653.
3. UK National Case-Control Study Group (1989) Oral contraceptive use and breast cancer risk in young women. Lancet I: 973-982.
4. White E, Malone KE, Weiss NS, Daling JR (1994) Breast cancer among young US women in relation to oral contraceptive use. J Natl Cancer Inst 86: 505-514.
5. Wingo PA, Lee NC, Ory HW et al (1991) Age-specific differences in the relationship between oral contraceptive use and breast cancer. Obst Gynecol 78: 161-170.
6. McPherson K (1993) Childbearing, oral contraceptive use and breast cancer. Lancet 341: 1604.
7. Lund E, Meirik O, Adami H-O et al (1989) Oral contraceptive use and premenopausal breast cancer in Sweden and Norway: Possible effects of different pattern of use. Int J Epidemiology 18: 527-532.
8. Thorogood M, Vessey MP (1990) Trends in use of oral contraceptives in Britain. Brit J Family Planning 16: 41-53.
9. Werner B, Chalk S (1986) Projections of first, second, third and later births. Population Trends 45: 26-34.

11. Tarone RE, Chu KC (1992) Implications of birth cohort patterns in interpreting trends in breast cancer rates. J Natl Cancer Inst 84: 1402-1410.
12. Early Breast Cancer Trialists' Collaborative Group (1992) Systemic treatment of early breast cancer by hormonal, cytotoxic, or immune therapy. Lancet 339: 1-15.
13. Doll R, Peto R (1981) The causes of cancer. J Natl Cancer Inst 66: 1191-1308.
14. Peto R, Lopez AD, Boreham J et al (1994) Mortality from smoking in developed countries 1950-2000. Oxford University Press. Oxford.

7

Gene Expression in Familial Breast Cancer: A Genetic-Epidemiologic Study of Premenopausal Bilateral Breast Cancer

Robert W. Haile, Sue Ingles, Victoria K. Cortessis,
Robert C. Millikan, Anh Diep, and Robert S. Sparkes

Introduction

The objectives of our study are to identify and characterize genes involved in the etiology of breast cancer, to define sources of heterogeneity, and to identify factors that may affect expression of inherited susceptibility genes (i.e., gene-gene and gene-environment interactions).

Methods

For this analysis, families were identified as part of a large, ongoing genetic-epidemiologic study of bilateral breast cancer. Cases of bilateral breast cancer diagnosed prior to 50 years of age were identified from the Los Angeles County Tumor Registry (1970-89), the Connecticut Tumor Registry (1935-89), and the major hospitals in Montreal and Quebec City, which identify 95% of all cases in these cities and the southern Quebec Province (1975-89). Index cases were sent questionnaires regarding family history of breast cancer. We received 434 completed questionnaires: 200, 156, and 78 from the Los Angeles, Connecticut, and Montreal study sites, respectively. From the questionnaires, we identified multiple-case families potentially informative for genetic linkage. Blood samples were obtained from as many living members as possible. DNA was extracted and stored for future analysis. Sixty-eight multiple-case families were identified, and DNA collected from 428 members. The families range in size from 4 members with 2 cases (both living) to 84 members (44 dead, 40 living) with 10 cases of breast cancer (5 dead, 5 living). Sixty-six of the families were studied for this analysis.

We determined alleles for the following non-17q markers: DRD2 (11q22-23), D1S160, D1S170; and the following 17q markers: MFD188, MFD191,

UT67, OF2, UT185, UT8, UT62, SCG43, D17S791, UT573, D17S800, CMM86, UT8, HSD-DEL, HSD-A3T, UT8182, UT8184, UT855, according to standard methods described elsewhere (1-3). We also conducted immunohistochemical staining for overexpression of HER-2 and p53, according to standard methods (4-6).

The 1992 version of the computer program MENDEL (6) was run on an IBM/9000 model 900 under MVS to calculate lod scores for a single recombination parameter, θ. The program was modified to incorporate the age-at-onset step function model for penetrance reported by Claus et al. (7). The model gives cumulative incidence as genotype-dependent step functions with steps every 10 years from age 30 to 80. For gene carriers, the cumulative incidence begins at 0.0167 for those under age 30 and increases to 1.000 for those age 80 or over. For noncarriers, the cumulative incidence begins at 0.0002 and increases to 0.1254 at age 80. We estimated the density at each step by dividing the increase in step height by the step length. The penetrance function is given by the density for affected individuals or by 1-cumulative incidence for unaffected persons. The frequency of the high-risk allele was set to 0.0006, and probabilities for males were set equal to those for females age 20-29. Individuals were treated as affected if they had a diagnosis of breast cancer, ovarian cancer or both. For individuals with either ovarian cancer only or both breast and ovarian cancer, age was coded as 35 years, while for those with breast cancer only, age was coded as the actual age at diagnosis.

Lod scores were calculated for 21 values of θ: 0.000001, 0.25, 0.50, 0.075, . . . 0.450, 0.475, 0.500. Vectors of lod scores were summed for all families and for subgroups defined by age at onset (families with all cases younger than 45 years at diagnosis versus families with one or more cases 45 years or older at diagnosis), family history of ovarian cancer, synchronicity (proband had both primaries diagnosed within one year vs. greater than one year), span of ages at diagnosis (families with all breast cancer diagnoses within 10 years of age vs. families with a difference in age at diagnosis greater than 10 years for at least two breast cancer cases), and histopathologic diagnosis. Histopathologic diagnosis was based on histopathologic findings (8) described in the original pathology report or rapid report form of the Los Angeles County Tumor Registry. Families were included in subgroups based on histopathologic diagnosis only if confirmed diagnosis was available for tumors of both breasts of the proband and for a tumor of at least one additional case in the family. Four groups of families were defined: families in which all cases with confirmed diagnoses had tumors classified as ductal carcinoma *in situ*, invasive ductal carcinoma or spatial variants thereof (ductal); families in which all cases with confirmed diagnoses had tumors displaying only lobular elements (lobular); families in which all cases with confirmed diagnoses had at least one tumor displaying both lobular and ductal elements (mixed); and families in which cases with confirmed diagnoses were discordant with respect to histopathologic diagnosis (discordant).

Predivided sample and admixture tests of heterogeneity (9) were performed on a Dell 486 microcomputer using 1988 versions of HOMOG and MTEST. Input consisted of lod scores calculated for all values of θ enumerated above except 0.5.

BRCA1 genotypes were inferred from 17q haplotypes. Odds ratios were estimated for oral contraceptive (OC) use, for the high risk BRCA1 allele, and for the OC-BRCA1 interaction by fitting a logistic model in which familial linkage status was modeled as a random effect.

Results

The results were focused on issues of heterogeneity and GxG and GxE interactions.

Using the predivided sample test, we found no significant evidence of heterogeneity when we stratified families by the following variables: Ages at diagnoses of cases within a family (all cases under 45, one or more cases over 45) ($p = 0.290$), age span between diagnoses of cases within a family (span $\leq$ 10 years, span > 10 years) ($p = 0.529$), and histopathologic diagnosis (ductal, lobular, mixed, discordant) ($p = 0.756$).

We also stratified the families by presence or absence of family history of ovarian cancer. Ten of 12 families with ovarian cancer are consistent with linkage, but, because the informativeness of some of these families is quite low, the summary lod scores are not very high. For families with positive history of ovarian cancer, the summary lod score at $\theta = 0.10$ is 0.430, while for those with negative family history, the summary lod score at the same point is 1.841 ($p = 0.927$). Notably, we found that a reasonably informative family with history of ovarian cancer (BCS072) had a negative lod score of -0.227 at $\theta = .000001$.

We did observed suggestions of heterogeneity by synchronicity, and by presence/absence of overexpression of HER-2 and p53 (2) (Figure 1).

With respect to interactions, we focus here on 1) associations between HER-2 and p53, as well as between BRCA1 and reproductive factors; and 2) associations between BRCA1 and use of OCs. Results in Tables 1 and 2 suggest that reproductive variables and probable presence of BRCA1 (as inferred from haplotypes) are positively associated with overexpression of HER-2 and p53 (4).

Results in Table 3 suggest that OCs interact with BRCA1 (10).

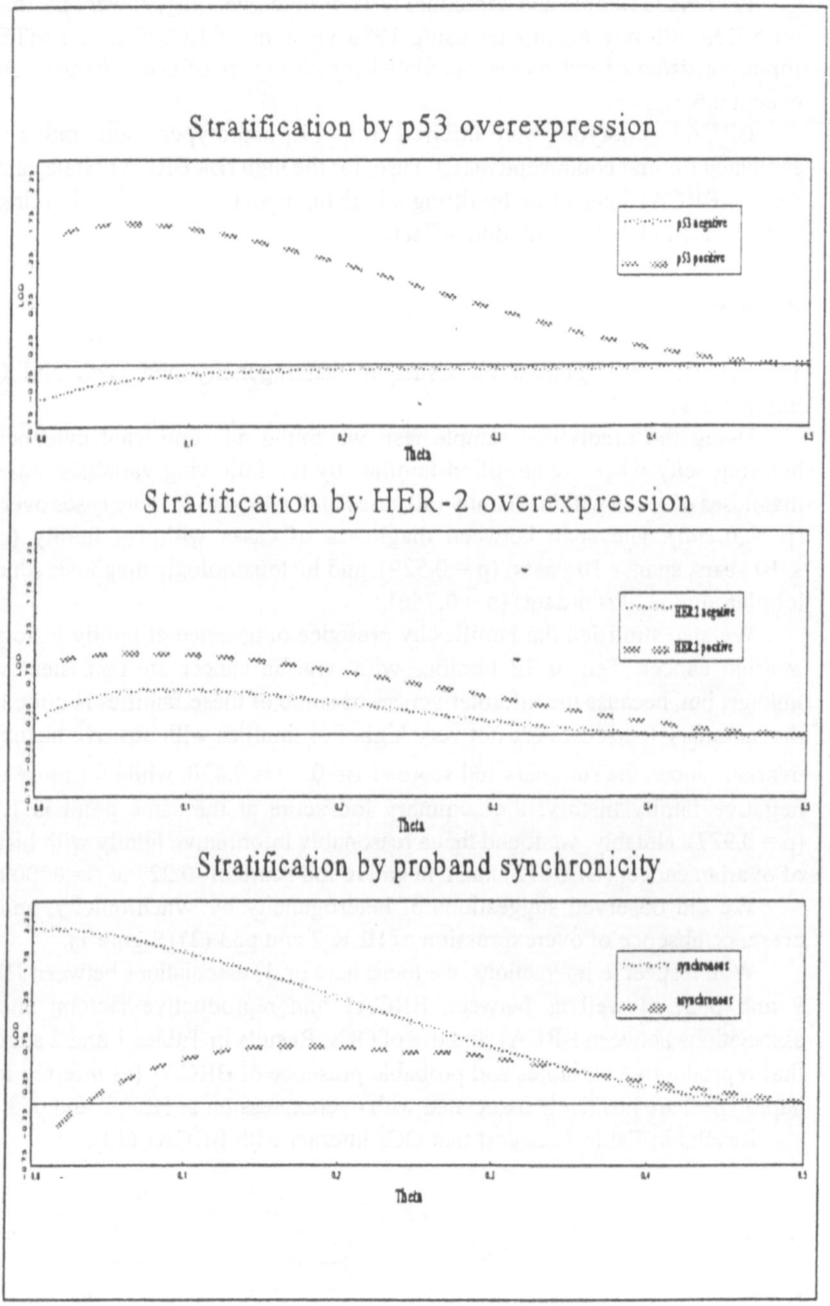

Figure 1. Linkage analysis of MFD 188 by all breast cancer.

Discussion

Our linkage results suggest heterogeneity by synchronicity and possibly by overexpression of HER-2 and p53. The results of the random effects model suggest OCs interact with BRCA1. Finally, the analyses of tumor blocks suggest that presence of BRCA1 and reproductive factors may be associated with overexpression of HER-2 and p53. Details of results are presented elsewhere (1-4). Although we find the results intriguing, all are based on relatively small numbers and warrant further consideration in our own and other data sets.

Table 1. HER-2 Overexpression and Epidemiologic Risk Factors.

	Odds ratio	(95% CI)
OCs use	2.81	(0.50, 15.77)
Smoking	1.28	(0.23, 5.49)
Menarche < age 12	1.70	(0.24, 11.94)
FFT > age 25	4.00	(0.55, 33.33)
Age at Dx < 44 years	0.80	(0.26, 2.50)
17q haplotype	2.00	(0.38, 10.61)
17q family	0.43	(0.08, 2.13)
GST M1 null	2.50	(0.82, 7.69)

Abbreviations: FFT = First Full Term Dx = Diagnosis

Table 2. p53 Overexpression and Epidemiologic Risk Factors.

	Odds ratio	95% CI
OCs use	6.19	(0.68, 56.07)
Smoking	0.50	(0.11, 2.19)
Menarche < age 12	0.89	(0.08, 9.69)
FFT > age 25	2.22	(0.30, 16.67)
Age at Dx < 44 years	1.96	(0.61, 6.25)
17q haplotype	3.29	(0.38, 28.85)
17q family	3.98	(1.07, 14.76)
GST M1 null	1.57	(0.52, 4.76)

Table 3. BRCA1 and OC Use Odds Ratios Estimated from Logistic Random-Effects Model.

BRCA1 aa, never OC	1.00
BRCA1 aa, ever OC	3.10
BRCA1 Aa, never OC	10.55
BRCA1 Aa, ever OC	64.10

References

1. Cortessis V, Ingles S, Millikan R, et al. (1993) Linkage Analysis of DRD2, a marker linked to the ataxia-telangiectasia gene, in 64 families with premenopausal bilateral breast cancer. Cancer Res 53:5083-5086.
2. Cortessis V, Ingles S, Millikan R et al (1995) A linkage analysis of MFD188 in 66 families with premenopausal bilateral breast cancer, with a focus on heterogeneity. Submitted.
3. Millikan RC, Ingles S, Diep A, et al., (1995) Linkage analysis of chromosome 1P using 66 multiple case breast cancer pedigrees. Submitted.
4. Ingles S, Millikan RC, Slamon D, et al., (1995) Descriptive epidemiology of HER-2 and p53 overexpression in bilateral breast cancer prevalence, concordance, and associations with BRCA1 and reproductive factors. Submitted.
5. Slamon DJ, Godolphin W, Jones LA et al (1989) Studies of the HER-2/neu proto-oncogene in human breast and ovarian cancer. Science 244:707-712.
6. Lange K, Boehnke M, Weeks D (1988) Programs for Pedigree Analysis: MENDEL, FISHER, and dGENE. Genetic Epidem 5:471-472.
7. Claus E, Risch N, Thompson WD (1991) Genetic analysis of breast cancer in the Cancer and Steroid Hormone Study. Am J Hum Genet 48:232-242.
8. Page D, Anderson T (1987) Diagnostic histopathology of the breast. New York: Churchill Livingston.
9. Ott J (1983) Linkage analysis and family classification under heterogeneity. Ann Hum Genet 47:311-320.
10. Ingles S, Diep A, Haile RW (1995) A logistic random-effects model for gene-environment interaction under linkage heterogeneity: Application to early-onset bilateral breast cancer data. Submitted.

PART 2. BREAST/MAMMARY GLAND

Introduction

The Importance of Human Breast Development in Mammary Carcinogenesis

Barry A. Gusterson, Michael Stratton, and
Ramaswamy Anbazhagan

Breast cancer is a heterogeneous disease, in terms of both its clinical course and its underlying molecular abnormalities. Its etiology is likely to involve a combination of congenital and acquired abnormalities. The introduction of this section focuses on the significance of human breast development and its possible contribution to our understanding of the roles of hormones and other factors in mammary carcinogenesis. The accompanying chapters will deal with the use of breast cancer animal model systems and their susceptibility to carcinogens, the definition of early events in mammary carcinogenesis, and the pathogenesis of the disease in relation to hormones and growth factors.

Genetic Predisposition

During development, the *in-utero* maternal milieu and unidentified prenatal factors may contribute to an increased breast cancer risk through their possible modulation of breast development, or through the inheritance of predisposing genes. In addition, prenatal exposure of embryonic fetal somatic cells or prezygotic exposure of the germ cells to carcinogens could also contribute to the development of breast cancer (1). Such prenatal influences have been well established for diethylstilbestrol (DES) in vaginal adenocarcinoma (2), and for ionizing radiation in a range of tumor types (3).

Estrogens have been shown to play an important role in mammary carcinogenesis. There is compelling evidence suggesting that the hormonal environment *in utero* may predispose to some types of breast cancer (4). Variation in the levels of maternal hormones, leading to differences in the proliferative rates of target cell population(s), may be a possible mechanism by which such influences could take place. Moreover, any change in proliferation of the epithelium *in utero* could induce an increase in the number of potential mutations *in utero* (5). Through an imprinting mechanism, estrogens may also induce altered responses in developing breast epithelial cells, resulting in increased susceptibility to carcinogens (4). Growing evidence suggests that genetic factors may affect breast development, thus predisposing the breast to neoplastic transformation.

Family history is a well-recognized risk factor in breast cancer. This risk has been explained as one or more autosomal dominant genes conferring a high risk of developing the disease. Approximately 40% of breast cancer families are linked to a gene localized on 17q21, which is now labeled as BRCA1 (6). Recently, however, it has become apparent that there are other genes for breast cancer susceptibility that have not been identified (7). Of particular interest for the relationship between hormones and breast carcinogenesis has been the finding that, within the BRCA1-linked breast cancer families, there is a high incidence of ovarian cancers. Currently, it is difficult to recognize the connection between breast and ovarian cancer in terms of identifying the responsible gene(s). However, recent data on mutations in the androgen receptor of male breast cancer samples (8, 9) may strengthen the argument that hormonal influences could play an important role in the development of breast cancer. Moreover, as part of their phenotype, some non-BRCA1 families have an increased incidence of male breast cancer (10). These observations raise the question of whether certain androgen/estrogen concentration(s) may predispose to breast cancer through mechanism(s) not yet identified. So far, human samples examined from normal breast of female adult carriers of the BRCA1 gene do not exhibit distinctive, histologically identifiable features in either the epithelial or the stromal component of the breast. It is definitely possible that these women may exhibit significant differences in the total number of breast lobules, or in other functional characteristics that cannot be recognized histologically. However, on histological examination of breast cancer specimens from eight members of the same BRCA1 family, all tumors were of high grade, suggesting that the BRCA1 gene may confer increased susceptibility to subsequent DNA damage (Gusterson and Bishop, personal communication).

Human Breast Development

Currently, we can only speculate about the possible significance of hormones and their effects on breast development in relation to mammary carcinogenesis; however, it is clear from clinical studies that the majority of breast carcinomas are hormone responsive whatever the complex molecular abnormalities involved. Therefore, it is likely that, in breast cancer, the effects of steroid hormones interface with many of the well-recognized oncogenic intracellular signal transduction pathways. This proposition could partially account for the tumor type association of some of the dominant and recessive oncogenes. In such circumstances, it is possible that the same genes may play significant roles in normal development of the mammary gland. It is therefore important to understand normal human breast development and the genes involved in this process. We consider here some recent findings in relation to the early stages of human breast development, and some data related to the formation of specialized breast stroma.

The human breast develops from the epithelium of the chest wall. This development is detectable macroscopically after 16 weeks of gestation. At this early stage, a strong expression of the BCL2 protein is detected in the epithelium and in the stromal fibroblasts surrounding the bud (11). Since the expression of BCL2 is known to inhibit apoptosis, its presence in breast epithelial and stromal progenitor cells enables the expansion of this precursor cell population. In the uterus, it has been clearly shown that BCL2 expression is cyclical; down-regulation occurs with the appearance of apoptosis in the late secretory phase, suggesting that BCL2 expression is hormonally regulated (12). This finding is consistent with our observation in breast carcinomas that the expression of BCL2 is linked with estrogen-receptor positivity (13). Concurrently with the expression of BCL2, TGFα is expressed in the epithelium of the bud and also in the same stromal population that exhibits BCL2 positivity (Anbazhagan and Gusterson, personal communication).

At birth, the degree of breast development does not correlate with chronological age, suggesting variations in the intrauterine hormonal environment as referred to above (4). As the breast lobules begin to develop, two populations of fibroblasts are easily recognized by their expression of a cell surface peptidase, dipeptidyl-peptidase IV (DPPIV) (14). One population of cells, referred as the intralobular stroma, is adjacent to the epithelium, and the other, the interlobular stroma, composes the bulk of the breast stroma. The identification of these specialized stroma is retained in the adult breast. It has been shown that a breast-specific tumor (phyllodes) is derived from the intralobular stroma (15). DPPIV is of particular interest because it belongs to a family of enzymes that may be involved in a range of functions, from growth control (16) to extracellular matrix degradation, all of which are required in tissue remodeling during morphogenesis. At about the same time, the enzyme neutral endopeptidase (NEP) is also expressed on the basal cell population (14, 17). This enzyme hydrolyzes bombazine, a growth-regulatory peptide that is a crucial factor in lung morphogenesis. NEP may have a similar regulatory role in the breast through the same or a similar substrate. It is important to mention that NEP is also expressed cyclically in the endometrium, raising the possibility that it may be hormonally regulated.

Breast stroma is important in breast carcinogenesis for a number of reasons. In the stroma of some breast carcinomas, the expression of TGFβ1 increases in response to tamoxifen treatment (18), and in some breast cancer cells TGFα expression is unregulated (19). During development, TGFα is strongly expressed in the presumptive interlobular fibroblasts, as well as in a subpopulation of breast epithelial cells. In all stages of breast development, the stroma stains strongly for TGFβ1. Therefore, it is possible that broad autocrine and paracrine interactions take place between the epithelial cells and the stroma. The significance of these data is unclear.

Future Directions

Regarding the heterogeneity in human breast carcinomas, it has recently been demonstrated that the two major types of human breast carcinoma, namely ductal and lobular, exhibit different expression of specific oncogenes. In particular, c-*erb*B-2 is expressed only in breast carcinomas of ductal origin (20), while p53 mutations, rare in lobular carcinomas (21), are seen in over 40% of ductal carcinomas. Specific tumor phenotypes may arise from different functional subpopulations of cells in the breast, or from a time-dependent exposure or susceptibility to events that may result in specific molecular abnormalities. In the case of lobular carcinomas, it is also possible that the high incidence of bilaterality in LCIS may be due to a familial component, although reports indicate that this may not be the case.

References

1. Tomatis L (1979) Prenatal exposure to chemical carcinogens and its effect on subsequent generations. Natl Cancer Inst Monog 51:159-184.
2. Greenwald P, Barlow JJ, Nasca PC, Burnett WS (1971) Virginal cancer after maternal treatment with synthetic estrogens. N Engl J Med 285:390-392.
3. Kato H, Yoshimoto Y, Schull WJ (1986) Risk of cancer among children exposed to atomic bomb radiation *in utero*: A review. In Napalkov NP, Rice JM, Tomatis L, Yamasaki H (eds): Perinatal and Multigeneration Carcinogenesis. IARC Scientific Publications No. 96. Lyon International Agency for Research on Cancer, pp 365-374.
4. Anbazhagen R, Gusterson BA (1994) Prenatal factors may influence predisposition to breast cancer. Eur J Cancer 30A(1):1-3.
5. Millar WR (1993) Breast cancer. Hormonal factors and risk of breast cancer. Lancet 341:25-26.
6. Hall JM, Lee MK, Morrow J, et al. (1990) Linkage of early onset familial breast cancer to chromosome 17q21. Science 250:1684-1689.
7. Easton DF, Bishop DT, Ford D, et al. (1993) Genetic linkage analysis in familial breast and ovarian cancer: Results from 214 families. Am J Hum Genet 52:678-701.

8. Wooster R, Mangion J, Eeles R, et al. (1992) A germline mutation in the androgen receptor gene in two brothers with breast cancer and Reifenstein syndrome. Nature Genet 2:132-134.
9. Lobaccaro J-M, Lumbroso S, Belon C, et al. (1993) Male breast cancer and the androgen receptor gene. Nature Genet 5:109-110.
10. Stratton MR, Ford D, Neuhausen S, et al. (1994) Familial male breast cancer is not linked to the *BRCA1* locus on chromosome 17q. Nature Genet 7:103-107.
11. Nathan B, Anbazhagan R, Clarkson P, et al. (1994) Expression of BCL2 in the developing human fetal and infant breast. Histopath 24:73-76.
12. Otsuki Y, Misaki O, Sugimoto O, et al. (1994) Cyclic *bcl*-2 gene expression in human uterine endometrium during menstrual cycle. Lancet 344:28-29.
13. Nathan B, Gusterson B, Jadayel D, et al. (1994) Expression of BCL-2 in primary breast cancer and its correlation with tumour phenotype. Ann Oncol 5:409-414.
14. Atherton AJ, Anbazhagan R, Monaghan P, et al. (1994) Immunolocalisation of cell surface peptidases in the developing human breast. Differentiation 56:101-106.
15. Atherton AJ, Monaghan P, Warburton MJ, et al. (1992) Dipeptidyl peptidase IV expression identifies a functional subpopulation of breast fibroblasts. Int J Cancer 50:15-19.
16. Kenny AJ, O'Hare MJ, Gusterson BA (1989) Cell surface peptidases as modulators of growth and differentiation. Lancet 785-787.
17. Gusterson BA, Monaghan P, Mahendran R, et al. (1986) Identification of myoepithelial cells in human and rat breasts by anti-common acute lymphoblastic leukaemia antigen antibody A12. J Natl Cancer Inst 77:343-349.
18. Butta A, MacLennan K, Flanders KC, et al. (1992) Induction of transforming growth factor β_1 in human breast cancer *in vivo* following tamoxifen treatment. Cancer Res 52:4261-4264.
19. Dickson RB, Lippman ME (1988) Control of human breast cancer by estrogen, growth factors and oncogenes. In Dickson RB, Lippman ME (eds): Breast Cancer: Cellular and Molecular Biology. Boston: Kluwer Academic Publishers, pp 119-165.
20. Ramachandra S, Machin L, Ashley S, et al. (1990) Immunohistochemical distribution of c-*erb*B-2 in *in situ* breast carcinoma--A detailed morphological analysis. J Pathol 161:7-14.
21. Gusterson BA, Shipley J, Crew J (1994) Applications of molecular genetics and cytogenetics to breast cancer and soft tissue sarcomas. Annals Oncol, in press.

8

Breast Susceptibility to Carcinogenesis

Jose Russo, Irma H. Russo, Gloria Calaf, Pei-Li Zhang, and Nandita Barnabas

Introduction

The understanding of the human breast has been a major biological puzzle, mainly due to the fact that the mammary gland seems to be the only organ that is not fully developed at birth (1, 2). No other organ presents such dramatic changes in size, shape and function as does the breast during growth, puberty, pregnancy and lactation (2-4). It is agreed that the developmental phase of the human breast starts as early as the stage of nipple epithelium during embryonic development, continuing steadily with body growth, and undergoing a spurt of growth with lobule formation at puberty. Four different lobular structures have been characterized in the breast of postpubertal women, each one representing sequential developmental stages (3). Lobules type 1 are the most undifferentiated ones. They are also called virginal lobules because they are present in the immature female breast before menarche; and they are composed of clusters of 6 to 11 ductules per lobule. Lobules type 2 evolve from the previous ones and have a more complex morphology, being composed of a higher number of ductular structures per lobule. They progress to lobules type 3 which are characterized by having an average of 80 ductules or alveoli per lobule. They are frequently seen in the breast of women under hormonal stimulation or during pregnancy. Lobules type 4 have been described to be present during the lactational period of the mammary gland, but they are not found in the breast of nulliparous post-pubertal women. Lobules type 4 are considered to be the maximal expression of development and differentiation (3).

The fact that the breast is the source of the most frequent malignancy in the female population, and the knowledge that breast cancer is heavily influenced by the reproductive history of the individual, requires a thorough understanding of the developmental pattern of the breast during the life-span of a woman. It is known that women with a history of early full term pregnancy are at a lower risk of developing breast cancer than nulliparous women (3, 5-11). The protective effect of pregnancy has been attributed to differences in the degree of differentiation of the breast (12-14). In the comparative study of the pathogenesis of chemically induced mammary carcinomas in experimental animal models with the pathogenesis of human breast cancer (13), it was concluded that the initiation of the neoplastic process, which was inversely related to the degree of differentiation of the mammary gland in the experimental animal, might occur under similar circumstances in women (11-16). The study of the pathogenesis of human breast cancer indicated that the lobules type 1 are the site of origin of preneoplastic lesions such as atypical ductal hyperplasia, which evolve to ductal carcinoma in situ, progressing to invasive carcinoma. Lobules type 2 are postulated to be the origin of atypical lobular hyperplasia (ALH) and lobular carcinoma in situ (13), and lobules type 3 to be the site of origin of secretory adenomas, fibroadenomas, sclerosing adenosis, and apocrine cysts (13). These observations indicate that the degree of differentiation or lobular development of the mammary gland influence the type of tumors developed in the human breast (13). Our studies have been aimed at answering the following questions: How does lobular differentiation affect the susceptibility of the breast epithelial cells to neoplastic transformation by chemical carcinogens *in vitro*? and which are the different steps of neoplastic transformation of human breast epithelial cells (HBEC) *in vitro*?

Influence of Degree of Mammary Gland Developed and Cell Proliferation on the Growth Rate of Breast Primary Culture

The susceptibility of the mammary gland to neoplastic transformation is related to its degree of development and proliferative activity (17-19). The degree of breast development is measured upon the type and number of lobules present (3), which allowed us to classify the human breast into two types: a) poorly differentiated breasts, those composed of lobules type 1 and 2 and b) well differentiated breasts, those composed almost exclusively of lobules type 3 (19). The breasts composed predominantly of lobules type 1 and 2, have a DNA-Labeling Index (LI) of 1.03 ± 0.48. Cells derived from these samples attached to the culture dish immediately with a high number of doublings (0.64 ± 0.47). Breasts composed almost exclusively lobules type 3, have a DNA-LI of 0.05 ± 0.05, that is markedly lower than the DNA-LI of cells derived from poorly differentiated breasts. The cells' number of doubling was 0.11 ± 0.06. There was a correlation coefficient ($r = 0.802$) between the number of doubling of cells in primary cultures and the percentage of type 1 and 2 lobules present in the specimen, whereas differentiation inversely correlated with this parameter ($r = 0.722$) (19).

Response of Human Breast Epithelial Cells (HBEC) in Primary Cultures to Carcinogen Treatment

Lobules type 1, 2, or 3, when plated in DMEM:F12 culture medium containing 1.05 mM calcium (high Ca^{++}) formed monolayers. The cells were passaged and then treated with the following carcinogens: 7,12-dimethylbenz(a)anthracene (DMBA), N-methyl-N-nitrosourea (NMU), methyl-N-nitro-nitrosoguanidine (MNNG) and benzo(a)pyrene (BP). BP and DMBA require metabolic activation, whereas NMU and MNNG are direct acting carcinogens (13, 17, 18). After the second or third passage post-treatment, the cells were seeded in agar methocel in order to measure anchorage independent growth, a phenotype usually expressed by transformed cells which correlates with tumorigenicity (20). Ten out of 20 samples tested expressed an increased survival efficiency in agar methocel when compared with the DMSO-treated control cells. BP and MNNG were the carcinogens that affected the largest number of samples, 6 and 5, respectively, whereas DMBA and NMU had the lowest effect, since only 2 samples expressed increased survival efficiency after treatment. The samples that reacted with the carcinogens were mainly derived from breast tissue containing lobules type 1 or 2, singly or combined. Those cells that did not show response were derived from samples that contained lobules type 3. We concluded that the developmental stage of the gland *in vivo* correlates with increased survival efficiency in agar methocel, which is an indicator of increased life-span *in vitro* and an early transformation phenotype.

Isolation and Characterization of Spontaneous Immortalized Human Breast Epithelial Cells

Normal human breast epithelial cells senesce after 10-20 passages *in vitro* (19, 21, 22). We have reported that a mortal human breast epithelial cell line derived from a subcutaneous mastectomy of a 36-year-old woman was cultured in medium containing a 0.04 mM calcium (low Ca^{++}) for over 2 years; it spontaneously gave rise to an immortal cell line MCF-10 which grew as attached cells (MCF-10A), or as floating cells (MCF-10F). Immortalization of these cells was characterized by their continuous growth in culture medium containing the conventional calcium levels (1.05 mM, high Ca^{++}) without entering into senescence. The availability of these mortal cells and their derived immortal cell lines led us to analyze growth characteristics that might explain the process of immortalization. For example, the growth curves of S#130, MCF-10A and MCF-10F cells grown in either low or high Ca^{++} media were similar. They differed, however, in the fact that S#130 cells were unable to continue growing in high Ca^{++} after the 20th passage, whereas the immortal cells MCF-10A and MCF-

10F continued growing indefinitely. None of the three cell types formed colonies in agar or in agar-methocel. We demonstrated that both the mortal and the immortalized cells were bonafide human breast epithelial in nature, expressing genetic, cytogenetic, ultrastructural and phenotypic characteristics of normal breast epithelium. These characteristics made this cell line the most near to a normal breast epithelial cell line available (22).

Response of Immortalized HBECs to Carcinogen Treatment

Since *in-vitro* treatment of HBECs in primary cultures with chemical carcinogens did not succeed in inducing the full expression of malignant transformation, we decided to use the protocols developed for primary cultures of breast epithelial cells with the immortalized cell line MCF-10F in order to elucidate whether immortalization is required for the expression of the full malignant phenotype. Upon treatment of MCF-10F cells with DMBA, MNNG, NMU, or BP for 24 hrs treated cells differed from controls cells in their altered morphology, altered pattern of growth, increased growth rate, and anchorage independence in agar-methocel after 8-10 passages (around 157 days post-treatment). Control cells never developed colonies, whereas carcinogen-treated cells formed colonies from which the following clones were derived: D1, D2, and D3, from DMBA, M4 from MNNG, and BP1, BP2, BP5, BP6, BP7, and BP10 from BP treated cells. Colonies isolated from NMU treated cells did not form clones (23).

All clones grew at a faster rate than their respective parental cells. Based upon this growth advantage, selected cell populations of clones BP1 and BP2 were isolated at approximately 446 days post-treatment; they were named BP1-E and BP2-B, respectively. BP1 and BP1-E cells showed increased anchorage-independent growth, chemotaxis and invasiveness. From the D3 clone the D3-1 cell line was originated. It showed increased chemotactic and invasive capabilities, but to a lesser degree than BP1-E. The tumorigenic potential of the cells was tested by inoculation into SCID mice. After 101 days of injection, mammary tumors developed from the BP1-E cell line, but none of the other BP clones or those derived from DMBA or MNNG formed tumors (23).

These results led us to conclude that the final malignant phenotype of tumorigenesis may be induced in HBEC by carcinogen treatment, provided the treated cells are previously immortalized and enough time and clone selection are allowed for its expression. This model allowed us to isolate clones of cells expressing different stages of progression to malignant transformation which are useful for determining whether specific phenotypes are the result of specific genotypic alterations.

Molecular Changes Induced by Chemical Carcinogens

In order to determine which are the genes that are affected in chemically induced transformation of HBEC, we studied the genes that have been linked to human breast cancer, namely, *ras*, TGF-α, *Erb*B/2, and the tumor suppressor gene p53 (24-27).

The *ras* gene has been reported to be involved in chemically induced mammary carcinomas, in breast cancer cell lines and in primary breast cancers, in which point mutations, loss of alleles and amplification of the gene product p21 protein have been reported (27-32). However, it is not known whether the mutationally activated *ras* oncogene is capable of inducing full transformation of normal HBEC *in vitro*. We studied the role of *ras* gene in the transformation of HBEC by using the following approaches: a) determination of whether chemical carcinogens induce alterations in the *c-Ha-ras* oncogene and b) what is the effect of *c-Ha-ras* gene transfected to immortalized cells.

Role of *Ras* Oncogene in the Transformation of Human Breast Epithelial Cells by Chemical Carcinogens

MCF-10F cells present two alleles of the *c-Ha-ras* gene identified by 1.0Kb and 1.2Kb restriction fragments. These alleles were analyzed at different passages of MCF-10F cells post DMSO and post-carcinogen treatments. Control cells contained the two alleles at all the passages; no point mutations or transformation phenotypes were expressed by these cells. Treatment with carcinogens resulted in the loss of one of the alleles (1.0Kb) in almost all the treated cells except in those treated with DMBA, in which a rearrangement was observed (33). To determine whether the loss of a *c-Ha-ras* allele in transformed cells was associated with duplication of the retained allele, we quantified the hybridization signals. Autoradiographs of these blots were scanned with a densitometer and analyzed to determine the ratios of hybridization of restriction fragments obtained from the normal and transformed cells DNA. The hybridization signal of the remaining *c-Ha-ras* allele in BP, BP1-E, BP2, BP2-B, and MNU cells was significantly increased in density. Thus, the loss of a *c-Ha-ras* allele in these transformed cells was associated with amplification of the retained allele, while a simple loss of *c-Ha-ras* allele occurred in BP1, BP5, BP6, D1, D3, MNNG, and M4 cells. Densitometric analysis also confirmed the marked reduction in intensity in the remaining allele in D1 and D2 cells. Amplification of the remaining *Ha-ras* allele is confirmed by using the DRD2 gene (dopamine D2 receptor located in chromosome 11q22-23), which did not change in carcinogen treated cells, indicating that equal amounts of DNA were loaded. To obtain evidence that the changes we observed were specific for *c-H-ras* and not simply widespread random alterations, we examined DNA markers on chromosomes 1q21 (PEM), 17q21 (*neu/Erb*B2/HER 2);13q14 (*RB* gene) and 17p13.1 (p53). We did not observe loss of any chromosome marker in these cells. However, these data do not allow us to rule out the possibility that other losses have occurred in other chromosomes or in other regions of the chromosomes tested. We also showed evidence that the loss of alleles of *c-Ha-ras* is followed by point mutations in codons 12 and 61 of this gene as demonstrated by the study of PCR-amplified DNA from all carcinogen transformed cells (33). All the carcinogens induced a mutation at the first position of codon 12 (GGCϒAGC). Another frequent mutation occurred at the first position of *c-Ha-ras* codon 61 (CAGϒGAG) (33). The loss of one allele was a common event in carcinogen treated cells and it occurred before point mutations were observed, indicating that it may be an early event in the initiation of cell transformation. The clonal expansion detected by the formation of colonies in agar methocel seemed to be associated with the loss of an allele but no specific changes of this gene correlated with the emergence of invasiveness or tumorigenesis, indicating that other genes may be involved in these two last processes. Whereas loss of alleles is present in 27% of primary breast cancers (34), point mutations of *c-Ha-ras* genes in primary breast cancer is mcuh less frequent (30). Interestingly enough, our work is the first one demonstrating that a chemical carcinogen induces alterations in the *ras* gene in human breast epithelial cells *in vitro* (33). Although the exact mechanism of *c-Ha-ras* action is not clear, it has been suggested that a normal suppressor effect of cell growth is present in the *c-Ha-ras* allele and that the loss of this suppressor locus leads to cell transformation (35,36). It is possible that this is the mechanism in operation, although a dosage effect cannot be ruled out, since amplification of the remaining allele has been observed in some of the transformed cells such as BP1-E, the tumorigenic cell line. A dosage effect is shown by transfection of the *c-Ha-ras* gene in BP and DMBA treated cells, which increases the tumorigenic effect as a function of the increased number of copies of the *c-Ha-ras* gene (37).

Introduction of the Mutated *c-Ha-ras* Oncogene in MCF-10 Cells by Transfection

Mutated *c-Ha-ras* oncogene was capable of inducing malignant transformation of the immortalized MCF-10 cells (37-40). Transfection to HBEC transformed by chemical carcinogens enhances the tumorigenic phenotype (37). MCF-10 cells and the clones BP1 and D3-1 were transfected with T24(T) activated *Ha-ras* oncogene. Transfected cells, called MCF-10AneoT, BP1-*ras* and D3-1-*ras* exhibited all the array of malignant changes observed in breast cancer cells (37). All the transfected cells were tumorigenic in nude mice, but BP1-*ras* and D3-1-*ras* cells induced tumors which grew faster than those induced by MCF-10AneoT*ras* cells (37). These results indicated that the mutated *ras* gene was able to induce all the malignant phenotypes in immortalized HBEC and it has an enhancing effect in the transformation phenotypes in chemically transformed cells (37). These data suggest a strong dosage effect of *ras* and is in agreement with what is published in the literature (41,42). The mechanism by which *c-Ha-ras* works, and how the loss of alleles and/or its amplification change the transformation phenotype is under investigation.

MCF-10AneoT*ras* cells exhibited anchorage independent growth in the absence of EGF. This phenomenon suggested the presence of an autocrine loop. This hypothesis was confirmed by inhibiting the growth of MCF-10AneoT*ras* cells growth with an anti-TGFα antibody and by the detection of TGFα production by the same cells. Although the expression of TGFα gene was associated with higher expression of the p21 protein of the *c-Ha-ras* gene when the TGFα gene was transfected to MCF-10A cells, it failed to induce any invasive or tumorigenic phenotypes, which indicated that TGFα was not responsible for the expression of those malignant phenotypes in *c-Ha-ras* transfected cells, although they exhibited a 4-8 fold increase in TGFα mRNA, and a 15 fold increase in the secretion of TGFα (43, 44).

Role of *erb*B/2 Gene in the Transformation of Breast Epithelial Cells

It has been reported that amplification of the *erb*B-2 gene correlates with the aggressive behavior of breast cancer (25), although it has not been determined whether it plays a role in the initiation of mammary carcinogenesis. Our studies had the purpose of determining whether treatment of the immortal cell line MCF-10F with chemical carcinogens known to induce their malignant transformation, results in amplification of *erb*B-2 oncogene. EcoRI digested DNA of MCF-10F cells, carcinogen-treated cells, and of all the clones derived from these cells, was studied by Southern blot analysis. Nine out of 18 treated cells exhibited 2-fold amplification of *erb*B/2 oncogene. The D_1 clone showed a 3-fold estimated level of amplification of the 9.0 kb fragment and the D_3 clone showed about a 6-fold amplification of both the 13 and 9.0 kb fragments, when compared with MCF-10F cells. MNNG, BP_5, D_2, and NMU-treated cells, did not show amplification in the number of copies (45). Immunocytochemical detection of the c-*erb*B/2 oncoprotein in MCF-10F, DMBA and BP treated cells and their derived clones revealed a similar pattern of reactivity, with a tendency to higher number of copies with increased survival efficiency in agar methocel. However, there seemed not to exist a direct relationship between the number of copies and the expression of malignant phenotypes such as invasiveness and tumorigenesis (45).

Alterations in p53 Gene During Transformation of HBEC by Chemical Carcinogens

The tumor suppressor gene p53 is thought to be involved in regulation of the normal cell cycle and, when inactivated, leads to uncontrolled cell proliferation. Abnormalities of p53 are particularly common in colorectal and breast cancer (24, 46, 47). Recent studies have shown that the expression of mutant p53 increases from 13% to 50% as breast cancers progress from early in situ to advanced metastatic lesions (48, 49). In our experimental *in-vitro* system Southern blot analysis of EcoR1, Hind III, and Bg III digested genomic DNA of MCF-10F control and carcinogen treated cells was conducted using a full-length human p53 cDNA probe. MCF-10F cells, the carcinogen-treated cells, and the clones derived from these cells did not exhibit alterations in Hind III, EcoRI, or BglII digested DNA (50).

Mutations too small to be resolved by Southern blotting were identified by single strand conformation polymorphism (SSCP) analysis. Almost none of the cell lines studied exhibited mutations in exons 2-11 of the p53 gene. Only one cell line, D3-1, showed in exon 7 an additional band which migrated faster than the normal bands observed. This shift was conserved in its *ras*-transformed progeny D3-1T*ras* and represents a conformational shift suggesting the presence of a mutation in this exon. Since this was not observed in the parental cell line D3, this finding indicates the acquisition of a p53 mutation in the progression of DMBA treated cells from D3 to D3-1 (50). We are presently in the process of sequencing these clones to find out the exact nature of the mutations in p53.

DNA from 17 cell lines was examined by Southern blot analysis for allelic loss on chromosome 17p using probes p144D6 (17p13) and pYNZ22 (17p13.3) to detect variable number of tandem repeats (VNTR). Using p144D6 as a probe, loss of heterozygosity was observed in the benzo(a)pyrene transformed tumorigenic cell line BP1E, whereas MCF-10F cells showed the presence of both alleles. T24, a bladder carcinoma cell line, and human pancreatic carcinoma cell line MiA1PaCa2 and PANC-1 were used as positive controls for loss of heterozygosity and rearrangements (50). In these experiments, we have established that the transformation of these cell lines and the progression in the expression of neoplastic phenotypes which culminates in tumorigenesis in SCID mice by BP1-E cells, are associated with the loss of the telomeric portion of chromosome 17. Our data also suggest the presence of another tumor suppressor gene on the short arm of chromosome 17, which may act alone or in conjunction with p53 (52-53).

Conclusions and Future Directions

Our results demonstrate that human breast epithelial cells can be transformed *in vitro* by chemical carcinogens. The response to chemical carcinogens is related to the presence of lobules type 1 and 2 in the host tissue. However, transformed cells express *in vitro* the full spectrum of neoplastic phenotypes progressing to tumorigenesis in a heterologous host if the cells appear only immortalized. In this experimental system we have developed clones of cells which manifest specific transformation phenotypes, making this system unique for asking specific questions, such as what genes are activated for each specific phenotype, whether the expression of the tumorigenic phenotype is the result of a cascade of events that have to occur sequentially or cumulatively, and more importantly, provides the adequate experimental system to test the functional role of these genes in the progression or reversion of the neoplastic phenotype.

Acknowledgments

We thank Jo Anne Bowman for typing this manuscript and Maria Rocio Rivera, Ana Lilian Romero, Sandino Estrada and Ricardo Moraes for their technical assistance. Supported by Public Health Service Grants CA38921 and CA48927.

References

1. Russo J, Russo IH, Rogers AE (1990) Tumors of the mammary gland. In Turusov V, Mohr H (eds): Pathology of Tumors of Laboratory Animals, Vol 1, pp 47-78.
2. Dabelow A (1957) Die Milchdrüse. In Bargmann W (ed): Handbuch der Mikroskopischen Anatomic des Menschen. Berlin: Springer-Verlag, pp 277-485.
3. Russo J, Russo IH (1987) Development of the human mammary gland. In Neville MD, Daniel C (eds):The Mammary Gland. New York: Plenum Publishing, Inc, pp 67-93.
4. Vorherr H (1974) The Breast, Academic Press, New York, pp 1-18..
5. MacMahon B, Cole P, Liu M, (1970) Age at first birth and breast cancer risk. Bull WHO 43:209-221.
6. MacMahon B, Cole P, Brown J (1973) Etiology of human breast cancer: A review. J Natl Cancer Inst 50:21-42.
7. Wynder EL, Bross IJ, Hirayama T (1960) A study of the epidemiology of cancer of the breast. Cancer 13:559-601.
8. Bonser GM, Dossett JA, Jull JW (1961) Human and Experimental Breast Cancer, Charles C. Thomas, Springfield, IL. pp. 316-363.

9. Russo J, Russo IH (1980) Susceptibility of the mammary gland to carcinogenesis II. Pregnancy interruption as a risk factor in tumor incidence. Am J Pathol 100:497-511.
10. Monaghan P, Perusinghe NP, Cowen P, Gusterson BA (1990) Peripubertal human breast development. The Anat Rec 226:501-508.
11. Russo J, Rivera R, Russo IH (1992) Influence of age and parity on the development of the human breast. Breast Cancer Res Treat 23:211-218.
12. Russo J, Russo IH (1980) Influence of differentiation and cell kinetics on the susceptibility of the mammary gland to carcinogenesis. Cancer Res 40:2677-2687.
13. Russo J, Gusterson BA, Rogers AE, (1990) Comparative study of human and rat mammary tumorigenesis. Lab Invest 62:1-32.
14. Russo IH, Russo J (1978) Developmental stage of the rat mammary gland as determinant of its susceptibility to 7,12-dimethylbenz(a)anthracene. J Natl Cancer Inst 61:1439-1449.
15. Russo J, Russo IH (1978) DNA labeling index and structure of the rat mammary gland as determinants of its susceptibility to carcinogenesis. J Natl Cancer Inst 61:1451-1459.
16. Russo J, Wilgus G, Russo IH (1979) Susceptibility of the mammary gland to carcinogenesis. I. Differentiation of the mammary gland as determinant of tumor incidence and type of lesion. Am J Pathol 96:721-734.
17. Russo J, Reina D, Frederick J, Russo IH (1988) Expression of phenotypical changes by human breast epithelial cells treated with carcinogens in vitro. Cancer Res 48:2837-2857.
18. Russo J, Calaf G, Russo IH (1993) A critical approach to the malignant transformation of human breast epithelial cells. CRC critical reviews in oncogenesis 4:403-417.
19. Russo J, Mills MJ, Moussalli MJ, Russo IH (1989) Influence of breast development and growth properties in vitro. In Vitro Cell Develop Biol 25:643-649.
20. Traul KA, Takayama K, Kachevsky V, Hink RJ, Wolff JS (1981) A rapid in vitro assay for carcinogenicity of chemical substances in mammalian cells utilizing an attachment independence endpoint-2-assay validations. J Appl Toxicol 1:190-195.
21. Soule HD, Maloney TM, Wolman SR (1990) Isolation and characterization of a spontaneously immortalized human breast epithelial cell line, MCF-10. Cancer Res 50:6075-6086.
22. Tait L, Soule H, Russo J (1990) Ultrastructural and immunocytochemical characterizations of an immortalized human breast epithelial cell line MCF-10. Cancer Res 50:6087-6099.

23. Calaf G, Russo J (1993) Transformation of human breast epithelial cells by chemical carcinogens. Carcinogenesis 14:483-492.
24. Stanbridge EJ (1990) Human tumor suppressor genes Ann Rev Genet 24:615-657.
25. Slamon DJ, Godolphin W, Jones LA, (1989) Studies of the HER-2/neu proto-oncogene in human breast and ovarian cancer. Science 244:707-712.
26. Sporn MB, Todaro GJ (1980) Autocrine secretion and malignant transformation of cells. N Engl J Med 303:878-880.
27. Bishop JM (1987) The molecular genetics of cancer. Science 235:305-311, 1987.
28. Zarbl H, Sukumar S, Arthur AV, (1985) Direct mutagenesis of *Ha-ras*-1 oncogenes by N-nitroso-N-methylurea during initiation of mammary carcinogenesis in rats. Nature 315:382-385.
29. Balmain A, Pragnell IB (1983) Mouse skin carcinoma induced in vivo by chemical carcinogens have a transforming Harvey-*ras* oncogene. Nature 303:72-74.
30. Barbacid M (1987) *Ras* genes. Annu Rev Biochem 56:779-827.
31. Bos JL (1988) The *ras* gene family and human carcinogenesis. Mutation Res 195:255-271.
32. Sukumar S (1989) *Ras* oncogenes in chemical carcinogenesis. In: Current Topics in Microbiology and Immunology, Vol. 148. Berlin: Springer-Verlag, pp 93-114.
33. Zhang PL, Calaf G, Russo J (1992) Point mutation in codons 12 and 61 of the *c-Ha-ras* gene in carcinogen-treated human breast epithelial cells (HBEC). Proc Am Assoc Cancer Res 33:669a.
34. Theillet C, Lidereau R, Escot C, et al. (1986) Loss of a *c-H-ras*-1 allele and aggressive human primary breast carcinomas. Cancer Res 46:4776-4781.
35. Comings DE (1973) A general theory of carcinogenesis. Proc Natl Acad Sci USA, 70:3324-3328.
36. Fearon ER, Feinberg AP, Hamilton SM, Vogelstein B (1985) Loss of genes on the short arm of chromosome 11 in bladder cancer. Nature 318:377-380.
37. Calaf G, Zhang PL, Estrada S, Alvarado M, Russo J (1993) *c-Ha-ras* oncogene enhances the tumorigenic phenotype human breast epithelial cells transformed with chemical carcinogens in vitro. Proc Am Assoc Cancer Res 34:711a.
38. Basolo F, Elliott J, Tait L, (1991) Transformation of human breast epithelial cells by *c-Ha-ras* oncogene. Mol Carcin, 4:25-35.
39. Russo J, Tait L, Russo IH (1991) Morphologic expression of cell transformation induced by *c-Ha-ras* oncogene in human breast epithelial cells. J Cell Science 99:1-10.

40. Ochieng J, Basolo F, Albini A, J (1991) Increased invasive, chemotactic and locomotive abilities of *c-Ha-ras* transformed human breast epithelial cells. Invasion and Metastasis 11:38-46.
41. Stacey DW, Kung HF (1985) Transformation of NIH 3T3 cells by microinjection of *Ha-ras* p21 protein. Nature 310:508-511.
42. Horand-Hand PH, Thor A, Wunderlich D (1984) Monoclonal antibodies of predefined specificity detect activated *ras* gene expression in human mammary carcinomas. Proc Natl Acad Sci USA 81:5227-5231.
43. Ciardello F, McGeady ML, Kim N (1990) Transforming growth factor alpha expression is enhanced in human mammary epithelial cells transformed by an activated *c-Ha-ras* protooncogene but not by the c-neu protooncogene, and overexpression of the TGFαcomplementary DNA leads to transformation. Cell Growth Diff 1:407-420.
44. Russo J, Calaf G, Ochieng J, (1992) Role of *Ras* Oncogene in Human Breast Cancer. In Dogliotti L, Sapino A, Busolatti G (eds): An Experimental Approach. Dordrecht: Kluwer, pp 105-117.
45. Russo J, Calaf G, Sohi N, Alvarado ME, Estrada S, Russo IH (1993) Critical steps in carcinogenesis. Breast Cancer: From biology to therapy. New York Acad Sci 698:1-20.
46. Biggs PJ, Warren N, Venitt S, Stratton MR (1993) Does a genotoxic carcinogen contribute to human breast cancer? Mutagenesis 8:275-283.
47. Coles C, Thompson AM, Elder PA, Cohen BB, Mackenzie IM, Cranston G, Chetty U, Mackay J, Macdonald M, Nakamura Y, Hoyheim B, Steel CM (1990) Evidence implicating at least two genes on chromosome 17p in breast cancer. Lancet 336:761-763.
48. Tsuda H, Iwaya K, Fukutomi T, Hirohashi S (1993) p53 mutations and c-*erb*B-2 amplification in intraductal and invasive breast carcinomas of high histologic grade. Jp J Cancer Res 84:394-401.
49. Lohmann D, Ruhri C, Schmitt M (1993) Accumulation of p 53 protein as an indicator for p53 gene mutation in breast cancer. Diagn Mol Patho 2:36-41.
50. Barnabas-Sohi N, Moraes RCB, Calaf G, Estrada S, Alvarado ME, Russo J (1993) Overexpression of p53 protein and loss of heterozygosity in chemically transformed human breast epithelial cells. Proc Am Assn Cancer Res 34:649a.
51. Sato T, Tanijani A, Yamokowa K, (1990) Allelotype of breast cancer: Cumulative allele losses promote tumor progression in primary breast cancer. Cancer Res 50:7184-7189.
52. Coles C, Thompson A, Elder PA (1990) Evidence implicating at least two genes on chromosome 17p in breast carcinogenesis. Lancet 336:761-763.

9

Hormones, Growth Factors, and Gene Expression in Preneoplasias of the Mouse Mammary Gland

Daniel Medina, Anne White, and Thenaa K. Said

Introduction

Breast cancer, in humans as in mice and rats, is thought to be the result of sequential changes in the epithelial cells of the mammary gland (1-3). In mice, the intermediate stages are well-defined and their biological properties extensively studied (2, 4). These intermediate stages are visualized as ductal or alveolar hyperplasias, both of which contain cells with an enhanced probability for tumor formation (2, 4, 5). These cell populations, termed preneoplasias, have a common set of biological properties. The essential properties are considered to be the following: 1) immortality, 2) morphological hyperplasia, and 3) enhanced tumorigenicity. The most frequently studied, the alveolar hyperplasias, can be induced by mouse mammary tumor virus (MMTV) (2), chemical carcinogens (2), hormones (2), or spontaneously, as a result of cell culture (6). Alveolar hyperplasias can be transplanted into the cleared fat pads of syngeneic mice, resulting in stable hyperplastic outgrowth lines (1, 4). Regardless of the etiology of the alveolar hyperplastic outgrowth lines, the essential biological properties are similar for the different lines. The outgrowth lines are essentially clonal derivatives of the primary hyperplastic alveolar nodules, as was demonstrated by examination of MMTV integration sites (7). Although some of the critical molecular changes associated with MMTV-induced and chemical carcinogen-induced tumorigenesis have been defined in terms of *wnt* gene (8) and *ras* gene activation (9), respectively; these molecular events do not explain the biological properties associated with most mammary alveolar hyperplasias, particularly those induced in the absence of these oncogenic agents. Recently, we have examined a new series of transformed mouse mammary cells for which the biological properties of immortality, hyperplasia and tumorigenicity are independently assortable. The *in-situ* outgrowth lines and their *in-vitro* cell line counterparts provide a means to examine the molecular events associated with each of the essential changes acquired during the evolution to the tumor phenotype (6, 10, 11).

Sequential Events in Mouse Mammary Tumor Development

The scheme below describes the sequential changes observed in mouse mammary tumorigenesis. This scheme suggests that the acquisition of the immortalized phenotype occurs as an early event and before the acquisition of morphological hyperplasia. The EL 5,7,11,12 outgrowth lines represent ductal outgrowths that can be serially transplanted for prolonged periods (currently more than five years), thus demonstrating indefinite proliferation potential (11). The progression from a ductal morphology to a morphological hyperplasia is observed in the EL12 line, which generates alveolar cells that have segregated subsequently into a stable alveolar hyperplasia, termed TM12, with a significant tumor-producing capability. In contrast, the EL11 line has remained a purely ductal outgrowth and nontumorigenic. Secondly, the scheme suggests that the emergence of morphological hyperplasia is independent from the ability to undergo further transformation into the neoplastic phenotype. This conclusion is based on the observations that the outgrowth lines TM2L and TM3, although composed entirely of alveolar cells, are essentially nontumorigenic in untreated virgin mice (10). In particular, the TM3 outgrowth line has failed to produce tumors either spontaneously with prolonged hormone stimulation or after a single dose of 7,12-dimethylbenzanthracene. Whether it can progress to tumors directly or indirectly via a stable hyperplasia II phenotype is unclear at this time. In contrast, most of the alveolar outgrowth lines (i.e., TM2H, TM10, TM12) fall into the hyperplasia II category and have significant tumorigenic potential in untreated virgin mice (10). If this model is valid, then it would suggest a minimum of 4 molecular changes is necessary for tumor development.

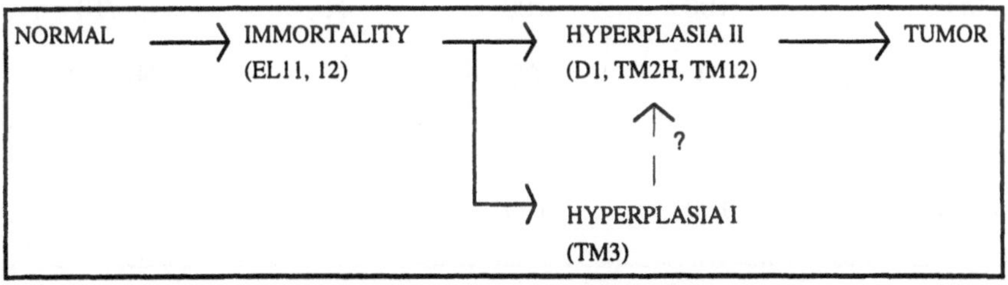

Figure 1. Sequential changes observed in mouse mammary tumorigenesis.

Hormones and Growth Factors

The majority of preneoplastic hyperplasias developed and examined in BALB/c and C3H mice are ovarian hormone-independent for growth and morphogenesis (2). However, the immortalized ductal lines and the nontumorigenic hyperplasia line TM3 are markedly ovarian hormone-dependent for growth and morphogenesis (Table 1), in contrast to the tumorigenic hyperplastic outgrowth lines TM2H and TM12. The TM2L exhibits a relative ovarian-hormone dependency for growth but not morphogenesis. The close correlation between hormone dependency and tumorigenic potential is probably not significant or causal, since earlier studies demonstrated that weakly tumorigenic hyperplasias can be ovarian-hormone dependent (12). However, the absolute ovarian dependency of the ductal EL lines and the fact that the alveolar hyperplasia in TM3 is transitory, as the TM3 outgrowth spontaneously involutes, suggests that changes in the estrogen-progesterone regulated pathways are important for the stable establishment of the alveolar hyperplastic morphology.

Table 1. Hormone and Growth Factor Dependency of Mammary Outgrowth Lines.

Line	Percent Fat Pad Filled		EGF-Dependence *in vitro*	Tumor-Producing Capability[b]
	Ovariectomized	Intact		
EL5	19	90	ND	NIL/12
EL11	5	50[a]	Y	NIL/12
EL12	5	60[a]	Y	11/12
TM12	100	100	N	56/10
TM3	11	46	Y	NIL/12
TM2L	45	90	N	<3/12
TM2H	100	100	N	90/5

a These samples were evaluated 6 weeks after transplantation, in contrast to the others, which were evaluated at 7-8 weeks after transplantation.

b The tumor-producing capabilities are stated in terms of percent tumors by months after transplantation.

Most of these *in-situ* outgrowth lines have been established as cell lines in monolayer cell culture. An assessment of the growth factor dependencies in cell culture revealed that the TM cell lines were independent of epidermal growth factor (EGF) for growth, in comparison to the cell lines established from the EL ductal lines or the TM3 hyperplastic outgrowth (10). This raised the possibility that the ovarian hormone and/or EGF independence was due to the activation of other genes in these pathways. An examination of the *in-situ* hyperplastic outgrowths and their tumors failed to demonstrate consistent mRNA overexpression and/or detectable protein for the following factors: *erb*-2, TGF amphiregulin, cripto. The dependence on EGF for growth in cell culture may reflect a fundamental property of mammary cells which is linked to ovarian hormone dependence *in vivo* and/or to the preneoplastic phenotype. This hypothesis is supported by the following observations. First, the FSK3 cell line, the *in-vitro* progenitor to the TM3 outgrowth *in situ*, retains EGF dependence up to late passages (passage 62), a passage history long after the FSK3 cell line progressed to bFGF independence (passage 50) and the ability to form colonies on plastic (passage 35). Second, the TM3 outgrowth line established in cell culture as the TM3 cell line retains EGF dependence into late passages (passages 35-40). Third, a TM3 tumor that arose as a consequence of oncogene transfection, is EGF-independent *in vitro*. The ovarian hormone and EGF dependencies of the rare TM3 tumors appearing after various oncogenic insults are currently being examined.

Oncogene Expression

Several oncogenes and tumor suppressor genes known or suspected to be operative in mammary gland tumorigenesis were examined for their mRNA expression levels. The oncogenes *Ha-ras*, *Ki-ras*, c-*myc*, *wnt*-1, *wnt*-3, *int*-2, and the tumor suppressor gene *RB* were not overexpressed in the TM hyperplasias and tumors (13). The mammary developmental genes, *wnt*-2, *wnt*-4, *wnt*-5b, which are expressed at different stages in the developmental cycle of the normal mammary gland was also examined by Northern blots (Table 2). Levels of expression for the three genes ranged from zero to 25% of the control levels in the tumorigenic hyperplasias and were undetectable in the nontumorigenic hyperplasias. These data do not support the hypothesis that these *wnt* genes may be persistently activated in the hyperplastic phenotype. The only gene consistently showing dysregulation was the tumor suppressor gene p53. Dysregulation of this gene was detected as mutations in the TM2H, TM2L, TM3, TM4 outgrowth lines and as overexpression of wild-type gene in the TM9, D1 and D2 outgrowth lines (14, 15). Dysregulation was consistently detected at the hyperplastic stage and not detected in the immortalized EL ductal lines. The mutations observed in the tumors were the same as those observed in the progenitor hyperplasias. There was no correlation between the mutations detected in the hyperplasias and their tumorigenic potentials. At the present time, the functional properties of these mutations are being analyzed.

Cell Cycle Genes

The factors responsible for the tumorigenic potential of the hyperplastic outgrowths remain obscure. Recently, we have examined the patterns of cell cyclin-dependent kinases (cdk's) in these outgrowths. Initiation of these experiments resulted from an observation that extent of tyrosine phosphorylation of proteins of 34kd differed widely between the weakly tumorigenic outgrowths and the highly tumorigenic outgrowths (Said and Medina, personal communication). There was a strong inverse correlation between the degree of tyrosine phosphorylation and the tumorigenic capabilities of the outgrowths.

Table 2. *wnt* Gene Expression in Mammary Hyperplasias.

Mammary Gland Stage	*wnt* Gene Levels (% of Control) *wnt*-2	*wnt*-4	*wnt*-5b
Virgin	100	100	8
Pregnant	0	8	100
TM2H, 9,12	0-10	4-10	25
TM2L, 3	0	0	0

Tumors resulting from the outgrowths exhibited low levels of tyrosine phosphorylation at p34kd proteins. Immunoprecipitation with monoclonal antibody to phosphorylated tyrosine followed by incubation of the precipitated proteins with specific antibodies indicated the proteins were cdc2 and cdk2. Since tyrosine dephosphorylation is an essential step prior to activation of cdc2 kinase activity and kinase activity of the cdc2 and cdk2 proteins is essential for cells to traverse the cell cycle (16), we hypothesized that the kinase activity of these cell cycle proteins was one difference between the nontumorigenic and tumorigenic hyperplasias (Table 3).

The data were complex but overall supported the hypothesis that an increase in cdc2 and cdk2 protein and kinase activity as well as in the cdc2 labeling index reflected the neoplastic phenotype and the incipient neoplastic phenotype, the latter represented in TM2H hyperplasias. TM2L hyperplasia has low values for each of these activities and was comparable to normal pregnant mammary gland. TM10 and TM12 hyperplasias, which have long latent periods for tumor formation, have levels comparable to TM2L, indicating a relatively quiescent period for these hyperplasias before tumors start to emerge. In contrast, TM2H hyperplasia, which produces a high incidence of tumors at 4 to 5 months, has increased activities in all the five parameters, some of which are comparable to the values found in tumors. For example, in TM2H hyperplasia, cdc2 and cdk2 protein increased significantly along with their kinase activities. In all three types of tumors (TM2, TM10, TM12), the kinase activities increased even further. The data suggest that, at 12 weeks after transplantation, the TM2H hyperplasias, although morphologically well differentiated and organized as hyperplasias, had the cell cycle control characteristics observed in frank neoplasias. The TM2H hyperplasias would produce tumors within a further 4 to 6 weeks.

Table 3. Cell Cycle Activities in Mammary Hyperplasias and Neoplasias.

Outgrowth Line	TPC/ Latency[a]	cdc2-L.I.[b] (% of tumor)	cdc2 prot./k.a.[c]	cdc2 prot./k.a..[c]	cycli n E[d]
TM2L HOG	0/12	29	1/1	1/1	65
TM2H HOG	90/5	90	23/4	7/8	110
TM10,12 HOG	60/10	35-41	N.D./1	N.D./2	5-45
TM2 Tumor	N.A.	100	49/17	8/22	180
TM12 Tumor	N.A.	100	N.D./10	N.D./6	100

a Numbers reflect the percent tumors by months after transplantation.

b Labeling index represents the percent of cells staining with antibody to cdc2.

c Protein is measured as OD from Western blots; kinase activity is measured as phosphorylation of histone H1. In all 4 assays, the values for TM2L were designated as one and the values for the other tissues are expressed as a fold increase of that in TM2L.

d Protein as measured from cyclin E bound to cdk2. N.D. = Not determined

The cyclin-dependent kinase proteins are regulated by cell cyclins, whose concentrations peak at different stages of the cell cycle. The cyclin E content was examined using p13 beads to bind cdk2 and its associated proteins, then the proteins incubated with antibodies to cyclin E. As in the analysis of kinase activities, only the TM2H hyperplasia and tumors exhibited high levels of cyclin E. The difference in the amount of cdk2-bound cyclin E between the hyperplasias (TM2L, TM10, TM12) and their respective tumors was only 3.5-fold; however, the TM2H hyperplasia clearly had levels comparable to the tumors. Current studies should clarify the activities of cyclins D1 and D2 and cdk4, the cyclin-dependent kinase partners involved in the early stages of G_1.

The central question that emerges from these results concerns the nature of the molecular changes triggering the cells of the hyperplasias into a highly proliferative state. Several possibilities emerge. First, dysregulation of the cyclin genes D or E (17) could lead to high levels of expression, which would then favor high cell cycle kinase activity and progression through the cell cycle. Second, the loss of specific inhibitory proteins such as p21 (18) and p27 (19) would remove a checkpoint for cdk proteins. Presumably, the dysregulation of the cyclins or the inhibitory proteins could be achieved via mutation of these genes (17) or deactivation of proteins that regulate these genes, such as TGF (20, 21). An alternative regulatory loop would be the dephosphorylation of the kinases, a prerequisite for kinase function. The strongest association observed so far in these mammary outgrowths was the high level of tyrosine phosphorylation observed in the TM2L and TM3 hyperplasias, in contrast to the low levels detected in the TM2H hyperplasia and tumors. The tyrosine phosphorylation level in TM10 hyperplasia was intermediate between TM2L and TM2H hyperplasias. The factors important for dephosphorylation of cdc2 and cdk2 in the mammary tissues have not been determined; however, one would predict that activity of the relevant phosphatases is directly correlated to tumorigenic potential of the hyperplasias. It is conceivable that dysregulation of specific phosphatase activity is a critical event which would relieve one of the checkpoints on cell cycle kinase activity, and that increased levels of cell cycle proteins and cyclins would be a consequence of this event.

Conclusions

The EL and TM mammary outgrowth lines provide a model system in which the biological phenotypes of immortality, hyperplasia and tumorigenicity are independent and assortable characteristics. The outgrowth lines arose in the absence of specific viral or chemical carcinogen treatment; thus, the contributions of *wnt* and/or *ras* genes to the development of these phenotypes is negligible. Of the oncogenes and tumor suppressor genes examined in these outgrowth lines, only alteration of p53 function/expression has been identified as having a possible causative role. Two other pathways have been hypothesized to have a significant role in this system of tumorigenesis. First, the factors regulating expression of EGF responsiveness may be important in hormone dependence and/or alveolar hyperplasia, although factors such as TGF, *erb*-2, cripto and amphiregulin cannot be shown to be overexpressed. However, other factors such as the EGF and estrogen receptors need to be examined. Second, the factors regulating cyclin-dependent kinase levels and functional activities may be important, as there is preliminary evidence that only in the incipient neoplastic phenotype and frank neoplasias are high levels of cdk and cyclin activity apparent. Whether the high cdk activity is due to activated cyclins, down-regulated kinase inhibitors, and/or dephosphorylation are questions being addressed.

References

1. Page DL, Dupont WD (1993) Anatomic indicators of increased breast cancer risk. Breast Cancer Res Treat 28:157-166.
2. Medina D (1988) The preneoplastic state in mouse mammary tumorigenesis. Carcinogenesis 9:1113-1119.
3. Russo J, Gusterson BA, Rogers AE (1990) Comparative study of human and rat mammary tumorigenesis. Lab Invest 62:244-278.
4. DeOme KB, Faulkin LJ Jr, Bern HA, Blair PB (1959) Development of mammary tumors from hyperplastic alveolar nodules transplanted into gland-free mammary fat pads of female C3H mice. Cancer Res 19:515-520.
5. Medina D (1976) Mammary tumorigenesis in chemical carcinogen-treated mice. VI. Tumor-producing capabilities of mammary dysplasias occurring in BALB/c mice. J Natl Cancer Inst 57:1185-1189.
6. Kittrell FS, Oborn CJ, Medina D (1992) Development of mammary preneoplasias *in vivo* from mouse mammary epithelial cells *in vitro*. Cancer Res 52:1924-1932.
7. Cardiff RD (1984) Protoneoplasia: The molecular biology of mouse mammary hyperplasia. Adv Cancer Res 42:167-190.
8. Tsukamoto AD, Grosschedl R, Guzman RC (1988) Expression of the int-1 gene in transgenic mice is associated with mammary gland hyperplasia and adenocarcinomas in male and female mice. Cell 55:619-625.
9. Kumar R, Medina D, Sukumar S (1990) Activation of H-*ras* oncogenes in preneoplastic mouse mammary tissues. Oncogene 5:1271-1277.
10. Medina D, Kittrell FS, Liu Y-J, Schwartz M (1993) Morphological and functional properties of TM preneoplastic mammary outgrowths. Cancer Res 53:663-667.
11. Medina D, Kittrell FS (1993) Immortalization phenotype dissociated from the preneoplastic phenotype in mouse mammary epithelial outgrowths *in vivo*. Carcinogenesis 14:25-28.
12. Medina D (1973) Preneoplastic lesions in mouse mammary tumorigenesis. Adv Cancer Res 7:1-53.

13. Medina D, Kittrell FS, Oborn CJ, Schwartz M (1993) Growth factor dependency and gene expression in preneoplastic mouse mammary epithelial cells. Cancer Res 53:668-674.
14. Jerry DJ, Ozbun MA, Kittrell FS (1993) Mutations in p53 are frequent in the preneoplastic stage of mouse mammary tumor development. Cancer Res 53:3374-3381.
15. Jerry DJ, Butel JS, Donehower LA (1994) Infrequent p53 mutations in 7,12-dimethylbenzanthracene-induced mammary tumors in BALB/c and p53 homozygous mice. Mol Carcin 9:175-183.
16. Norbury C, Nurse P (1992) Animal cell cycles and their control. Ann Rev Biochem 61:441-470.
17. Keyomarsi K, Pardee AB (1993) Redundant cyclin overexpression and gene amplification in breast cancer. Proc Natl Acad Sci, USA 90:1112-1116.
18. Harper JW, Adami GR, Wei N, (1993) The p21 cdk-interacting protein Cip1 is a potent inhibitor of G1 cyclin-dependent kinases. Cell 75:805-816.
19. Polyak K, Kato J-Y, Solomon MJ (1994) p27kip1, a cyclin-cdk inhibitor links transforming growth factor-αand contact inhibition to cell cycle arrest. Genes and Develop 8:9-22.
20. Koff A, Ohtsuki M, Polyak K (1993) Negative regulation of G1 in mammalian cells: Inhibition of cyclin E-dependent kinase by TGF. Science 260:536-538.
21. Ewen ME, Sluss HK, Whitehorse LL, Livingston DM (1993) TGF inhibition of cdk4 synthesis is linked to cell cycle arrest. Cell 74:1009-1020.

10

Pathogenesis of Ductal and Lobular Progestin-Induced Mammary Carcinomas in BALB/c Mice

Alfredo A. Molinolo, Patricia Pazos, Fernanda Montecchia, Edith C. Kordon, Graciela Dran, Fabiana Guerra, Patricia Elizalde, Isabel Luthy, Eduardo H. Charreau, Christiane Dosne Pasqualini, and Claudia Lanari

Introduction

Several years ago, we demonstrated that medroxyprogesterone acetate (MPA)-induced mammary adenocarcinomas in female BALB/c mice with an incidence close to 80% and a mean latency of around 13 months (1). These tumors were mostly ductal, progestin-dependent (PD) adenocarcinomas with high levels of estrogen and progesterone receptors (ER and PR) (2). We later found that progesterone (P) also induced mammary carcinomas, but this time the tumors were mostly lobular, progestin-independent (PI) adenocarcinomas with lower levels of ER and PR (3). There was a constant correlation between progestin dependence and morphology, that is, lobular carcinomas were always PI and ductal carcinomas PD (3). To extend this study further, we designed a model of co-carcinogenesis using medroxyprogesterone (MPA) with N-methyl N-nitrosourea (MNU) in BALB/c mice. We obtained a high incidence of mammary adenocarcinomas similar to the hormone-induced lobular tumors, and showed that MPA can act as a potent promoter.

Among the different factors that influence progestin-induced tumors, we have especially studied EGF, an interest that was first triggered by the finding of hyperplastic salivary glands in female mice treated with MPA. EGF is one of several growth factors implicated in the induction and progression of cancer (4), and in mice, salivary glands are the main source of this factor. Its synthesis can be stimulated by different hormones, among them MPA (5), via androgen receptors, but the correlation between variations in salivary gland EGF (SG-EGF) and serum levels (S-EGF) remains controversial (6-8).

In the present study, we present our findings regarding the possible influence of EGF on the induction of these tumors, and the *in-vitro* growth regulation of hormone-dependent tumor lines.

Effect of Sialoadenectomy on Tumor Induction

MPA-Induced Carcinogenesis

Virgin female BALB/c mice were sialoadenectomized (Sx) or remained intact and received one 40-mg pellet of MPA every 3 months for 18 months. After this time, those mice that had not developed clinically evident tumors were sacrificed. Sx significantly reduced ($p < 10^{-4}$) the incidence of MPA-induced mammary tumors (Table 1); tumor latencies were similar in Sx and intact MPA-treated animals. A similar ratio of ductal to lobular and PD to PI tumors was registered; ductal PD tumors/lobular PI: 6/2 in Sx and 18/9 in controls. In syngeneic passages, all ductal tumors, with one exception, were PD and all lobular tumors were PI. The correlation between histology and hormone-responsiveness was significant ($p < 10^{-5}$), confirming previous results (3). There were no morphological differences between mammary glands of untreated normal and Sx female mice 45 days after surgery in whole mounts. However, there was an important difference in branching in mammary glands from Sx and control mice treated for 2 months with MPA.

(NMU + MPA)-Induced Carcinogenesis

Sx MPA-treated female BALB/c mice, sham-operated, and controls were inoculated with a single NMU dose (50mg/kg, i.p) 45 days after the MPA inoculum. MPA treatment was repeated every 3 months, and the animals were followed for 10 months. The experiment was terminated after 10 months to avoid interference with MPA-induced tumors (average latency of 13 months). Sx significantly inhibited tumor incidence (Table 1). The tumors originated were histologically acinar or papillary adenocarcinomas. Most of them showed extensive squamous metaplasia, with areas that closely resembled the histology of the human calcifying epithelioma (Malherbe tumor or pilomatricoma).

Serum, Salivary Gland and Tumor EGF Levels

Sx or intact female mice were treated sc with 40 mg MPA (one week after surgery), or remained untreated. Pooled sera from 6-10 mice were used. Salivary glands and/or tumors were excised and frozen at -70° C. EGF was measured in serum, salivary glands, and tumor samples by RIA as previously described (2). MPA induced an increase in both S-EGF and SG-EGF in intact mice (Table 2).

S-EGF's increment started to be evident approximately 20 days after MPA inoculum. In Sx animals, MPA did not increase S-EGF levels (median: 1.051 ng/ml, range: 0.68-1.69), whereas in sham-operated animals, S-EGF values were higher (median: 3.6 ng/ml; range 3-4.9 ng/ml; $p < 0.05$). EGF was not detected in 6 ductal (3 PD and 3 PI) and 3 lobular progestin-induced tumors.

Table 1. Effect of sialoadenectomy on tumor incidence.

Treatment	#tumors / #animals	Tumor Actuarial Incidence	Tumor Latency (weeks)
Sx-MPA	11 / 48	[a]32%	52.5 ± 3.8
MPA	34 / 47	[a]98%	50.1 ± 2.1
Sx - MPA (MNU)	11 / 40	[b]28%	31 ± 5.4
MPA (MNU)	41 / 83	[b]50%	30 ± 5.4
MNU	0 / 40	0%	0

[a] $p < 0.0001$
[b] $p < 0.001$

Table 2. EGF levels in mice treated or not with MPA.

EGF	MPA-treated	Untreated	
Salivary glands	761 ± 157 ng/mg fresh tiss. n - 14	35 ± 11 ng/mg fresh tiss. n = 7	$p < 0.01$
Serum	9.13 (2.58-120) median (range) n = 20	1.68 (0.95 - 1.95) median (range) n = 6	$p < 0.001$
MPA-induced Tumors 3 lobular / 3 ductal	Not detectable	Not detectable	

EGF Receptors (EGF-R)

Tumor growth rate, hormone dependence, ER, PR, and EGF-R levels of syngeneic transplants of 8 lobular, 5 ductal PD with their PI variants, and 3 tumors originated in animals treated with NMU and MPA tumors are shown in Figure 1. The difference in EGF-R between lobular progestin-induced and (MNU + MPA)-induced vs. ductal progestin-induced was significant ($p<0.01$). Lobular progestin-induced and (MNU + MPA)-induced tumors express neither ER nor PR in syngeneic passages, while PD or PI ductal tumors retain high levels of ER and PR.

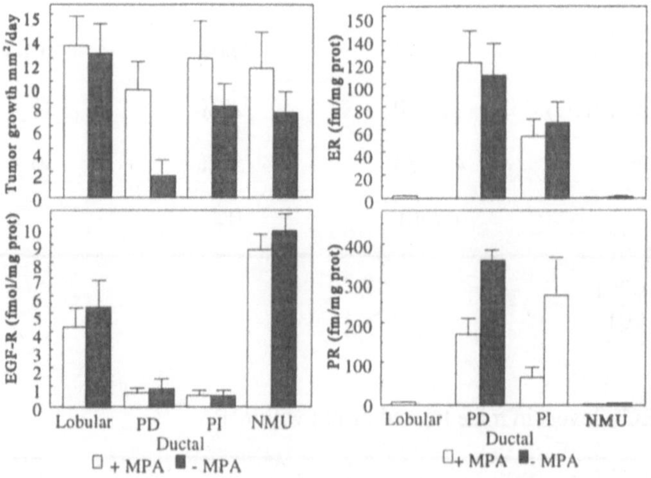

Figure 1. Tumor growth, EGF-R, ER and PR in lobular and ductal progestin-induced tumors, and in MPA+MNU-induced tumors. Differences in EGF-R, ER and PR expression are significant ($p<0.01$) between lobular and MNU+MPA vs. ductal tumors. *: $p<0.05$.

EGF-R also correlates with absence of steroid receptors (ER: correlation coefficient = -0.5105, $p < 0.01$, PR: correlation coefficient= -0.5217, $p < 0.01$) but not with tumor growth rate (correlation coefficient = 0.0929, ns). In 7/8 of the different lobular tumor lines characterized, ER and PR were detected in the original tumor and in the first passage, (ER: 5-87 fmoles/mg protein, PR: 7-159 fmoles/mg protein).

Effect of Sialoadenectomy (Sx) on Tumor Growth

Three ductal [EGF-R$^-$] tumor lines (2 PD and 1 PI) and two lobular PI tumor lines, both [EGF-R$^+$], were transplanted into Sx or sham-operated animals; half of the animals were treated with MPA, and half remained untreated. No correlation was found between growth and Sx in either lobular [EGF-R$^+$] or ductal tumors [EGF-R$^-$].

Effect of MPA and EGF on *In-Vitro* Growth of Primary Cultures of a Ductal Tumor Line

Epithelial and fibroblast-enriched cultures were obtained from MPA-induced PD ductal tumors, incubated 48 h with MPA and EGF, and grown in D MEM/ F12 with 5% charcoal-treated fetal calf serum. Cell growth was evaluated using thymidine uptake. MPA significantly stimulated epithelial cell growth in concentrations ranging from 10^{-6} to 10^{-11} M, while it did not affect fibroblasts. Conversely, EGF significantly stimulated fibroblast proliferation in a concentration of 5 ng/ml, but had no effect on epithelial cells (Figure 2).

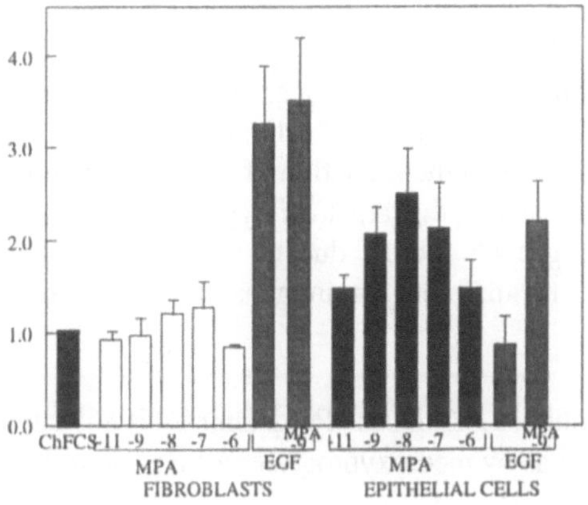

Figure 2. Effects of MPA on [^{3}H]-thymidine uptake of subcultures of primary epithelial or fibroblastic enriched-cultures (C4-HD). MPA was tested at concentrations ranging from 10^{-11} to 10^{-6} M. EGF was used at 5 ng/ml. The results represent a mean value of 3-4 experiments. Proliferation index is defined as the ratio between the total cpm counting of each experimental group and the control. Significant differences between MPA-treated groups vs. control group were observed only in epithelial cells ($p < 0.05$).

Discussion

The results reported herein demonstrate that the incidence of mammary tumors induced by MPA and by MNU + MPA in BALB/c mice is significantly reduced by sx, however, this procedure does not significantly affect the growth of tumors already established. Similar results have been reported in different models (7, 8). These findings were attributed to the lack of stimulatory activity of SG-EGF, although the contribution of SG-EGF to S-EGF remains controversial (6). In our model, we demonstrated a sustained increase in SG-EGF, followed by an increase in S-EGF, evident 3-4 weeks after starting MPA treatment. EGF was not detected in sx animals, and the incidence of mammary tumors in this group was significantly lower. These data suggest that SG-EGF may play an important role in the development of mammary carcinomas. Additional support for this hypothesis is provided by the lack of mammary gland development in Sx MPA-treated animals. On the other hand, SG-EGF does not seem to play a vital role in the growth of tumors already established. The lack of growth stimulation by EGF in *in-vitro* cultures also supports this finding.

Regardless of the presence of salivary glands, no morphological or biological differences were found in the developing tumors; MPA induced mostly ductal tumors, and MNU + MPA lobular adenocarcinomas with squamous metaplasia. EGF-R were only detected in lobular and (NMU + MPA)-induced tumors, while ductal tumors expressed high levels of ER and PR.

MPA may induce mammary cancers by acting directly on the gland and indirectly through the stimulation of salivary gland EGF synthesis. This factor seems to play a key role in the induction of mammary carcinomas by MPA and MNU, probably expanding the target-cell type. The deregulation of the synthesis and secretion of growth factors due to hormone imbalances may create a common pathway for different carcinogens to act on similar cell targets.

References

1. Lanari C, Molinolo AA, Dosne Pasqualini C (1986) Induction of mammary adenocarcinomas by medroxyprogesterone acetate in BALB/c female mice. Cancer Lett 33:215-223.
2. Molinolo AA, Lanari C, Charreau EH (1987) Mouse mammary tumors induced by medroxyprogesterone acetate: Immunohistochemistry and hormonal receptors. J Natl Cancer Inst 79:1341-1350.
3. Kordon E, Molinolo AA, Pasqualini CD (1993) Progesterone induction of mammary carcinomas in BALB/c female mice. Correlation between hormone dependence and morphology. Breast Cancer Res Treat 28:29-39.
4. Stoschek CM, King LE (1986) Role of epidermal growth factor in carcinogenesis. Cancer Res 46:1030-1037.
5. Bullock LP, Barthe TH, Mawzowicz I, CW (1975) The effect of progestin

on submaxillary gland epidermal growth factor: Demonstration of androgenic, synandrogenic and anti-androgenic actions. Endocrinology 97:189-195.

6. Perheentupa J, Lakshmanan J, Hoath SB, Fisher DA (1984) Hormonal modulation of plasma concentration of epidermal growth factor. Acta Endocrinol 107:571-576.
5. Kurachi H, Okamoto S, Oka T (1985) Evidence for the involvement of the submandibular gland epidermal growth factor in mouse mammary tumorigenesis. Proc Natl Acad Sci USA 82:5940-5943
6. Inui T, Tsubura A, Morii S (1989) Incidence of precancerous foci of mammary glands and growth rate of transplantable mammary cancers in s ialoadenectomized mice. J Natl Cancer Inst 81:1660-1663.

PART 3. ENDOMETRIUM / UTERUS

11

Is the hsp90 Connection Between Steroid Receptors and Immunosuppressant Binding Immunophilins Involved in the Control of Gene Transcription and Cell Growth?

Etienne-Emile Baulieu

The very first experiments related to the title had apparently a very limited objective. We wanted to determine the protein composition of the so-called "8S," unliganded form of steroid hormone receptors that does not bind to DNA, while it was known that there was a smaller ("4S") hormone- and DNA-binding receptor form. We found that the "8S" form is a heterooligomer, which includes the same non-DNA-binding protein typical of all steroid receptors (1, 2). Cloning revealed that it was the heat-shock protein (MW ~ 90,000) hsp90, an abundant (~ 1%) soluble protein (3) present in nearly all cells of all organs and remarkably well conserved through evolution. Associated with steroid receptors, hsp90 exerts a "chaperon" function, stabilizing the conformation of the ligand-binding domain of gluco- and mineralocorticosteroid receptors, as well as exerting negative control on steroid-regulated gene transcription. Further analysis indicated that hsp90 increases in stress situations, is regulated during the cell cycle [maximum at the G1/S transition (4)], and interacts with many other proteins (pp60v-src and other viral oncogenic protein kinases, dioxin receptor, myoD I, casein kinase II, eIF2 kinase, calmodulin, actin, etc.), whose activity it may regulate. In other words, hsp90 may be involved in a number of cellular functions (5) besides interacting with steroid receptors.

Putting together the heterooligomeric receptor with a monoclonal antibody obtained by Faber, we formed a complex that we described as a "new" protein of apparent MW 59,000 (6). Cloning revealed that its N-terminal region was highly homologous to the one previously described as FKBP12, which binds the immunosuppressant FK506 (thus the name of FKBP) and rapamycin (an immunosuppressant). FKBP12 complexed with FK506 (but not with rapamycin) is immunosuppressive via its interaction with the protein phosphatase calcineurin. The function of FKBP12 in non-immunological cells is poorly understood. It has

The function of FKBP12 in non-immunological cells is poorly understood. It has been proposed that its peptidyl proline isomerase (PPIase) activity plays a role in the control of protein substrate conformation. FKBP59 displays the same immunosuppressant binding and enzymatic activities as FKBP12, but even when complexed with FK506, it does not interact with calcineurin (7) as FKPB12 does (8, 9). In contrast to FKBP12, FKBP59 binds to hsp90 (hence the acronym HBI for heat-shock protein 90-binding immunophilin) (10, 11) via a triple tetratricopeptide sequence located at the N-terminal extremity of the C-terminal of the calmodulin binding site (12). We have demonstrated that hsp90 and unliganded steroid receptors do interact *in vivo* (13). However, the interaction between hsp90 and FKBP5, although likely, has not so far been directly confirmed in intact cells. There is little doubt that the three proteins, FKBP59-hsp90-steroid receptors, form *in-vivo* transitory heterooligomeric complexes, as represented schematically in Figure 1.

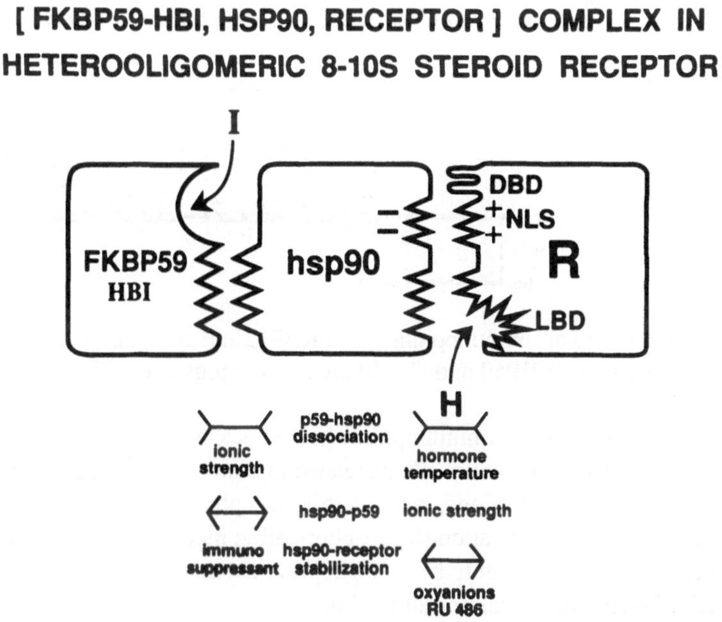

Figure 1. A proposed schematic representation of interactions between the immunophilin FKBP59-HBI, the heat-shock protein hsp90, and a steroid receptor R. DBD: DNA-binding domain; LBD: ligand-binding domain; NLS: nuclear localization signal; I: immunosuppressant drug; H: steroid hormone. Several maneuvers are mentioned that tend to reinforce ( ) or dissociate (>__<) protein-protein interactions.

Recently, two research groups (14, 15) have identified a cyclophilin of MW ~ 40,000 (Cyp40). Cyp40 includes an N-terminal domain exceptionally analogous to the cyclophilin Cyp18, the first identified immunophilin (16), which binds cyclosporin (CsA) and is also a PPlase. The rest of the Cyp40 molecule, as in FKBP59-HBI, consists of a tetratricopeptide triple repeat (Figure 2). Cyp40 has been found associated to the 8S form of the estrogen receptor; it probably also binds to hsp90 (unpublished data), and thus is also an HBI.

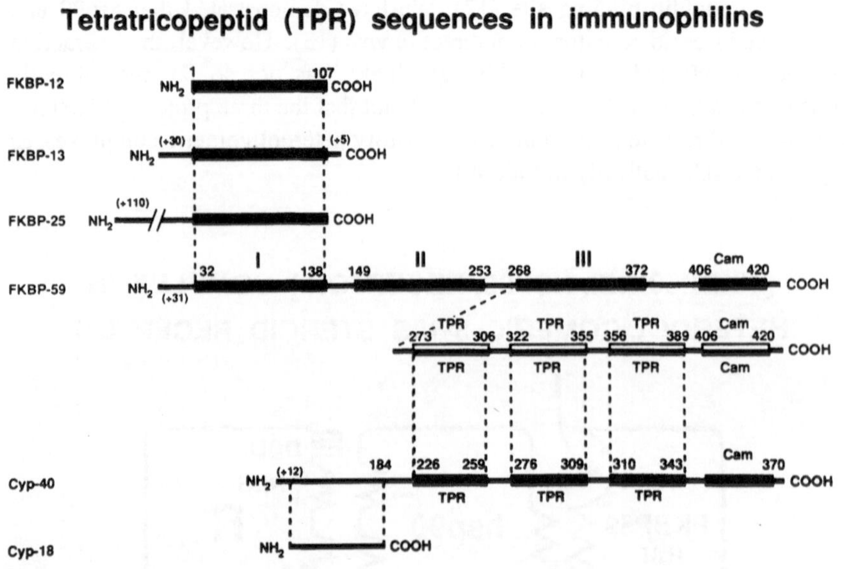

Figure 2. The "classical" immunophilins are FKBP12 and cyclophilin 18. The TPR sequences present in FKBP59 and Cyp40 are homologous but not identical.

The labile structure of the immunophilin-heat shock protein-receptor may be involved in a number of potential effects related to immunosuppressant activities, heat-shock protein interactions, and steroid receptor functions, as well as connections with phosphorylation-dephosphorylation mechanisms (receptor and heat-shock protein are known to be phosphoproteins) and the Ca2+-calmodulin duet [FKBP59-HBI binds calmodulin (17)].

A limited (20%) but consistent increase in the binding affinity of steroids to their receptors (agonists such as progesterone, dexamethasone, and estradiol, or antagonists, such as RU486) has been found in preparations containing immunophilins and the appropriate receptor, when FK506, rapamycin, or cyclosporin was added (18). There is no evidence for a direct affinity of immunosuppressants to the "4S" steroid-receptor molecule, although the three-protein sequence is a likely candidate for the effect of immunosuppressants (consistent with hirsutism observed in cyclosporin-treated patients after heart transplants).

We have searched for the biological effects of immunosuppressants upon hormonal activities. In L-cells (used to study a purely endogenous system with no transfected component), dexamethasone decreased cell growth via the glucocorticosteroid receptor and RU486, an antiglucocorticosteroid, regularly reestablished the normal growth. Remarkably, FK506 abolished the RU486 effect, and, under these circumstances, cell growth was inhibited as with dexamethasone alone. In control studies, FK506 directly potentiated neither the dexamethasone effect nor those of FK506, RU486, or their combination (FK506 did not affect L-cell growth). Whether this effect on proliferation is directly related to the molecular association of the three proteins has yet to be demonstrated. Nevertheless, the antihormone effect of RU486 is dose-dependent. Rapamycin was not tested because it inhibits cell growth (19).

We have also studied another remarkable activity of FK506, rapamycin, and cyclosporin: all increase MMTV transcription in L-cells permanently transfected with an MMTV construct (LMCAT), potentiating the dexamethasone-induced transcription of the virus sequences in a dose-dependent manner (20, 21). Again, these results suggest a connection between immunosuppressants and immunophilins on the one hand, and steroids and their receptors on the other. However, there is as yet no direct evidence in favor of a mechanism acting through the transitory association of these three proteins as indicated in Figure 1. We should also mention that, for both growth and transcription control, we obtained identical results using several different cell types.

The principal lesson to be learned from these complex results is to perform experiments using these three protein types together. Already, we have observed that steroids and their analogs, frequently administered during the course of cancer treatments, may suffer modifications in their activity by immunosuppressant drugs. Indeed, studies with some non-immunosuppressive immunosuppressant derivatives have provided preliminary evidence indicating that they may have the same activity in growth and transcription as the *bona fide* immunosuppressants. This opens the way to new drug design and to novel drug associations of therapeutic interest, particularly in the field of cell growth. Even though immunophilins, hsp90, and steroid receptors are widely distributed among cells, as already mentioned, the same results are not obtained with all cell types, giving hope for cellular specificities in these new pharmacological approaches. The role of hsp90 is not fully understood, but more and more this protein appears to participate in networks vital to cell function.

References

1. Baulieu E-E, Binart N, Buchou T, et al. (1983) Biochemical and immunological studies of the chick oviduct cytosol progesterone receptor. In Eriksson H & Gustafsson J-A (eds.): Steroid Hormone Receptors: Structure and Function. Nobel Symposium 57. Amsterdam: Elsevier, pp 45-72.
2. Joab I, Radanyi C, Renoir JM, et al. (1984) Immunological evidence for a common non hormone-binding component in "non-transformed" chick oviduct receptors of four steroid hormones. Nature 308:850-853.
3. Catelli MG, Binart N, Jung-Testas I, et al. (1985) The common 90-kd protein component of non-transformed "8S" steroid receptors is a heat-shock protein. EMBO J 4:3131-3135.
4. Jérôme V, Vourc'h C, Baulieu EE, et al. (1993) Cell cycle regulation of the chicken hsp90 expression. Exptl Cell Res 205:44-51.
5. Jakob U, Buchner J (1994) Assisting spontaneity: The role of hsp90 and small hsps as molecular chaperones. TIBS 19:205-211.
6. Lebeau MC, Massol N, Herrick J, et al. (1992) P59, an hsp90 binding protein: Cloning and sequencing of its cDNA. Preparation of a peptide-directed polyclonal antibody. J Biol Chem 267:4281-4284.
7. Lebeau MC, Myagkikh I, Rouvière-Fourmy N, et al. (1994) Rabbit FKBP-59/HBI does not inhibit calcineurin activity *in vitro*. Biochem Biophys Res Commun 203:750-755.
8. Liu J, Farmer JD Jr, Lane WS, et al. (1991) Calcineurin is a common target of cyclophilin-cyclosporin A and FKBP-FK506 complexes. Cell 66:807-815.
9. Fruman DA, Klee CB, Bierer BE, et al. (1992) Calcineurin phosphatase activity in T lymphocytes is inhibited by FK506 and cyclosporin A. Proc Natl Acad Sci USA 89:3686-3690.
10. Renoir JM, Radanyi C, Faber LE, et al. (1990) The non-DNA binding heterooligomeric form of mammalian steroid hormone receptors contains a hsp90-bound 59 kDa protein. J Biol Chem 265:10740-10745.
11. Callebaut I, Renoir JM, Lebeau MC, et al. (1992) An immunophilin that binds Mr 90,000 heat shock protein: Main structural features of a mammalian p59 protein. Proc Natl Acad Sci USA 89:6270-6274.
12. Radanyi C, Chambraud B, Baulieu E-E (1994) The ability of the immunophilin FKBP59-HBI to interact with the 90-kDa heat shock protein is encoded by its tetratricopeptide repeat domain. Proc Natl Acad Sci USA 91:11197-11201.
13. Kang KI, Devin J, Cadepond F, et al. (1994) *In vivo* functional protein-protein interaction: Nuclear targeted hsp90 shifts cytoplasmic steroid receptor mutants into the nucleus. Proc Natl Acad Sci USA 91:340-344.

14. Ratajzack T, Carrello A, Mark PJ (1993) The cyclophilin component of the inactivated estrogen receptor contains a tetratricopeptide repeat domain and shares identity with p59 (FKBP59). J Biol Chem 268:13187-13192.
15. Kieffer LJ, Seng TW, Li W (1993) Cyclophilin-40, a protein with homolog to the p59 component of the steroid receptor complex. Cloning of the cDNA and further characterization. J Biol Chem 268:12303-12310.
16. Harding MW, Handschumacher RE, Speicher DW (1986) Isolation and amino acid sequence of cyclophilin. J Biol Chem 261:8547-8555.
17. Massol N, Lebeau MC, Renoir JM, et al. (1992) Rabbit FKBP59-heat shock protein binding immunophilin (HBI) is a calmodulin binding protein. Biochem Biophys Res Commun 187:1330-1335.
18. Renoir JM, Le Bihan S, Mercier-Bodard C, et al. (1994) Effects of immunosuppressants FK506 and rapamycin on the heterooligomeric form of the progesterone receptor. J Steroid Biochem Mol Biol 48:101-110.
19. Jung-Testas I, Delespierre B, Baulieu E-E (1993) Inhibition par l'immunosuppresseur FK506 de l'effet antiglucocorticostéroïde du RU486 sur la croissance de fibroblastes de souris en culture. C R Acad Sci Paris 316:1495-1499.
20. Ning YM, Sanchez ER (1993) Potentiation of glucocorticoid receptor-mediated gene expression by the immunophilin ligands FK506 and rapamycin. J Biol Chem 268:6073-6076.
21. Renoir JM, Mercier-Bodard C, Hoffman K, et al. (1995) Cyclosporin A, as FK506, potentiates the dexamethasone-induced MMTV-CAT activity in LMCAT cells. A possible role for different heat shock protein binding immunophilins (HBIs). In preparation.

12

Expression and Regulation of 17β-Hydroxysteroid Dehydrogenase Type 1 in Steroidogenic Cells and Estrogen Target Tissues

Veli V. Isomaa, E. Hellevi Peltoketo, Matti H. Poutanen, and Reijo K. Vihko

Introduction

The biological actions of estrogens in target tissues are regulated by the concentration of nuclear estrogen receptors and by the intracellular estrogen concentration. The latter concentration is dependent on estrogen plasma levels, as well as local intracellular production and metabolism in the tissue. The ovarian follicles are the main source of estrogens during the reproductive years. However, in postmenopausal women most of the estrogens are produced in peripheral target tissues, which contain the required enzymes to synthesize estrogens from inactive steroid precursors of adrenal and ovarian origins. A key enzyme for the synthesis of estradiol is 17β-hydroxysteroid dehydrogenase (17HSD), which catalyzes the reaction between estrone and estradiol. This review deals mainly with the expression and regulation of 17HSD type 1 in steroidogenic organs and estrogen target tissues.

Properties of Human 17HSD Isoenzymes

17HSD activity represents a group of enzymes catalyzing the reversible reaction between 17-keto- and 17-hydroxysteroids. So far, three separate human17HSD isoenzymes with different tissue distribution, subcellular localization, substrate and cofactor specificity, and preference for catalysis of oxidation versus reduction reactions have been characterized (Table 1). Type 1 enzyme is cytosolic, is highly estrogen-specific, and favors the reduction of estrone to estradiol. This isoenzyme is expressed in the placenta, in ovarian granulosa cells, and in several estrogen target tissues (Figure 1), as well as in some other tissues (1). 17HSD type 2 is a membrane-bound isoenzyme that catalyzes reactions between both estrogens and androgens; it favors the oxidation of estradiol and testosterone to estrone and androstenedione, respectively (2). The tissue distribution of type 2 enzyme is not known, but it is expressed in at least the human placenta and prostate. Type 3 isoenzyme is a microsomal enzyme that is expressed predominantly in the testis and prefers androstenedione as a substrate (3). The diversity of the enzymatic properties of these isoenzymes explains their different roles in steroid hormone metabolism. Also, the type and relative abundance of each 17HSD isoenzyme expressed in a particular cell type may significantly regulate its intracellular concentration in target tissues (4).

Table 1. Properties of the Cloned Human 17HSD Isoenzymes.

Type	Chromosomal Location of Gene	Sub-Cellular Location	Tissue Distribution	Substrate Specificity	Preferred Reaction Direction
1[a]	17q12-q21	cytosol	placenta, ovary, breast, endometrium	estrogens	reductive
2[b]	16q24	microsome s	placenta, prostate	estrogens androgens	oxidative
3[c]	9q22	microsome s	testis	androgens estrogens	reductive

[a] Peltoketo, et al. (5)
[b] Wu, et al. (2).
[c] Geissler, et al. (3).

17HSD in Steroidogenic Tissues

Reductive 17HSD catalyzes the last step of estradiol biosynthesis in the granulosa cells, and testosterone biosynthesis in the Leydig cells. Thus, the enzyme activity is essential for the production of the most important circulating sex steroids in both females and males. The tissue distribution of 17HSD isoenzymes clearly shows that different enzymes are mainly responsible for the production of male and female sex steroids from steroidogenic tissues.The crucial role of 17HSD type 3, testosterone biosynthesis and development of normal male sexual characteristics, has been confirmed by the demonstration that mutations in 17HSD type 3 gene result in male pseudohermaphroditism (3).

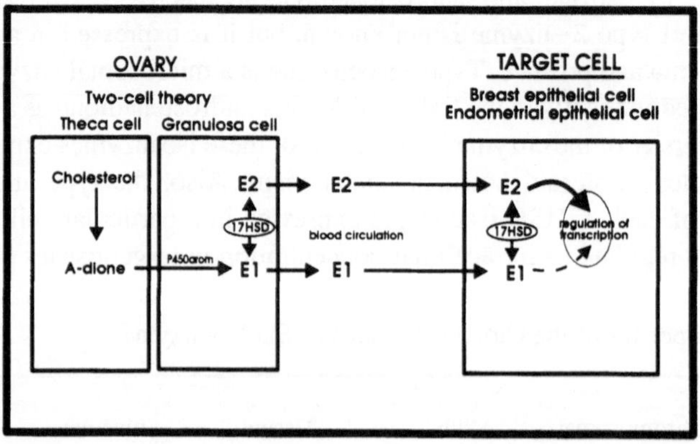

Figure 1. Schematic representation of the role of 17HSD in estrogen metabolism in the ovary and estrogen target cells.

In both human and rat granulosa cells, 17HSD type 1 expression is related to the differentiation stages of the follicles (6-8). A good correlation between 17HSD type 1 and aromatase P450 expression has been found in small and large follicles obtained from patients undergoing *in-vitro* fertilization (IVF) (6). In rat ovaries, 17HSD type 1 expression is low in antral follicles, up-regulated during follicular maturation, and highest in Graafian follicles, which also have the highest capacity for estradiol biosynthesis (Figure 2). Thereafter, 17HSD type 1 expression decreases in the rat ovary during luteinization and is almost undetectable in corpora lutea (7). In contrast, the enzyme is expressed in human corpora lutea and granulosa luteal cells obtained from IVF patients (6, 8).

The difference in the regulation of 17HSD type 1 between humans and rats is probably related to the fact that, in contrast to the rat, human corpora lutea produce estrogens. It is evident from these data that, in addition to P450 aromatase, 17HSD type 1 plays an important role in the regulation of ovarian estradiol biosynthesis.

In humans, the placenta is the major source of estradiol synthesis during pregnancy, and as seen in granulosa cells, type 1 17HSD is highly expressed in human placenta (9) and in three choriocarcinoma cell lines: JAR, JEG-3 and BeWo (10). The 17HSD type 1 and type 3 enzymes are particularly important during gestation, a function that is consistent with their essential roles in estradiol and testosterone synthesis, respectively. In contrast, 17HSD type 2 predominantly catalyzes oxidative activity; therefore, its main function is related to the inactivation of highly reactive sex steroids. It is interesting to note that, at least in the placenta, both type 1 and 2 isoenzymes are expressed. Recent data suggest that in steroidogenic cells, including granulosa cells, trophoblasts and choriocarcinoma cells, 17HSD type 1 is regulated by a cAMP-dependent pathway. However, opposite effects have been observed in other cell types. In short-term cultured human granulosa-luteal cells, the enzyme expression decreased after exposure to Br-cAMP; however, in cultured normal human trophoblasts and choriocarcinoma cells, it was up-regulated by cAMP-analogs (11).

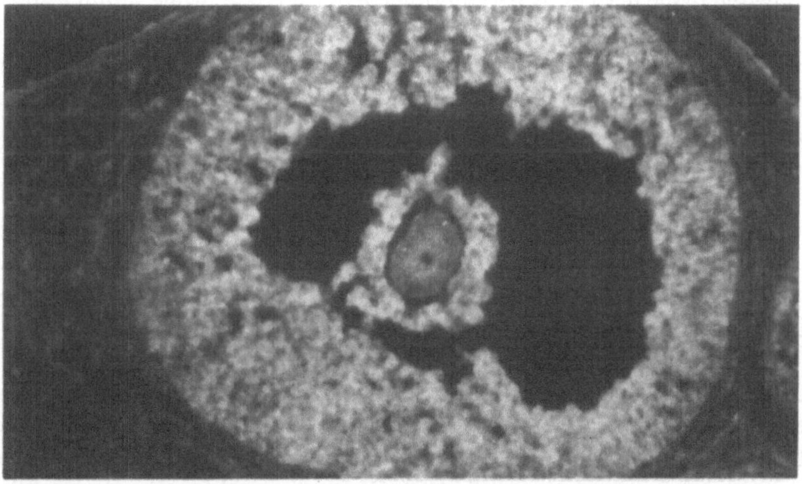

Figure 2. Fluorescence micrograph showing staining for 17HSD type 1 in a frozen section of rat ovary. Strong staining is seen in a growing Graafian follicle after treatment with pregnant mare serum gonadotrophin.

17HSD Type 1 in Human Endometrium and Breast Tissue

The morphological and physiological changes in the human endometrium during the menstrual cycle are primarily due to the cyclic variation in concentrations of estradiol and progesterone. The mitogenic activity of estradiol induces DNA synthesis, resulting in a sharp increase in cell proliferation around the tenth day of the cycle (12). In the endometrium, progesterone plays an essential role in differentiation; it antagonizes estradiol action by reducing estradiol receptor levels and by inducing 17HSD activity. Growth and differentiation of the breast are also mostly determined by estrogens and progesterone. Estrogens lead to the growth of the ductal and lobular systems, periductal and perilobular stroma, and fat tissue. The proper differentiation of these structures is regulated by the action of progesterone. There is, however, a distinct difference in hormone response between endometrial and breast epithelia, as the mitotic rate of the breast epithelium reaches its peak during the luteal phase following the serum progesterone surge (13).

Intracellular metabolism's role in estrogen action has been clearly established in human endometrium and breast tissue. The concentration ratio of estradiol/estrone in the endometrium differs significantly from that in plasma (14). The tissue concentration of estradiol peaks at mid-cycle and drops after ovulation without rising again during the luteal phase. Estrone concentration in endometrial tissue reaches its peak during the luteal phase, while serum estrone concentration peaks during the follicular phase (14). During the secretory phase, estradiol is partially metabolized to sulfate conjugates, but, in the endometrium, 17HSD is the major enzyme metabolizing estradiol to estrone (15). Analogous findings have been reported in breast cancer samples with high tissue concentrations of estradiol after menopause. This increase leads to a significant concentration gradient (10- to 100-fold) of estradiol between plasma and breast cancer tissues (16). Since both tissues express estrogen-metabolizing enzymes, it has been suggested that estrogen biosynthesis and metabolism in peripheral tissues strongly affect their biological activity, which is not truly reflected by their plasma concentrations.

We have studied the expression of 17HSD type 1 in normal and malignant human endometrium and breast tissue using immunological methods (Figure 3). In the endometrium, 17HSD type 1 was expressed in the gland and surface epithelial cells only when serum progesterone was present at concentrations characteristic of an ovulatory cycle (17). During the proliferative phase of the menstrual cycle, 17HSD type 1 staining was absent; however, it was maximal during the mid-secretory phase, in concert with an increased serum progesterone concentration. In tissue sections, the staining for 17HSD type 1 was not seen indiscriminately in all glands; however, all the cells of responsive glands were stained (17). Other 17HSD isozymes are apparently also expressed in the human endometrium. This is evident since 17HSD's major role in the endometrium is oxidation of estradiol to estrone (15), an activity typical of the type 2 enzyme. At present, however, no direct evidence for the expression of 17HSD type 2 at the mRNA or protein level is available.

It has been suggested that several forms of the 17HSD enzymes are expressed in breast tissue. We have shown one of the isoenzymes of 17HSD type 1 in breast cancer tissue, as well as in breast cancer cell lines (18).

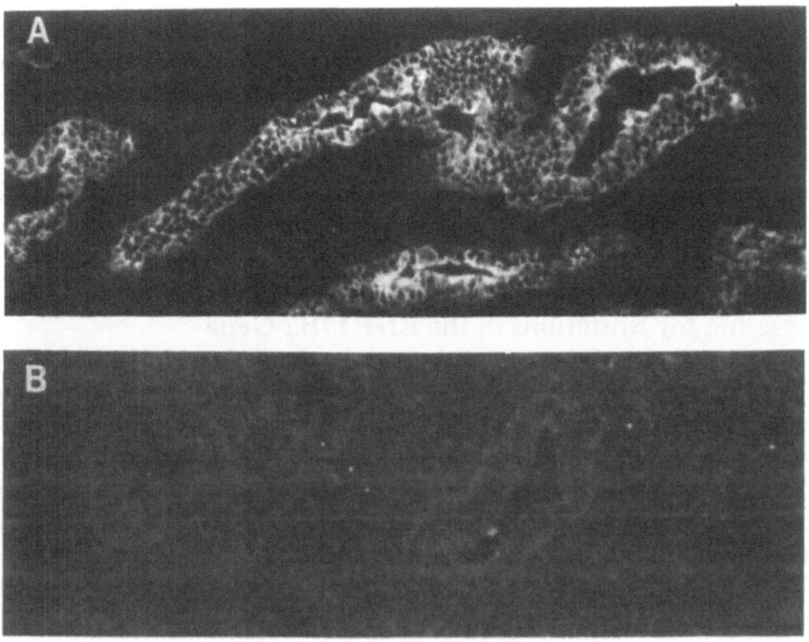

Figure 3. Micrographs showing the effect of mifepristone on the fluorescence staining of 17HSD type 1 in normal human endometrium during the luteal phase of the menstrual cycle. (A) Intensive immunostaining for 17HSD type 1 during the control cycle. (B) Negative staining during the luteal phase on day LH +6 after the administration of 200 mg mifepristone on day LH +2.

In the cell line T47D, 80% of the total enzyme activity was recovered in the cytosol fraction, suggesting that the major activity represents type 1 isoenzyme (19). Immunohistochemistry was used to clarify further the cellular distribution of 17HSD, and the enzyme was localized in the cytoplasm of epithelial cells of benign breast lesions and in breast cancer tissue. Detectable 17HSD type 1 was found in about 50% of the tumors; the immunofluorometric concentration of the enzyme was highly variable in both tumor specimens, as well as in the different breast cancer cell lines (18). Similarly, about 50% of the endometrial carcinoma specimens were 17HSD type 1-positive (20).

The concentration of 17HSD type 1 in the normal endometrium varied with plasma progesterone concentrations, and increased expression of 17HSD type 1 was seen in progesterone receptor-positive breast cancer specimens (18). In addition, an antiprogestin, mifepristone, blocked progestin-induced expression of 17HSD type 1 in the endometrium, as well as in the breast cancer cell line T47D in culture (19, 21). All these data support the notion of the possible progesterone receptor-mediated induction of the enzyme. Other factors may also regulate the expression of 17HSD type 1 in malignant tissues. About 40% of the 17HSD-positive endometrial adenocarcinoma specimens were estrogen and progesterone receptor-negative. In addition, a negative correlation was observed between serum progesterone and 17HSD type 1-positive endometrial adenocarcinoma specimens (20). Most of the endometrial carcinoma patients were postmenopausal and had low serum progesterone concentrations. Therefore, in these tumors the positive expression of 17HSD type 1 is regulated by factors other than progesterone.

Screening for Mutations in the EDH17B2 Gene

The gene encoding 17HSD type 1 (EDH17B2 gene) has been identified in the border region of loci 17q12-q21 (22). Because of its location, as well as the involvement of 17HSD type 1 in the regulation of estrogen action, the EDH17B2 gene may be a plausible candidate for the breast-ovarian cancer susceptibility gene BRCA1, mapped to 17q21 (23). To test the possible linkage between BRCA1 and EDH17B2, we screened for mutations of the coding regions of the EDH17B2 and of the region expected to contain the promoter for the transcription of the 1.3-kb 17HSD type 1 mRNA.

Altogether, single-strand conformation polymorphism screening and sequencing revealed three frequent (named vI, vIII, and vVI) and three rare (vII, vIV, and vV) nucleotide variations (Figure 4) in the specimens from both breast cancer patients and control individuals (24, 25). An A to G transition at exon six (vVI) was the only altered sequence detected in the coding regions of the EDH17B2 gene. This frequent nucleotide change led to an amino acid substitution of serine to glycine at position 312 in the 17HSD type 1 protein. The substitution did not, however, alter the activity of the protein, demonstrated with [312(Gly)]-mutated recombinant protein (Puranen, et al., personal communication).

The mutation vIV, an alteration from A to C, was located in the putative TATA-box of the EDH17B2 gene, 27 bp upstream from the transcription start point of the 1.3-kb 17HSD type 1 mRNA (25). Interestingly, the mutation vIV changed the putative TATA-box sequence of the EDH17B2 from ATATCAA to ATATCCA, which is also the analogous sequence in the EDH17B1, suggested to be a pseudogene or to code a protein other than 17HSD type 1 (26, 27).

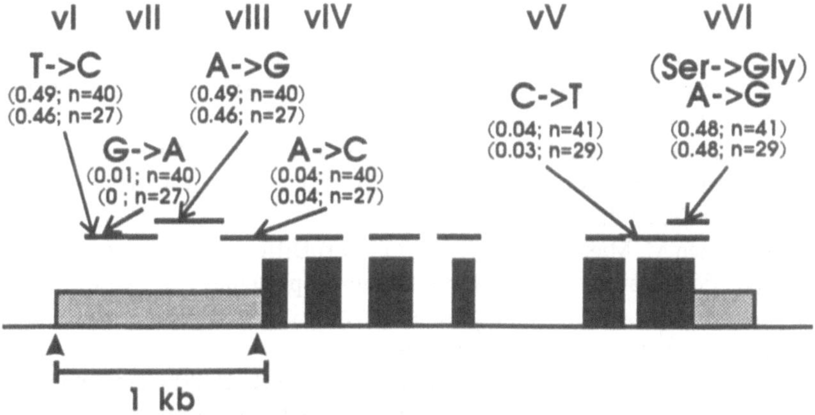

Figure 4. Schematic exon-intron structures of the EDH17B2 gene and PCR-amplified regions investigated by SSCP analyses. Coding and noncoding regions are, respectively, depicted as black and hatched boxes, and the lines above the boxes represent the amplified regions. The arrows indicate the location of the nucleotide variants vI-vVI. The frequency of each transition or transversion, and the number of patients (upper line) and control individuals (lower line) are presented below the nucleotide change. The two transcription start points are indicated by arrowheads.

Due to the position of the mutation vIV, its effect was characterized by reporter gene assays using fragments from the promoter region of the wild-type and mutated EDH17B2 genes and the analogous region from the EDH17B1 gene. The results indicated that the fragments derived from intact EDH17B2 had a promoter activity greater by an average of 80% in Cos-m6 and choriocarcinoma JAR cells than those constructed from the EDH17B1 or mutated forms of EDH17B2.

Neither mutations in the EDH17B2 gene enriched in patients affected with familial breast cancer nor functional alterations apart from the mutation vIV were detected. Thus, the data do not support the identification of EDH17B2 as the BRCA1 gene.

References

1. Isomaa VV, Ghersevich SA, Mäentausta OK, et al. (1993) Steroid biosynthetic enzymes: 17β-hydroxysteroid dehydrogenase. Ann Med 25:91-97.

2. Wu L, Einstein M, Geissler WM, et al. (1993) Expression cloning and characterization of human 17β-hydroxysteroid dehydrogenase type 2, a microsomal enzyme possessing 20α-hydroxysteroid dehydrogenase activity. J Biol Chem 268:12964-12969.
3. Geissler WM, Davis DL, Wu L, et al. (1994) Male pseudohermaphroditism caused by mutations of testicular 17β-hydroxysteroid dehydrogenase 3. Nature Genet 7:34-39.
4. Poutanen M, Miettinen M, Vihko R (1993) Differential estrogen substrate specificities of transiently expressed human placental 17β-hydroxysteroid dehydrogenase and an endogenous enzyme expressed in cultured COS-m6 cells. Endocrinology 133:2639-2644.
5. Peltoketo H, Isomaa V, Mäentausta O, Vihko R (1988) Complete amino acid sequence of human placental 17β-hydroxysteroid dehydrogenase deduced from a DNA. FEBS Lett 239:73-77.
6. Ghersevich S, Poutanen M, Martikainen, et al. (1994) Expression of 17β-hydroxysteroid dehydrogenase in human granulosa cells: Correlation with follicular size, cytochrome P-450 aromatase activity and estradiol production. J Endocrinol, in press.
7. Ghersevich S, Poutanen M, Rajaniemi H, et al. (1994) Expression of 17β-hydroxysteroid dehydrogenase in the rat ovary during follicular development and luteinization induced with pregnant mare serum gonadotrophin and human chorionic gonadotrophin. J Endocrinol 140:409-417.
8. Sawetawan C, Milewich L, Word RA, et al. (1994) Compartmentalization of type I 17β-hydroxysteroid oxidoreductase in the human ovary. Mol Cell Endocrinol 99:161-168.
9. Mäentausta O, Peltoketo H, Isomaa V, et al. (1990) Immunological measurement of human 17β-hydroxysteroid dehydrogenase. J Steroid Biochem 36:673-680.
10. Jantus-Lewintre E, Orava M, Peltoketo H, et al. (1994) Characterization of 17β-hydroxysteroid dehydrogenase type 1 in choriocarcinoma cells: Regulation by basic fibroblast growth factor. Mol Cell Endocrinol, in press.
11. Tremblay Y, Beaudoin C (1993) Regulation of 3β-hydroxysteroid dehydrogenase and 17β-hydroxysteroid dehydrogenase messenger ribonucleic acid levels by cyclic adenosine 3',5'-monophosphate and phorbol myristate acetate in human choriocarcinoma cells. Mol Endocrinol 7:355-364.
12. Ferenczy A, Bertrand G, Gelfand MM (1979) Proliferation kinetics of human endometrium during the normal menstrual cycle. Am J Obstet Gynecol 133:859-867.

13. Going JJ, Anderson TJ, Battersby S, et al. (1988) Proliferative and secretory activity in human breast during natural and artificial menstrual cycles. Am J Pathol 130:193-203.
14. Pollow K, Schmidt-Gollwitzer M, Pollow B (1980) Progesterone- and estradiol-binding proteins from normal human endometrium and endometrial carcinoma: A comparative study. In Wittliff JL, Dapunt O (eds): Steroid Receptors and Hormone Dependent Neoplasia. New York: Masson Publishing USA, pp 69-94.
15. Tseng L (1980) Hormonal regulation of steroid metabolic enzymes in human endometrium. In Thomas RL, Singhal JA (eds): Advances in Sex Steroid Hormone Research. Baltimore: Urban & Schwarzenbirg, pp 329-361.
16. Vermeulen A, Deslypere JP, Paridaens R, et al. (1986) Aromatase, 17β-hydroxysteroid dehydrogenase and intratissular sex hormone concentrations in cancerous and normal glandular breast tissue in postmenopausal women. Eur J Cancer Clin Oncol 22:515-52.
17. Mäentausta O, Sormunen R, Isomaa V, et al. (1991) Immunohistochemical localization of 17β-hydroxysteroid dehydrogenase in the human endometrium during the menstrual cycle. Lab Invest 65:582-587.
18. Poutanen M, Isomaa V, Lehto V-P, et al. (1992) Immunological analysis of 17β-hydroxysteroid dehydrogenase in benign and malignant human breast tissue. Int J Cancer 50:386-390.
19. Poutanen M, Isomaa V, Kainulainen K, et al. (1990) Progestin induction of 17β-hydroxysteroid dehydrogenase enzyme protein in the T-47D human breast-cancer cell line. Int J Cancer 46:897-901.
20. Mäentausta O, Boman K, Isomaa V, et al. (1992) Immunohistochemical study of the human 17β-hydroxysteroid dehydrogenase and steroid receptors in endometrial carcinoma. Cancer 70:1551-1555.
21. Mäentausta O, Svalander P, Gemzell Danielsson K, et al. (1993) The effects of an antiprogestin, mifepristone, and an antiestrogen, tamoxifen, on endometrial 17β-hydroxysteroid dehydrogenase, and on progestin and estrogen receptors during the luteal phase of the menstrual cycle: An immunohistochemical study. J Clin Endocrinol Metab 77:913-918.
22. Winqvist R, Peltoketo H, Isomaa V, et al. (1990) The gene for 17β-hydroxysteroid dehydrogenase maps to human chromosome 17, bands q12-q21, and shows an RFLP with ScaI. Hum Genet 85:473-476.
23. Hall JM, Lee MK, Newman B, et al. (1990) Linkage of early-onset familial breast cancer to chromosome 17q21. Science 250:1684-1689.
24. Mannermaa A, Peltoketo H, Winqvist R, et al. (1994) Human familial and sporadic breast cancer: Analysis of the coding regions of the 17β-hydroxysteroid dehydrogenase 2 gene (EDH172) using a single-strand conformation polymorphism-assay. Hum Genet 93:319-324.

25. Peltoketo H, Piao Y, Mannermaa A, et al. (1994) A point mutation in the putative TATA-box, detected in non-diseased individuals and patients with hereditary breast cancer, decreases promoter activity of the 17β-hydroxysteroid dehydrogenase gene 2 (EDH17B2) *in vitro*. Genomics, in press.
26. Luu-The V, Labrie C, Simard J, et al. (1990) Structure of two in tandem human 17β-hydroxysteroid dehydrogenase genes. Mol Endocrinol 4:268-275.
27. Peltoketo H, Isomaa V, Vihko R (1992) Genomic organization and DNA sequences of human 17β-hydroxysteroid dehydrogenase genes and flanking regions. Localization of multiple Alu sequences and putative *cis*-acting elements. Eur J Biochem 209:459-466.

13

Growth Factors in Endometrial Cancer

Naoki Terakawa

Introduction

Progressive tumor growth has been associated with neovascularization induced by an angiogenic factor(s) that the tumor secretes (1). Therefore, inhibition of neovascularization could be effective in suppressing tumor growth (2). The antiproliferative effects of progestins in human endometrial cancer cells have been demonstrated by several investigators. Previously, we reported that the growth of human endometrial adenocarcinoma cells in primary culture was significantly suppressed by medroxyprogesterone acetate (MPA) (3); a similar growth-inhibitory effect has been reported in a nude mouse system (4). We have suggested that the antiproliferative effect of progestin on adenocarcinoma cells is mediated through progestin receptors (PR) present in the cells (5). In contrast, using the rabbit cornea assay, Gross et al. have shown that MPA inhibits the angiogenesis induced by several tumors of laboratory animals (6). Recent studies have shown that human endometrial adenocarcinoma cells produce angiogenic factors (7). Although MPA seems to act directly on endometrial cancer cells to inhibit their growth, the present data suggest that its inhibition of angiogenesis may share the same mechanism by which it inhibits the growth of endometrial adenocarcinoma. In the present study, to gain further information on MPA's mechanism of action in human endometrial cancer, we examined MPA's effect on angiogenesis induced by adenocarcinoma. We also investigated the synthesis and secretion of angiogenic growth factor(s) in a human endometrial cancer cell line.

Angiogenesis of Endometrial Adenocarcinoma

Tumor specimens were obtained from a patient with primary endometrial adenocarcinoma. Angiogenesis was determined by a rabbit cornea bioassay previously described (8). Human tumor fragments (1.5 mm^3 pieces) were transplanted into both corneas of a rabbit. One eye was implanted with an MPA pellet, and the other, without MPA, was used as control. Ten days after the transplant, angiogenesis was evaluated from the limbus. The tumors induced angiogenesis in 50/71 (70.4%) control corneas, but only in 17/79 (21.5%) of corneas implanted with MPA. This difference was statistically significant. These results indicate that MPA inhibited the angiogenesis induced by endometrial adenocarcinoma. We have suggested that in a cell culture system the antiproliferative effect of progestin on adenocarcinoma cells is mediated through PR present in the cells (5). Clinical studies have shown that endometrial adenocarcinomas positive to PR are likely to respond to progestin therapy (9). In the present study, however, MPA inhibited angiogenesis of both PR-positive and -negative tumors (Table 1), suggesting that the antiangiogenic effect observed was not mediated through PR.

Table 1. Effect of MPA on Angiogenesis Induced by PR-positive and PR-negative Endometrial Adenocarcinomas.

Progestin Receptors	Treatment[a]	No. Corneal Grafts[b]	No. Corneas w/ Angiogenesis[c]	% Corneas w/ Angiogenesis
Positive (n = 17)	Control	38	24	63.2
	MPA	44	10	22.7[d]
Negative (n = 11)	Control	30	23	76.7
	MPA	32	5	15.6[d]

[a] A 1-mm^3 pellet of ethylene vinyl acetate polymer with or without 200 μg of MPA was placed at the base of the corneal pocket; a 1.5-mm^3 piece of tumor was then introduced.

[b] Tumor fragments obtained from a patient were transplanted into six corneas of three rabbits. In each rabbit, one eye was implanted with MPA and the other was not.

[c] Ten days after the transplantation, angiogenesis was evaluated macroscopically from the limbus. The results were confirmed by reexamining the excised corneas under a dissection microscope.

[d] Mantel-Haenszel and Breslow-Day tests were used for statistical analysis. Significant difference from control, $p < 0.01$.

A statistically significant relationship has been demonstrated between tumor grading of endometrial adenocarcinomas and their response to MPA therapy (9). A previous *in-vitro* study showed that its cytocidal action is more pronounced in well-differentiated adenocarcinomas (10). In the present study, however, MPA had antiangiogenic effects on both well-differentiated (Grade 1) and poorly-differentiated (Grades 2 and 3) adenocarcinomas (Table 2). Furthermore, microscopic examination of tumor grafts showed that adenocarcinoma cells were viable in the presence and absence of MPA. Therefore, the antiangiogenic effect on adenocarcinomas was unlikely to be due to a cytocidal action of the compound.

Table 2. Effect of MPA on Angiogenesis Induced by Endometrial Adenocarcinomas of Various Histologic Grades

Histologic Grade	Treatment	No. Corneal Grafts	No. Corneas w/Angiogenesis	% Corneas w/ Angiogenesis
Grade 1 (n = 22)	Control	54	39	72.2
	MPA	60	12	20.2[a]
Grades 2 and 3 (n = 7)	Control	17	11	64.7
	MPA	19	5	26.3[a]

[a] Mantel-Haenszel and Breslow-Day tests were used for statistical analysis. Significant difference from control, $p < 0.01$.

Progestins and Neovascularization Induced by Angiogenic Factors

Several studies have shown that human endometrial adenocarcinoma cells produce angiogenic factors such as fibroblast growth factor (FGF) and transforming growth factors (TGFα and TGFβ) (7, 11-13). MPA also reduced the activity of these angiogenic factors *in vitro* in an extract of endometrial adenocarcinoma cells (12). These results suggest that MPA may prevent the progressive growth of endometrial adenocarcinoma by inhibition of angiogenic factor(s). Moreover, MPA inhibited angiogenesis and collagenolysis induced by an extract of a human hepatoma (6). Moreover, it also significantly suppressed the activity of a plasminogen activator produced by bovine endothelial cells *in vitro* (14). These findings suggest that MPA may also act on the host endothelial cells to inhibit angiogenesis.

Pellets containing the angiogenic factors, acidic FGF (aFGF) or TGFα, without tumor tissue, induced angiogenesis in all the corneas examined (Table 3). MPA almost completely inhibited the neovascularization induced by the angiogenic factors; however, progesterone had no effect. These findings support the conclusion that the angiogenic action of MPA is not mediated through PR. Previous studies have shown that the antiangiogenic effect induced by tumors may be the inactivation of the endothelial cells responsible for neovascularization (6, 14). In the current study, MPA inhibited neovascularization induced by the angiogenic factors aFGF and TGFα. Thus, these data and previous findings (6, 14) suggest that MPA may act on endothelial cells and inhibit angiogenesis in endometrial adenocarcinomas, although the possibility that it may reduce the production of other angiogenic factor(s) present in adenocarcinomas cannot be ruled out. The antiangiogenic effect of MPA shown herein may be another mechanism by which this compound inhibits the progressive growth of adenocarcinomas.

Table 3. Effects of Progestins on aFGF- or TGFα-induced Angiogenesis

Angiogenic Factor	Treatment	No. Corneal Transplants[a]	No. Corneas w/ Angiogenesis	% Corneas w/ Angiogenesis
aFGF	Control	16	16	100
	MPA	9	1	11.1[b]
	Progesterone	9	9	100
TGFα	Control	5	5	100
	MPA	5	0	0[b]

[a] A pellet containing 1 μg aFGF or 1μg TGFα instead of a piece of tumor was transplanted into the corneal pocket, with or without a pellet containing MPA or progesterone.

[b] The χ^2 test was used for statistical analysis. Significant difference from control, $p < 0.01$.

Autocrine Growth Mechanism by TGFβ$_1$ in Endometrial Cancer

In a variety of cells, TGFβ$_1$ consisting of two 12.5 kDa subunits is a potent modulator of growth. TGFβ was first isolated from human platelets, and five distinct cDNAs of TGFβ have currently been cloned, each of which is approximately 65-80% homologous to the others (15, 16). Considerable amounts of TGFβ and its mRNA have been detected in different cell types, especially transformed ones (15). TGFβ$_1$ can promote growth of mesenchymal cells, such as normal rat kidney and fibroblasts. In contrast, TGFβ has been reported to inhibit the growth of epithelial cells, including human endometrial cancer cells (17). In estrogen-dependent breast cancer cell lines, tamoxifen stimulated TGFβ$_1$ synthesis and inhibited cell growth (18). These data suggest that growth-inhibitory effects of tamoxifen may be mediated by a TGFβ$_1$ autocrine mechanism. However, a recent report failed to confirm the inhibitory effect of TGFβ$_1$ on the growth of endometrial cancer cells that expressed relatively large amounts of TGFβ$_1$ (19). Furthermore, it has been shown that TGFβ had a stimulatory effect on the growth of estrogen-responsive endometrial as well as breast cancer cells (20). We investigated the synthesis and secretion of TGFβ$_1$ in a progestin-sensitive human endometrial cancer cell line IK-90 (5), and evaluated the effects of TGFβ$_1$ on cell growth

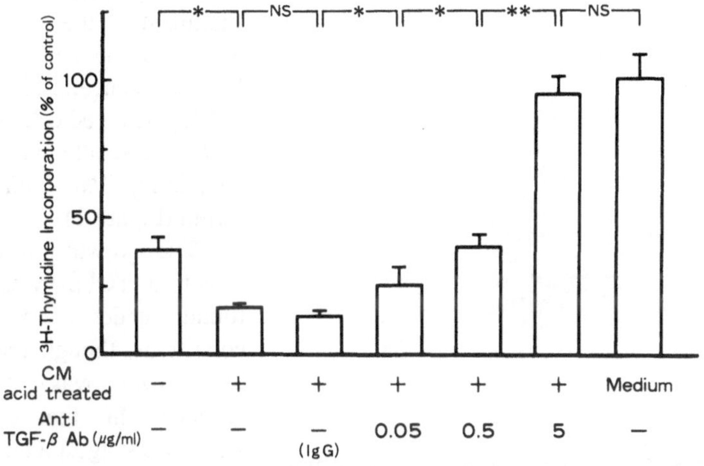

Figure 1. TGFβ$_1$ activity in conditioned medium of IK-90 cells. Results are expressed as % of the amount of [^{3}H]-thymidine incorporated into CCl64 cells incubated with the conditioned media (CM)/those incubated without the conditioned media. Values are expressed as the mean ± SEM. Unpaired t-test was used for statistical analysis. Significant differences from control medium, *$p < 0.05$, **$p < 0.01$.

TGFβ$_1$ activity in conditioned medium from IK-90 cells was examined by the growth-inhibition assay using CCl64 mink-lung epithelial cells (21). TGFβ$_1$ activity was 156 ± 19 pg/ml (value interpolated from standard curves prepared with authentic TGFβ$_1$ and neutralized with anti-TGFβ IgG). TGFβ was predominantly produced as a latent form that could be activated by acid or heat treatment. Acid-treated conditioned medium showed more profound inhibition of DNA synthesis in CCl64 cells, and the level TGFβ$_1$ activity quantified was more than 1.5 ng/ml (Figure 1).

This activity was also neutralized by rabbit anti-TGFβ$_1$-IgG in a concentration-dependent manner, but not by non-immune IgG. Northern blot analysis consistently revealed that the mRNA species hybridized with [^{32}P]-labeled TGFβ$_1$ cDNA were 2.5 and 4.0 kb in size (Figure 2).

Thus, it was demonstrated that IK-90 cells produced and secreted abundant amounts of bioactive TGFβ$_1$ protein (22). IK-90 cells expressed a high-affinity TGFβ$_1$ receptor (Kd; 74 pm, 2,000 sites/cell). Cross-linking studies showed that the cells had two specific TGFβ$_1$ receptors with molecular weights of 280 kDa and 65 kDa, which were consistent with two of the three types of TGFβ$_1$ receptor described previously (17). Finally, we examined the effects of TGFβ$_1$ on cell growth. Various concentrations of TGFβ$_1$ increased cell growth dose-dependently in medium containing 1% fetal calf serum (Figure 3).

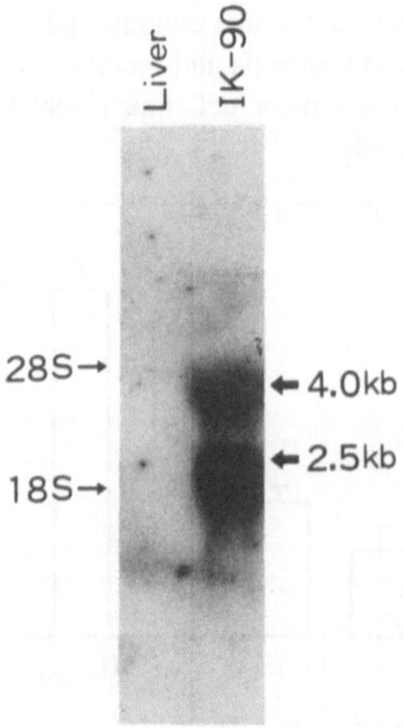

Figure 2. Northern blot analysis showing TGFβ$_1$ mRNA in 20 μg of total RNA from IK-90 cells and BALB/c mouse liver. Total RNA samples from mouse liver were used as a negative control.

The growth-promoting effect of TGFβ$_1$ was also found under serum-free conditions, though the rate of growth was slightly reduced. In the current study, we suggest a possible autocrine growth mechanism by TGFβ$_1$ in IK-90 cells. Our observations and previous conflicting results (17, 19, 20) suggest that the sensitivity of endometrial cancer cell lines to exogenous TGFβ$_1$ is variable and is inversely correlated to the level of TGFβ$_1$ production by cancer cells.

Conclusions

Using the rabbit cornea assay, MPA inhibited the angiogenesis induced by both PR-positive and PR-negative endometrial adenocarcinomas, and by both well- and poorly-differentiated adenocarcinomas. The current study, therefore, suggests that the antitumor effect of MPA on adenocarcinoma cells, which plays an important role in the treatment of endometrial cancer, may act in concert with its inhibitory action on angiogenesis. We have also shown that human endometrial adenocarcinoma cells produce angiogenic factors such as FGF, TGFα, and TGFβ (7, 11-13), and that MPA reduced the activity of these angiogenic factors in an extract of adenocarcinoma cells *in vitro* (12). Moreover, we have shown a growth-promoting effect of $TGF\beta_1$ in a progestin-sensitive human endometrial cancer cell line, IK-90, which produces and secretes abundant amounts of $TGF\beta_1$. Autocrine regulation of cancer cell growth by $TGF\beta_1$ is a complex process, and further research is required to elucidate the detailed autocrine mechanism by which $TGF\beta_1$ affects endometrial cancer growth. In progestin-sensitive Ishikawa cells, the parent cell line of IK-90 cells, TGFα mRNA was detected. A significant decrease in its mRNA level was shown after exposure to progestin (13). Moreover, a growth-promoting effect of TGFα was demonstrated in these cells.

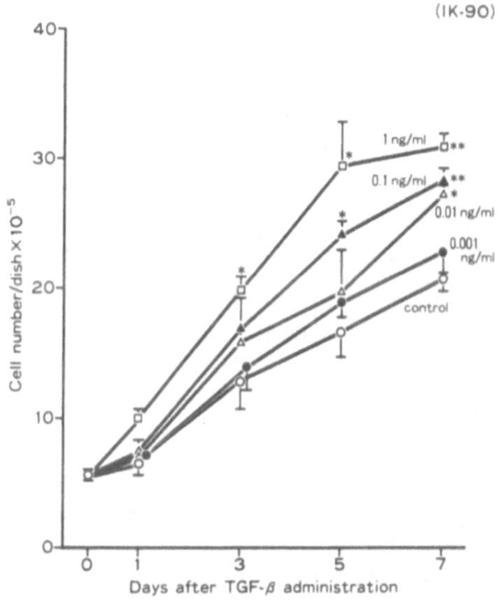

Figure 3. Effects of $TGF\beta_1$ on IK-90 cell growth in culture. Cells ($5\ 10^5$) were plated in culture dishes. Two days after plating (day 0), the medium was replaced by medium containing 1% dextran-coated charcoal-treated fetal calf serum without (control) or with indicated concentrations (0.001-1 ng/ml) of $TGF\beta_1$. Media were changed every 48 h. Cells were removed at the times indicated, and counted. Significant differences from control cultures, *$p < 0.05$, **$p < 0.01$.

It is possible that TGFα may also be an autocrine modulator in the growth regulation of IK-90 cells.

References

1. Folkman J, Klagsbrun M (1987) Angiogenic factors. Science 235:442-447.
2. Folkman J (1990) What is the evidence that tumors are angiogenesis dependent? J Natl Cancer Inst 82:4-6.
3. Terakawa N, Ikegami H, Shimizu I et al (1988) Inhibitory effects of danazol and medroxyprogesterone acetate on [^{3}H]thymidine incorporation in human endometrial cancer cells. J Steroid Biochem 31:131-135.
4. Zaino RJ, Satyaswaroop PG, Mortel R (1985) Hormonal therapy of human endometrial adenocarcinoma in a nude mouse model. Cancer Res 45:539-541.
5. Terakawa N, Hayashida M, Shimizu I (1987) Growth inhibition by progestins in a human endometrial cancer cell line with estrogen-independent progesterone receptors. Cancer Res 47:1918-1923.
6. Gross J, Azizkhan RG, Biswas C (1981) Inhibition of tumor growth, vascularization, and collagenolysis in the rabbit cornea by medroxyprogesterone. Proc Natl Acad Sci USA 78:1176-1180.
7. Ishiwata I, Ishiwata C, Soma M (1988) Tumor angiogenic activity of gynecologic tumor cell lines on chorioallantoic membrane. Gynecol Oncol 29:87-93.
8. Jikihara H, Terada N, Yamamoto R (1992) Inhibitory effect of medroxyprogesterone acetate on angiogenesis induced by human endometrial cancer. Am J Obstet Gynecol 167:207-211.
9. Kauppila AJI, Isotalo HE, Kivinen ST, Vihko RK (1986) Prediction of clinical outcome with estrogen and progestin receptor concentrations and their relationships to clinical and histopathological variables in endometrial cancer. Cancer Res 46:5380-5384.
10. Grönroos M, Mäenpää J, Kangas R (1987) Steroid receptors and response of endometrial cancer to hormones *in vitro*. Ann Chir Gynecol 76:76-79.
11. Presta M, Maier JAM, Rusnati (1988) Modulation of plasminogen activator activity in human endometrial adenocarcinoma cells by basic fibroblast growth factor and transforming growth factor-β. Cancer Res 48:6384-6389.
12. Fujimoto J, Fujita H, Hosoda S (1989) Effect of medroxy-progesterone acetate on secondary spreading of endometrial cancer. Invasion Metastasis 9:209-215.
13. Gong Y, Anzai Y, Murphy LC (1991) Transforming growth factor gene expression in human endometrial adenocarcinoma cells: Regulation by progestins. Cancer Res 51:5476-5481.

14. Ashino-Fuse H, Takano Y, Oikawa T (1989) Medroxy-progesterone acetate, an anti-cancer and anti-angiogenic steroid, inhibits the plasminogen activator in bovine endothelial cells. Int J Cancer 44:859-864.
15. Derynck R, Jarrett JA, Chen EY (1985) Human transforming growth factor-β complementary DNA sequence and expression in normal and transformed cells. Nature 316:701-705.
16. Kondaiah P, Sands MJ, Smith JM (1990) Identification of a novel transforming growth factor-β (TGF-β5) mRNA in Xenopus laevis. J Biol Chem 265:1089-1093.
17. Murray K, Haussler CA, Trookman NS (1987) Divergent effects of epidermal growth factor and transforming growth factors on a human endometrial carcinoma cell line. Cancer Res 47:4909-4914.
18. Knabbe C, Lippman ME, Wakefield LM (1987) Evidence that transforming growth factor β is a hormonally regulated negative growth factor in human breast cancer cells. Cell 48:417-428.
19. Boyd JA, Kaufman DG (1990) Expression of transforming growth factor β_1 by human endometrial carcinoma cell lines: Inverse correlation with effects on growth rate and morphology. Cancer Res 50:3394-3399.
20. Croxtall JD, Jamil A, Ayub M (1992) TGF-β stimulation of endometrial and breast-cancer cell growth. Int J Cancer 50:822-827.
21. Jikihara H, Ikegami H, Sakata M (1991) Epidermal growth factor attenuates cell proliferation by down-regulating transforming growth factor-β receptor in the osteoblastic cell line MC3T3-E1. Bone Mineral 15:125-136.
22. Sakata N, Kurachi H, Ikegami H (1993) Autocrine growth mechanism by transforming growth factor (TGF)-β_1 and TGF-β_1 receptor regulation by epidermal growth factor in a human endometrial cancer cell line IK-90. Int J Cancer 54:862-867.

PART 4. KIDNEY

Introduction

Estrogens and Genitourinary Cancers of Syrian Hamsters

James S. Norris

Although estrogens have been implicated in tumor development for over fifty years, we are still struggling to understand their mechanisms of action in this regard. Investigations in diverse areas including receptor activity, mutagenesis, steroid metabolism, cell cycle control and genetics have been carried out, leading to hypotheses including estrogen-induced mitogenesis, epigenetic control of differentiation, and direct chemical interaction at genomic and/or protein loci with resultant genomic instability. However, no unifying hypothesis has been sustained. Thus, in spite of model systems and an exponentially increasing core of basic knowledge, we still await the "leap" whereby mechanistic insights will solve the fundamental questions of estrogen-induced carcinogenesis. In the following, I will present a brief review of where we stand today.

Syrian Hamster Kidney Model

A 1950 report documented the induction of renal tumors by prolonged administration of either 17β-estradiol (E_2) or diethylstilbestrol (DES) to male animals (1). Subsequent studies analyzing induced tumors have created a controversy as to their cell of origin, reported to be embryonal rests (2, 3), primitive interstitial cells (3), tubular epithelium (4), stromal elements (5), or arterial smooth muscle (6). This complicated classification issue may have occurred because at least two cell types are histologically and histochemically identifiable in most tumors.

Induction of tumors in this model is complicated. Both intact and castrated males are 100% susceptible to induction within 9-10 months using a variety of estrogens (1). However, ethinylestradiol (EE), a potent estrogen, has little to no induction capacity (0-10%), although it will induce a "florid dysplasia" (4). Curiously, administration of EE with DES blocks tumor formation (3), as does androgen or progesterone.

The situation is more complicated in female hamsters. Generally, no tumors are observed in females treated with DES or E_2, although dysplastic areas are occasionally observed. It was initially considered that the ovary might contribute to this resistance, but ovariectomy does not significantly enhance the rate of tumor formation. Neither hypophysectomy nor hysterectomy significantly alters this result (JJ Li, personal communication).

At the genetic level, the only known difference between males and females is the number of copies of the X chromosome. How this may affect susceptibility is unknown. The lack of spontaneous kidney tumors in both sexes suggests that a mutational event must occur as a result of steroid treatment. Two copies of the X chromosome seem to ameliorate the impact of such an event. Because the tumors have the characteristics of nephroblastoma, the possibility of a Wilms tumor-like gene defect may be relevant. There is no evidence that imprinting plays a role in susceptibility.

Receptor Biology

Clearly, the biological activities of estrogens are primarily manifested through the estrogen receptor. This receptor belongs to a large family of transcription factors; thus, one can predict that transcriptional control plays a role in estrogen carcinogenesis. Which genes in the carcinogenesis pathway are controlled by estrogen, and in what cell type, are legitimate, but unanswered, questions. Protooncogenes have been demonstrated to be estrogen-regulated, but despite speculation, no direct linkage to the estrogen receptor has been established (7). These transcription factor pathways are clearly very complex, with both negative and positive regulatory proteins balancing cellular outcomes, for example, the redox cycling control of the AP1 transcription factor (8). Redox cycling has been implicated in estrogen carcinogenesis (9), but not necessarily in this context (see below). Another equally important question involves the cascade of signals evoked by estrogen-receptor activation of peptide hormones and growth factors. Specifically, the control of induction of endocrine, paracrine, and autocrine peptides and how they signal control of cellular growth and death is poorly understood. These peptide signals must play a major role in controlling promotion and progression following the still undefined initiation events.

Cytogenetic Analysis of Hormone-Induced Tumors

Cytogenetic abnormalities have been known to accompany neoplastic transformation. Abnormalities include aneuploidy, translocations, formation of double minutes, gene amplification events, and various other effects. Frequently associated with cytogenetic changes are either mutations (*ras*, p53) or altered expression of proto-oncogenes (*jun, myc, fos*). This subject is now being defined expression of proto-oncogenes (*jun, myc, fos*). This subject is now being defined in terms of genomic instability measurable at both the cytogenetic and molecular levels.

Chromosomal changes in Syrian hamster embryo cells exposed to carcinogens have been noted by Oshimura, et al. (10), who demonstrated that nonrandom loss of chromosomes occurred during v-*ras*/v-*myc*-induced cellular transformation. In rat fibroblasts, van den Berg, et al. (11), demonstrated that overexpression of *fos* could induce genomic instability, as monitored cytogenetically. Baseline levels of *jun* expression in hamster kidney tumors were shown to be persistently elevated long after estrogen stimulation (7). Since *jun* is a member of AP1 (as is *fos*), it is possible to speculate that a sequela of this activity is the propensity to genomic instability during altered expression of these oncogenes.

The mechanisms for inducing aneuploidy are not clear, but studies by Metzler's group suggest a mechanism of microtubular disruption by covalent interaction of steroid metabolites with cysteinyl residues on tubulin (12). Since aneuploidy is frequently observed in estrogen-treated hamster proximal tubule cells or in estrogen-induced hamster kidney tumors (13), disruption of microtubular function by estrogens is a plausible mechanism to explain aneuploidy and correlates well with metabolic activation of these steroids.

Metabolic Activation of Estrogens and Carcinogenesis

In some models, estrogens are both initiators and promoters of carcinogenesis as, for example, in the Syrian hamster kidney model. Metabolic activation of estrogens leading to formation of DNA and protein adducts has been reviewed (14, 15). Estrogen metabolites have been analyzed following both microsomal and *in-vivo* activation, and the Syrian hamster kidney is clearly exposed to these metabolites from both endogenous activities and as an end organ clearing such molecules from the circulation. It is reasonable to expect estrogen metabolites to act as carcinogens, but no specific DNA targets have been identified. Conversely, adducts have been described on tubulin whose function can be implicated in the carcinogenic process (induction of aneuploidy). Nevertheless, estrogen metabolites do seem to interact with DNA, and efforts by Roy, et al. (16), have demonstrated that redox cycling of diethylstilbestrol will hydroxylate DNA on quinine residues in a kidney-specific manner if hamsters are exposed to the steroids for 3 months. Demonstrations of other types of DNA adducts presumably associated with catechol estrogens have been published (17, 18). In no case to date has the adduct been chemically identified or its genetic location identified. Thus, although estrogen seems to act as an initiator of carcinogenesis in the hamster model, relevant DNA molecular targets remain to be identified.

Genomic Instability Studied With Retrotransposons

As discussed above, genomic instability is the hallmark of cancer cells, but is poorly understood mechanistically. My laboratory has been interested in this problem and, as described below, we have instituted studies in transgenic animals to quantify events measuring genomic instability.

To understand our approach, one must first realize that 5-10% of primate and rodent genomes are composed of transposable elements, including retroviruses, LINES, SINES, and others (19). It has been further estimated that 0.04% of the genome are retroviruses (20). If one estimates that each retrovirus is 8 Kb long (probably the maximum), then there are at least 15,000 copies of full-length retrovirus in the genome, equivalent to one every 2×10^5 bases. One class of these viruses, intracisternal A particles (IAP), consist of defective endogenous retroviral elements, of which at least 1,000 copies exist in the mouse and hamster genomes (21). IAPs are classified into groups of truncated 3.5-Kb and 7.0-Kb sequences. Reverse transcriptase activity has been purified from these particles, but there is no evidence for an extracellular infectious phase. There is clear evidence for expression of IAPs during embryogenesis (22), in specific tissues (23), and in cancers (24). The extent of IAP retrotransposition within cells remains unknown. However, there are instances in which IAP sequences have been found inserted near specific genes, i.e., c-*mos*, IL3, Hox 2.4, and others (25). From those studies, the Heidmanns proposed a way to monitor transposition of IAPs by tagging them with a neomycin resistance gene that required splicing for activity. Their studies unequivocally demonstrated retrotransposition.

IAPs appear to have enhanced expression in tumor cells. We submit this increased expression is likely a harbinger of genomic instability, since increased transcription of retroelements is frequently associated with neoplastic progression. However, the IAP genome, due to its high copy number, is difficult to monitor in this respect. To circumvent this problem, we have utilized a corrected murine IAP genome [MIA60 is the corrected version of MIA14, (26)] and inserted a modified neomycin resistance gene replacing part of the envelope region. Our marker gene is a neomycin resistance gene (neo) with a hamster intron inserted in the translated region of the phosphotransferase (27). This construct results in a neo gene that no longer confers resistance to G418 (a neomycin analogue) in prokaryotes (no RNA splicing mechanism), but still functions in eukaryotes. When introduced into an animal as a transgene, the unspliced version incorporates into the genome. Subsequently during retrotransposition, splicing of the intron occurs. The reconstituted (i.e., spliced) neo gene regains function in prokaryotes; thus, any new integration of the spliced virus can be cloned by direct selection on G418 as either a cosmid or a plasmid. This will allow us to analyze the DNA surrounding the new integration site and also provide a means for statistical analysis of transposition frequency. *In-situ* PCR should define locations of retrotranspositional events. If successful, generating transgenic Syrian hamsters, it will allow introduction of a series of these transgenes into hamsters, and test their mobility during hormone-induced carcinogenesis in both the kidney and vas deferens models.

We believe this will give us several levels of knowledge regarding degree of transposition as a function of genomic instability and access to potentially important growth control genes possibly interrupted by viral activation or inactivation as a function of insertion.

Summary

As stated before, much has been done, but little is understood about the mechanism of estrogen-induced kidney tumors in Syrian hamsters. We now believe estrogens are initiators and promoters in this model. Their actions are at least partially through receptor-mediated mechanisms, as well as by direct chemical activity on DNA and protein targets that remain largely unidentified. Aneuploidy and chromosome gains or losses are integral in the process and may occur via perturbation of such transcription factors as *myc* and AP1, which acts to promote genomic instability by unknown mechanisms. Even the target cells on which estrogens act as carcinogens are still debated. In today's series of lectures, we will hear updates on several of these issues. Perhaps new insights will be forthcoming as we digest this new information.

Acknowledgments

I would like to thank Dr. Lea Sekely for editorial comments, and Janis Davis and Anne Donaldson for typing and formatting. This work has been supported by NIH research grants CA52085, CA49949, RR09885, and CA58030.

References

1. Kirkman H, Bacon RL (1950) Malignant renal tumors in male hamsters (Cricetus auratus) treated with estrogens. Cancer Res 10:122-124.
2. Llombart-Bosch A, Peydro A (1975) Morphological, histochemical and ultrastructural observations of diethylstilbestrol-induced kidney tumors in the Syrian golden hamster. Eur J Cancer 11:403-412.
3. Oberley TD, Gonzalez A, Lauchner LJ, et al. (1991) Characterization of early kidney lesions in estrogen-induced tumors in the Syrian hamster. Cancer Res 51:1922-1929.

4. Goldfarb S, Pugh TD (1990) Morphology and anatomic localization of renal microneoplasms and proximal tubule dysplasia induced by four different estrogens in the hamster. Cancer Res 50:113-119.
5. Hacker HJ, Vollmer G, Chiquet-Ehrismann R, et al. (1991) Changes in the cellular phenotype and extracellular matrix during progression of estrogen-induced mesenchymal kidney tumor in Syrian hamsters. Virchows Arch B Cell Pathol 60:213-223.
6. Hacker HJ, Bannasch P, Liehr J (1988) Histochemical analysis of the development of estradiol-induced kidney tumors in male Syrian hamsters. Cancer Res 48:971-976.
7. Liehr JG, Chiappetta C, Roy D, Stancel GM (1992) Elevation of protooncogene messenger RNA in estrogen-induced kidney tumors in the hamster. Carcinogenesis 13:601-604.
8. Abate C, Patel L, Rauscher FJ III, Curran T (1990) Redox regulation of *fos* and *jun* DNA-binding activity *in vitro*. Science 249:1157-1161.
9. Liehr JG, Roy D (1990) Free radical generation by redox cycling of estrogens. Free Radical Biol Med 8:415-423.
10. Oshimura M, Gilmer TM, Barrett JC (1985) Nonrandom loss of chromosome 15 in Syrian hamster tumors induced by V-Ha-*ras* plus v-*myc* oncogenes. Nature 316:636-639.
11. Van den Berg, Kaina B, Rahmsdorf HJ, et al. (1991) Involvement of *fos* in spontaneous and ultraviolet light-induced genetic changes. Molec Carcinogen 4:460-466.
12. Li JJ, Gonzalez S, Banerjee SK, et al. (1993) Estrogen carcinogenesis in the hamster kidney: Role of cytotoxicity and cell proliferation. Environ Health Perspect 101:254-264.
13. Epe B, Harttig U, Stopper H, Metzler M (1990) Covalent binding of reactive estrogen metabolites to microtubular proteins as a possible mechanism of aneuploidy induction and neoplastic cell transformation. Environ Health Perspect 88:123-127.
14. Li JJ, Li SA (1990) Estrogen carcinogenesis in hamster tissues: A critical review. Endocrine Rev 11:524-531.
15. Liehr JG (1990) Genotoxic effects of estrogens. Mutation Res 238:269-276.
16. Roy D, Floyd RA, Liehr JG (1991) Elevated 8-hydroxy deoxyquanosine levels in DNA of diethylstilbestrol-treated Syrian hamsters: Covalent DNA damage by free radicals generated by redox cycling of diethylstilbestrol. Cancer Res 51:3882-3885.
17. Liehr JG, Han X, Bhat HK (1993) ^{32}P-postlabelling in studies of hormonal carcinogenesis. IARC Sci Publ 124:149-155.

18. Di Augustine RP, Walker M, Li SA, et al. (1992) DNA adduct profiles in hamster kidney following chronic exposure to various carcinogenic and non-carcinogenic estrogens. In Li JJ, Nandi S, Li SA (eds): Hormonal Carcinogenesis. New York: Springer-Verlag, pp 280-284.
19. Baltimore D (1985) Retroviruses and retrotransposons: The role of reverse transcription in shaping the eukaryotic genome. Cell 40:481-482.
20. Callahan R, Todaro GJ (1978) Four major endogenous retrovirus classes each genetically transmitted in various species of Mus. In Morse CH III (ed): Origins of Inbred Mice. New York: Academic Press, pp 689-713.
21. Lueders KK, Kuff EL (1977) Sequences associated with intracisternal A particles are reiterated in the mouse genome. Cell 12:963-972.
22. Piko L, Hammons MD, Taylor KD (1984) Amounts, synthesis and some properties of intracisternal A-particle-related RNA in early mouse embryos. Proc Nat Acad Sci USA 81:488-492.
23. Kuff EL, Fewell JW (1985) Intracisternal A-particle gene expression in normal mouse thymus tissue: Gene product and strain-related variability. Mol Cell Biol 5:474-483.
24. Shen-Ong GLC, Cole MD (1982) Differing population of intracisternal A-particle genes in myeloma tumors and mouse subspecies. J Virol 42: 411-421.
25. Heidmann O, Heidmann T (1991) Retrotransposition of a mouse IAP sequence tagged with an indicator gene. Cell 64:159-170.
26. Meitz JA, Grossman Z, Leuders KK, Kuff EL (1987) Nucleotide sequence of a complete mouse intracisternal A-particle genome: Relationship to known aspects of particle assembly and function. J Virol 61:3020-3029.
27. Schwartz DA, Dahm M, Bai L, et al. (1993) Construction of a retrotransposition indicator sequence using a neomycin resistance-encoding gene containing a functional intron. Gene 127:233-236.

14

Interstitial Cell Origin of Estrogen-Induced Kidney Tumors in the Syrian Golden Hamster

Antonio Llombart-Bosch, Amando Peydro-Olaya,
Antonio Cremades, Virginia Cortes, and Maria Aurelia Gregori

Introduction

It is well established that the Syrian golden hamster (SGH) possesses a unique capability for developing kidney tumors under prolonged estrogen administration (1-3).

Several morphological and histological patterns of this tumor have been reported by various authors (4, 5), including ourselves (6). Its histogenesis, however, remains controversial and unresolved. Several hypotheses have been proposed, namely, that it is: of epithelial-tubular origin (5); of mesenchymal-interstitial nature (7); from vascular smooth muscle cell proliferation (8); of mixed epithelial and mesenchymal composition (4, 9); of juxtaglomerular, renin-positive cell origin (10). As recently proposed, the blastema nature of the stem cell has been implicated in the origin of such a heterogeneous neoplasm (11, 12).

It is possible that DES administration results in consecutive stages within the SGH kidney, causing an interstitial hyperplasia of blastema remnants, which may lead to small tumors (tumorlets) and confluent invasive tumors. In the first stage, this neoplasm is hormone-dependent; however, when serially transplanted into SGH, it eventually loses its hormonal dependency (6, 13).

This estrogen-induced lesion of the SGH kidney bears some structural similarities to human blastema proliferation reported in infantile nephroblastomatosis and in Wilms' tumors (14, 15).

In the present report, we postulate that the induction of kidney tumors by DES in the SGH provides an experimental model for human nephroblastoma, and that the blastema cells develop a divergent phenotype capable of maturation toward neuroectodermal cells of Schwannian differentiation, among others. We have published numerous studies on this subject (6, 9, 13, 16).

Our studies are performed in groups of castrated male and female SGH treated with DES alone or with DES + N-nitroso-N-ethylurea (ENU). This well-known carcinogen induces peripheral nervous tumors transplacentally. This SGH/ENU transplacentally-induced nervous tumor has been proposed as a model for von Recklinghausen's (NFM-1) disease (17). In order to detect early lesions and renal tumors, the treated animals are sacrificed at periods between 1.5 and 10.0 months after treatment After prolonged treatment, large metastatic neoplasms have been reported (9). Our studies include conventional histology, electron transmission microscopy, immunohistochemistry of cell secretory products or intermediate filaments, biochemical detection of DT-diaphorase enzymes with immunohistochemical support, and estrogen-receptor analysis. These techniques allow us to examine the effects of estrogen treatment in the biological and morphological behavior of hormone-dependent, as well as hormone-independent transplanted tumors (6). We have also used short-term cultures and cytogenetic studies to determine the clonicity of the neoplasm. Our most significant results are discussed below.

Types of Microscopic Lesions

Shortly after estrogen treatment, two main types of lesions have been observed in the SGH kidneys: interstitial cell hyperplasia and tubular cell dysplasia.

As the treatment continues, microscopically detectable neoplasms (tumorlets) are found, followed by widespread multiple tumoral nodules in the cortex (0.1-0.3 mm) and invasive confluent neoplasms (up to 4-5 mm). The interstitial cell hyperplasia and the tubular dysplasia may appear isolated or together within neighboring fields associated or not with tumors or tumorlets.

Histologically, interstitial cells appear as small, dark, oval groups with scant cytoplasm and delicate intertubular projections. They are located in the intertubular fields but not necessarily connected to vascular type smooth muscle cells. They proliferate, forming nests or cords protruding into the tubules and capillaries.

Tubular dysplasia consists of various degrees of atypical changes at the nuclei of the tubular cells. These cells tend to proliferate (increase mitotic activity) and show pseudostratification.

No continuity could be determined between these two types of lesions, but both coexist in close proximity.

Histology of the Neoplasms

The histology of the neoplasms is highly heterogeneous and complex. We have distinguished four different stages:

Tumorlets, small microscopic foci of interstitial cells grouped in the intertubular spaces.

Small multiple tumoral nodules, macroscopically detectable, pinhead-sized groups of cells. They were bilaterally located and did not disturb kidney contour or size.

Tumors, macroscopic neoplasms (up to several cm in size) that coexist with previous lesions within the kidney, occasionally extending into the cortex and invading the peritoneum.

Distant metastasis (vascular metastasis), infrequent lesions observed after prolonged treatment.

Several histological subtypes of structures have been observed. The most common is an epithelial pattern, with hypernephroid clear cell type. It coexists with areas of adenoid, glandular or papillary configuration and foci of neuroendocrine differentiation. It exhibits ribbon-like organoid texture.

The second histological pattern, which is in continuity with the former, has been considered blastema It is composed of undifferentiated interstitial cells with scarce maturation. Some of these cells possess sarcomatoid patterns of spindle cell type.

Finally, there are small interstitial cell groups, in Homer-Wright-like rosettes, that display a neuroectodermal appearance. This is the least frequent structure.

In large tumors, as well as in hormone-dependent transplant tumors, the dominant microscopic pattern is epithelial, clear cell type, or adenoid, solid-glandular. After several transplants, the cells become more undifferentiated, adopting a sarcoma-type configuration.

Administration of large amounts of DES to animals carrying hormone-dependent tumors results in histologic changes of the transplanted tumors, which acquire a sarcoma-type configuration. Also, DES withdrawal results in reduced growth rate and the appearance of involute figures within the tumor characterized by the presence of abundant lipid-laden deposits in the cell cytoplasm.

Immunohistochemistry

Tumor cells are positive for epithelial (CAM 5.2, cytokeratin, CEA, and EMA), as well as for mesenchymal and muscle cell differentiation markers (vimentin, desmin, myoglobin). Furthermore, neuroectodermal markers such as S-100 protein, HNK-1, and neuron-specific enolase are positive only in some cells. Coexpression of these specific markers was found in several groups of tumors. Our results are in agreement with previous reports (8, 10, 11, 17). Coexpression of several mesenchymal, epithelial, and neural markers has also been reported in human Wilms' tumor (18-20).

ENU + DES treatment does not induce substantial changes at the immunohistochemical level. The effects of such treatment are similar to those observed after DES treatment alone.

Estrogen and Progesterone Receptors

Attempts to establish correlations between hormonal receptor status and DES-induced neoplasms, using both the dextran charcoal method (21) and solid-phase enzymatic immunoassay (the latter being more sensitive), have been described (22-25).

Both types of receptors were present in 50% of the neoplasms studied. When the receptor levels were related to tumor size and grade of histological differentiation, we found that the estrogen receptor (ER) values increased progressively with the size of the tumor and the time of DES exposure.

Immunohistochemistry analysis, using an affinity-purified ER antibody (ER-715), revealed the presence of ER protein exclusively in the nuclei of both interstitial tumoral cells and other nontumorous renal cells (arterial wall, capillaries, podocytes). These data provide further support for the interstitial cell origin of the neoplasms (26).

DT Diaphorase and Antioxidant Enzymatic Activities

We have recently reported (27, 28) a marked decrease in microsomal DT diaphorase activity in preneoplastic and tumoral lesions detected by biochemical and immunohistochemical means. Of special interest in relation to this finding is the observation that Wilms' tumor samples have shown a similar decrease in microsomal DT under similar experimental conditions.

In addition, increases were observed in the activity levels of superoxide dismutase (SOD), an antioxidant enzyme, and glutathione peroxidase (GSH-Px), while glutathione disulphide reductase was decreased in both SGH tumors and human carcinomas. These findings confirm previous reports (27) indicating that DES may be involved in free radical-mediated carcinogenesis.

Electron-Microscopic Results

Our findings using both DES- and DES + ENU-treated SGH, as well as transplanted tumors, are consistent (6, 9, 11, 13, 29). In these neoplasms, epithelial differentiation features are dominant, with distinct basal poles, limiting basal membranes, and well-oriented adenoid patterns. Intercellular and intracellular lumina are frequently seen crowned by microvilli or abundant ciliary structures. There is secretory activity within the cells, with large Golgi fields and active REG profiles. Mucinous material and secretory granules with crystalloid are seen.

The blastema cells are more undifferentiated, nonsecretory, and provided with intermediate filaments and lamellar bodies. They are fusiform in contour, lack cilia or microvilli, and possess interdigitations with neighboring cells. Nuclei are fusiform, irregular and cleaved, showing a dense heterochromatin pattern. Nests of interstitial cells display neural appearance with processes scalloping or involving neural structures, mimicking Schwannian cell differentiation.

Furthermore, pseudorosettes with cell processes filling the central cores are also present. Membrane-bound dense-core granules are present not only within cell processes but in the cytoplasm. Close contact exists between all of the above-mentioned cell types and the interstitial cell hyperplastic foci, as well as the neoplastically transformed foci.

Tubular structures of the remaining kidney show degenerative lesions but no neoplastic transformation.

Cytogenetic Findings

A number of tumors have been assayed for cytogenetic analysis. After short-term tissue culture, metaphases have been karyotyped and classified using trypsin GTG banding (30). The chromosome pattern is similar to that proposed for human chromosomes (ISCN, 37).

Our findings demonstrate the absence of clonal abnormalities, structural and/or numerical, in these tumors (primary DES-induced and transplanted). Isolated numerical abnormalities are consistent with chromosomal gains or losses. Several degrees of aneuploidy and polyploidy with chromosomal endoreduplications have been observed.

Present findings are consistent with the hypothesis postulated by Tsutsui, et al. (31), stressing the significance of aneuploidy in the mechanism of genetic transformation in DES-induced neoplasms.

Acknowledgment

Supported by a grant from the Spanish Association Against Cancer (AECC of Valencia, Spain).

References

1. Kirkman H, Bacon RL (1952) Estrogen-induced tumors of the kidney. I. Incidence of renal tumors in intact and gonadectomized male golden hamsters treated with diethylstilbestrol. J Natl Cancer Inst 13:745-755.

2. Kirkman H, Bacon RL (1952) Estrogen-induced tumors of the kidney. II. Effect of dose, administration, type of estrogen, and age on the induction of renal tumors in intact male golden hamster. J Natl Cancer Inst 13:757-771.
3. Kirkman H (1959) Estrogen-induced tumors of the kidney in the Syrian hamster. III. Growth characteristics in the Syrian hamster. Natl Cancer Inst Monogr 1, pp 1-57.
4. Kirkman H, Robbins M (1959) Estrogen-induced tumors of the kidney in the Syrian hamster V. Histology and histogenesis in the Syrian hamster. Natl Cancer Inst Monogr 1, pp 93-139.
5. Horning ES, Wittick JW (1957) The histogenesis of stilboestrol-induced renal tumors in the male golden hamster. Brit J Cancer 8:451-457.
6. Llombart-Bosch A, Peydro-Olaya A (1986) Estrogen-induced malignant kidney tumors in the Syrian hamster. In Jones TC, Mohr U, Hunt RD (eds.): Urinary System. ILS Monographs on Pathology of Laboratory Animals, pp 141-152.
7. Dontenwill W, Eder M (1959) Histogenese und biologische Verhaltensweise hormonell ausgelöster Geschwülste. Beitr Path Anat 120:270-301.
8. Hacker HJ, Bannasch P, Liehr JG (1988) Histochemical analysis of the development of estradiol-induced kidney tumors in male Syrian hamsters. Cancer Res 48:971-976.
9. Llombart-Bosch A, Peydro-Olaya A (1975) Morphological, histochemical and ultrastructural observations of DES-induced kidney tumors in the Syrian golden hamster. Eur J Cancer 11:403-412.
10. Dodge AH, Brownfield M, Reid IA, Inagami T (1988) Immunohistochemical renin study of DES-induced renal tumor in the Syrian hamster. Am J Anat 182:347-352.
11. Gonzalez A, Oberley TD, Li JJ (1989) Morphological and immunohistochemical studies of the estrogen-induced Syrian renal tumor: Probable cell of origin. Cancer Res 49:1020-1028.
12. Oberley TD, Gonzalez A, Lauchner LJ, et al. (1991) Characterization of early kidney lesions in estrogen-induced tumors in the Syrian golden hamster. Cancer Res 51:1922-1929.
13. Cortes V, Peydro-Olaya A, Llombart-Bosch A (1995) Morphological and immunohistochemical support for the interstitial cell origin of estrogen-induced kidney tumors in the Syrian golden hamster. Carcinogenesis, in press.
14. Cardesa A, Ribalta T (1986) Nephroblastoma, kidney, rat. In Jones TC, Mohr U, Hunt RD (eds.): Urinary System. ILSI Monographs on Pathology of Laboratory Animals, pp 71-80.

15. Llombart-Bosch A, Peydro-Olaya A, Cerda-Nicolas M (1980) Presence of ganglion cells in Wilms' tumors: A review of the possible neuroepithelial origin of nephroblastoma. Histopathology 4:321-330.
16. Cortes V, Perez M, Llombart-Bosch A (1988) Immunohistochemical characterization of diethylstilbestrol-induced neoplasms in the golden hamster. Abst Verhandl Deuts Ges Pathol 23:544.
17. Nakamura T, Hara M, Kasuga T (1989) Transplacental induction of peripheral nervous tumors in the Syrian golden hamster by N-nitroso-N-ethylurea. A new animal model for von Recklinghausen's neurofibromatosis. Am J Pathol 135:251-259.
18. Altmannsberger M, Osborn M, Schäfer H (1984) Distribution of nephroblastomas from other childhood tumors using antibodies to intermediate filaments. Virchows Arch B (Cell P) 45:113-124.
19. Yeger H, Baumal R, Bailey D, (1985) Histochemical and immunohistochemical characterization of surgically resected and heterotransplanted Wilms' tumor. Cancer Res 45:2350-2357.
20. Wick MR, Manivel C, O'Leary TP, Cherwitz DL (1986) Nephroblastoma. A comparative immunocytochemical and lecitin-histochemical study. Arch Pathol Lab Med 110:630-635.
21. Cortes-Vizcaino V, Llombart-Bosch A (1993) Estrogen and progesterone receptors in the diethylstilbestrol kidney neoplasms of the Syrian golden hamster: Correlation with histopathology and tumoral stages. Carcinogenesis 14:1215-1219.
22. Steggles AW, King RJB (1972) Oestrogen receptors in hamster tumors. Eur J Cancer 8:323-334.
23. Li JJ, Talley DJ, Li SA, Villee CA (1976) Receptor characteristics of specific estrogen binding in the renal adenocarcinoma of the golden hamster. Cancer Res 36:1127-1132.
24. Li SA, Li JJ, Villee CA (1977) Significance of the progesterone receptor in the estrogen-induced and -dependent renal tumor of the Syrian golden hamster. Ann NY Acad Sci 286:369-383.
25. Liehr JG, Dague BB, Ballatore AM, Sirbasku DA (1982) Multiple roles of estrogen in estrogen-dependent renal clear-cell carcinoma of Syrian hamster. Cold Spring Harbor Conf Cell Prol 9:445-458.
26. Bhat HK, Hacker HJ, Bannasch E (1993) Localization of estrogen receptors in interstitial cell of hamster kidney and in estradiol-induced renal tumors as evidence of the mesenchymal origin of this neoplasm. Cancer Res 53:5447-5451.

27. Segura-Aguilar J, Cortes-Vizcaino V, Llombart-Bosch A (1990) The levels of quinone reductase, superoxide dismutase and glutathione-related enzymatic activities in diethylstilbestrol-induced carcinogenesis in the kidney of male Syrian golden hamster. Carcinogenesis 11:1727-1732.
28. Segura-Aguilar J, Cremades A, Llombart-Bosch A (1995) Activity and immunohistochemistry of DT-diaphorase in hamster and human kidney tumors. Carcinogenesis, in press.
29. Dodge AH (1974) Fine structural HaLVgs antigen and reverse transcriptase study of the Syrian hamster stilbestrol-induced renal carcinoma. Lab Invest 31:250-257.
30. Gregori-Romero M, Cremades A, Gil-Benso R (1995) Cytogenetic characterization of kidney tumors induced by diethylstilbestrol and N-nitroso-N-ethylurea in the Syrian hamster. Submitted.
31. Tsutsui T, Maizumi H, McLachlan JA, Barrett JC (1983) Aneuploid induction and cell transformation by diethylstilbestrol. A possible chromosomal mechanism in carcinogenesis. Cancer Res 43:3814-3821.

15

Effects of Estrogens on Microtubule Assembly: Significance for Aneuploidy

Manfred Metzler, Erika Pfeiffer, Maik Schuler, and Brigitte Rosenberg

Introduction

The molecular mechanisms of hormonal carcinogenesis are not yet fully understood. For carcinogenic estrogens, it is likely that multiple mechanisms act together, namely, stimulation of cell proliferation, heritable reprogramming of cellular differentiation, and induction of genetic changes (1). The latter may involve numerical and/or structural chromosomal aberrations, or gene mutations. Despite efforts from several laboratories, the evidence for the induction of DNA damage and gene mutations by estrogens remains elusive. On the other hand, an increasing number of observations suggests an important role for chromosomal alterations, in particular numerical aberrations, in estrogen carcinogenesis (2). Several mechanisms have been proposed to explain the induction of aneuploidy in mitotic cells. The proposed cellular targets include spindle microtubules (MT), MT-associated proteins and regulatory molecules, kinetochores and centromeres, centrioles and centrosomes, as well as the cytoplasmic membrane and DNA (3). Several laboratories have reported colchicine-like effects when the carcinogenic estrogen diethylstilbestrol (DES) was tested in several cell systems (4-6). These effects cause inhibition of MT polymerization in cell-free conditions, as well as disruption of the mitotic spindle, and induction of aneuploidy and micronuclei (MN) in various mammalian cells. Colchicine (COL) is known to elicit its effects through noncovalent binding to the major MT protein, tubulin. We have reported the interaction of various estrogens with tubulin and its putative role in the induction of MN and aneuploidy (6-8). Herein, we summarize our recent findings on the biochemical mechanisms of known aneuploidogenic estrogens, and on the ability to induce MN and aneuploidy of recently-identified environmental estrogens and other compounds.

Binding of DES and Estradiol-17ß to Tubulin

Although it is well established that DES has COL-like effects, the binding of DES to tubulin has not been carefully analyzed. We have determined the binding of COL, DES, and 17ß-estradiol (E_2) to tubulin using the equilibrium gel filtration method described by Hummel-Dreyer (6). COL was used as the positive control, and E_2 was used as an example of a compound that does not inhibit MT assembly. Briefly, samples containing 10 nmol of tubulin (obtained from phosphocellulose-chromatographed mixtures of MT proteins devoid of MT-associated proteins) and a known concentration of the ligand (10-20 μM) were immediately applied to gel columns (Bio Gel P4) equilibrated with assembly buffer at identical ligand concentrations (8). The time allowed for binding was equal to the mean chromatographic elution time of the protein (void volume). When labeled ligands were used, the ligand concentration was measured by determining either the radioactivity level or the UV absorbance. Binding was indicated by a peak of ligand eluting with the protein, as well as a trough of the free ligand at the retention time.

Table 1. Binding of COL, DES, and E_2 to Tubulin.

Compound	Binding (mol ligand per mol tubulin) after *x* min 10	30	60
COL	0.28	0.42	0.71
DES	1.70	1.72	1.70
E_2	<0.10	0.25	0.35
COL/DES	n.d.[a]	0.37/1.20	0.65/0.60
E_2/DES	n.d.	0.23/1.32	n.d.
COL/E_2	n.d.	0.45/0.25	n.d.

[a] n.d., not determined

The results obtained at various binding times are summarized in Table 1. Incubation times longer than 60 min were precluded because of the instability of tubulin. COL treatment resulted in slow binding approaching a molar ratio of ligand to tubulin of 1. After DES treatment, more rapid binding with a molar ratio of ~2 was observed. Surprisingly, E_2 treatment resulted in a small but significant binding. When tubulin was preincubated with an excess of COL or DES and later chromatographed with a ligand-free buffer, COL but not DES was

DES and later chromatographed with a ligand-free buffer, COL but not DES was detected in the void volume, suggesting that the tubulin-COL-complex is stable, whereas the tubulin-DES-complex dissociates under these experimental conditions. When similar experiments were carried out with two ligands, DES/COL and DES/E_2, the binding of DES was markedly decreased (Table 1). The binding to tubulin after treatment with E_2/COL was not significantly different from that obtained when these compounds were tested individually. These data suggest that DES may have two binding sites on tubulin, one of which may be at the COL binding site, where it may affect the functional integrity of the MT assembly. The tubulin binding and dissociation were faster after DES treatment than after COL. While E_2 binds slowly to tubulin, this binding does not seem to be directed to the COL site. It therefore does not seem to affect MT polymerization, though it does interfere with DES binding.

Inhibition of MT Assembly by Environmental Estrogens

The *in-vitro* MT polymerization assay is relatively simple, and its value in the detection of potential aneuploidogen compounds has been demonstrated (9). Using this assay, we have studied the ability of various environmental estrogens to inhibit MT assembly under cell-free conditions. The results of such studies are summarized in Table 2.

Table 2. Ability of Various Environmental Estrogens to Inhibit Mt Assembly under Cell-Free Conditions.

Inhibitors of MT assembly[a]	
Bisphenol A	4-Hydroxy-2',4',6'-trichlorobiphenyl
p-Nonylphenol	4-Hydroxy-2,2',5'-trichlorobiphenyl
Pentachlorophenol	3-Hydroxy-2',5'-dichlorobiphenyl
Non-inhibitors of MT assembly[b]	
Coumestrol (50 μM)	Zearalenone (50 μM)
Genistein (50 μM)	Zeranol (50 μM)
Daidzein (50 μM)	Enterolactone (100 μM)
Equol (100 μM)	Enterodiol (100 μM)

[a] Concentration range: 50 to 200 μM

assembly with an activity comparable to DES. However, only some analogs of BP-A exhibited similar properties. BP-A disrupted the cytoplasmic MT complex and induced MN in cultured Chinese hamster V79 cells, as reported by Pfeiffer, et al., in this volume (pp 450-453). p-Nonylphenol, another MT inhibitor, has recently been identified as an environmental estrogen (10). It is an antioxidant and plasticizer that is released from alkylphenol polyethoxylates. The latter compounds are widely used as lubricants and detergents in toiletries, and in the textile and paper industries.

Pentachlorophenol (PCP) has been used extensively as a fungicide to preserve wood and, on a smaller scale, as a herbicide and bactericide. It has been shown to be carcinogenic in mice, and in a cell-free assay, PCP caused a concentration-dependent inhibition of MT assembly, reaching 52% at 200 μM. This level of activity corresponds to about one third of that observed for DES. Recently, we have shown that 200 μM PCP clearly disrupts the cytoplasmic MT complex in cultured Chinese hamster V79 cells (unpublished results).

In collaboration with Drs. John A. McLachlan and Ken Korach (National Institute of Environmental Health Sciences, Research Triangle Park, NC), we have started a study on the aneugenic potential of hydroxylated polychlorinated biphenyls, weak estrogens and metabolites of PCBs that accumulate in the environment (11). Several congeners of this large class of substances clearly inhibited MT polymerization with an activity resembling that of DES.

Whereas most compounds with an inhibitory effect on the *in-vitro* assembly of MT produce aneuploidy and MN induction in intact cells, the reverse is not necessarily true. The cell-free assay may give "false negative" results if the concentration of the compound is too low (which may be the case with lipophilic and poorly water-soluble substances), if the compound requires metabolism to affect the MT, or if cellular targets other than MT are involved. As discussed below, E_2 is a clear example of a false negative result using the MT assembly assay.

Conformational Changes of Tubulin Evoked by Estrogens

Binding of MT inhibitors to tubulin most likely results in alteration(s) of the protein conformation. Tubulin is a heterodimer consisting of an α- and a ß-chain with a total of twenty cysteine moieties, twelve at the α-chain and eight at the ß-chain. Due to the presence of one or more intramolecular disulfide bridges in the ß-subunit, as well as for sterical reasons, only twelve of the twenty sulfhydryl groups are accessible to reagents such as 5,5'-dithiobis-(2-nitrobenzoic acid) (DTNB, Ellman's reagent) or N-ethylmaleimide. Moreover, these twelve thiol groups exhibit different reaction rates. Under suitable conditions, a triphasic "titration" curve is obtained with DTNB in the presence of native tubulin

(Figure 1, left chart). The graph indicates that three of the thiols react rapidly, five are less reactive, and the remaining four react at a slower rate. If the conformation of the tubulin dimer is changed as a result of this treatment, given the reaction rates obtained, it is possible that the steric conditions of the thiol groups may also
be modified. In fact, it has been reported that the well-known MT inhibitor, nocodazol, causes the number of the highly-reactive sulfhydryl groups to increase (12).

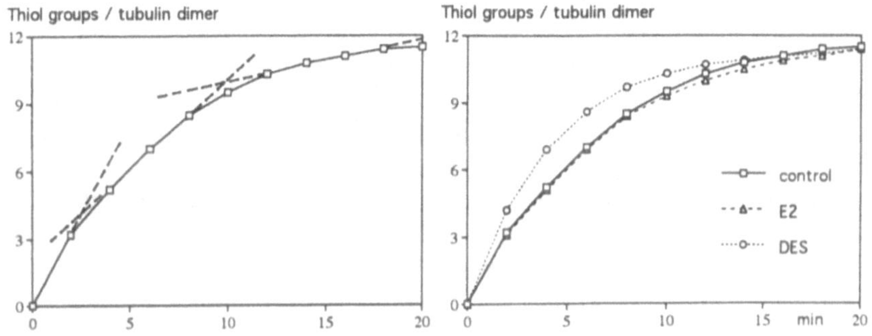

Figure 1. DTNB titration of tubulin (left) and effects of DES and E_2 (right)

When DES was added to MT proteins at a concentration leading to a marked inhibition of MT assembly, a significant change was observed in the DTNB titration curve (Figure 1, right chart): the number of highly reactive thiol groups was increased by about two. The same effect was achieved with meso-hexestrol, bisphenol-A (BP-A) and 3',3"-dimethoxy-BP-A (data not shown). When E_2, an estrogen capable of binding to MT proteins without causing inhibition of MT assembly (see above), was examined, no effect on the DTNB titration curve was noted (Figure 1). Two other non-inhibitors of MT, bis-(4'-hydroxyphenyl)-methane and bis-(4'-hydroxyphenyl)-sulfone (see Pfeiffer, et al., pp. 450-453, in this volume), behaved like E_2 and did not alter the number of DTNB-reactive thiol groups (data not shown). These studies, which have yet to be extended to other compounds, imply a correlation between inhibition of MT assembly and conformational changes of the tubulin dimer as indicated by the accessability of the thiol groups. In this respect, the DTNB titration method may become a useful supplement to the MT polymerization assay.

Aneuploidogenic Effects of E_2

E_2 is an example of an estrogen that lacks MT-inhibiting activity but able to induce mitotic arrest, aneuploidy, and MN in various mammalian cell systems (for details see paper by Schuler et al., this volume). More recently, we have reported that E_2 induces MN in cultured Syrian hamster embryo (SHE) and ovine seminal vesicle (OSV) cells (13). Characterization of these MN with CREST antibodies showed that only 44-48% were kinetochore-positive. The same result (45%) was obtained in human chorionic villi (HCV) cells (Schuler et al., this volume). However, when the MN in the HCV cells was characterized by fluorescence *in-situ* hybridization with the centromere-specific DNA probe p82H, the percentage of signal-positive MN increased to 70%. This observation and the absence of MT assembly suggest that E_2 and DES induce MN by different mechanisms. It is proposed that E_2 causes kinetochore damage and/or chromosome breakage within the centromeric region.

It is interesting to note that coumestrol, although it lacks MT-inhibiting potential, induces high incidences of MN in HCV cells (Schuler et al., this volume). However, these MN are both CREST- and p82H-negative and may arise through a clastogenic effect.

Conclusion

Among the large number of estrogens so far studied, a significant proportion exhibits the ability to interfere with MT assembly and thereby cause the induction of aneuploidy and micronuclei. The fact that recently identified environmental estrogens, such as bisphenol A and p-nonylphenol, also show MT-inhibiting properties suggests that the category of aneuploidogenic estrogens will expand in the future. However, the molecular mechanisms and the structural requirement for MT inhibition have still to be elucidated.

Acknowledgment

Studies carried out in our laboratory were supported by the Deutsche Forschungsgemeinschaft [Grant Me 574(9-1)].

References

1 Barrett JC (1992) Molecular mechanisms of hormonal carcinogenesis. In Li JJ, Nandi S, Li SA (eds): Hormonal Carcinogenesis. New York: Springer-Verlag, pp 159-163.

2. Oshimura M, Barrett JC (1986) Chemically induced aneuploidy in mammalian cells: Mechanisms and biological significance in cancer. Environ Mutagen 8:129-159.
3. Liang JC, Brinkley BR (1985) Chemical probes and possible targets for the induction of aneuploidy. In: Dellarco VL, Voytek PE, Hollaender A (eds): Aneuploidy: Etiology and Mechanisms, Plenum Press, New York, pp 491-505.
4. Degen GH, Metzler M (1987) Sex hormones and neoplasia: Genotoxic effects in short term assays. Arch Toxicol Suppl 10:264-278.
5. Wheeler WJ, Cherry LM, Downs T, Hsu TC (1986) Mitotic inhibition and aneuploidy induction by naturally occurring and synthetic estrogens in Chinese hamster cells *in vitro*. Mutation Res 171: 31-41.
6. Pfeiffer E, Metzler M (1992) Effects of steroidal and stilbene estrogens and their peroxidative metabolites on microtubular proteins. In Li JJ, Nandi S, Li SA (eds): Hormonal Carcinogenesis. New York: Springer-Verlag, pp 313-317.
7. Metzler M, Pfeiffer E, Köhl W, Schnitzler R (1992) Interactions of carcinogenic estrogens with microtubular proteins.In Li JJ, Nandi S, Li SA (eds): Hormonal Carcinogenesis. New York: Springer-Verlag, pp 86-94.
8. Epe B, Harttig U, Stopper H, Metzler M (1990) Covalent binding of reactive estrogen metabolites to microtubular protein as a possible mechanism of aneuploidy induction and neoplastic cell transformation. Environ Health Perspect 88:123-127.
9. Wallin M, Friden B, Billger M (1988) Studies of the interaction of chemicals with microtubule assembly *in vitro* can be used as an assay for detection of cytotoxic chemicals and possible inducers of aneuploidy. Mutation Res 201:303-311.
10. Raloff J (1994) The gender benders. Are environmental hormones emasculating wildlife? Science News 145:24-27.
11. Korach KS, Sarver P, Chae K, McLachlan JA, McKinney JD (1988) Estrogen receptor-binding activity of polychlorinated hydroxybiphenyls: Conformationally restricted probes. Molec Pharmacol 33:120-126.
12. Lee JC, Field DJ, Lee LLY (1980) Effects of nocodazole on structures of calf brain tubulin. Biochemistry 19:6209-6215.
13. Schnitzler R, Foth J, Degen GH, Metzler M (1994) Induction of micronuclei by stilbene-type and steroidal estrogens in Syrian hamster embryo and ovine seminal vesicle cells *in vitro*. Mutation Res, in press.

16

Estrogen Carcinogenesis: A Sequential, Epi-Genotoxic, Multi-Stage Process

Sara Antonia Li, Xiaoying Hou, and Jonathan J. Li

Introduction

The estrogen-induced renal neoplasm in the Syrian hamster has emerged, in the last decade, as one of the primary experimental models in hormonal carcinogenesis. Observed as an early event, cell proliferation is believed to play an essential role in the tumorigenic process in the hamster kidney (1-3). Recent advances in our laboratory have suggested that estrogen-induced tumorigenesis is a consequence of a sequential epi-genotoxic multi-stage process. Evidence is presented herein to support this concept. An epi-genotoxic agent is defined as one not involved in direct (covalent) or indirect interactions with genetic material, but nevertheless able to elicit heritable changes by alternative mechanisms (4).

Cell Proliferation

Since diethylstilbestrol (DES) and 17β-estradiol (17β-E_2) have been shown to induce estrogen receptor (ER) as well as progesterone receptor (PR) in the untransformed hamster kidney *in vivo* (1, 5, 6), it is not necessarily unexpected that these hormones are able to elicit renal cell proliferation both *in vivo* and *in vitro* (2, 3, 7). Using either bromodeoxyuridine (BrdU) or proliferative cell nuclear antigen (PCNA) methods, studies were performed to assess S-phase labeled kidney cell distribution in various regions in the kidney, as well as the temporal sequence of labeled renal cells in response to estrogen treatment. These data indicate a significant increase in labeled kidney cells after 2.0 to 3.0 months of estrogen treatment, particularly in the cortico-medullary region of the cortex (8). The relatively early enhanced S-phase labeled renal cells after estrogen treatment suggest the presence of preexisting estrogen-sensitive cells, shown by these immunolabeling techniques.

Renal Cell Damage and Regenerative Cell Proliferation

Although renal tubular injury was observed as early as 1.5 months after estrogen treatment (2, 9), it was uncommon and not severe and could only be seen at the electron microscopic level. The kidney tubular damage was, however, progressive, increasing in severity following continuous estrogen treatment, particularly after 3.0-6.0 months of estrogen treatment (8-10). It is likely that it is an ER-mediated cell proliferative process. The increasing regenerative or reparative hyperplasia (S-phase labeled cells) seen after estrogen treatment coincided with growing evidence of renal tubular damage.

Aneuploidy

One consequence of estrogen-driven renal tubular damage and regenerative cell proliferation, both in proximal tubules and in nascent developing renal interstitial cell lesions is aneuploidy. Studies indicate that a certain level of estrogen-driven cell proliferation is required to elicit aneuploidy since aneuploid cells could not be found in hamster kidney sections following either 17α-E_2 or β-dienestrol treatment (9). Aneuploid frequency with different estrogens was DES $\geq$ 17β-E_2 >> Ethinylestradiol (EE). The frequency of near diploid (between 40 and 43, and 45 and 48 chromosomes) was 6%; and the near tetraploid frequency (>51) was less than 2% in untreated control kidneys; normal diploid kidney chromosomes $2n = 44$ (2). Between 1.5 and 3.5 months of estrogen treatment, the near diploid frequency rose to 38-39% and the near tetraploid frequency to 6-7% (2).

Gene Expression

Employing Northern blot analyses, the overexpression of nuclear protooncogenes (c-*myc*, c-*fos*, and c-*jun*) (Figure 1) as well as tumor suppressor genes p53 and *WT-1* (Figure 2) was assessed at different intervals during estrogen-induced renal tumorigenesis and in the renal tumor. For the first time, significant elevations of early estrogen-responsive protooncogenes have been observed in 5.0- and 6.0-month DES-treated hamster kidneys (Figure 1). Compared with age-matched controls, c-*myc* expression increased 2.8- and 4.1-fold in kidneys of 5.0- and 6.0-month estrogen-treated animals, respectively. Similarly, c-*fos* expression was elevated 4.6- and 4.8-fold in kidneys of 5.0- and 6.0-months of DES-treated hamsters, respectively. Following the same pattern, c-*jun* expression increased 2.8- and 5.1-fold after 5.0 and 6.0 months' hormone treatment, respectively, compared with untreated kidneys. The level of c-*myc*, c-*fos*, and c-*jun* expression rose 6.0-, 10.0-, and 7.0-fold, respectively, in the renal tumor, compared with levels in the untreated kidney. These latter renal tumor gene expression data are largely comparable to those previously reported (11).

Although *WT-1* was evidently not overexpressed during estrogen-induced renal tumorigenesis, it was elevated 8.1-fold in the renal tumor compared with the untreated kidney (Figure 2). Interestingly, there may be two *WT-1* transcripts or homologues in the hamster kidney. The higher molecular weight transcript, less dominant in the normal kidney, is more prominent in the tumor, as is seen in the newborn mouse kidney. These data suggest that the late overexpression of *WT-1,* which is found only in the renal tumor, may be more related to tumor progression than tumor development. In contrast, p53 expression in the kidney was elevated 1.8-fold after 6.0 months of DES treatment and 2.0-fold in the renal tumor (Figure 2).

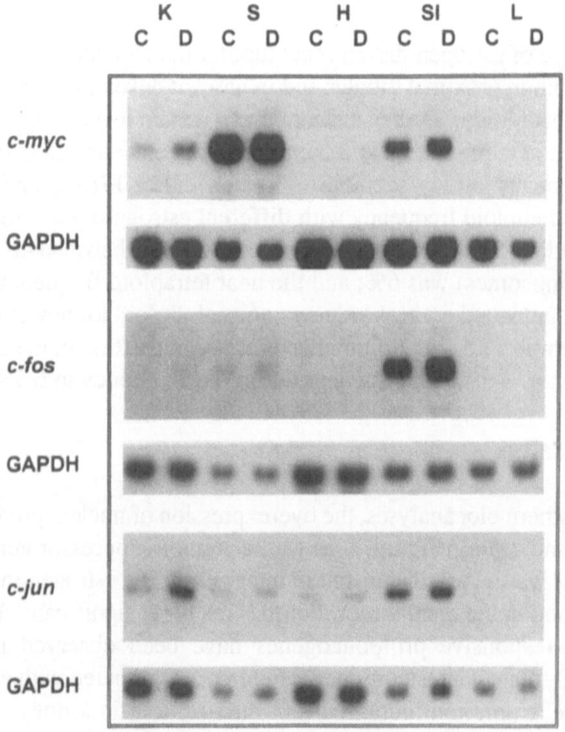

Figure 1. Effect of DES on the expression of c-*fos,* c -*myc*, and c-*jun* in hamster kidney and estrogen -induced tumor tissue. Total RNA (pooled from at least five hamsters) was prepared from age-matched untreated hamsters and hamsters treated with DES for various periods of time. Aliquots (10 μg) of total RNA were examined by Northern blot analysis. (M) months. RNA prepared from: (C) kidneys of age-matched untreated animals, (D) kidneys of DES-treated animals, (T) estrogen-induced kidney tumors.

Based on the findings presented herein, one may anticipate that *in-situ* hybridization studies using immunocytochemical methods for these gene expressions would reveal overexpression of these gene transcripts earlier (< 5.0 months) in estrogen-induced renal tumorigenesis.

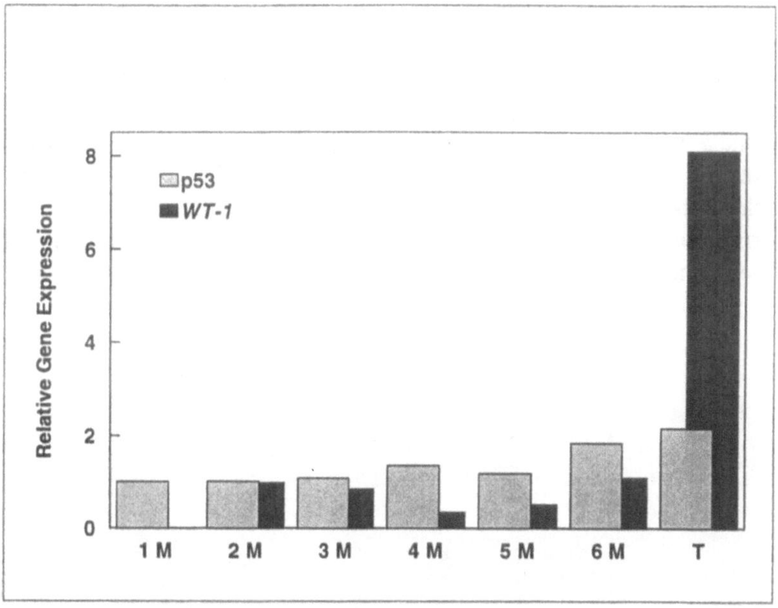

Figure 2. Densitometric quantification of the relative expression of *WT-1* and p53 in the kidneys of hamsters treated with DES from 1.0 to 6.0 months and estrogen-induced tumor tissue (T). Total RNA (pooled from at least five hamsters) was prepared from age-matched untreated hamsters and hamsters treated with DES for various periods of time. Aliquots (10 μg) of total RNA were examined by Northern blot analysis and quantified by densitometric scanning. (M) months. The relative expression represents the ratio of the densitometric values for treated versus control kidney samples.

Studies presented in this section further support the notion that overexpression of these early-response protooncogenes and perhaps p53 gradually confers on a subset of altered estrogen-dependent renal cells growth advantages that eventually contribute to the transition from normal to neoplastically transformed phenotype.

Chromosome Instability and Aberrations

In our laboratory, nonrandom numerical chromosomal changes have been observed which are common to both DES- and 17β-E_2-induced hamster kidney tumors (12). Gains in chromosomal number (trisomies, tetrasomies) were seen in chromosomes 1, 2, 3, 11, 15, 16, 20, and 21, and losses in chromosomal number were found in chromosome 20. Recent studies in kidneys treated for 3.0 and 5.0 months have also shown nonrandom or consistent numerical chromosomal changes in some of the same chromosomes seen in estrogen-induced renal neoplasms (Li JJ, et al., unpublished data).

Previous reports have indicated that both steroidal and nonsteroidal estrogens can induce unscheduled DNA synthesis in HeLa cells, mouse germ cells, and Syrian hamster embryo cells (13-15). These estrogens are also capable of inducing sister chromatid exchange (SCE) in other cell lines, and in fibroblasts and lymphocytes (16-18). It should be noted that these studies were performed *in vitro* in either non-epithelial or transformed cells exposed to estrogens and thus may have little bearing on *in-vivo* epithelial neoplastic transformation processes.

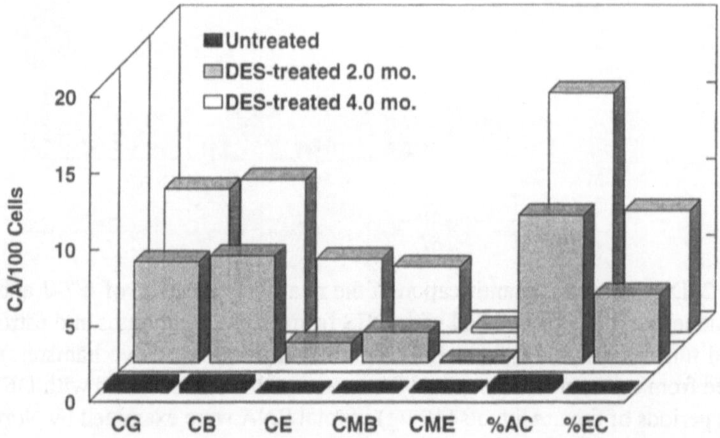

Figure 3. Frequency of chromosomal aberrations induced by DES in Syrian hamster kidney proximal tubules. Groups of five to seven hamsters were treated with DES for 2.0 and 4.0 months. At least 200 well-spread metaphases were analyzed for each animal. Chromatid and isochromatid gaps were not included as aberrations. The results are expressed as chromosomal aberrations for 100 cells. (CG) chromosome gaps, (CB) chromosome breaks, (CE) chromosome exchanges, (CMB) chromatid breaks, (CME) chromatid exchanges, (AC) aberrant cells, (EC) endoreduplicated cells (24).

Recently, however, both DES and 17β-E_2 have been shown to induce SCE in cultured neonatal mouse uterine cervical epithelium (19). Nevertheless, chromosomal aberrations (CA) in various other *in-vitro* cell systems exposed to either steroidal or nonsteroidal estrogens have been essentially negative (20-23). Banerjee, et al. (24), in our laboratory, reported that DES induced substantial increases in CA, attaining a maximal level after 5.0 months of continuous hormone treatment (Figure 3). The frequency of CA was evidently cumulative, increasing from 0.5 to 5.0 months of estrogen treatment (24). These CA include chromatid gaps (CG), chromatid breaks (CB), chromatid exchanges (CE), chromosome breaks (CME), and endoreduplicated cells (EC). There were no significant differences in the frequency of these parameters (all of which remained at nearly undetectable levels) in age-matched controls as a result of increasing age corresponding to estrogen-treatment periods. Additionally, the frequency of CA in hamster kidney cells has been studied after 5.0 months of treatment employing a variety of potent and weak estrogens (Figure 4).

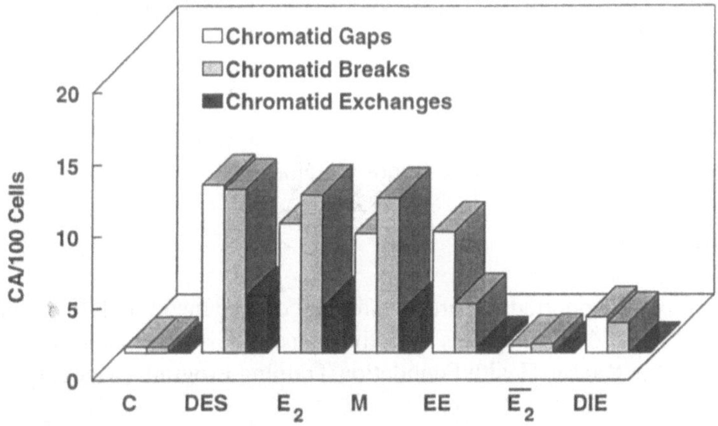

Figure 4. Frequency of chromosomal aberrations induced by different estrogens in Syrian hamster kidney proximal tubules. Groups of five and seven hamsters were treated for 5.0 months with different estrogen implants. Age-matched control hamsters were used for the control groups. At least 200 well-spread metaphases were analyzed for each animal. The results are expressed as chromosomal (1983), aberrations per 100 cells. (C) control, (DES) diethylstilbestrol, (E_2) 17β-estradiol, (M) Moxestrol, (E_2) 17α-estradiol, (DIE) dienestrol (24).

Not surprisingly, Moxestrol (MOX), a very potent estrogen and carcinogenic agent, but only poorly metabolized in the kidney, exhibited CA frequencies similar to those of DES and 17β-E_2. In contrast, estrogens with very weak estrogen potencies, such as 17α-E_2 and β-dienestrol, elicited only low CA frequencies, not appreciably different from untreated age-matched control levels

(Figure 4). Moreover, treatment with EE, a normally potent estrogen but only weakly carcinogenic in the hamster kidney, did result in a significant rise in CG; however, these did not evolve into significant increases in either CB, CE, CMB, or CME (24). Although EE treatment resulted in a modest but significant increase in the frequency of aberrant cells in hamster renal cortical cells, the level was approximately 4.0-fold lower than the frequency observed for the other potent estrogens examined. Neither progesterone nor androgen treatment, for the same 5.0-month period, was found to induce any CA in the hamster kidney, indicating that the CA generated by estrogens are hormone specific (data not shown).

Summary

As a result of studies from our laboratory, both those described here and those presented previously, an estrogen-driven, epi-genotoxic, multi-stage sequence is proposed which begins to elucidate the broad mechanism(s) of estrogen carcinogenesis in this experimental hormonally-induced model. We believe that, once the details of each stage in this sequence are identified, a clearer understanding of the cellular and molecular mechanisms of estrogen carcinogenesis in the hamster kidney will be established and prove useful for other solely hormonally-induced tumor systems.

Acknowledgments

This study was supported in part by National cancer Institute, NIH grants CA 58030 and CA 22008, and grants from the Kansas Masonic Oncology Research Center and the Kansas Health Foundation Training Program (Xiaoying Hou).

References

1. Li JJ, Li SA (1990) Estrogen carcinogenesis in hamster tissues: A critical review. Endocrine Rev 11:524-531.
2. Li JJ, Gonzalez A, Banerjee S, et al. (1993) Estrogen carcinogenesis in the hamster kidney. Role of cytotoxicity and cell proliferation. Environ Health Perspect 101:259-264.
3. Li JJ, Li SA, Oberley TD, Parsons JA (1995) Carcinogenic activities of various steroidal and nonsteroidal estrogens in the hamster kidney: Relation to hormonal activity and cell proliferation. Cancer Res, in press.
4. Li JJ (1995) Perspectives in hormonal carcinogenesis: Animal model to human disease. In Huff J, Boyd J, Barrett JC (eds): Cellular and Molecular Mechanisms of Hormonal Carcinogenesis: Environmental Influences. New York: Wiley-Liss, in press.

5. Li JJ, Li SA (1993) Estrogen-induced tumorigenesis in the Syrian hamster: Roles for hormonal and carcinogenic activities. Arch Toxicol 55:110-118.
6. Li SA, Li JJ, Villee CA (1977) Significance of the progesterone receptor in the estrogen-induced and -dependent renal tumor in the Syrian golden hamster. Ann NY Acad Sci 286:369-383.
7. Oberley TD, Lauchner LJ, Pugh TD, et al. (1989) Specific estrogen-induced cell proliferation of cultured Syrian hamster renal proximal tubular cells in serum-free chemically-defined media. Proc Natl Acad Sci USA 86:2107-2111.
8. Banerjee S, Banerjee SK, Li SA, Li JJ (1992) Cell proliferation studies during estrogen-induced renal tumorigenesis in Syrian hamsters. Proc Am Assoc Cancer Res 33:115.
9. Banerjee SK, Banerjee S, Li SA, Li JJ (1992) Regenerative cell proliferation and aneuploidy during estrogen-induced renal tumorigenesis in Syrian hamsters. Proc Am Assoc Cancer Res 33:129.
10. Li JJ, and Li SA (1995) Estrogen carcinogenesis in the hamster kidney: A hormone-driven multi-step process. In Huff J, Boyd J, Barrett JC (eds): Cellular and Molecular Mechanisms of Hormonal Carcinogenesis: Environmental Influences. New York: Wiley-Liss, in press.
11. Liehr JG, Chiappetta C, Roy D, Stancel GM (1992) Elevation of protooncogene messenger RNAs in estrogen-induced kidney tumors in the hamster. Carcinogenesis 13:601-604.
12. Banerjee SK, Banerjee S, Li SA, Li JJ (1990) Cytogenetic changes in renal neoplasms during estrogen-induced renal tumorigenesis in hamsters. In Li JJ, Nandi S, Li SA (eds): Hormonal Carcinogenesis. New York: Springer-Verlag, pp 247-250.
13. Martin CN, McDermid AC, Gaines RC (1978) Testing of known carcinogens and noncarcinogens for their ability to induced unscheduled DNA synthesis in HeLa cells. Cancer Res 38:2721-2627.
14. Racine RR, Schmid BP (1983) DNA-damaging potential of diethylstilbestrol evaluated in the germ cells unscheduled DNA synthesis assay. Environ Mutagen 6:211-218.
15. Tsutsui T, Degan GH, Schiffmann D, et al. (1984) Dependence on exogenous metabolic activation for induction of unscheduled DNA synthesis in Syrian hamster embryo cells by diethylstilbestrol and related compounds. Cancer Res 49:184-189.
16. Kochhar TS (1985) Inducibility of chromosome aberrations by steroid hormones in cultured Chinese hamster ovary cells. Toxicol Lett 29:201-206.
17. Rudigar HW, Haenisch F, Metzler M, et al. (1979) Metabolites of diethylstilbestrol induce sister chromatid exchange in human cultured fibroblasts. Nature 5730:392-394.

18. Hill A, Wolff S (1983) Sister chromatid exchanges and cell division delays induced by diethylstilbestrol, estradiol, and estriol in human lymphocytes. Cancer Res 43:4114-4118.
19. Hillbertz-Nilsson K, Forsberg J-G (1989) Genotoxic effects of estrogens in epithelial cells from the neonatal mouse uterine cervix. Modification by metabolic modifiers. Teratogen Carcinogen Mutagen 9:97-110.
20. Mehnert K, Speit G, Vogel W (1985) Effect of diethylstilbestrol on the frequencies of sister chromatid exchange *in vitro* and *in vivo*. Cancer Res 49:3626-2630.
21. Husum B, Wulf HC, Niebuhr E (1982) Normal sister chromatid exchanges in oral contraceptive users. Mutation Res 103:161-164.
22. Drevon C, Piccoli C, Montesano R (1981) Mutagenicity assay of estrogenic hormones in mammalian cells. Mutation Res 89:83-90.
23. Chrisman CL, Baumgartner AP (1979) Cytogenetic effects of diethylstilbestrol-diphosphate (DES-dp) on mouse bone marrow monitored by the micronucleus test. Mutation Res 67:157-160.
24. Banerjee, SK, Banerjee S, Li SA, Li JJ (1994) Induction of chromosome aberrations in Syrian hamster renal cortical cells by various estrogens. Mutation Res 311:191-197.

PART 5. LIVER

Introduction

Role of Estrogens in Liver Carcinogenesis

James D. Yager and Joanne Zurlo

In women, while inappropriate and/or excessive exposure to estrogens is generally associated with breast, endometrial, and vaginal cancer, an increased risk for developing hepatic neoplasms also represents a public health issue. One of the first reports of this association was by Baum, et al. (1), who in 1973 reported an increased incidence of liver neoplasms in women with a history of long-term use of oral contraceptives (OC). Since then, there have been numerous reports confirming an association between OC use and the development of hepatocellular adenomas and carcinomas (2, 3). A recent report of a case-control study in Milan presents data showing that the increased relative risk was directly related to duration of use and, interestingly, persisted for longer than 10 years after exposure ended (4).

The early clinical reports prompted investigation of this association. Several early experimental studies employing initiation-promotion protocols demonstrated that the synthetic steroidal estrogens are promoters of hepatocarcinogenesis in both male and female rats (5-8). In fact, synthetic steroidal estrogens such as ethinylestradiol (EE) are very strong promoters and, although not directly genotoxic, are also weak complete hepatocarcinogens (9).

The mechanisms of estrogen-induced effects on liver, including hepatocarcinogenesis, continue to be studied by a small group of investigators in three animal models: the Armenian hamster (10), the α-naphthoflavone-treated Syrian hamster (2), and the rat (9, 11-13). Based on our past and recent findings in the rat model, we would like to highlight three aspects of the female rat liver's complex response to chronic exposure to EE. As shown in Figure 1, these aspects are the transient induction of hepatic DNA synthesis, reflecting additive or adaptive hyperplasia; the subsequent onset of a mitosuppressed state in the hyperplastic liver; and the continuous metabolism of EE to catechol metabolites, which, through further enzymatic and non-enzymatic processes, leads to increased oxidative DNA damage.

EE-Induced Liver Growth

A common response to exposure to nonhepatotoxic doses of most hepatic promoters, including phenobarbital (14, 15), various peroxisome proliferators (16), and EE (9, 15) is a rapid, transient, dose-dependent increase in DNA synthesis, reflecting additive or adaptive growth. As a result, the liver increases in size, i.e., becomes hyperplastic, and remains in that state until exposure to the xenobiotic ceases (17). Our investigation of the mechanisms of EE-induced growth has shown that EE is not a complete hepatic mitogen. Rather, in addition to enhancing the levels of a serum factor stimulatory for hepatocyte DNA synthesis, EE and other estrogens have co-mitogenic effects (9). As such, the estrogens enhance hepatocyte DNA synthesis in response to several complete hepatic mitogens by a mechanism which results in an increase in the fraction of responding hepatocytes (18).

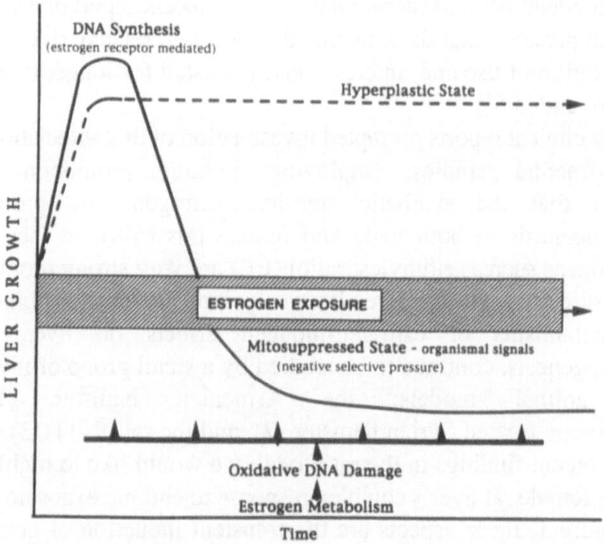

Figure 1. A scheme showing rat liver growth and oxidative DNA damage associated with estrogen exposure.

The observations that tamoxifen inhibits EE-induced DNA synthesis *in vivo* (15) and co-mitogenesis in culture (19) suggests that these effects are mediated through the estrogen receptor. The involvement of the estrogen receptor is further supported by the observation that a nonestrogenic metabolite of estradiol, 2-methoxyestradiol, does not have comitogenic activity (20). Thus, the initial, transient stimulation of liver growth by EE appears to occur through indirect effects which originate from signal transduction pathways leading from the estrogen receptor, and which have the effect of increasing hepatocyte growth factor responsiveness.

EE-Induced Mitosuppression

Studies by others on the effects of several hepatic promoters, including phenobarbital (21-23) and clofibrate (16), have shown that at nonhepatotoxic doses, following the initial, transient stimulation of hyperplastic growth, continued exposure causes an inhibition in basal and/or induced liver growth. Using 7-day osmotic minipumps to deliver BrdU to allow cumulative labeling of replicating hepatocytes, we have found that chronic exposure to EE also results in the appearance of a mitosuppressed state (24). Thus, after 28 and 42 days of exposure to EE, hepatocyte nuclear labeling indices were decreased 72% and 88%, respectively, compared to control. In addition, regenerative growth following partial hepatectomy was reduced in animals previously exposed to EE for 21 days. These results show that chronic exposure of female rats to EE leads to the appearance of a mitosuppressed state characterized by reduced cell turnover and decreased growth responsiveness. It is in this growth-suppressed environment that altered hepatic foci representing preneoplastic lesions undergo clonal expansion during promotion. Thus, while promoters may have additional effects such as suppression of apoptosis, these foci must be resistant to the growth suppressive effects, lending support to the resistant hepatocyte model of Farber (25) as a common but not necessarily exclusive pathway of hepatic tumor promotion. In addition, similar findings with several hepatic promoters suggest that their initial growth stimulatory effects may have little bearing on their promoting potential. Jirtle, et al. (26), have found that the regulation and expression of TGF-β and mannose 6-phosphate/insulin-like growth factor-II receptors are altered in association with mitosuppression induced by phenobarbital. EE's effects on this growth regulatory pathway are unknown. In addition, it becomes important to identify other gene products whose expression is altered during mitosuppression in order not only to gain insight into the mechanisms of growth suppression, but also to identify potential molecular biomarkers/dosimeters for mitosuppression.

Oxidative DNA Damage as a Result of Estrogen Metabolism

The oxidative metabolism of estrogens is complex, but the major pathways for phase I metabolism of estradiol, mediated by various cytochrome P-450s, are shown in Figure 2 (27, 28). Fishman, Bradlow, and co-workers have reported an association between increased risk of breast cancer and excessive formation of the 16α-hydroxy estrone which can, through formation of a Schiff base, covalently bind to the estrogen receptor (29) and can also cause DNA damage (30). However, metabolism of EE at the 16α position is blocked by the 17α-ethinyl group and does not occur (27). On the other hand Liehr and coworkers have proposed that redox cycling of catechol metabolites, mediated by microsomal cytochrome P-450 oxidase/reductase activities, results in the generation of free radicals and reactive oxygen species which cause increased oxidative damage to protein and DNA (31, 32). That this pathway may contribute to the etiology and/or progression of estrogen carcinogenesis is suggested by the detection of increased 8-OH-deoxyguanosine (8-OH-dG) levels and DNA strand breaks in Syrian hamster kidney DNA following DES and 4-OH-estradiol treatment (32, 33), as well as increased levels of 8-OH-dG in human breast tumor tissue (34).

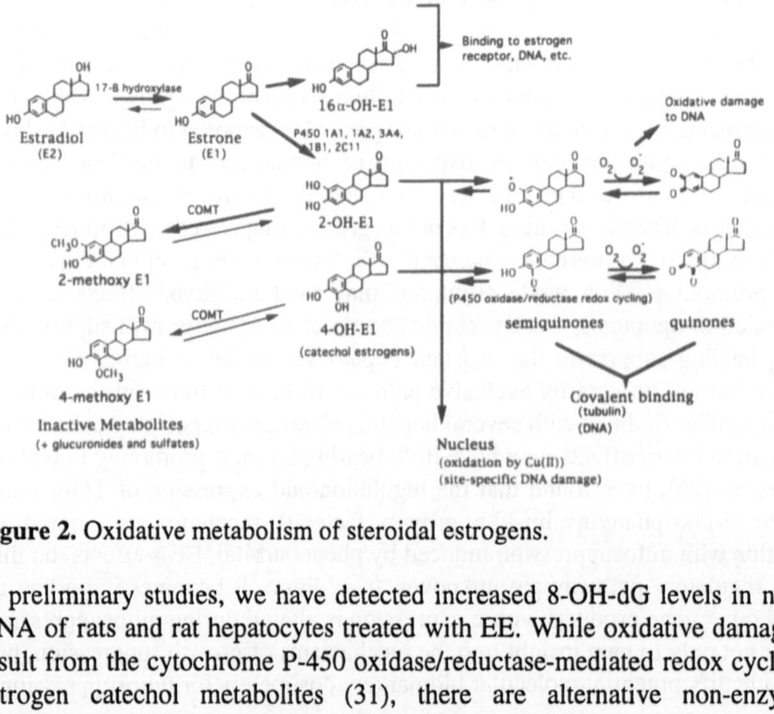

Figure 2. Oxidative metabolism of steroidal estrogens.

In preliminary studies, we have detected increased 8-OH-dG levels in nuclear DNA of rats and rat hepatocytes treated with EE. While oxidative damage can result from the cytochrome P-450 oxidase/reductase-mediated redox cycling of estrogen catechol metabolites (31), there are alternative non-enzymatic mechanisms that cause oxidation of xenobiotics. One such mechanism involves copper, which has been shown to bind DNA, particularly in association with guanines. We have shown that Cu (II) can oxidize 2-OH- and 4-OH-estradiol, resulting in the generation of a copper redox cycle and the formation of both single- and double-strand breaks in ϕx-174 phage DNA (35). Our results suggest that an alternative non-enzymatic, site-specific mechanism exists for the formation of oxidative DNA damage as illustrated in Figure 3.

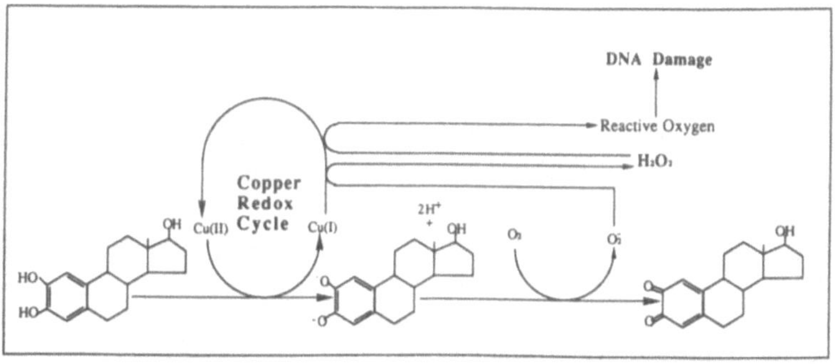

Figure 3. A scheme for the oxidative metabolism of estrogen catechols by copper, leading to a copper redox cycle and site-specific DNA damage.

In this model, an estrogen catechol which has entered the nucleus is oxidized by DNA-bound copper, setting the stage for a Cu(II)/Cu(I) redox cycle. Our results indicate that a free hydroxyl radical is not involved in formation of the DNA damage, suggesting a role for a bound reactive species (34). The estrogen quinone may also bind to DNA and proteins, contributing to the carcinogenic process. Together with those of Liehr and colleagues, our results suggest that increasing tissue P-450s, which catalyze the metabolism of estrogens to catechol metabolites, results in an increased risk for oxidative DNA damage through P-450 oxidase/reductase redox cycling, or through site-specific Cu(II)/Cu(I) redox cycling. Hence, environmental agents that induce xenobiotic metabolism may increase the level of oxidative DNA damage in normal or tumor tissue and thus contribute indirectly to the carcinogenic process through such a mechanism.

References

1. Baum JK, Holtz F, Bookstein JJ, Klein EW (1973) Possible association between benign hepatomas and oral contraceptives. Lancet 2:926-929.
2. Li JJ, Kirkman H, Li SA (1992) Synthetic estrogens and liver cancer: Risk analysis of animal and human data. In Li JJ, Nandi S, Li SA (eds): Hormonal Carcinogenesis. New York: Springer-Verlag, pp 217-224.
3. Palmer JR, Rosenberg L, Kaufman D, et al. (1989) Oral contraceptive use and liver cancer. Am J Epidemiol 130: 878-882.
4. Tavanti A, Negri E, Parazzini F, et al. (1993) Female hormone utilisation and risk of hepatocellular carcinoma. Brit J Cancer 67:636-637.
5. Taper HS (1978) The effect of estradiol-17-phenylpropionate and estradiol benzoate on N-nitrosomorpholine-induced liver carcinogenesis in ovariectomized female rats. Cancer 42:462-467.

6. Yager JD Jr, Yager R (1980) Oral contraceptive steroids as promoters of hepatocarcinogenesis in female Sprague-Dawley rats. Cancer Res 40:3680-3685.
7. Cameron R, Imaida K, Ito N (1981) Promotive effects of ethinyl estradiol in hepatocarcinogenesis initiated by diethylnitrosamine in male rats. Gann 72:339-340.
8. Wanless IR, Medline A (1982) Role of estrogens as promoters of hepatic neoplasia. Lab Invest 46:313-320.
9. Yager JD, Zurlo J, Ni N (1991) Sex hormones and tumor promotion in liver. Proc Soc Exptl Biol Med 198:667-674.
10. Coe JE, Ishak KG, Ross MJ (1990) Estrogen induction of hepatocellular carcinomas in Armenian hamsters. Hepatology 11:570-577
11. Schulte-Hermann R, Ochs H, Bursch W, Parzefall W (1988) Quantitative structure-activity studies on effects of sixteen different steroids on growth and monooxygenases of rat liver. Cancer Res 48:2462-2468.
12. Vickers AEM, Lucier GW (1991) Estrogen receptor, epidermal growth factor receptor and cellular ploidy in elutriated subpopulations of hepatocytes during liver tumor promotion by 17 α-ethinylestradiol in rats. Carcinogenesis 12:391-399.
13. Mayol X, Neal G.E, Davies R, et al. (1992) Ethinyl estradiol-induced cell proliferation in rat liver. Involvement of specific populations of hepatocytes. Carcinogenesis 13:2381-2388.
14. Peraino C, Fry RJM, Staffeldt E, Christopher JP (1975) Comparative enhancing effects of phenobarbital, amobarbital, diphenylhydantoin, and dichlorodiphenyltrichloroethane on 2-acetylaminofluorene-induced hepatic tumorigenesis in the rat. Cancer Res 35:2884-2890.
15. Yager JD, Roebuck BD, Paluszcyk TL, Memoli VA (1986) Effects of ethinyl estradiol and tamoxifen on liver DNA turnover and new synthesis and appearance of gamma glutamyl transpeptidase-positive foci in female rats. Carcinogenesis 7:2007-2014.
16. Tanaka K, Smith PF, Stromberg PC, et al. (1992) Studies of early hepatocellular proliferation and peroxisomal proliferation in Sprague-Dawley rats treated with tumorigenic doses of clofibrate. Toxicol Appl Pharmacol 116:71-77.
17. Schulte-Hermann R (1974) Induction of liver growth by xenobiotic compounds and other stimuli. CRC Crit Rev Toxicol 3:97-158.
18. Ni N, Yager JD (1994) Co-mitogenic effects of estrogens on DNA synthesis induced by various growth factors in cultured female rat hepatocytes. Hepatology 19:183-192.
19. Yager JD, Shi YE (1991) Synthetic estrogens and tamoxifen as promoters of hepatocarcinogenesis. Preventive Med 20:27-37.

20. Ni N, Yager JD (1994) The co-mitogenic effects of various estrogens for TGF-α-induced DNA synthesis in cultured female rat hepatocytes. Cancer Lett, 841:133-140.
21. Jirtle RL, Meyer SA (1991) Liver tumor promotion: Effect of phenobarbital on EGF and protein kinase C signal transduction and transforming growth factor-β expression. Digestive Dis Sci 36:659-668.
22. Barbason H, Rassenfosse C, Betz EH (1983) Promotion mechanism of phenobarbital and partial hepatectomy in DENA hepatocarcinogenesis cell kinetics effect. Brit J Cancer 47:517-525.
23. Abanobi SE, Lombardi B, Shinozuka H (1982) Stimulation of DNA synthesis and cell proliferation in the liver of rats fed a choline-devoid diet and their suppression by phenobarbital. Cancer Res 42:412-415.
24. Yager JD, Zurlo J, Sewall C, et al. (1994) Growth stimulation followed by growth inhibition in livers of female rats treated with ethinyl estradiol. Carcinogenesis 15:2117-2123.
25. Farber E (1990) Clonal adaptation during carcinogenesis. Biochem Pharmacol 39:1837-1846.
26. Jirtle RL, Hankins GR, Reisenbichler H, Boyer IJ (1994) Regulation of mannose 6-phosphate/insulin-like growth factor-II receptors and transforming growth factor beta during liver tumor promotion with phenobarbital. Carcinogenesis 15:1473-1478.
27. Martucci CP, Fishman J (1993) P450 enzymes of estrogen metabolism. Pharmacol Ther 57:237-257.
28. Guengerich FP (1991) Oxidation of estrogens and other steroids by cytochrome P-450 enzymes: Relevance to tumorigenesis. In Li JJ, Nandi S, Li SA (eds): Hormonal Carcinogenesis: Proceedings of the first international symposium. New York: Springer-Verlag, pp 104-109.
29. Fishman J, Swaneck G (1991) Differential interactions of estradiol metabolites with the estrogen receptor: Genomic consequences and tumorigenesis In: Li JJ, Nandi S, Li SA (eds) Hormonal Carcinogenesis, New York, Springer-Verlag, pp 95-103.
30. Telang NT, Suto A, Wong G, et al. (1992) Induction by estrogen metabolite 16α-hydroxyestrone of genotoxic damage and aberrant proliferation in mouse mammary epithelial cells. J Natl Cancer Inst 84:634-638.
31. Liehr JG, Roy D (1990) Free radical generation by redox cycling of estrogens. Free Rad Biol Med 8: 415-423.
32. Han X, Liehr JG (1994) DNA single-strand breaks in kidneys of Syrian hamsters treated with steroidal estrogens: Hormone-induced free radical damage preceding renal malignancy. Carcinogenesis 15:997-1000.

33. Roy D, Liehr JG (1991) Elevated 8-hydroxydeoxyguanosine levels in DNA of diethylstilbestrol-treated Syrian hamsters: Covalent DNA damage by free radicals generated by redox cycling of diethylstilbestrol. Cancer Res 51:3883-3885.
34. Malins DC, Holmes EH, Polissar NL, Gunselman SJ (1993) The etiology of breast cancer: Characteristic alterations in hydroxyl radical-induced DNA base lesions during oncogenesis with potential for evaluating incidence risk. Cancer 71: 3036-3043.
35. Li Y, Trush MA, Yager JD (1994) DNA damage caused by reactive oxygen species originating from a copper-dependent oxidation of the 2-hydroxy catechol of estradiol. Carcinogenesis 15:1421-1427.

17

Hormonal Regulation of Hepatic Cell Proliferation and Apoptosis: Implications for Carcinogenesis

Rolf Schulte-Hermann, W. Bursch, B. Grasl-Kraupp, L. Müllauer, H. Ochs, W. Parzefall, and B. Ruttkay-Nedecky

Introduction

Steroid hormones have been implicated in hepatocarcinogenesis in humans and in experimental animals. The use of oral contraceptives (OC) increases the risk of adenomas and possibly carcinomas in human liver; androgenic/anabolic steroid use has also been associated with hepatic neoplastic lesions (1-4). Furthermore, a possible role of sex steroids in hepatocarcinogenesis has been suggested by the 2.0-fold increase in male:female cancer risk ratio in areas with high exposure to aflatoxin B1, or high prevalence of hepatitis B infections (5). Induction of liver tumors has been reported in rodents after prolonged treatment with relatively high doses of several estrogens and progestins. Anabolic steroids have also been shown to induce or promote tumors in rat or mouse livers (1, 4, 6, 7). As to the possible mechanisms underlying these hepatocarcinogenic effects, two major hypotheses have been considered. One hypothesis presumes that sex steroids have tumor-initiating or complete carcinogenic activity. However, when sex steroids have been tested in assays for genotoxic or mutational effects, the results have been generally negative. Recently, two synthetic compounds, tamoxifen and cyproterone acetate (CPA), have been shown to have genotoxic effects, as well as initiating and complete carcinogenic activity in rat liver (8-13). Using sensitive techniques that detect DNA adducts, weak genotoxic effects have been reported for other steroids (14); however, the relevance of these weak effects is unknown. Indeed, for most steroidal hormones the "complete carcinogen" mechanism appears unlikely. The alternative hypothesis assumes a nongenotoxic mechanism for steroid carcinogenicity. In initiation/promotion studies, several steroids were found to be liver tumor promoters (7, 13, 15). Therefore, hepatocarcinogenesis may result from promotion of initiated cells formed spontaneously in the liver (16, 17).

In the present paper, we review the effects of some steroids on liver cell proliferation and cell death, as well as the possible implications that these effects may have in carcinogenesis.

Effects of Steroids in Rat Liver Cell Proliferation and in Cultured Rat and Human Hepatocytes

Using CPA and ethinylestradiol (EE) as prototype compounds, we have studied the effects of a variety of steroids on liver growth, as well as the change of activity of associated enzymes in the rat. Several of the steroids selected are tumorigenic in rodent liver. CPA and EE rapidly induced liver enlargement, as well as increases in total liver DNA, DNA synthesis, and mitosis. DNA synthesis of the sinus wall cells was also increased (18-23). The increase in cell proliferation was restricted to the first days of treatment and subsequently returned to normal; while the increased level of total liver DNA was maintained throughout the duration of the treatment (19, 21). The DNA rise, studied with the CPA model, was exclusively due to hyperplasia since no increase in cellular ploidy could be demonstrated. There was, however, a considerable increase in nuclear ploidy and a simultaneous dramatic decline in hepatocyte binuclearity (20). Other agents, unrelated to steroid hormones (phenobarbital, hexachlorocyclohexane, peroxisome proliferators, etc.) have shown similar effects (12).

The effects of the two prototype compounds exhibited important differences. CPA induced a significant increase in the activity of cytochrome P450-dependent microsomal liver enzymes, while 17β-estradiol (E_2) and EE have little or no effect (18, 23). The extrapolated threshold for the liver growth induced by E_2 and EE_2 was approximately 1 μg/kg, 3-4 orders of magnitude lower than that of CPA; however, the dose-response curve for CPA was steeper (23). The natural analogs, progesterone and E_2, had similar effects but of lower magnitude (21-23). Further structure-activity studies using steroids predominantly estrogenic in rats (norethynodrel and norethisterone) resulted in EE-like effects. Other progestins and spironolactone had effects similar to those observed with CPA and progesterone, while cortisol, dexamethasone, and pregnenolone-16α-carbonitrile (PCN) were less effective. All these steroids have a saturated alkyl substituent with at least two carbon atoms at position C_{17}. Progestins, gestodene, and levonorgestrel, as well as testosterone and methyltestosterone, which do not possess this alkyl side chain, had little or no effect on the parameters studied in the liver (23). These findings support the notion that estrogenic compounds exert their hepatic effects via estrogen receptors. On the other hand, the effects of the C_{17}-alkylated steroids did not appear to be associated with any "classical" endocrine activity or mechanism.

The induction of DNA synthesis by CPA was confirmed in primary cultures of adult rat hepatocytes in serum-free medium. These data raise the possibility that CPA may directly affect the liver *in vivo* (8, 26). Under similar conditions, human hepatocytes isolated from seven donors did not respond to CPA with increased DNA synthesis, although treatment with epidermal growth factor (EGF) stimulated DNA synthesis in these cells (27).

Apoptosis After Steroid Withdrawal

Within a few days of CPA withdrawal, the size of the liver returned to almost normal proportions. The enhanced DNA content rapidly returned to normal levels, although regression was not complete (19, 20, 28). We wondered how DNA was eliminated so quickly from the liver, and analyzed for the occurrence of apoptosis, a type of active cell death (29-31).

In normal resting liver, the incidence of apoptosis is very low (in adult rats approximately 2-4/10,000 hepatocytes). During treatment with CPA, we noted a moderate increase in apoptotic activity. After the treatment stopped, a dramatic increase in apoptosis occurred. CPA and other liver mitogens prevented apoptosis and sustained hyperplasia (28, 30, 32). This inhibition of apoptosis by mitogens is a general property of active cell death, in contrast to the necrotic process that occurs after severe tissue damage. We used this inhibitory effect induced by CPA to estimate the duration of apoptosis. From the kinetics of disappearance of apoptotic bodies in regressing hyperplastic liver, we calculated an average duration of 3 hours (33). This short duration explains why relatively few cells are seen undergoing apoptosis even in states of extensive tissue regression. With the duration of apoptosis known, rates of apoptosis from histological counts were determined. In the liver regressing after CPA withdrawal, we observed a rate of 0.5% per hr and, in regressing putative preneoplastic foci, 5% per hr.

Also, the feeding/fasting state of the animals appears to regulate apoptosis. After some days of complete starvation, the incidence of apoptosis strongly increased in the liver, coinciding with a fall in liver DNA content. Fasting (food restriction to 40-60% of the normal amount) facilitated induction of apoptosis in hyperplastic liver. Apoptosis in fasting animals was suppressed by feeding carbohydrates, proteins, or fat (34). It is possible glucocorticoids, glucagon, adrenalin, insulin, etc., may play a role in the regulation of liver apoptosis.

Transforming Growth Factor ß1 (TGF ß1) in Hepatocyte Apoptosis

We have been interested in TGF β1 because it inhibits DNA synthesis in regenerating liver *in vivo*, as well as in hepatocytes in culture. We found that apoptotic hepatocytes in regressing rat liver showed positive immunostaining for TGF β1 precursor protein (antibodies provided by M. Sporn, Bethesda, MD) (35). TGF β1 also induced apoptosis *in vitro* in cultured hepatocytes and *in vivo* when injected a few hrs before sacrifice (36). The induced-apoptosis effect of TGF β1 was potentiated in rats whose livers were in the state of regression due to CPA withdrawal treatment. We concluded that TGF β1 may act preferentially upon cells already committed to apoptosis by preexisting hyperplasia. Therefore, TGF β1 may not be a primary signal triggering apoptosis, but instead serve as a permissive factor. Moreover, hepatocytes preparing for apoptosis show preexisting TGF β1 a few hrs before the start of chromatin condensation. Staining for preexisting TGF β1 may provide a marker for preexisting apoptotic cells, and may also be useful to discriminate apoptosis from other types of cell death since necrotic hepatocytes induced with carbon tetrachloride (CCl_4) were not stained (35).

Recently, we determined whether TGF β1 peptide would increase in the liver after CPA treatment. For this purpose, the acid-ethanol procedure of Roberts, et al. (37), was used. The results suggested that neither in the growth period nor during involution following CPA treatment and withdrawal did a detectable increase in the concentration of TGF β1 occur. In contrast, after intoxication with CCl_4 used as control, a marked increase in liver TGF β1 was observed (Ruttkay-Nedecky, et al., personal communication). Any changes in TGF β1 levels during regression are restricted to cells undergoing apoptosis and their immediate environment, and these changes may be undetectable in whole liver homogenates.

Also, activin, a member of the TGF β1 family of polypeptides, induced apoptosis in rat liver and in cultured hepatocytes. It was one tenth as active as TGF β1, and its presence was required for 24 hrs to induce apoptosis (38). In conclusion, TGF β1 and related peptides inhibit hepatic cell proliferation and enhance cell death in the liver. This suggests they may be involved in the rat liver regulation of the balance between cell proliferation and death in the liver.

Effect of Steroid Hormones on Different Stages of Liver Carcinogenesis

Carcinogenesis includes initiation, promotion, and progression. Initiation can be induced through genotoxic carcinogens, and may result from mutational events on crucial genes or chromosomes. Initiation can also occur "spontaneously", as has been demonstrated in rat liver (16, 39). This "spontaneous" initiation is reflected in most human cancers, particularly in hormone-dependent organs, where the factors responsible for initiation are usually unknown. Promotion results in growth stimulation of initiated cells, and preneoplastic cell clones give rise to accelerated tumor development (17).

In rodent liver, histological, histochemical, and immunocytochemical techniques allow the identification of putative initiated cells and clones as single cells or cell foci. Some estrogenic and progestational steroids, as well as many other nongenotoxic hepatocarcinogens have been shown to accelerate the growth rate of these foci, leading to early appearances of neoplastic nodules and/or frank carcinomas (7, 13, 17). We have shown that high doses of CPA and progesterone effectively promote liver tumor development (15).

Cell Proliferation and Cell Death in Hepatic Foci: Implications for Liver Carcinogenesis

Early studies on growth kinetics of putative preneoplastic (ppn) liver foci revealed rates of DNA synthesis and mitosis 5.0- to 10.0-fold higher than those observed in surrounding unaltered liver tissue. Interestingly, the growth rate of ppn liver foci was much lower than predicted from their high proliferation rate. This apparent contradiction was resolved when we found that ppn foci exhibited a high incidence of apoptosis. This largely counterbalanced the high proliferative activity so that little net growth of foci occurred (28, 40). Promoting steroids like CPA or progesterone, as well as nonsteroid promoters such as phenobarbital, increased the proliferative activity of ppn liver foci above their already enhanced level and simultaneously inhibited apoptosis (15, 28, 40, 41). As a result, the accumulation of preneoplastic cells led to a dramatic acceleration of ppn liver focal growth. Thus, liver ppn cells respond to tumor promoters in a manner qualitatively similar to normal liver cells, but stronger.

Fasting induced effects opposite to those observed after treatment with tumor promoters. When the daily amount of food was restricted for 95 days to 60% of *ad libitum* consumption, normal liver DNA synthesis was reduced and apoptosis slightly enhanced. Similar responses occurred in ppn liver foci but were much more pronounced; total liver cells (measured as DNA content) decreased by 15%, while ppn focal cells decreased by 85%. Fasting, by favoring a shift from cell proliferation to cell death, inhibited promotion and enhanced apoptosis. These effects resulted in the elimination of numerous initiated clones, as suggested by the reduced number of ppn foci, and by the reduced tumor response observed after subsequent treatment with the potent promoter nafenopin (a peroxisome proliferator) (34). These data show, for the first time, that food restriction may induce preferential apoptosis in preneoplastic cell clones. This may contribute to the well-known fasting-induced protective effect against cancer development in experimental animals and humans.

We also studied whether endogenous signals were involved in the control of cell proliferation of preneoplastic cells. Although tamoxifen treatment did not inhibit DNA synthesis in normal or ppn liver cells, this observation by itself does not provide sufficient evidence to demonstrate the involvement of endogenous estrogens in the control of liver focal cell proliferation. Similarly, TGF β1 induced the same level of apoptosis in normal and ppn liver focal cells, suggesting again that altered responsiveness to this agent may not be crucially important in ppn cells (Müllauer, et al., personal communication).

Conclusion

These studies suggest that, in the liver, the significantly higher levels of cell proliferation and cell death observed in initiated and preneoplastic cells, when compared with normal hepatic cells, may be the result of an intrinsic defect in growth control. A similar shift, but even more pronounced, occurs in liver tumor cells (unpublished data). This defect renders preneoplastic and neoplastic cells highly susceptible to signals stimulating either replication or apoptosis.

Because of the enhanced levels of cell proliferation and apoptosis, the mitotic effects of steroid hormones, as well as of any other liver-tumor promoters, on initiated (preexisting) neoplastic cells may be more pronounced than those observed in normal liver cells. Promoters or overfeeding protect preneoplastic cells from death and favor their replication; these effects result in selective growth advantage. Antipromoting and anti-initiating effects may be exerted by the metabolic and hormonal alterations induced by food restriction.

References

1. International Agency for Research on Cancer (1987) Evaluation of carcinogenic risks to humans. IARC Monograph Suppl 7.
2. Rosenberg L (1991) The risk of liver neoplasia in relation to combined oral contraceptive use. Contraception 43:643-652.
3. Tao L-C (1991) Oral contraceptive-associated liver cell adenoma and hepatocellular carcinoma. Cancer 68:341-347.
4. Shimoji N, Imaida K, Hasegawa C, et al. (1990) Enhanced effect of oxymetholone, an anabolic steroid, on development of liver cell foci in rats initiated with N-diethylnitrosamine. Cancer Lett 49:165-168.
5. International Agency for Research on Cancer (1993) Evaluation of carcinogenic risks to humans. IARC Monograph Vol 56.

6. Schuppler J, Schulte-Hermann R, Timmermann-Trosiener I, et al. (1982) Proliferative liver lesions and sex steroids in rats. Toxicol Pathol 10:132-144.
7. Yager JD, Campbell HA, Longnecker DS, et al. (1984) Enhancement of hepatocarcinogenesis in female rats by ethinyl estradiol and mestranol but not estradiol. Cancer Res 44:3862-3869.
8. Neumann I, Thierau D, Andrae U, et al. (1992) Cyproterone acetate induces DNA damage in cultured rat hepatocytes and preferentially stimulates DNA synthesis in gamma-glutamyltranspeptidase-positive cells. Carcinogenesis 13:373-378.
9. Topinka J, Andrae U, Schwarz LR, et al. (1993) Cyproterone actetate generates DNA adducts in rat liver and in primary rat hepatocyte cultures. Carcinogenesis 14:423-427.
10. Deml E, Schwarz LR, Oesterle D (1993) Initiation of enzyme-altered foci by the synthetic steroid cyproterone actetate in rat liver foci. Carcinogenesis 14:1229-1231.
11. Han X, Liehr JG (1992) Induction of covalent DNA adducts in rodents by tamoxifen. Cancer Res 52:1360-1363.
12. Hard GC, Iatropoulos MJ, Jordan K, et al. (1993) Major difference in the hepatocarcinogenicity and DNA adduct forming ability between toremifene and tamoxifen in female Crl:CD(BR) rats. Cancer Res 53:4534-3541.
13. Ghia M, Mereto E (1989) Induction and promotion of gamma-glutamyltranspeptidase-positive foci in the liver of female rats treated with ethinyl estradiol, clomiphene, tamoxifen and their associates. Cancer Lett 46:195-202.
14. Shimomura M, Higashi S, Mizumoto R (1992) ^{32}P-postlabeling analysis of DNA adducts in rats during estrogen-induced hepatocarcinogenesis and effect of tamoxifen on DNA adduct level. Jpn J Cancer Res 83:438-444.
15. Schulte-Hermann R, Schuppler J, Ohde G (1982) Effect of tumor promoters on proliferation of putative preneoplastic cells in rat liver. Carcinogenesis 7:99-104.
16. Schulte-Hermann R, Timmermann-Trosiener I, Schuppler J (1983) Promotion of spontaneous preneoplastic cells in rat liver as a possible explanation of tumor production by nonmutagenic compounds. Cancer Res 43: 839-844.
17. Schulte-Hermann R (1985) Tumor promotion in the liver. Arch Toxicol 57:147-158.
18. Schulte-Hermann R, Parzefall W (1980) Adaptive responses of rat liver to the gestagen and anti-androgen cyproterone acetate and other inducers. I. Induction of drug-metabolizing enzymes. Chem Biol Interact 31:279-286.

19. Schulte-Hermann R, Hoffmann V, Parzefall W, et al. (1980) Adaptive responses of rat liver to the gestagen and anti-androgen cyproterone acetate and other inducers. II. Induction of growth. Chem Biol Interact 31:287-300.
20. Schulte-Hermann R, Hoffmann V, Landgraf H (1980) Adaptive responses of rat liver to the gestagen and anti-androgen cyproterone acetate and other inducers. III. Cytological changes. Chem Biol Interact 31:301-311.
21. Ochs H, Düsterberg H, Günzel P, et al. (1986) Effect of tumor promoting contraceptive steroids on growth and drug metabolizing enzymes in rat liver. Cancer Res 46:1224-1232.
22. Ochs H, Düsterberg B, Schulte-Hermann R (1986) Induction of monooxygenases and growth in rat liver by progesterone. Arch Toxicol 59:146-149.
23. Schulte-Hermann R, Ochs H, Bursch W, et al. (1988) Quantitative structure-activity studies on effects of sixteen different steroids on growth and monooxygenases of rat liver. Cancer Res 48:2462-2468.
24. Schulte-Hermann R (1974) Induction of liver growth by xenobiotic compounds and other stimuli. Crit Rev Toxicol 3:97-158.
25. Gerlyng P, Grotmol T, Seglen PO (1994) Effect of 4-acetylaminofluorene and other tumour promoters on hepatocellular growth and binucleation. Carcinogenesis 15:371-379.
26. Parzefall W, Monschau P, Schulte-Hermann R (1989) Induction by cyproterone acetate of DNA synthesis and mitosis in primary cultures of adult rat hepatocytes in serum free medium. Arch Toxicol 63:456-461.
27. Parzefall W, Erber E, Sedivy R, et al. (1991) Testing for induction of DNA synthesis in human hepatocyte primary cultures by rat liver tumor promoters. Cancer Res 51:1143-1147.
28. Bursch W, Lauer B, Timmermann-Trosiener I, et al. (1984) Controlled death (apoptosis) of normal and putative preneoplastic cells in rat liver following withdrawal of tumor promoters. Carcinogenesis 5:453-458.
29. Kerr JFR, Wyllie AH, Currie AR (1972) Apoptosis: A basic biological phenomenon with wide-ranging implications in tissue kinetics. J Cancer 26:239-257.
30. Bursch W, Taper HS, Lauer B, et al. (1985) Quantitative histological and histochemical studies on the occurrence and stages of controlled cell death (apoptosis) during regression of rat liver hyperplasia. Virchows Arch 50:153-166.
31. Bursch W, Oberhammer F, Schulte-Hernann (1992) Cell death by apoptosis and its protective role against disease. Trends in Pharmacol Sci 13:245-251.
32. Bursch W, Düsterberg B, Schulte-Hermann R (1986) Growth, regression and cell death in rat liver as related to tissue levels of the hepatomitogen cyproterone acetate. Arch Toxicol 59:221-227.

33. Bursch W, Paffe S, Putz B, et al. (1990) Determination of the length of the histological stages of apoptosis in normal liver and in altered hepatic foci of rats. Carcinogenesis 11:847-853.
34. Grasl-Kraupp B, Bursch W, Ruttkay-Nedecky B, et al. (1995) Food restriction eliminates preneoplastic cells through apoptosis and antagonizes carcinogenesis in rat liver. Proc Nat Acad Sci USA, in press.
35. Bursch W, Oberhammer F, Jirtle RL, et al. (1993) Transforming growth factor β1 as a signal for induction of cell death by apoptosis. Brit J Cancer 67:531-536.
36. Oberhammer F, Pavelka M, Sharma S, et al. (1992) Induction of apoptosis in cultured hepatocytes and in regressing liver by transforming growth factor-β1. Proc Natl Acad Sci USA 89:5408-5412.
37. Robert AB, Lamb LC, Newton DL, et al. (1980) Transforming growth factors: Isolation of polypeptides from virally and chemically transformed cells by acid/ethanol extraction. Proc Nat Acad Sci USA 77:3494-3498.
38. Schwall RH, Robbins K, Jardieu P, et al. (1993) Activin induces cell death in hepatocytes *in vivo* and *in vitro*. Hepatology 18:347-356.
39. Kraupp-Grasl B, Huber W, Taper H, et al. (1991) Increased susceptibility of aged rats to hepatocarcinogenesis by the peroxisome proliferator nafenopin and the possible involvement of altered liver foci occurring spontaneously. Cancer Res 51:666-671.
40. Schulte-Hermann R, Timmermann-Trosiener I, Barthel G, et al. (1990) DNA synthesis, apoptosis and phenotypic expression as determinants of growth of altered foci in rat liver during phenobarbital promotion. Cancer Res 50:5127-5135.
41. Schulte-Hermann R, Ohde G, Schuppler J, et al. (1981) Enhanced proliferation of putative preneoplastic cells in rat liver following treatment with the tumor promoters phenobarbital, hexachlorocyclohexane, steroid compounds and nafenopin. Cancer Res 41:2556-2562.

18

Mechanisms of Tamoxifen-Induced Genotoxicity and Carcinogenicity

Ian N.H. White and Lewis L. Smith

Introduction

Tamoxifen is an anti-estrogen widely used with proven efficacy in the treatment of breast cancer in women (1). It is well tolerated and has potentially beneficial actions in reducing serum cholesterol (2) and the incidence of fatal heart attacks (3). Preliminary results suggest that tamoxifen will reduce the incidence of breast cancer in such women (2). However, several epidemiological studies in women with breast cancer have shown that tamoxifen can also lead to increased incidence of endometrial tumors (4-6). Although tamoxifen is primarily an anti-estrogen, it also has some estrogenic properties. At present, it is not clear whether the effect on the endometrium is due to estrogen-like stimulation by tamoxifen, causing cell proliferation and the promotion of endogenous lesions, or a direct mutagenic effect on the affected cells.

Tamoxifen is not positive in Ames *Salmonella* tests for genotoxicity (7). However, further concerns about its safety in healthy women were raised when it was found that long-term tamoxifen administration to rats at dose levels considerably higher than those used therapeutically (up to 35 mg/kg/day) gave rise to hepatocellular carcinomas (8-10). In these studies, there were no reports of tumors of the reproductive system. Differences in response to tamoxifen between species were confounded by differences in its hormonal actions. In mice and dogs, tamoxifen is estrogenic; in rats and humans, it has anti-estrogen/estrogen actions; and in the chick, it is a pure anti-estrogen (11). Long-term treatment of mice with tamoxifen showed no liver tumors (7). At the present time, rats are the only laboratory animal species in which liver tumors have been reported.

More evidence is required concerning the mechanism of carcinogenic action of tamoxifen in experimental systems so that women taking this drug may be better informed of risk-benefit analysis. Since the discovery that tamoxifen itself was not genotoxic but could be activated in the rat liver to produce genotoxic intermediates (12), the strategy for studies initiated in the MRC Toxicology Unit has been to define the nature of the activating enzyme systems, factors influencing DNA damage, and the development of liver tumors.

The Mechanisms of Tamoxifen Genotoxicity and its Carcinogenic Effects in Rats

Work in this field has been divided into four specific areas: 1. Studies on the genotoxic potential of tamoxifen and related analogues. Toremifene and droloxifene (Figure 1), chemical analogs of tamoxifen that share similar therapeutic actions, do not induce liver tumors in rats in lifetime dosing studies (9, 10, 13). 2. Studies on the induction of cytochrome P-450 drug-metabolizing enzymes and the isoenzymic forms of P-450 involved in the metabolism and metabolic activation of tamoxifen in rats, mice, and women. 3. Determination of the differential induction of tamoxifen-induced lesions in three strains of rat and mouse. 4. Studies on the response of women treated with tamoxifen.

The prime objective of the present study was to establish animal models that will allow us to identify those factors contributing to development of hepatic tumors in rat but not in mouse. Having established these, it may then be possible to determine whether they also operate in humans and, if so, establish the quantitative relationship between potential responses to tamoxifen of human and rat liver.

All of the evidence so far supports the view that tamoxifen itself is not genotoxic. It needs to be metabolized to reactive intermediate(s) that will bind irreversibly with DNA. This

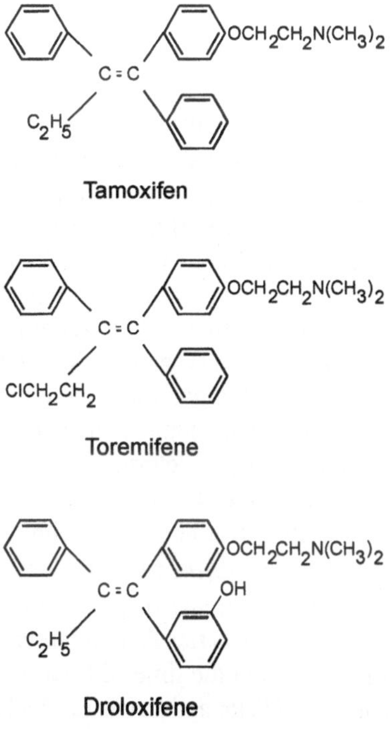

Figure 1. Chemical structures of tamoxifen, toremifene, and droloxifene.

interaction is quite distinct from tamoxifen binding to the estrogen receptor and is not related to the estrogen/anti-estrogen potency of the drug.

DNA Damage and Pathological Changes in the Liver

We have investigated the ability of tamoxifen and its sister anti-estrogen drugs, toremifene and droloxifene (Figure 1), to induce DNA damage using [^{32}P]-post-labeling. It was established that tamoxifen could cause such damage, even after a single dose, and that this damage was selective for the liver. No damage could be detected in DNA extracted from lung, kidney, or peripheral lymphocytes. In liver DNA, the degree of damage was dependent on dose and length of exposure. Toremifene caused only trace levels, and droloxifene no detectable labeling (12). Tamoxifen-induced [^{32}P]-post-labeling was detected in DNA extracted from mouse livers, but the level of adduct formation after seven days of treatment was about one third that observed in rats. There is an apparent correlation between [^{32}P]-post-labeling of DNA in the liver and production of tumors. It is recognized, however, that such DNA damage is only the first step in a multi-sequence pathway and may not necessarily lead to tumor formation.

Following tamoxifen treatment, there are two features of the DNA lesions observed in rat liver that may illuminate its hepatocarcinogenic effects. First, following withdrawal, the adducts are only slowly repaired, so that after thirty days some loss is detectable (12), but it takes three months for a 50% reduction of the initial levels (14). Second, in rats continuously exposed to tamoxifen, there appears to be no equilibrium between DNA damage and repair. The extent of DNA damage continues to increase for many months (Figure 2).

There is strong evidence that cell proliferation plays an important role in the promotion and progression of DNA damage. In a study involving female Fischer, Wistar, and Lewis rats given dietary tamoxifen (420 ppm) corresponding approximately to 40 mg/kg/day, there were only small differences between the parent compound and its major metabolites in the liver. Similarly, the extent of hepatic DNA damage, determined by [^{32}P]-post-labeling after six months of treatment in all three strains was about 3000 adducts/10^8 nucleotides (15). At three months after treatment, using either conventional histochemical staining or the more selective enzyme markers, γ-glutamyl-transpeptidase or glutathione S-transferase P, the number of positive liver foci was about 10.0-fold higher in both Wistar and Lewis rats than in the Fischer animals. There were also marked strain differences in the time of liver tumor development. After six months of treatment, both Wistar and Lewis rats had liver tumors, while the Fischer animals had none.

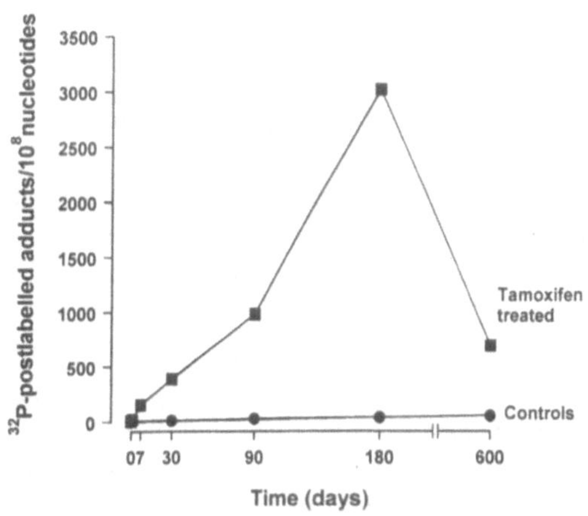

Figure 2. Effects of time on the accumulation of [^{32}P]-post-labeled adducts in the liver DNA of female Fischer rats exposed to tamoxifen. (Data modified from 12 and 15).

After eleven months, all of the Wistar and Lewis rats had liver carcinomas, while the Fischer animals had none. Carcinomas were seen in Fischer animals only when they were killed at twenty months. A comparison of the extent of hepatic parenchymal cell division (determined either by bromodeoxyuridine incorporation or immunohistochemically by proliferating cell nuclear antigen), relative to control untreated animals after six months of tamoxifen exposure, showed a decrease in Fischer rats but an increase in Wistar and Lewis rats. Therefore, the increase in cell proliferation induced by tamoxifen is consistent with its promotion of liver tumor foci, and with the subsequent tumor progression in the latter two strains.

If one action of tamoxifen is to promote initiated cells, can other tumor promoters have a similar effect? This seems to be the case. When Wistar rats were exposed to the same dose of dietary tamoxifen for only three months, the level of hepatic DNA damage determined by [^{32}P]-post-labeling was about 400 adducts/10^8 nucleotides. Development of liver tumors was also observed when these animals were returned to a normal diet and subsequently promoted with phenobarbitone in the drinking water. Even in about one third of those animals receiving no additional promotion after the initial three months, tamoxifen treatment induced liver tumors in a lifetime study (16).

The relative contributions of DNA damage initiation by tamoxifen, and the subsequent promotion by tamoxifen itself or by phenobarbitone remains to be established. Following [^{32}P]-post-labeling, there were many DNA adducts separable by HPLC or two-dimensional PEI-cellulose chromatography. From previous studies with other liver carcinogens, it appears highly probable that the adducts detected will not all be equally effective in causing liver neoplasias.

Effects on Drug Metabolism and Cytochrome P-450

When rats were given tamoxifen (40 mg/kg/day for 4 days), total microsomal cytochrome P-450 was slightly reduced. However, compared with controls, the rate of benzyloxy- or pentoxyresorufin metabolism by liver microsomal preparations increased 30.0- to 60.0-fold (17). Smaller increases were seen in the 6β- and 16α-hydroxylation of testosterone, as well as the oxidation of testosterone to androstenedione. These results suggest that tamoxifen induces predominantly CYP2B1/CYP2B2 and CYP3A families. This was confirmed by immunoblotting experiments which showed a 2.0- to 3.0-fold increase in CYP2B1, CYP2B2, and 3A1 proteins in liver microsomal fractions. The chemically related anti-estrogens, toremifene and droloxifene, also increased these liver cytochromes, although droloxifene was less effective. Treatment of female C57Bl/6 mice with tamoxifen, however, caused the rate of pentoxyresorufin dealkylase activity to decrease.

We have studied the metabolism of tamoxifen in naive and pretreated Fischer rats and C57Bl/6 mice. Control mice were found to metabolize tamoxifen at a much greater rate than rats (18). Tamoxifen pretreatment stimulated the rate of N-demethylation in rat liver microsomes but not in mouse microsomes, whereas 4-hydroxylation was depressed by pretreatment with the drug in both species. This depression was more significant in the mouse than the rat. We have concluded that, by stimulating its own metabolism, tamoxifen may accelerate the rate of its own excretion and could also increase the production of a reactive genotoxic metabolite(s). Some support for this view comes from studies looking at the effects of tamoxifen on unscheduled DNA synthesis in isolated rat hepatocytes. In a standard OECD-guideline unscheduled DNA synthesis assay, tamoxifen was negative. When tamoxifen was administered to rats for four days and the isolated hepatocytes exposed to tamoxifen *in vitro*, a small but significant increase in unscheduled DNA synthesis was observed (12). This could suggest that tamoxifen was stimulating its own metabolism. Alternatively, it could be that after one exposure *in vitro*, damage to the DNA was so mild that it takes several exposures before the sensitivity of the assay is sufficient to detect DNA damage. Toremifene, over a similar range of concentrations, gave negative results in hepatocytes from naive animals nor those from tamoxifen-pretreated animals.

[^{32}P]-Post-Labeling of DNA from Women Taking Tamoxifen Therapeutically

Liver DNA obtained from seven women receiving 20 or 40 mg/kg/day tamoxifen or from a "control group" not receiving this drug was analyzed using [^{32}P]-post-labeling. DNA damage was detected in both groups, but the patterns of post-labeled spots were not the same as that detected in tamoxifen-treated rat liver DNA. No significant difference in the level of DNA damage (18 to 80 adducts/10^8 nucleotides) was observed between the two groups (19). The marked difference in levels of hepatic DNA damage between rats that develop liver tumors (3000 adducts/10^8 nucleotides) and women suggests that the hazard may be considerably less in humans. Several factors may affect this finding. First, in the present study, only a few human samples were obtained compared with the very large numbers of women treated with this drug. Second, many factors are likely to influence individual susceptibility to the carcinogenicity of tamoxifen treatment. These may include genetic polymorphisms in Phase 1 or Phase II enzymes responsible for the activation and detoxification of tamoxifen and for the balance between these pathways; the efficiency of DNA repair; and the extent of cell proliferation.

Metabolism of Tamoxifen

We have investigated tamoxifen metabolism in liver microsomal systems to compare the qualitative and quantitative differences among mice, rats, and humans (18). Comparison of the metabolites produced by rat and human microsomal preparations showed the patterns to be very similar. However, there is a considerable quantitative difference. The rat liver metabolizes tamoxifen more rapidly than does the human liver. This can also be seen in terms of the half-life in the serum of rats and women treated with this drug. Tamoxifen is cleared in a matter of hours from rat serum with a t½ of »14h, whereas the half-life in women treated with this drug is measured in days (terminal t½ of »7 days) (20). If the parent compound tamoxifen itself were responsible for liver tumors, then, based on these results, women exposed to the drug would be likely to be at greater risk than the rat. However, if it is shown that a metabolic activation of tamoxifen is required to induce DNA damage and tumor development, then the much more rapid metabolism of tamoxifen to a reactive intermediate by the rat would indicate that the risk to women is quantitatively much less than in the rat. In contrast to the similar patterns of metabolites in humans and rats, a comparison of tamoxifen metabolites in human and mouse liver indicates qualitative differences. Although similar metabolites are produced, the mouse much more effectively hydroxylates tamoxifen, suggesting that the drug may be cleared more rapidly from the liver. However, more work is required before the precise differences among rats, mice, and humans are established.

Isoenzymic Forms of Cytochrome P-450 Involved in Tamoxifen Activation

Clastogenicity Using Crespi Cell Lines

Using a human lymphoblastoma derived from an MCL-5 cell line in which the human isoenzymes CYP1A1, 1A2, 2A6, 3A4, and 2E1 are functionally expressed together with P-450 reductase and epoxide hydrolase, tamoxifen was positive in a micronucleus assay (12). In such a test system, where the majority of the Phase II detoxification systems are absent and there are no pharmacokinetic considerations such as biliary elimination, toremifene also gave a positive result, though with a lower activity (21) (Figure 3). We have extended these observations using similar cell lines that express individual cytochrome P-450s. Our results suggest that the isoenzymes CYP2E1 and CYP3A4 may be capable of metabolizing tamoxifen to genotoxic intermediates, as judged by a positive micronucleus test. The Crespi cells expressing CP1A1, CYP1A2, CYP2D6, CYP2A6, or CYP2B6 did not give positive results over the range of concentrations used with the former two isoenzymes. It should be noted that, because the extent to which these various isoenzymes are expressed within the cells has not been established, one cannot say categorically that negative results reflect the response in either rodent or human tissues. These results do show that human P-450s can activate tamoxifen, and at concentrations normally found in the serum of women taking tamoxifen therapeutically (»300 ng/ml).

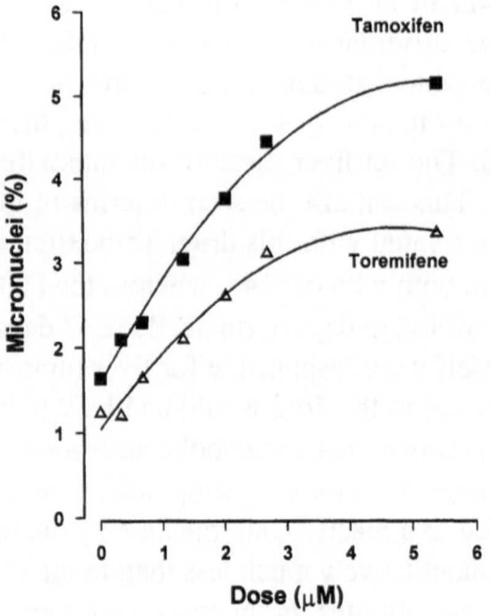

Figure 3. Effects of tamoxifen or toremifene dose on micronucleus formation in Crespi MCL-5 cells (21).

Irreversible Binding to Microsomal Protein

Human and rat liver microsomal preparations are able to activate [^{14}C]-tamoxifen in the presence of NADPH to bind irreversibly to microsomal proteins. When the microsomal proteins were run on SDS-gels, a major protein involved in such covalent interaction appeared to be a 54-kDa component that may be the isoenzymic form of cytochrome P-450 involved in the activation (22). Protein binding must be used as an index of metabolic activation and as a surrogate for DNA binding. The extent of binding to DNA appears to be of the order of 50.0-fold lower than to protein and at the limit of detection using conventional liquid scintillation methods for radioactive detection. Using a panel of twelve human microsomal preparations characterized for the cytochrome P-450 content of nine CYP isoenzyme forms by Western blotting, we showed that CYP3A4 and CYP2B6 were involved with the metabolic activation of tamoxifen to metabolites that covalently bound to protein (23). This study suggests that the same isoenzyme forms involved in the N-demethylation of tamoxifen are also involved in covalent binding. The 4-hydroxylation reaction is catalyzed by a distinct CYP2C9 isoenzyme. It is proposed that how well the tamoxifen molecule fits at the CYP3A active site determines whether the activated oxygen species, derived from molecular oxygen, bring about a detoxification reaction, such as N-demethylation, or an activation reaction resulting in irreversible binding. In this study, no correlation with the presence of CYP2E1 was found. Consistent with the involvement of CYP3A4 and CYP2B6, pretreatment of rats with either dexamethasone or phenobarbitone caused a significant increase in the protein binding and N-demethylation rates, whereas 4-hydroxylation was not significantly affected. As yet, we cannot distinguish whether the active metabolites involved in protein and in DNA binding are the same.

The binding of tamoxifen to protein *in vitro* in rats is 3.8-fold, and in mice 17.0-fold higher than in human liver microsomes (23). In this respect, the greater activity in the mouse microsomal preparations reflects their higher levels of overall metabolism. Although the liver was the main site of activation, binding in microsomal preparations from normal human breast tissue could also be detected at rates 7.0-fold lower than in human liver. A major difference in the response of mice to tamoxifen from those of rats or humans was the apparent Km value. The mouse enzyme is saturated at relatively low tamoxifen concentrations, whereas apparent Km values are over an order of magnitude higher in rat and human liver microsomal preparations. Such studies suggest that the rat is a better model than the mouse for human liver microsomal activation of tamoxifen, with respect to both kinetic parameters and the pattern of metabolic products. The genetic diversity of the drug-metabolizing enzymes in the human population and of the induction of the P-450 enzymes will affect an individual's response to this drug's harmful effects. The present results show that, while the studied *in-vitro* systems demonstrate the *potential* for activation, they do not reflect the actual extent to which this will occur *in vivo*. In the case of mice, they fail to explain the low level of DNA damage resulting when this species is treated with tamoxifen.

In the cell, DNA damage will depend on not only the rate of tamoxifen activation of but also its rate of detoxification by other routes such as N-demethylation, N-oxidation, or Phase II pathways. Once formed, the active metabolite can be detoxified by pathways such as conjugation with reduced glutathione or hydrolysis by epoxide hydrolases (Figure 4). Most of these pathways are absent from an *in-vitro* microsomal system; therefore, it is not surprising that these systems give a poor representation of the situation seen in the whole animal.

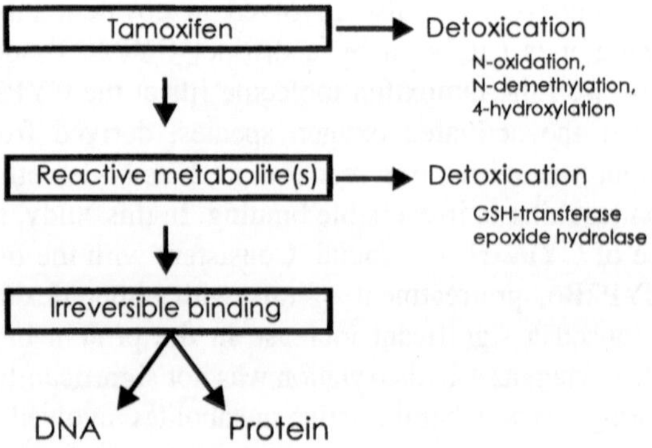

Figure 4. Routes of detoxification of tamoxifen and its reactive metabolite.

Activation of Tamoxifen by Peroxidases

Peroxidase activation of tamoxifen has recently been demonstrated (24, 25). Horseradish peroxidase and peroxidase-related enzymes, such as prostaglandin synthase in the presence of H_2O_2, can N-oxidize, N-demethylate, and activate tamoxifen to induce protein binding, causing damage to exogenous calf thymus DNA. The peroxidase system has a lower substrate selectivity than the P-450s since tamoxifen and toremifene are activated to a similar extent (25). The pattern of [^{32}P]-post-labeled adducts following two-dimensional PEI-cellulose chromatography appears similar to the major adducts seen following tamoxifen or toremifene treatment of rats. With respect to DNA damage using the peroxidase activation system, 4-hydroxytamoxifen appears to be a much better substrate than tamoxifen.

Identification of Tamoxifen Active Metabolite(s): Role of Epoxides?

It had been shown that a model ethylenic epoxide of tamoxifen, prepared chemically, would react slowly with calf thymus DNA in the absence of activating enzymes to give adducts detectable by [^{32}P]-post-labeling (26). These adducts had similar Rf values to those of adducts formed upon tamoxifen treatment of rats. Because of the high chemical stability of this epoxide, it is unlikely that it could be a significant metabolite formed by liver microsomes and it has not been detected in such a system. Studies using liquid chromatography with on-line electrospray mass spectrometry have detected the presence of several metabolites formed from tamoxifen in microsomal incubation mixtures corresponding formally to the addition of oxygen to the drug. A number of metabolites with known retention times on HPLC, such as tamoxifen N-oxide, can be assigned to the peaks observed. Two peaks believed to represent aromatic 3,4-epoxide and 3',4'-epoxide of tamoxifen have been described. In the presence of acid, they are converted to the corresponding dihydrodiols. Formation of these epoxides is detected using rat, mouse, and human liver microsomal preparations (18). In the peroxidase activation system, such 3,4-epoxytamoxifen metabolites can also be detected (25). Other putative active metabolites have been proposed, such as α-hydroxy-ethyl-tamoxifen (27) or an intermediate resulting from further activation of 4-hydroxytamoxifen (24).

Conclusions

Three factors have been found to be important in the development of tamoxifen-induced liver tumors in rats. These are: 1) the nature and quantity of metabolism; 2) the initiation of DNA damage; and 3) the expression of this damage by the degree of cell proliferation with altered DNA. We have examined the difference in relative rates of metabolism and metabolic activation in the rat and human liver, and in a small number of women. Results show that the possible risk of liver tumors in tamoxifen therapy is probably low.

References

1. Cummings FJ, Gray R, Tormey DC, et al. (1993) Adjuvant tamoxifen versus placebo in elderly women with node-positive breast cancer: Long-term follow-up and causes of death. J Clin Oncol 11:29-35.
2. Powles TJ, Jones AL, Ashley SE, et al. (1994) The Royal Marsden Hospital pilot tamoxifen chemoprevention trial. Breast Cancer Res Treat 31:73-82.
3. McDonald CC, Stewart HJ (1991) Fatal myocardial infarction in the Scottish adjuvant tamoxifen trial. Brit Med J 303:435-437.

4. Marshall E (1994) Tamoxifen: Hanging in the balance. Science 264:1524-1527.
5. Jordan VC (1988) Tamoxifen and endometrial cancer. Lancet 2:1019
6. Fornander T, Rutqvist LE, Wilking N (1991) Effects of tamoxifen on the female genital tract. Ann N Y Acad Sci USA 622:469-473.
7. Tucker MJ, Adams HK, Patterson JS (1984) Tamoxifen. In Laurence DR, McLean AEM, Weatherall M (eds): Safety Testing of New Drugs. New York: Academic Press, pp 125-161.
8. Greaves P, Goonetilleke R, Nunn G, et al. (1993) Two-year carcinogenicity study of tamoxifen in Alderly Park Wistar-derived rats. Cancer Res 53:3919-3924.
9. Hard GC, Iatropoulos MJ, Jordan K, et al. (1993) Major differences in the hepatocarcinogenicity and DNA adduct forming ability between toremifene and tamoxifen in female Cr1:CD(BR) rats. Cancer Res 53:4534-4541.
10. Hirsimaki P, Hirsimaki Y, Nieminen L, et al. (1993) Tamoxifen induces hepatocellular carcinoma in rat liver--A 1-year study with 2 antiestrogens. Arch Toxicol 67:49-54.
11. Jordan VC, Robinson SP (1987) Species-specific pharmacology of antiestrogens: Role of metabolism. Fed eration Proc 46:1870-1874.
12. White INH, De Matteis F, Davies A, et al. (1992) Genotoxic potential of tamoxifen and analogues in female Fischer F344/N rats, DBA/2 and C57Bl/6 mice and in human MCL-5 cells. Carcinogenesis 13:2197-2203.
13. Dahme E, Rattel B (1994) Unlike tamoxifen, droloxifene produces no hepatic tumors in the rat. Onkologie 17:6-16.
14. Carthew P, Martin EA, White INH, et al. (1995) Tamoxifen induces short-term cumulative DNA damage and liver tumors in rats: Promotion by phenobarbital. Cancer Res 55:544-547.
15. Carthew P, Rich KJ, Martin EA, et al. (1995) DNA damage as assessed by ^{32}P-postlabelling in three rat stains exposed to dietary tamoxifen: The relationship between cell proliferation and liver tumour formation. Carcinogenesis, in press.
16. Carthew P, Martin EA, White INH, et al. (1995) Tamoxifen induces short-term cumulative DNA damage and liver tumors in rats: Promotion by phenobarbital. Cancer Res 55:544-547.
17. White INH, Davies A, Smith LL, et al. (1993) Induction of CYP2B1 and 3A1, and associated monooxygenase activities by tamoxifen and certain analogues in the livers of female rats and mice. Biochem Pharmacol 45:21-30.
18. Lim CK, Yuan Z, Lamb JH, et al. (1994) A comparative study of tamoxifen metabolism in female rat, mouse and human liver microsomes. Carcinogenesis 15:589-593.

19. Martin EA, Rich K, White INH, et al. (1995) ^{32}P-Postlabelled DNA adducts in liver obtained from women treated with tamoxifen. Carcinogenesis, in press.
20. Fromson JM, Pearson S, Bramah S (1973) Metabolism of tamoxifen (ICI 46,474) Part II: In female patients. Xenobiotica 3:711-714.
21. Styles JA, Davies A, Lim CK, et al. (1994) Genotoxicity of tamoxifen, tamoxifen epoxide and toremifene in human lymphoblastoid cells containing human cytochrome P450s. Carcinogenesis 15:5-9.
22. Mani C, Kupfer D (1991) Cytochrome P450 mediated action and irreversible binding of the antiestrogen tamoxifen to proteins in rat and human liver: Possible involvement of the flavin-containing monooxygenases in tamoxifen activation. Cancer Res 51:6052-6058.
23. White INH, De Matteis F, Gibbs AH, et al. (1995) Species differences in the covalent binding of [^{14}C]tamoxifen to liver microsomes and the forms of cytochrome P450 involved. Biochem Pharmacol, in press.
24. Pathak DN, Pongracz K, Bodell WJ (1995) Microsomal and peroxidase activation of 4-hydroxytamoxifen to form DNA adducts: Comparison with DNA adducts formed in Sprague-Dawley rats treated with tamoxifen. Carcinogenesis 16:11-15.
25. Davies AM, Martin EA, Jones RM, et al. (1995) Peroxidase activation of tamoxifen and toremifene to cause DNA damage and covalently bound protein adducts. Carcinogenesis 16:539-545.
26. Phillips DH, Hewer A, White INH, et al. (1994) Co-chromatography of a tamoxifen epoxide-deoxyguanylic acid adduct with a major DNA adduct formed in the livers of tamoxifen-treated rats. Carcinogenesis 15:793-795.
27. Phillips DH, Potter GA, Horton MN, et al. (1994) Reduced genotoxicity of [D5-ethyl]-tamoxifen implicates α-hydroxylation of the ethyl group as a major pathway of tamoxifen activation to a liver carcinogen. Carcinogenesis 15:1487-1492.

19

Regulation of Growth-Associated Nuclear Transcription Factors During Growth Hormone-Regulated, Sex-Differentiated Rat Liver Carcinogenesis

Inger Porsch Hällström, DeZhong Liao, Lena Ohlson, and Agneta Blanck

Introduction

The incidence of hepatocellular carcinoma (HCC) in man is markedly sex differentiated, with a higher frequency in men than in women, in both high-risk and low-incidence areas (1, 2). Furthermore, both androgens and synthetic estrogens are potential risk factors for liver cancer (3, 4). Also, the induction of liver tumors in experimental animals is often sex differentiated, and both gonadal and pituitary hormones play an important role in this process (5-7).

Hormone-Regulated Sex Differences in the RH-Model

Our studies focus primarily on the resistant hepatocyte model (RH model) for rat liver carcinogenesis (8). This model consists of initiation with a single dose of diethylnitrosamine (DEN) and promotion with a two-week exposure to dietary 2-acetylaminofluorene (2-AAF), with a partial hepatectomy (PH) as a proliferative stimulus in the middle of the period. This procedure leads to a rapid and synchronized outgrowth of initiated cell foci, which can be detected one week after the PH. A fraction of these early lesions progresses into persistent nodules, which represent the precursors of hepatocellular carcinomas. The induction of tumors in the RH model is markedly sex differentiated. The differences are manifested during promotion, with a greater rate of putatively preneoplastic focal growth in males than in females (Figure 1) (9, 10). We have shown that this dimorphism is regulated by the secretory pattern of growth hormone (GH) from the pituitary (9, 10), which is sex differentiated in the rat. The male pattern, which is imprinted neonatally and maintained in the adult by circulating androgens, is characterized by low basal GH serum levels with sharp peaks every 3-4 h, while females have a higher GH basal level with small irregular peaks (11). This pattern determines a number of sex-differentiated liver functions, e.g., steroid and xenobiotic metabolism. The secretory pattern in male rats can be feminized by several endocrine manipulations, such as continuous GH administration, implantation of ectopic pituitaries, and castration. Such manipulations also decrease the focal growth rate during promotion in the RH model toward the level observed in females (9, 10). The same effect is obtained during promotion of DEN-initiated lesions with two other regimens that are more efficient in males, deoxycholic acid and a choline-deficient diet (12, 13).

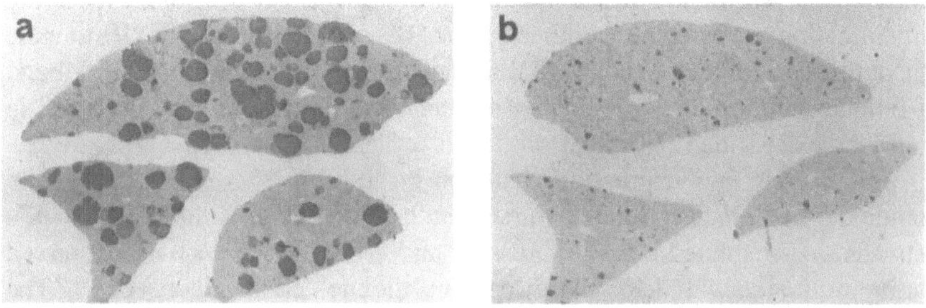

Figure 1. Liver focal size three weeks after PH in the RH model in a) male and b) female rats. Immunohistochemical staining for the placental form of glutathione-S-transferase.

Regulation of Transcription Factors Involved in Hepatocyte Growth Control

This sex differentiation, which can be manipulated hormonally, provides a tool to study growth control during preneoplasia and allows identification, at the molecular level, of factors important for the growth advantage of these lesions compared with surrounding normal hepatocytes. We have studied a set of nuclear transcription factors involved in the control of normal hepatocyte proliferation. In the liver, most hepatocytes reside in G_0, but a proliferative stimulus such as removal of 2/3 of the liver mass induces 95% of the hepatocytes to enter the cell cycle rapidly, followed by an increase in DNA synthesis 16-24 hr later (14). During the first hrs after the operation, genes contributing to this immediate-to-early response are transiently expressed, including transcription factors regulating other genes involved in subsequent G_1 events (14, 15). These include c-*myc*, commonly misregulated in human hepatocellular carcinomas (16); c-*fos*, c-*jun*, *jun*-B, and liver regeneration factor 1 (LRF-1), involved in growth signaling by binding to the AP-1 recognition sequence; and *ets*-2 of the rapidly growing *ets* family of transcription factors. *ets*-2 is shown to be overexpressed in human HCCs (16). We have studied C/EBPα, a liver-enriched factor engaged in the control of liver-specific genes such as albumin, but regarded as a negative growth regulator which is down-regulated in actively proliferating cells (17).

The expression of transcription factors during different stages of carcinogenesis in the RH model was studied, starting during promotion, on the assumption that the involvement of any critical factor for the dimorphism in growth of early lesions should be sex differentiated and regulated by GH. Our results indicate that during the early days after PH, when the sex differences in focal growth are first manifested, the expression of c-*myc* was increased in initiated, 2-AAF-treated males, but neither in females nor in animals receiving only initiation or promotion regimens alone (18). Administration of GH to male rats during 2-AAF treatment down-regulated c-*myc* expression to the level found in females (Figure 2a) (19). Increased nuclear transcription was observed in males three days after PH, which was abolished by the GH treatment (Figure 2b).

Increased c-*myc* expression was seen in nodules from males, but not in nodules from females isolated three weeks after PH. A similar correlation between GH-regulated c-*myc* expression and focal growth was also observed during promotion with deoxycholic acid or a choline-deficient diet (12, 13). The sex differences in the RH-model are maintained during most of the progression stage (20). The expression of c-*myc* increased in females in very advanced stages, and reached the male level in HCCs. Also, in these persistent nodules, the increased expression was due to an increased level of transcription (Figure 3a). About the time when expression began to increase in females, the advanced male lesions close to tumor development lost the GH control of the gene (Figure 3b) (20). c-*Myc*, which is regulated by the secretory pattern in normal liver (female 60-70% of the male level) (21), can be down-regulated by continuous GH treatment in both nodules and surrounding tissue eight months after initiation, but at eleven months, the nodular expression no longer responds to GH treatment. At this stage, not even hypophysectomy affected the nodular expression of c-*myc* (20).

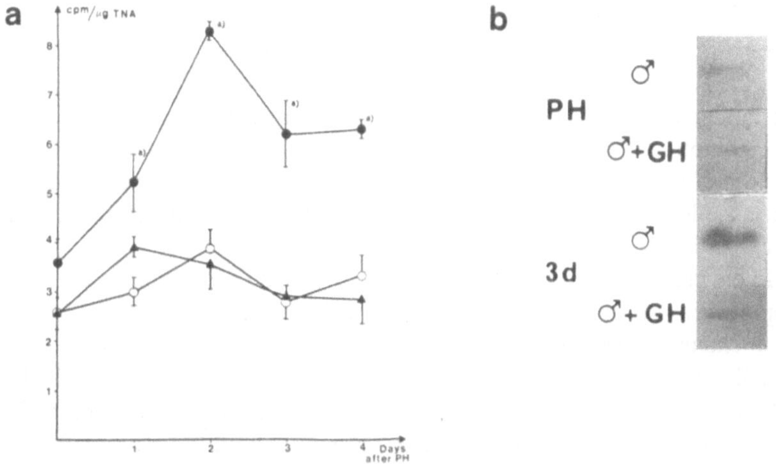

Figure 2. a) c-*myc* mRNA expression in males (•), females (○), and GH-treated males (▴) during promotion in the RH model, analyzed by hybridization of total nucleic acids (TNA) to [^{35}S]-labeled cRNA probe in solution. [a]Significantly higher ($p < 0.05$) than females and GH males. b) Transcription of the c-*myc* gene by nuclei from males and GH-treated males at PH and 3 days later. Reprinted by permission of Oxford University Press (19).

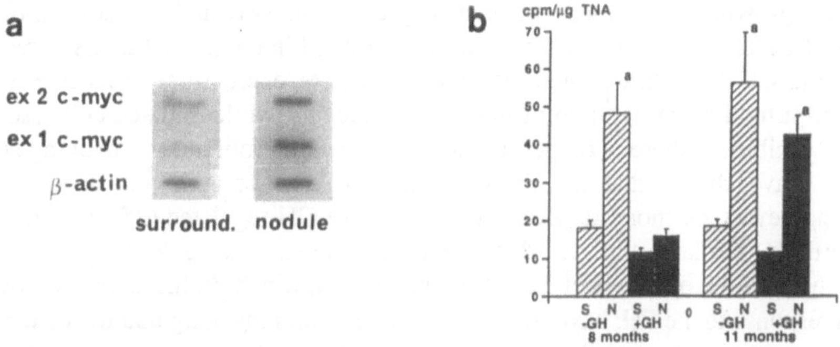

Figure 3. a) Transcription of the c-*myc* gene by nuclei from nodules and surrounding tissue 11 months after initiation. b) Effects of GH treatment on c-*myc* mRNA levels in males 8 and 11 months after initiation in the RH model. Reprinted by permission of John Wiley and Sons, Inc. (20). N, nodules; S, surrounding livers.

An inverse relation between C/EBPα and c-*myc* has been shown (22), and a recognition site for *myc/max* binding has been identified in the promoter region of the C/EBPα gene (23). We reported that the CEBPα expression was reduced in both preneoplastic and neoplastic lesions, in early nodules isolated two weeks after promotion (24). The decrease was due to a reduced transcription of the gene; therefore, one of the functions of c-*myc* during carcinogenesis may be to reduce the level of C/EBPα protein, shown to inhibit mitotic growth in target tissues.

In the group of transcription factors binding to the AP-1 site, we found that, except for a small short-lived increase one day after PH, the expression of c-*fos* was unchanged during early stages (18, 19); c-*jun*, *jun*-B, and LRF-1 mRNA gradually increased during the first days after PH 3.0- to 6.0-fold from the basal level. For LRF-1 and c-*jun*, the increase was more pronounced in males, and GH treatment down-regulated the expression. Increased transcription of LRF-1 and *jun*-B was observed, while the increase in c-*jun* seemed to depend on post-transcriptional modifications. To our surprise, however, similar increases were observed in uninitiated animals treated with 2-AAFt. Figure 4 shows the expression of LRF-1 one week after PH; similar results were obtained for the *jun* genes.

The similarity between foci-containing initiated animals, and uninitiated animals without foci suggests that these genes are not involved in the control of focal growth. Furthermore, the gene expression in early nodules isolated three weeks after PH did not differ from that observed in uninitiated 2-AAF-treated animals. 2-AAF, like many other liver promoters, exerts some of its effects by inhibiting the proliferation of normal hepatocytes, while initiated cells escape this inhibition, thereby rendering these cells with a proliferative advantage (8). We have shown that the mito-inhibitory effects of AAF on surrounding hepatocytes are more marked in males and are GH regulated (25), which may partially explain the sex-differentiated response to 2-AAF promotion. Consequently, the livers of 2-AAF-treated males, which proliferate very slowly in response to the PH, have the highest expression, indicating that the observed increases are not coupled to a proliferative response. To verify these results, we used gel mobility shift analysis to study the binding capacity of nuclear extracts to the AP-1 recognition sequence from the growth-related collagenase gene. We observed a decrease in binding after PH. Particularly, *jun*-containing complexes declined in all 2-AAF-treated animals.

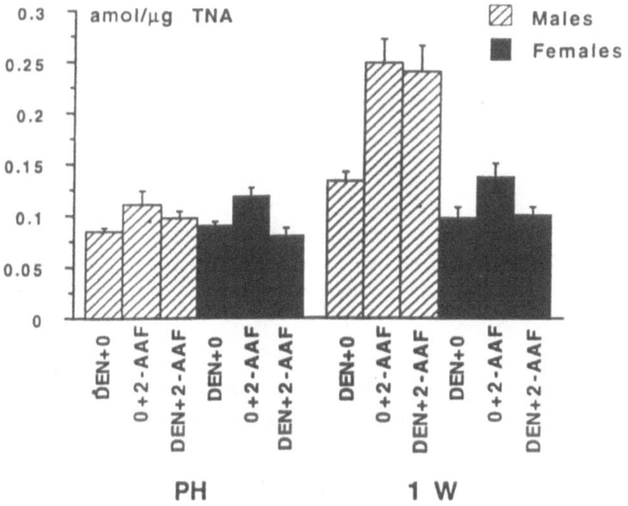

Figure 4. LRF-1 mRNA levels at PH and one week after PH in the RH model, analyzed by solution hybridization. [a]Significantly higher when compared with PH values.

In advanced lesions, expression of c-*fos* and LRF-1 was elevated 2.0- to 3.0-fold compared to corresponding surrounding livers. Similar results were obtained with *ets*-2, which was unchanged during promotion (*jun*-B, LRF-1, and *ets*-2 are shown in Figure 5).

The levels of *jun* mRNAs were elevated when compared with the levels observed in age-matched control rats, but were not significantly different from the surrounding tissue. The increased expression of several members of the AP-1 family may result in a higher transactivation of promoters containing this sequence. It should also be considered that synergistic effects have been observed between *ets* and *fos*/*jun* proteins in transactivation of viral and cellular promoters (26-28). Therefore, it is possible that cooperation between transcription factors contributes to the growth advantage of these lesions.

Finally, a brief mention of sex-differentiated liver carcinogenesis. Since it is known that the male lesions grow faster during promotion in the RH model,

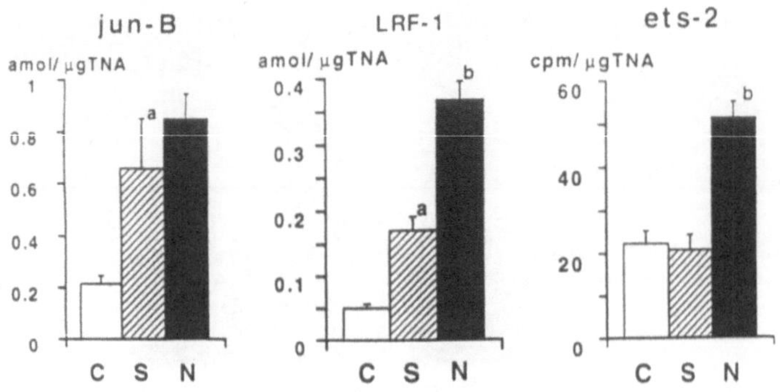

Figure 5. Expression of *jun*-B, LRF-1, and *ets*-2 mRNA analyzed by solution hybridization in: N, persistent nodules eight months after PH; S, surrounding liver; and C, age-matched control liver. [a]Significantly higher than C;[b] higher than S ($p < 0.05$).

we studied whether the same rate of growth continued during progression, or whether the differences in tumor outcome only reflect the advantage that the male lesions acquired during promotion. To analyze this question, nodule-bearing animals received [^{3}H]-thymidine ten months after initiation. The groups were sacrificed either one week or four weeks after administration. We compared the labeling index in male and female nodules and surrounding tissue, and determined the radioactivity incorporated in isolated DNA from both groups. We observed the same rate of cell division in lesions from both sexes; however, the loss of incorporated radioactivity from DNA was much higher in female nodules, indicating a higher cell loss and lower net growth in these lesions than in those observed in the males. These data suggest that hormone-mediated differences in programmed cell death contribute to sex differences during progression of liver cancer.

Conclusions

In conclusion, in the RH model, the expression of *c-myc* may be involved in the sex-differentiated response to promotion, while the AP-1 site did not show any

Conclusions

In conclusion, in the RH model, the expression of *c-myc* may be involved in the sex-differentiated response to promotion, while the AP-1 site did not show any obvious connection with focal growth. Moreover, during progression, synergism between transcription factors as well as hormone-regulated sex differences in programmed cell death may contribute to growth advantage in preneoplastic lesions.

Acknowledgments

This work has been supported by the NIEHS, NIH (1RO1 CA 57925-01), and the Swedish Cancer Society.

References

1. Schonland MM, Millward-Sadler GH, Wright DH, et al. (1979) In: R Wright, et al. (eds.): Liver and Biliary Disease. London: WB Saunders Co, Ltd, pp 886-925.
2. Lotze MT, Flickinger JC, Carr, BI (1993) Hepatobiliary neoplasms. In Devita VT, Hellman S, Rosenberg ST (eds.): Cancer: Principles and Practice of Oncology. J.B. Philadelphia: Lippincott Co, pp 83-91.
3. Westaby D, Portman B, Williams R (1984) Androgen related primary hepatic tumors in non-Fanconi patients. Cancer 51:1947-1962.
4. Shar SR, Kew MC (1982) Oral contraceptives and hepatocellular carcinoma. Cancer 49:407-410.
5. Wogan GN, Newberne PM (1967) Dose-response characteristics of aflatoxin B_1 carcinogenesis in the rat. Cancer Res 27:2370-2376.
6. Stasney J, Paschkis KE, Cantarow A, et al. (1947) Neoplasms in rats with 2-acetylaminofluorene and sex hormones. Cancer Res 7:356-362.
7. Weisburger JH, Pai SR, Yamamoto RS (1964) Pituitary hormones and liver carcinogenesis with N-hydroxy-N-2-fluorenylacetamide. J Natl Cancer Inst 32:881-904.
8. Solt D, Farber E (1976) New principle for the analysis of chemical carcinogenesis. Nature 262:701-703.

9. Blanck A, Hansson T, Eriksson LC, et al. (1987) Growth hormone modifies the growth rate of enzyme-altered hepatic foci in male rats treated according to the resistant hepatocyte model. Carcinogenesis 8:1585-1588.
10. Blanck A, Hansson T, Gustafsson J-Å, et al. (1986) Pituitary grafts modify sex differences in liver tumor formation in the rat following initiation with diethylnitrosamine and different promotion regimens. Carcinogenesis 7:981-985.
11. Eden S (1979) Age and sex related differences in episodic growth hormone secretion in the rat. Endocrinology 105:555-560.
12. Porsch Hällström I, Svensson D, Blanck A (1991) Sex differentiated deoxycholic acid promotion is under hypothalamo-pituitary control. Carcinogenesis 12:2035-2040.
13. Porsch Hällström I, Liao DZ, Blanck A (1994) Ectopic pituitary grafts modify the response of male rats to sex differentiated promotion of diethylnitrosamine-initiated hepatic lesions with a choline-deficient diet. Carcinogenesis 15:921-925.
14. Fausto N, Webber EM (1993) Control of liver growth. Crit Rev in Eucaryotic Gene Expression 3:117-135.
15. Hsu JC, Bravo R, Taub R (1992) Interactions among LRF-1, *jun*-B, c-*jun* and c-*fos* define a regulatory program in the G_1 phase of liver regeneration. Mol Cell Biol 12:4654-4665.
16. Gu JR (1988) Molecular aspects of human hepatic carcinogenesis. Carcinogenesis 9:697-703.
17. Xanthopoulos KG, Mirkowich J (1993) Gene regulation in rodent hepatocytes during development, differentiation and disease. Eur J Biochem 216:353-360.
18. Porsch Hällström I, Blanck A, Eriksson LC, et al. (1989) Expression of the c-*myc*, c-*fos* and c-ras^{Ha} protooncogenes during sex-differentiated rat liver carcinogenesis in the resistant hepatocyte model. Carcinogenesis 10:1793-1800.
19. Porsch Hällström I, Gustafsson J-Å, Blanck A (1989) Effects of growth hormone on the expression of c-*myc* and c-*fos* during early stages of sex-differentiated rat liver carcinogenesis in the resistant hepatocyte model. Carcinogenesis 10:2339-2343.

20. Porsch Hällström I, Gustafsson J-Å, Blanck A (1991) Role of growth hormone in the regulation of the c-*myc* gene during progression of sex-differentiated rat liver carcinogenesis in the resistant hepatocyte model. Mol Carcinogenesis 4:376-381.
21. Porsch Hällström I, Gustafsson J-Å, Blanck A (1990) Hypothalamo-pituitary regulation of the c-*myc* gene in rat liver. J Mol Endocrinol 5:267-274.
22. Freytag S, Geddes T (1992) Reciprocal regulation of adipogenesis by *myc* and C/EBPα. Science 256:379-382.
23. Legraverend C, Antonson P, Flodby P, et al. (1993) High level activity of the mouse CCAAT/enhancer binding protein (C/EBPα) gene promoter involves autoregulation and several ubiquitous transcription factors. Nucleic Acid Res 21:1735-1742.
24. Flodby P, Liao DZ, Blanck A, et al. (1994) Expression of the liver-enriched transcription factors C/EBPα, C/EBPβ, HNF-1 and HNF-4 in preneoplastic nodules and hepatocellular carcinoma in rat liver. Submitted.
25. Blanck A, Eriksson LC, Porsch Hällström I (1991) Sex differentiated and growth hormone regulated mitoinhibition in rat liver during treatment with 2-acetylaminofluorene and partial hepatectomy in the resistant hepatocyte model. Carcinogenesis 12:1259-1264.
26. Wasylyk B, Wasylyk C, Flores P, et al. (1990) The c-*ets* protooncogene encodes transcription factors that cooperate with c-*fos* and c-*jun* for transcriptional activation. Nature 346:191-193.
27. Wang CY, Bassuk AG, Boise LH, et al. (1994) Activation of the granulocyte-macrophage colony-stimulating factor promoter in T cells requires cooperative binding of Elf-1 and AP-1 transcription factors. Mol Cell Biol 14:1153-1159.
28. Wu H, Moulton K, Horvai A, et al. (1994) Combinatorial interactions between AP-1 and *ets* domain proteins contribute to the developmental regulation of the macrophage scavenger receptor gene. Mol Cell Biol 14:2129-2139.

PART 6. PROSTATE

20

Isolation of Differentially Expressed cDNAs from Prostate Cancer Cell Lines Using Differential Display PCR: Identification of an Androgen-Regulated Gene

Leen J. Blok, M. Vijay Kumar, and Donald J. Tindall

Introduction

The transition of prostate cancer cells from androgen-dependent growth, which can be treated by androgen ablation therapy, to androgen-independent growth, which is virtually beyond medical control (1), can be viewed in the broader context of carcinogenesis: Tumor development is associated with the activation of oncogenes and/or the inactivation of tumor suppressor genes (2-5). In the case of the transition of prostate cancer cells from androgen-dependent proliferation to androgen-independent proliferation, it is likely that the expression of additional genes is altered. In an attempt to isolate these genes, the mRNA expression patterns of different prostate cancer cell lines were compared. Using differential display PCR (DD-PCR) (6-11), 13 genes were isolated that were differentially expressed (12-13) between the androgen-dependent prostate carcinoma cell line LNCaP (13) and two androgen-independent prostate carcinoma cell lines, PC3 (14) and DU145 (15). In this manuscript the use of dd-PCR to isolate differentially expressed genes is discussed. Furthermore, one of the clones (TL5) is analyzed for its regulation by androgens, tissue-specific expression, and expression in different prostate carcinoma tissues.

Differential Display PCR

DD-PCR is a new method which compares mRNA expression patterns between different tissues, cell lines, or treatment groups (6-11). In brief, mRNA was isolated and reverse transcribed using one of 12 different poly T primers (3'NNT_{11}5'). The cDNAs produced were used as templates in an amplification reaction (40 cycles: 30 sec 94°C, 2 min 40°C, 30 sec 72°C [AmpliTaq Polymerase from Perkin Elmer, Norwalk, CT]) using the same poly T primer, a random 10-mer primer and radioactively labeled dATP, [^{35}S] or [^{32}P]. The annealing temperature was relatively low to allow binding of both primers. The PCR product was electrophoresed on a 6% sequencing gel (4h at 90w). The gel was fixed, dried and exposed to X-ray film for 10h. Bands of interest were excised from the gel (Figure 1), eluted with an electro-eluter (100V, 0.3mA, 1.5h using TBE buffers), and reamplified in two rounds of PCR under conditions described above. The PCR products were ligated into a

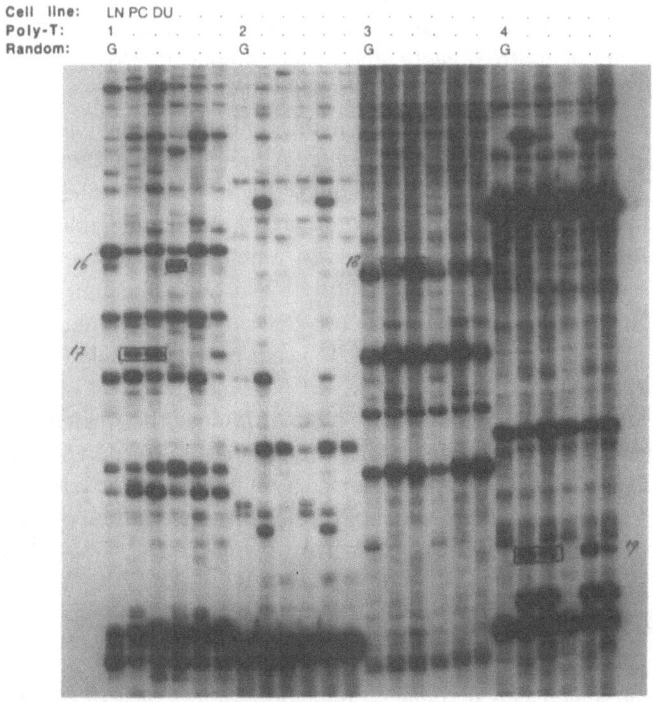

Figure 1. Identification of differentially expressed genes in LNCaP (LN), PC3 (PC) and DU145 (DU) cells. This figure gives an example of a DD-PCR amplification reaction performed in the presence of [^{32}P]-labeled dATP. Poly (A^+) RNA was reverse transcribed using poly T primers 1-4 (3'ACT_{11}5', 3'CGT_{11}5', 3'GGT_{11}5', 3'CCT_{11}5'). The synthesized cDNA was used as a template in the PCR amplification using the poly T primers together with a random primer G (GTACTCAGAC). For each primer combination, the six lanes indicate duplicate samples of LNCaP, PC3, and DU145 cells.

TA-cloning vector (Invitrogen Corporation, San Diego, CA) and used as probes for Northern blots, screening of cDNA or genomic libraries or as templates in sequencing reactions.

Using mRNA isolated from three prostate cancer cell lines (LNCaP, PC3, and DU145) as starting material, 46 cDNAs were isolated from sequencing gels, reamplified and ligated into the vectors. In a secondary screening, using the cDNA clones as probes on Northern blots containing mRNA from the three cancer cell lines, only thirteen cDNA clones showed a clear, differential expression pattern. Other clones could not be detected by Northern blot analysis, either because the cDNA probes were too short to yield substantial hybridization, or because their levels of expression were too low. The expression pattern on the Northern blots was in agreement with the expression pattern on the sequencing gels (DD-PCR) in approximately 50% of the samples (7/13). Thus the overall success rate of DD-PCR, in our hands, was about 15%.

The choice of using three cell lines in this study has substantially contributed to finding differentially expressed clones. Although all three cell lines were originally derived from metastatic lesions of the human prostate, the cell lines are different from each other in many respects. For example, the chromosome number differs between the cell lines (we identified 61-85 chromosomes in LNCaP cells, 55-60 in PC3 cells, and 59-62 in DU145 cells). Furthermore, although LNCaP cells express both androgen receptors and androgen-responsive genes such as PSA, neither PC3 nor DU145 cells have detectable levels of these transcripts (unpublished observations).

However, despite these and many other differences between the cell lines, four cDNAs were cloned because they were either up- or down-regulated by androgens in LNCaP cells and not regulated by androgens in PC3 or DU145 cells (12). TL5 is one of these cDNAs. Another clone isolated turned out to be PSA, representing an accurate positive control.

Clone TL5

Clone TL5 presented a clear signal on the Northern blot, was of expected base-pair length, and seemed to be up-regulated by androgens (Figure 2A). Therefore, the expression of this clone was further investigated. Sequence analysis (Figure 2B) of the PCR-amplified transcript revealed that a 58bp stretch was highly homologous (94%) to the mouse sequence *tag ml621* (16). To determine the kinetics of the androgen response, a time course experiment in the presence or absence of androgens was performed (Figure 3A). TL5 mRNA expression increased between 4h and 16h after the addition of androgens to the culture medium (RPMI 1640 + 7.5% charcoal stripped FCS), reaching maximum levels at 24h. Quantitation of the hybridization signal revealed a 3.0- to a 4.0-fold increase in TL5 mRNA after 24h of androgen treatment (Figure 3B). When LNCaP cells were cultured for 24h in the presence of other steroids (3.2 x 10^{-7} M 17ß-estradiol, 4 x 10^{-7} M dexamethasone, 10^{-9} M progesterone), only the cells grown in the presence of androgens (10^{-9} M mibolerone, or 10^{-5} M DHT) showed up-regulation of TL5 mRNA expression (data not shown).

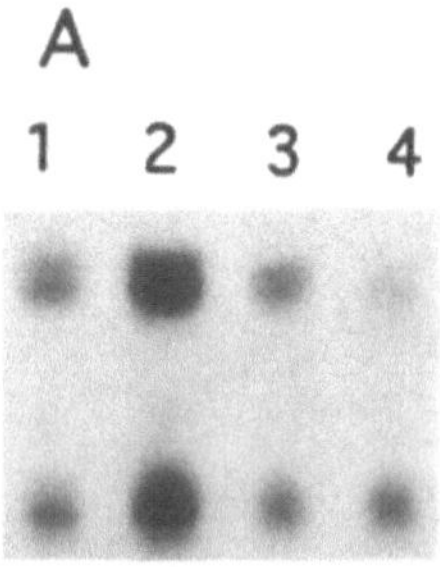

B

```
  1  tccagtatgg ccaacgagaa gcacagcaag aacatcaccc agcgcggcaa
 51  cgtcgccaag acctcgagaa agtgcccccg aagagaaggc gtctgtagga
101  ccctggttat tggctctctt cattttgtt gtctgtggtt ctgcaatttt
151  ccagattatt caaagtatca ggatgggcat gtgaagtgac tgaccttaag
201  atgtttccat tctcctgtga aaaaaaaaa
```

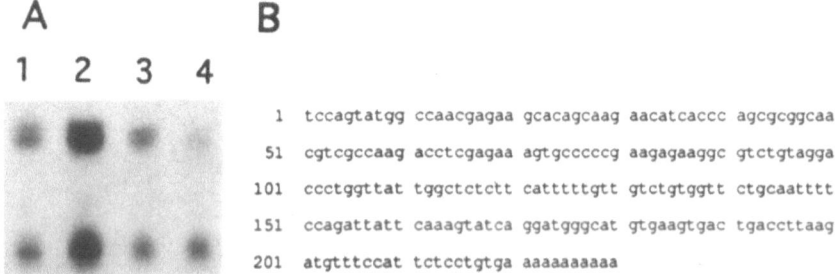

Figure 2. Northern blot analysis and sequence data of clone TL5. In panel A, RNA isolated from LNCaP, PC3 and DU145 cells was probed with TL5 cDNA. Lane 1 contains 20μg total RNA isolated from LNCaP cells cultured in RPMI 1640 medium supplemented with 7.5% FCS, without added hormone; Lanes 2, 3 and 4 represent LNCaP cells, PC3 cells, and DU145 cells cultured in the presence of 1 nM mibolerone. Panel B represents the sequence data. Northern blots and sequencing reactions were repeated and generated similar results as shown.

Next, the expression of TL5 in different human tissues was examined (Figure 4). The tissues were obtained as surgical waste material from the Department of Pathology at Mayo Clinic, and were all diagnosed as normal, except prostate and liver tissues, which were diagnosed as non-malignant. All RNA preparations from the tissues were of good quality. Of all the tissues examined, TL5 mRNA expression was highest in the prostate and mandibular gland and very low in skeletal muscle (Figure 4). Furthermore, examination of rat and mouse tissues revealed that, of the three mRNA bands (3.5-, 3.3-, and 1-kb), the 3.5-kb mRNA band is human- or primate-specific, and the 1-kb band was expressed at much lower levels in the rat and mouse tissues (data not shown), suggesting alternative splicing of TL5 pre-mRNA in different species. In order to analyze this clone further, the full-length cDNA will be isolated by using TL5 as a probe to screen a human prostate cDNA library.

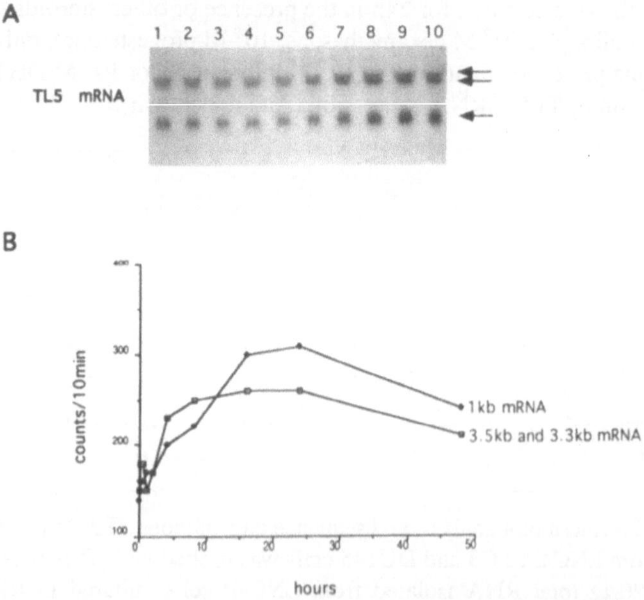

Figure 3. TL5 expression in LNCaP cells cultured in the presence of androgens (10 nM mibolerone) for 0, 15, and 30 min, and for 1, 2, 4, 8, 16, 24, and 48h (Lanes 1-10), respectively. Panel A shows the autoradiograph. In panel B, the androgen treatment time has been plotted against radioactive counts on the Northern blot after probing with [^{32}P]-labeled TL5 cDNA. Each lane of the gel was loaded with 20 μg of total RNA. The arrows in panel A indicate the 3.5-kb, 3.3-kb and 1-kb mRNA bands. Actin mRNA was probed to verify equal loading of the gel.

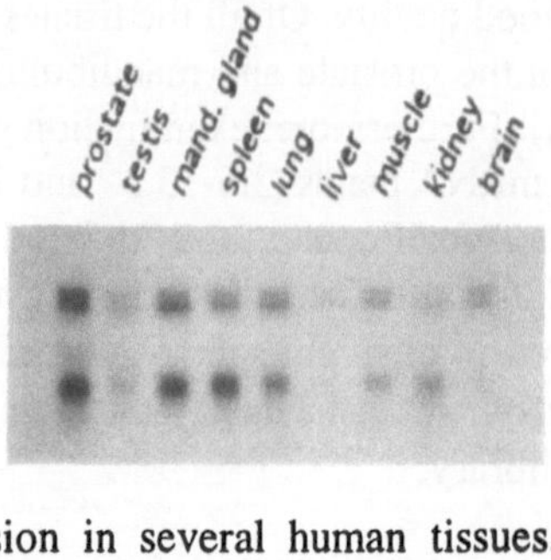

Figure 4. TL5 expression in several human tissues. Prostate (BPH), testis, mandibular gland, spleen, lung, liver (non-malignant), muscle, kidney, and brain. Each lane of the gel was loaded with 20 μg of total RNA.

To determine whether TL5 is actually involved in androgen-independent prostate cancer growth, clone TL5 was hybridized to RNA isolated from different human prostate carcinoma tissues (Figure 5). TL5 expression in grade D human prostate carcinoma tissues (Figure 5, Lanes 1 and 3) appeared markedly higher than in benign prostate tissues derived from the same patients (Figure 5, Lanes 2 and 4). Furthermore, GAPDH expression (Figure 5) showed a very similar distribution, while actin mRNA expression (Figure 5) appeared stably expressed. The observation that TL5 expression is similar to GAPDH expression is very interesting because of the association between GAPDH expression and metastatic potential in Dunning R-3327 rat prostatic adenocarcinoma cell lines (17). Furthermore, elevated GAPDH expression has been reported in lung cancer (18), pancreatic adenocarcinoma (19), and Hodgkin's disease (20), as well as in oncogene-transformed cell lines (21). The present study suggests an association between TL5 mRNA expression and the metastatic potential of human prostate carcinomas.

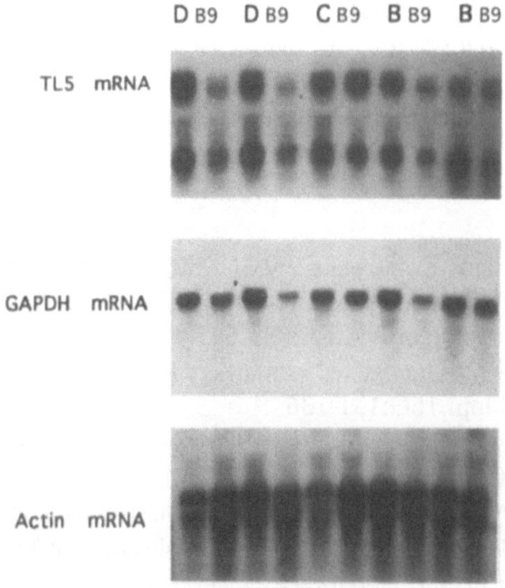

Figure 5. Northern blot analysis of TL5, GAPDH and actin expression in human prostate carcinomas. Total RNA (20 μg/lane). From left to right--Lanes 1 and 2: Tumor (D1, diploid) and benign (B9) prostate tissue, serum PSA level of 6.5 ng/l; Lanes 3 and 4: Tumor (D1, diploid) and benign prostate tissue, serum PSA level of 33.7 ng/l; Lanes 5 and 6: Tumor (C, diploid) and benign prostate tissue, serum PSA level of 19.3 ng/l; Lanes 7 and 8: Tumor (B2, aneuploid) and benign prostate tissue, serum PSA level of 17.6 ng/l; Lanes 9 and 10: Tumor (B2, diploid) and benign prostate tissue, serum PSA level of 15.6 ng/l. Tumor grading, A-D grading according to Hopkins. PSA levels in serum samples prior to surgery.

Summary

We have successfully used the mRNA DD-PCR technique to isolate several differentially expressed clones. Although, in our hands, this method gave only a 15% success rate, it is nonetheless a significant addition to the already existing methods for screening differentially expressed genes.

Acknowledgments

The authors are grateful for the skillful technical assistance of Ms. M. D. Blexrud, and for the secretarial help of Ms. K. Ambroson. The authors wish to thank Dr. D. G. Bostwick, Department of Pathology, Mayo Foundation, for providing the human prostate carcinoma tissues and Drs. S. M. Jalal and G. W. Dewald, Department of Laboratory Medicine, Mayo Foundation, for the karyotype analysis of the cells. This research was supported by NIH Grants CA58225, CA32387 and HD09140, awarded to DJT; a grant from CaPCure and the American Foundation for Urologic Diseases in cooperation with The Connaught Foundation, awarded to MVK; and a NATO-Science Fellowship (N 88-181.92) from the Netherlands Organization for Scientific Research, awarded to LJB.

References

1. Grayhack JT and Kozlowski JM (1980) Endocrine therapy in the management of advanced prostatic cancer: The case for early initiation of treatment. Urol Clin North Am 7:639-643.
2. Cooper GM (1992) Oncogenes as markers for early detection of cancer. J Cell Biochem Suppl 16G:131-136.
3. Fearon ER and Vogelstein BA (1990) Genetic model for colorectal tumorigenesis. Cell 61:759-767.
4. Godbout R, Dryja TP, Squire J (1983) Somatic inactivation of genes on chromosome 13 is a common event in retinoblastoma. Nature 304:451-453.
5. Knudson AG Jr (1985) Hereditary cancer, oncogenes and antioncogenes. Cancer Res 45:1437-1443.
6. Liang P and Pardee AB (1992) Differential display of eucaryotic messenger RNA by means of the polymerase chain reaction. Science 257:967-970.
7. Liang P, Averboukh L, Keyomarsi K (1992) Differential display and cloning of messenger RNAs from human breast cancer versus mammary epithelial cells. Cancer Res 52:6966-6968.
8. Liang P, Averboukh L and Pardee AB (1993) Distribution and cloning of eukaryotic mRNAs by means of differential display: Refinement and optimization. Nucleic Acids Res 14:3269-3275.

9. Sager R, Anisowicz A, Neveu M, (1993) Identification by differential display of alpha 6 integrin as a candidate tumor suppressor gene. FASEB J 7:964-970.
10. Zhang L and Medina D (1993) Gene expression screening for specific genes associated with mouse mammary tumor development. Mol Carcinogenesis 8:123-126.
11. Bauer D, Muller H, Reich J, et al. (1993) Identification of differentially expressed mRNA species by an improved display technique (DDRT-PCR). Nucleic Acids Res 21:4272-4280.
12. Blok LJ, Kumar MV, Tindall DJ (1995) Isolation of cDNAs that are differentially expressed between androgen-dependent and androgen-independent prostate carcinoma cells using differential display PCR. The Prostate 26:213-224.
13. Horoszewicz JS, Leong SS, Kawinski E, et al. (1983) LNCaP model of human prostatic carcinoma. Cancer Res 43:1809-1818.
14. Kaighn ME, Shankar Narayan K, Ohnuki Y, et al. (1979) Establishment and characterization of a human prostatic carcinoma cell line (PC3). Invest Urol 17:16-23.
15. Stone KR, Mickey DD, Wunderli H, et al. (1978) Isolation of a human prostate carcinoma cell line (DU145). Int J Cancer 21:274-281.
16. Warden CH, Mehrabian M, He KY, et al. (1993) Linkage mapping of 41 randomly isolated liver cDNA clones in the mouse. Sequence published in EMBL data base under Musest 621a.Gb-Est..
17. Epner DE, Partin AW, Schalken JA, et al. (1993) Association of glyceraldehyde-3-phosphate dehydrogenase expression with cell motility and metastatic potential of rat prostatic adenocarcinoma. Cancer Res 53:1995-1997.
18. Tokunaga K, Nakamura Y, Sakata K, et al. (1987) Enhanced expression of a glyceraldehyde-3-phosphate dehydrogenase gene in human lung cancers. Cancer Res 47:5616-5619.
19. Schek N, Hall BL, Finn OJ (1988) Increased glyceraldehyde-3-phosphate dehydrogenase gene expression in human pancreatic adenocarcinoma. Cancer Res 48:6354-6359.
20. Perfetti V, Manenti G, Dragani TA (1991) Expression of housekeeping genes in Hodgkin's disease lymph nodes. Leukemia 5:1110-1112.
21. Persons DA, Schek N, Hall BL, Finn OJ (1989) Increased expression of glycolysis-associated genes in oncogene-transformed and growth-accelerated states. Mol Carc 2:88-94.

21

Progression of LNCaP Human Prostate Carcinoma Cells: Androgen Receptor Activity and *c-myc* Gene Expression

John M. Kokontis, Nissim Hay, Richard A. Hiipakka, and Shutsung Liao

Introduction

Prostatic carcinogenesis is a multi-step process involving progression from precancerous cells to cells that uncontrollably proliferate and metastasize. Understanding the molecular events driving the progression is critical to the early detection of and choice of treatment for prostate cancer. The growth and development of prostate cancer appears to be initially androgen-dependent, making it vulnerable to androgen ablation and anti-androgen therapies (1). However, prostate cancer cells gradually lose androgen dependency, and tumor cells which are resistant to endocrine therapy ultimately proliferate. While loss of androgen receptor (AR) expression may accompany loss of androgen dependency and responsiveness (2), other prerequisite cellular events may occur which allow cell proliferation to bypass the androgen requirement. Loss of AR expression may either drive the selection for such events or occur secondarily (3).

Our recent studies (4) also suggest that alterations in the androgenic responses of prostate cancer cells may be caused by an adaptive increase in AR activity that enables cell proliferation to occur at very low concentrations of androgen. Thus, an increase in AR activity can be responsible for an apparent lack of androgen-dependent stimulation or repression of prostate cancer cell proliferation.

Strategy in the Use of Androgen-Sensitive LNCaP Cells

For this study, we employed androgen-responsive human prostatic carcinoma LNCaP cells, which have been used extensively as an *in-vitro* model for examining the role of AR in the control of prostatic carcinoma cell proliferation

(5). AR in LNCaP cells possess a mutation in the androgen binding domain which alters the specificity of ligand binding and steroid-induced transactivation (6, 7), and which is probably responsible for the aberrant proliferative response of LNCaP cells to anti-androgens (8, 9). Production of AR-mRNA in LNCaP cells is negatively controlled by androgen. Our strategy was to find out whether changes might occur in the proliferative response of LNCaP cells to androgen when cells are cultured over a long period of time in androgen-depleted medium and, if so, to identify accompanying molecular changes. A demonstration that changes in specific gene expression take place over time in a clonal isolate would support the idea that prostate tumor cells are able to adapt to lowered androgen concentration in their environment, and that androgen-independent cells do not necessarily arise during androgen ablation therapy through selection of a preexistent androgen-independent subpopulation (10).

Isolation of Parental Androgen-Sensitive 104-S Cells

LNCaP clonal subline cells displayed striking heterogeneity in the proliferative response to 0.1 nM 17β-hydroxy-17-methyl-estra-4,9,11-trien-3-one (R1881). The synthetic androgen R1881 was used in our study because it is metabolically much more stable than DHT. We and many others have shown that R1881 at a concentration of 0.1 nM is optimal for inducing proliferation in LNCaP cells in DMEM supplemented with 10% dextran-coated charcoal-stripped fetal bovine serum (CS-FBS). At concentrations higher than 0.1 nM, R1881 induces proliferation less effectively in medium containing CS-FBS and inhibits proliferation of LNCaP cells in medium containing untreated FBS (11). A subline designated 104-S exhibited the strongest proliferative response to androgen when the cells were grown in DMEM supplemented with CS-FBS. LNCaP 104-S cells at very early passage numbers (< 5) proliferated at a 12-fold higher rate in the presence of 0.1 nM R1881, while other sublines exhibited lower but significant growth responses to androgen.

Isolation of 104-I and 104-R Cell Lines Exhibiting Altered Proliferative Response to Androgen

LNCaP 104-S cells previously cultured for 40 passages in DMEM with 1 nM DHT and 10% FBS were subcultured every 7-10 days in medium supplemented with 10% CS-FBS. Starting from about passage 60 to passage 70, 0.1 nM R1881 induced proliferation only about twofold or less over cells grown in the absence of R1881. Cells at this "intermediate stage" were designated "104-I." At this time, the basal level of proliferation in 10% CS-FBS also began to increase. At about passage 80, the proliferation of these cells was significantly repressed by 0.1 nM R1881. LNCaP 104 cells at this third stage were designated "104-R."

The inhibition of 104-R cell proliferation was specific to androgen and steroids which exhibit androgenic activity in LNCaP cells. Progesterone and 17β-estradiol, which have been shown to utilize LNCaP AR in gene transactivation (6, 7), also exhibited some inhibitory activity. The relative inhibitory activities of steroids at 0.1 nM concentration were, in decreasing order, R1881 = DMNT > 5α-DHT >> progesterone > 17β-estradiol. Dexamethasone had no effect. LNCaP 104-S cells maintained in medium supplemented with 10% FBS and 1 nM DHT for more than 80 passages exhibited the same proliferative sensitivity to androgen as the parental 104-S cells.

Androgen Receptor Expression in LNCaP 104 Cell Lines

Northern blot analysis of total RNA from LNCaP 104-S and 104-R revealed the presence of the major form of AR mRNA, 9-10 kb in length. RNase protection assay also showed that AR mRNA levels in 104-R cells were over twofold higher than the levels detected in LNCaP 104-S cells. The increase in AR protein level was much more dramatic. The amounts of AR detected in whole cell and nuclear extracts of 104-I and 104-R cells not exposed to androgen were about 2.0- and 16.0-fold higher, respectively, than the amount of AR detected in the nuclear extract of 104-S cells. This result suggests that, as cells were cultured in the absence of androgen, the amount of AR in the cell increased.

Androgen Receptor Activity in LNCaP 104 Cell Lines

Changes in the transcriptional activity of AR in LNCaP 104 cells were assessed by two methods: (1) androgen induction of prostate-specific antigen (PSA) mRNA as measured by RNase protection assay and (2) androgen induction of chloramphenicol acetyltransferase (CAT) activity in cells transiently transfected with a CAT gene linked with the mammary tumor virus long terminal repeat (MMTV-LTR) (12). The 5'-regulatory region of the PSA gene and the MMTV-LTR have been shown to contain functional androgen response elements (13).

The induction of PSA mRNA above basal levels was about 8.0-fold higher in 104-R cells than in 104-S cells at every tested R1881 concentration. Similarly, androgen-induced CAT activity increased with LNCaP 104 passage number. Normalized CAT activity in androgen-induced 104-R cells transfected with MMTV-CAT was 20- to 27-fold higher than the CAT activity from androgen-induced 104-S cells transfected with MMTV-CAT.

Expression of *c-myc* in LNCaP 104 Cell Lines

Based on RNase protection analysis, the pattern of androgen-dependent stimulation and inhibition of the cellular *c-myc* mRNA level mirrored that of cell proliferation. At 0.1 nM, R1881 induced both cell proliferation and *c-myc* mRNA level about threefold or more in 104-S cells but reduced both of these parameters to about 10-20% of the control values in 104-R cells. 104-I cells showed intermediate responses. In the presence of untreated FBS, high concentration R1881 (20 nM) reduced the *c-myc* mRNA level to about 25% of control levels in both 104-S and 104-R cells. This result is again consistent with the inhibitory effect of androgen at high concentration on LNCaP cell proliferation. The level of *c-myc* expression in LNCaP 104-S cells grown in 10% CS-FBS supplemented with 0.1 nM R1881 was equivalent to the level in cells grown in 10% untreated FBS, suggesting that 0.1 nM R1881 compensated, at least with respect to *c-myc* expression, for a factor(s) or androgens removed from FBS by charcoal treatment.

Effect of Retroviral *c-myc* Expression on Androgen Response of LNCaP Cells

To examine the effect of overexpression of *c-myc* on androgen control of LNCaP 104 cell proliferation, LNCaP 104-S, 104-I and 104-R cells were infected with pMV7 retrovirus carrying the *c-myc* gene linked with the enhancer/promoter of the Moloney murine sarcoma virus long terminal repeat. R1881, at 20 nM, inhibited the proliferation of uninfected LNCaP 104-S, 104-I, and 104-R cells by about 50% of the control values, but this inhibitory effect was not observed in cells overexpressing *c-myc*. The protection from growth inhibition was *c-myc*-dependent, since proliferation of LNCaP 104-S cells infected with retrovirus not expressing *c-myc* was reduced to a similar level by 20 nM R1881. R1881 at 0.1 nM, however, significantly induced proliferation of retrovirally infected LNCaP 104-S, 104-I, and 104-R cells. This positive proliferative response to androgen by LNCaP 104-R cells, only observed when exogenous retroviral *c-myc* expression compensated for androgen-repressed endogenous *c-myc* levels, suggests that androgen most likely acts through a second proliferative pathway independent of *c-myc*.

Concluding Remarks

Shifting in the Biphasic Androgenic Response to Lower Concentrations of Androgen During Cell Progression

The effect of R1881 on LNCaP cell proliferation was biphasic. At the early stage of progression, R1881 stimulated 104-S cell proliferation in a dose-dependent manner at 0.01-0.1 nM but showed reduced stimulation at 1 nM or higher concentrations. After about 10-20 passages in androgen-depleted medium (104-I), cell proliferation in the absence of androgen progressively increased. After about 20 more passages (104-R cells), cell proliferation was many-fold higher than that of the 104-S cells in the absence of added androgen, and stimulation of cell proliferation by R1881 was less obvious than with 104-S cells. In contrast to 104-S cells, proliferation of 104-R cells was repressed by 0.1 nM or lower concentrations of R1881. The shift in the biphasic response may be due to the increased expression of proliferation-related genes in LNCaP cells that acquired enhanced sensitivity to low concentrations of androgen. This idea is supported by the lower threshold for induction of androgen-responsive genes in 104-R cells compared with those in 104-S and 104-I cells.

Factors That May Change AR Activity During Cell Progression

Enhanced sensitivity to R1881 may be the additive result of increased expression of the AR gene and increased transcriptional activity of AR in 104-R cells. Changes in AR phosphorylation, activity of accessory transcription factors, affinity of AR for ligands, nuclear translocation of AR, affinity of AR for chromatin, or AR recycling kinetics may also explain the increased transciptional activity (14). The increase in AR expression and activity, probably brought about by multiple mechanisms, may be as an adaptive response of LNCaP 104 cells to exceedingly low concentrations of available androgen (10). In contrast to LNCaP cells, Shionogi 115 mouse mammary tumor cells become androgen-insensitive after androgen withdrawal while maintaining AR expression (3).

Loss of AR Expression

The adaptive response of LNCaP cells to low androgen concentration offers no information on how some prostate tumor cells and established cell lines such as PC-3 and DU145 have lost AR expression. It is possible that loss of AR expression occurs after alterations in the expression of specific genes involved in control of cell proliferation take place, making AR expression nonessential for cell growth and survival. For example, PC-3 cells do not express p53 (15), and

DU145 cells express an inactive form of the RB gene product (16). In a study examining changes in gene expression in a *ras+myc*-induced mouse prostate carcinoma after transplantion into castrated hosts, Egawa, *et al.* (17), showed that AR mRNA expression dropped significantly after three weeks of growth when compared with AR mRNA expression in tumors grown in intact hosts. The growth of this tumor was shown to be androgen independent.

The Mechanism by Which Androgen Regulates *c-myc* mRNA Level in LNCaP Cells

Androgens have been reported to have no effect on the production of epidermal growth factor (EGF) and EGF-like peptides in LNCaP cells (18), while others have reported that TGFα, but not EGF production, is induced by androgen (19). Androgen induces EGF receptor in LNCaP cells (20) and a strong correlation of EGF receptor level with *c-myc* expression in human prostate carcinoma tissue samples has been reported (21). EGF induces *c-myc* mRNA in mouse fibroblasts via a cAMP-dependent pathway (22). While these observations lend support for an indirect mechanism by which androgen controls *c-myc* levels, conclusive data are lacking that demonstrate that positive modulation of *c-myc* expression by androgen is achieved through pathways involving peptide growth factors and their receptors. Because overexpression of *c-myc* has been associated with tumor progression in prostatic carcinoma (23, 24), and *v-myc* has been shown to cooperate with *v-Ha-ras* in the initiation of prostatic carcinoma in a reconstituted mouse prostate graft system (25), up-regulation of *c-myc* expression by androgen or estrogen in hormone-responsive tumors is a critical issue.

Positive and Negative Regulation of *c-myc* Expression by Androgen in the Prostate

In the normal rat prostate, *c-myc* expression is down-regulated by androgen; castration induces *c-myc* mRNA levels which are decreased by subsequent androgen administration (26). It is possible that in normal differentiated cells of a steroid-responsive tissue, negative regulation of *c-myc* by steroid hormone represents the normal state of proliferative control. A transition from negative to positive regulation may represent a crucial event in tumorigenesis. The LNCaP cell line may be an example of androgen-sensitive cells which display both positive and negative regulation of *c-myc* expression by androgen.

Clinical Implications

If the type of adaptation in AR activity exhibited by LNCaP 104 cells is a general phenomenon, the clinical implications are significant. Some prostatic tumor cells, under the selective pressure of anti-androgen and/or androgen ablation therapy, may be able to adapt to lower androgen concentration (such as the level supplied by the adrenals in castrated patients) by increasing the transcriptional activity or steroid affinity of AR, thereby circumventing the therapy. Additionally, adaptation to lower androgen availability may allow cancer cells to survive and proliferate during therapy until other events occur which enable cells to bypass the androgen requirement altogether. Indeed, proliferation of some prostate cancer cells that have adapted to low concentrations of androgen in a patient undergoing androgen blockade may be repressed by moderate concentrations of androgen, a counterintuitive form of endocrine therapy.

Acknowledgments

This work was supported by NIH grant DK 41670 and NCI grant CA 58073.

References

1. Huggins C, Hodges CV (1941) Studies on prostate cancer. I. The effect of castration, of androgen, and androgen injection on serum phosphatases in metastatic carcinoma of the prostate. Cancer Res 1:293-297.
2. Quarmby VE, Beckman JWC, Cooke DB (1990) Expression and localization of androgen receptor in the R-3327 Dunning rat prostatic adenocarcinoma. Cancer Res 50:735-739.
3. Darbre PD, King RJB (1987) Progression to steroid insensitivity can occur irrespective of the presence of functional steroid receptors. Cell 51:521-528.
4. Kokontis J, Takakura K, Hay N, Liao S (1994) Increased androgen receptor activity and altered *c-myc* expression in prostate cancer cells after long term androgen deprivation. Cancer Res 54: 1566-1573.
5. Horoszewicz JS, Leong SS, Kawinski E (1983) LNCaP model of human prostatic carcinoma. Cancer Res 43:1809-1818.
6. Veldscholte J, Ris-Stalpers C, Kuiper GGJM (1990) A mutation in the ligand binding domain of the androgen receptor of human LNCaP cells affects steroid binding characteristics and response to anti-androgens. Biochem Biophys Res Comm 173:534-540.
7. Kokontis J, Ito K, Hiipakka RA, Liao S (1991) Expression and function of normal and LNCaP androgen receptors in androgen-insensitive human prostatic cancer cells: Altered hormone and antihormone specificity in gene transactivation. Receptor 1:271-279.

8. Wilding G, Chen M, Gelman EP (1989) Aberrant response *in vitro* of hormone-responsive prostate cancer cells to anti-androgens. Prostate 14:103-115.
9. Olea N, Sakabe K, Soto AM, Sonnenscheim C (1990) The proliferative effect of "anti-androgens" on the androgen-sensitive human prostate tumor cell line LNCAP. Endocrinology 126:1457-1463.
10. Isaacs JT, Coffey DS (1981) Adaptation versus selection on the mechanism responsible for the relapse of prostate cancer to androgen ablation therapy as studied in the Dunning R-3327-H adenocarcinoma. Cancer Res 41:5070-5075.
11. Wolf DA, Schulz P, Fittler F (1991) Synthetic androgens suppress the transformed phenotype in the human prostate carcinoma cell line LNCaP. Brit J Cancer 64:47-53.
12. Parker MG, Webb P, Needham M , et al. (1987) Identification of androgen response elements in mouse mammary tumour virus and the rat prostate C3 gene. J Cell Biochem 35:285-292.
13. Riegman PHJ, Vietstra RJ, Van der Korput JAGM (1991) The promoter of the prostate-specific antigen gene contains a functional androgen responsive element. Mol Endocrin 5:1921-1930.
14. Liao S, Kokontis J, Sai T, Hiipakka RA (1989) Androgen receptors: Structures, mutations, antibodies and cellular dynamics. J Steroid Biochem Mol Biol 34: 41-5651.
15. Rubin SJ, Hallahan DE, Ashman CR (1991) Two prostate carcinoma cell lines demonstrate abnormalities in tumor suppressor genes. J Surg Oncol 46:31-36.
16. Bookstein R, Shew JY, Chen PL (1990) Suppression of tumorigenicity of human prostate carcinoma cells by replacing a mutated RB gene. Science 247:712-715
17. Egawa S, Kadmon D, Miller GJ (1992) Alterations in mRNA levels for growth-related genes after transplantation into castrated hosts in oncogene-induced clonal mouse prostate carcinoma. Mol Carcinogen 5:52-61.
18. Connolly JM, Rose DP (1990) Production of epidermal growth factor and transforming growth factor-alpha by the androgen-responsive LNCaP human prostate cancer cell line. Prostate 16:209-218.
19. Wilding G, Valverius E, Knabbe C, Gelmann, EP (1989) Role of transforming growth factor-α in human prostate cancer cell growth. Prostate 15:1-12.
20. Schuurmans ALG, Bolt J, Voorhorst MM (1988) Regulation of growth and epidermal growth factor receptor levels of LNCaP prostate tumor cells by different steroids. Int J Cancer 42:917-922.

21. Eaton CL, Davies P, Phillips MEA (1988) Growth factor involvement and oncogene expression in prostatic tumours. J Steroid Biochem 30:341-345.
22. Ran W, Dean M, Levine RA (1986) Induction of *c-fos* and *c-myc* mRNA by epidermal growth factor or calcium ionophore is cAMP dependent. Proc Natl Acad Sci USA 83:8216-8220.
23. Fleming WH, Hamel A, MacDonald R (1986) Expression of the *c-myc* proto-oncogene in human prostatic carcinoma and benign prostatic hyperplasia. Cancer Res 46:1526-1531.
24. Buttyan R, Sawchuk IS, Benson, MC (1987) Enhanced expression of the *c-myc* protooncogene in high-grade human prostate cancers. Prostate 11:327-337.
25. Thompson TC, Southgate J, Kitchener G, Land H (1989) Multistage carcinogenesis induced by *ras* and *myc* oncogenes in a reconstituted organ. Cell 56:917-930.
26. Quarmby VE, Beckman J W C, M WE, French FS (1987) Androgen regulation of *c-myc* messenger ribonucleic acid levels in rat ventral prostate. Mol Endocrinol 1:865-874.

22

Regulation of Prostate Growth and Gene Expression: Role of Stroma

Leland W. K. Chung, Chuan Gao, and Haiyen E. Zhau

Introduction

Cellular interaction is recognized as a continuing event, essential in multicellular organisms for acquiring preprogrammed structure and assuming differentiated functions. Aberrant cellular interaction could create misguided signals between cells, resulting in developmental defects and unregulated growth control in both vertebrate and invertebrate organisms. Examples found in the mammalian species are the mosaic Tfm/y mice, which produced offspring that might inherit regional developmental defects of their reproductive organs (1). Reactivation of the inductive potential of embryonic mesenchymes in the adult prostate gland was proposed as one underlying mechanism for the histogenesis of benign prostatic hyperplasia (2).

To understand the molecular and cellular basis of cellular interaction, we established a cell-cell interaction model *in vivo* (3) using well-characterized stromal cells (defined here as desmin- and cytokeratin-negative but vimentin-positive staining cells derived from the prostate and the bone) and relevant target epithelial cells (e.g., prostate or urothelial cells) as the model to evaluate the action of androgen and the roles of stromal cells in altering the behavioral parameters of the target epithelial cells. In this communication, we summarized our findings as follows:

1. Prostate stromal cells conferred androgen-induced growth responses to the target prostate and non-prostate epithelial cells *in vivo*.
2. Prostate and bone (the preferential site of prostate cancer metastasis) stromal cells markedly stimulated prostate epithelial tumor growth *in vivo*.
3. Stromal-epithelial interaction may "permanently" alter the behavioral parameters of the responding cancer epithelial cells. Evidence shows that through stromal induction of cancer growth *in vivo*, the interactive prostate cancer epithelial cells acquired androgen-independent and osseous metastatic potential.

4. Conditioned medium obtained from the relevant stromal cells presumably contains soluble growth factors that can stimulate cancer growth and metastasis. A novel molecular cloning strategy was established to isolate and characterize soluble growth factor(s) that may be responsible for the induction of epithelial growth. Many of the potential candidate molecules that are responsible for regulating prostate growth and gene expression are also examined.

Fetal Rat Urogenital Sinus Mesenchymal (rUGM) Cells Conferred Androgen Responsiveness to Target Prostate and Non-prostate Epithelial Cells

A number of studies conducted by our laboratories (4, 5) and others (6, 7) supported the hypothesis that androgen-induced growth and differentiation of the prostate epithelial cells *in vivo* may be indirect and possibly mediated by the mesenchyme. This hypothesis has led to the recent effort of searching for an *in-vivo* model of hormone-induced prostate growth and differentiation and to the identification of putative mesenchymal factors that may be responsible for stimulating epithelial growth and differentiation. In this regard, we summarize our recent work on the establishment of a cell-cell interaction model, and the identification of putative factors that could be important in affecting prostate epithelial growth and tumorigenesis.

Because fetal urogenital sinus mesenchymal tissues were demonstrated to be inductive when associated closely with either fetal or adult prostate epithelium *in vivo* as tissue recombinants, we separated the mesenchymal tissue compartments into cellular constituents and examined the effect of a cell line derived from rat urogenital sinus mesenchyme (rUGM) in inducing the growth of prostate and non-prostate cells *in vivo* (8). Figure 1 summarizes these findings. rUGM cells, a nontumorigenic fibroblast cell line derived from rat UGM tissues, were found to induce the growth of both "androgen-responsive" (e.g., a human androgen-responsive prostate cancer epithelial cell line, LNCaP, and a normal rat prostate epithelial cell line, NbE-1) and "androgen-nonresponsive" (e.g., a human androgen-insensitive prostate cancer epithelial cell line, PC3, and a human bladder transitional carcinoma cell line, WH) epithelial cell lines to grow as solid tumors *in vivo* in an androgen-responsive manner. It is likely that rUGM cells, presumably an androgen target cell line, conferred androgen-induced growth responsiveness to the prostate or non-prostate epithelial cells *in vivo* (9, 10) through paracrine mediators. This is further evidenced by the experiments where fibroblasts derived from androgen non-target organs, such as human bladder (WH cells) or fetal mouse tissues (3T3 cells) failed to confer androgen responsiveness to prostate and non-prostate (11) epithelial cells *in vivo*.

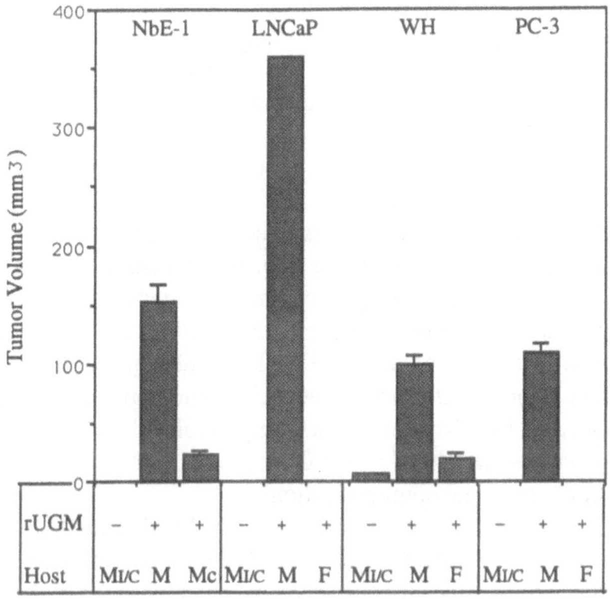

Figure 1. Mesenchymal cells derived from rat urogenital sinuses (rUGM) promoted the growth of prostate and non-prostate cells in intact male hosts. Chimeric tumor volumes were recorded in cell-cell recombinants composed of an equal number of the inductive rUGM cells and the responding NbE-1 (a normal rat prostate epithelial cell line), LNCaP (an androgen-dependent human prostate cancer epithelial cell line), WH (a human bladder transitional carcinoma cell line), or PC-3 (an androgen-independent human prostate cancer epithelial cell line) cells. Note rUGM cells induced the growth of prostate and non-prostate cells in the male (M) but not in the female (F) or castrated male (Mc) hosts. In the absence of rUGM cells, the inoculated epithelial cells alone failed to form tumors in either intact or castrated male hosts (Mi/c) during the observation period.

Prostate and Bone Stromal Cells Stimulate Prostate Carcinoma Growth *In Vivo*

To test the hypothesis whether the induction of prostate carcinoma growth by stromal cells *in vivo* may exhibit organ specificity, we co-inoculated five different nontumorigenic fibroblast cell lines separately with LNCaP cells and observed that only prostate and bone (the favorable site of prostate cancer metastasis) but not kidney, lung, or 3T3 fibroblasts stimulated LNCaP tumor growth *in vivo* (12). The growth-promoting effect of prostate and bone fibroblasts can be replaced by their serum-free and chemically-defined conditioned media collected from the cultured prostate and bone fibroblasts *in vitro*, suggesting that certain soluble factor(s), possibly in the form of paracrine growth factor(s), may be responsible for the stimulation of prostate cancer growth *in vivo*. We have begun to isolate and characterize this soluble factor(s) by conventional chromatographic procedures, using polyclonal antibody approaches to identify the existence of this factor (13). Results of these approaches are presented below.

We focused on the characterizations of soluble growth factors produced by a human bone osteoblast MS cell line capable of stimulating prostate tumor growth *in vivo* and anchorage-independent growth *in vitro*. A heparin sepharose affinity column-bound fraction, eluted by 1 M NaCl (or MS1), was determined to possess growth-stimulating activity. We have subsequently fractionated this material further by high-performance liquid chromatography and polyacrylamide gel electrophoresis. Results showed that a high molecular weight protein, with an apparent molecular weight of >200 kDa is active in stimulating prostate tumor growth (Li, et al., personal communication).

Another approach is to prepare a polyclonal antibody against the biologically-active growth factor fractions MS1 obtained from heparin affinity column. This antibody is then used as a reagent to screen against a cDNA library prepared from a human MS bone stromal cell line. Screening against this cDNA library with MS1 antibody resulted in the detection of several positive clones. One such clone, termed BPGF-1 (bone-prostate growth factor-1), was isolated and characterized (see below).

Cloning and Characterization of a Novel Growth Factor, BPGF-1 That Stimulates Prostate Tumor Growth *In Vitro*

If soluble paracrine growth factor(s) is responsible for stimulating prostate tumor growth *in vivo*, we reasoned that inductive prostate or bone fibroblasts must contain the respective mRNAs encoded for such growth factor(s). To test this hypothesis, we constructed a cDNA library of human bone fibroblasts, and screened this library with a polyclonal antibody prepared against the MS1 fraction, the biologically active protein fraction isolated from a human bone stromal cell (MS) conditioned medium fractionated by a heparin sepharose affinity column. After repeated screening of the expression cDNA library, we cloned a 3.2-kb novel paracrine growth factor cDNA, termed bone and prostate growth factor-1 (BPGF-1), with an estimated molecular weight of 70 kDa. The cDNA reading frame of this growth-promoting factor gene is 1.6 kb. Northern-blot hybridization revealed two BPGF-1 transcripts with their respective sizes of 3.3 and 2.5 kb. The transcripts were detected in bone and prostate fibroblasts, but not in the cell extracts of a host of other human cells and tissues (14). BPGF-1 cDNA, when expressed in COS cells, stimulates prostate cell growth and anchorage-independent growth *in vitro*. These results, taken together, support the hypothesis that certain novel and organ-specific mesenchymal factors may be responsible for stimulating prostate cancer epithelial growth *in vivo*.

Induction of Prostate Epithelial Tumor Growth *In Vivo* by Cellular Interaction with Organ-Specific Fibroblasts: A Hypothesis of "Genomic Adaptation" as the Mechanism for Tumor Cells to Acquire Androgen-Independent and Metastatic Potential

LNCaP cells were obtained from the lymph node of a patient with androgen-independent and metastatic prostate cancer (15). This is the only human prostate cancer cell line that is grown readily *in vitro*, responds to androgen, and synthesizes and secretes a clinically useful human prostate cancer marker, prostate-specific antigen (PSA). Presently, the genetics of LNCaP cells and their adaptive capability have not been thoroughly investigated. The "evolution" of this cell line *in vivo* that results in the acquisition of tumorigenic, androgen-independent, and metastatic phenotypes, however, appears predictable and nonrandom. LNCaP was initially an androgen-dependent (i.e., unable to grow in female or castrated male hosts), tumorigenic, but nonmetastatic cell line, as demonstrated by Horosewicz's laboratory. Subsequently, despite LNCaP cells' ability, after repeated *in vitro* passages and culture, to retain their cytogenetic marker chromosomes, lost their tumorigenicity (12, 16). Tumorigenicity of LNCaP cells in athymic nude mice, however, can be "recreated" by interaction with organ-specific fibroblasts, such as interaction with prostate and bone fibroblasts (12), interaction with reconstituted extracellular matrix (Matrigel) (17), or injecting the cells orthotopically, presumably interacting with host fibroblasts (16, 18, 19). Using bone fibroblasts as inducers, our laboratory cloned several LNCaP sublines that exhibited diverse tumorigenic, androgen-dependent, and metastatic properties (20). We attributed these changes of LNCaP cells *in vivo* as a consequence of "progression" through cellular interaction with fibroblasts in an androgen-deficient microenvironment (21). Mechanistically, how do organ-specific fibroblasts "instruct" LNCaP cells to progress from nontumorigenic to tumorigenic status, and from androgen-dependent to androgen-independent and osseous metastatic phenotypes? Because the acquisition of androgen-independent and metastatic phenotypes is rapid (~ 4 weeks) and occurs in the absence of massive cell death (22), it is unlikely that "clonal expansion" or "clonal evolution" alone can account for these experimental observations. A more likely explanation for the alterations in behavioral traits of the LNCaP cells is "adaptive mutation" (23), in which both dividing and non-dividing cells may undergo a "nonrandom genomic adaptation" so that certain favored genotypes and phenotypes are selected and evolved under the influence of host micro-environment. This hypothesis differed from the current concept that tumor cell heterogeneity results both from genomic changes that must occur only in cells that undergo mitosis (24) and from the selection and expansion of a previously existing cell clone(s) as the tumor progresses. Based on his experimental data, Rubin (25) has elegantly presented his view on neoplastic progression and revisited Foulds' rules.

Because LNCaP tumors grown in intact hosts have slow doubling times (~2 weeks) and failed to undergo apparent apoptosis in the absence of androgen (22), it is unlikely that the androgen-independent and metastatic LNCaP sublines are generated through the mechanism of "clonal expansion." Rather, our data appear to support the hypothesis that **host microenvironment** may play a crucial role in facilitating the emergence of "genomically stable" clone(s) of tumor cells (regardless of whether they are dividing or non-dividing, pre-existing or not) in a nonrandom fashion with distinct growth advantages. Since all cells, dividing or non-dividing, are known to be genetically unstable and thus could undergo "genomic adaptation," this could be the explanation of why both the parental and derived cell lines are clonal when subjected to cytogenetic analysis (20, 22). This hypothesis will fit the data from rapid generation of phenotypically and genotypically (different from the parental cells) yet stable clone(s) of LNCaP cells with defined behavioral parameters *in vivo*. Figure 2 lists the distinctions between "clonal expansion" as opposed to "genomic adaptation" as the potential theoretical basis for the generation of tumor cell heterogeneity.

Parameters	Clonal Expansion	Genomic Adaptation
Cell Division	Required	Not required
Cell Death	Required	Not required
Rate of generation of selective clones	Affected by the rate of cell growth and death	Not affected by the rate of cell growth or cell death
Genotype of the dominant clones	Must pre-exist	Either pre-existing or acquired

Figure 2. Distinction between "clonal expansion" and "genomic adaptation" as a potential mechanism for the generation of tumor cell heterogeneity.

Summary

Our results support the hypothesis that: (1) The mesenchymal cell compartment is responsible for conferring androgen sensitivity, enhancing growth of prostate epithelial cells *in vivo* (10). (2) The organ-specific mesenchymal cells (i.e., prostate and bone fibroblasts) accelerated human prostate cancer growth *in vivo* (12). (3) A novel soluble paracrine growth factor has been cloned and sequenced from a bone mesenchymal cDNA library. This growth factor and its receptor could serve as a mediator between stromal and epithelial cellular communication. (4) "Genomic adaptation" may be the mechanism underlying prostate cancer cells' "irreversible" progression and acquisition of tumor cell heterogeneity *in vivo*. That is, the influence of bone stromal cells in the castrated host microenvironment may allow the progression from androgen-dependent to androgen-independent and metastatic status. These results, taken together, suggest that reciprocal interaction between stromal and epithelial cells *in vivo* is not only important for supporting normal prostate development and conferring androgen-induced growth responsiveness to the epithelium but may also determine the rate of androgen dependency and progression of the cancer. This study provides a rational basis to identify induction and progression factors, as well as markers that could serve as diagnostic and prognostic factors, in order to design appropriate therapeutic interventions for prostate cancer.

Acknowledgments

The authors are indebted to the excellent secretarial and editorial assistance provided by Ms. Carolyn Davis and Ms. Sunita Patterson, respectively. This work was supported in part by financial support from National Institutes of Health grants CA56307 and DK38649 awarded to Leland W. K. Chung, and CA57361 awarded to Haiyen E. Zhau.

References

1. Schleicher G, Stumpf WE, Thiedemann K, Drews U (1988) Intersex mice composed of androgen insensitive Tfm and wild-type cells analyzed by ^{3}H dihydrotestosterone autoradiography. Anat Embryol 178:521-528.
2. McNeal, JE (1978) Origin and evolution of benign prostate enlargement. Invest Urol 15:340-345.
3. Chung LWK, Chang SM, Bell C, et al., (1989) Co-inoculation of tumorigenic rat prostate mesenchymal cells with non-tumorigenic epithelial cells results in the development of carcinosarcoma in syngeneic and athymic animals. Int J Cancer 43:1179-1187.

4. Cunha GR, Chung LWK (1981) Stromal-epithelial interaction: I. Induction of prostatic phenotype in urothelium of testicular feminized (TFm/y) mice. J Steroid Biochem 14:1317-1321.
5. Chung LWK and Cunha GR (1983) Stromal-epithelial interactions: II. Regulation of prostate growth by embryonic urogenital sinus mesenchyme. Prostate 4:503-511.
6. Lasnitzki I, Mizuno T (1980) Prostatic induction: Interaction of epithelium and mesenchyme from normal wild type mice and androgen-insensitive mice with testicular feminization. J Endocrinol 85:423-428.
7. Neubauer BL, Best K, Hoover DM, et al., (1986) Mesenchymal-epithelial interactions as factors influencing male accessory sex organ growth in the rat. Federation Proc 45:2618-2626.
8. Chung LWK, Gleave ME, Hsieh J, et al., (1991) Reciprocal mesenchymal-epithelial interaction affecting prostate tumour growth and hormonal responsiveness. Cancer Surveys 11:91-121.
9. Chung LWK, Chang SM, Bell C, et al., (1988) Prostatic carcinogenesis evoked by cellular interaction. Environ Health Perspect 77:23-28.
10. Zhau HYE, Hong SJ, Chung LWK (1994) A fetal rat urogenital sinus mesenchymal cell line (rUGM) accelerated growth and conferral of androgen-induced growth responsiveness to a human bladder cancer epithelial cell line in vivo. Int J Cancer 56:706-714.
11. Chung LWK, Hong SJ, Zhau HYE, et al., (1991) Fibroblast-mediated human epithelial tumor growth and hormonal responsiveness *in vivo*. Mol Cell Biol Prostate Cancer 19:91-102.
12. Gleave ME, Hsieh JT, Gao C, et al., (1991) Acceleration of human prostate cancer growth *in vivo* by prostate and bone fibroblasts. Cancer Res 51:3753-3761.
13. Chung LWK, Li W, Gleave ME, et al., (1992) Human prostate cancer model: Roles of growth factors and extracellular matrices. J Cell Biochem 16H:99-105.
14. Gao C (1994) Molecular mechanisms of prostate cancer growth and differentiation: Roles of BPGF-1 and extracellular matrices. Ph.D. thesis submitted to Graduate School of Biomedical Sciences, The University of Texas Health Science Center at Houston, Texas.
15. Horoszewicz JS, Leong SS, Chu TM, et al., (1980) The LNCaP cell line--A new model for studies on human prostatic carcinoma. In Murphy GP (ed): Models for Prostate Cancer. New York: Alan R Liss, Inc, pp 115-132.
16. Stephenson RA, Dinney CPN, Gohji K, et al., (1992) Metastatic model for human prostate cancer using orthotopic implantation in nude mice. J Natl Cancer Inst 84:951-957.

17. Pretlow TG, Delmoro CM, Dilley GG, et al., (1991) Transplantation of human prostatic carcinoma into nude mice in Matrigel. Cancer Res 51:3814-3817.
18. Gleave ME, Hsieh JT, von Eschenbach AC, Chung LWK (1992) Prostate and bone fibroblasts induce human prostate cancer growth *in vivo*: Implications for bidirectional stromal-epithelial interaction in prostate carcinoma growth and metastasis. J Urol 147:1151-1159.
19. Fu X, Herrera H, Hoffman RM (1992) Orthotopic growth and metastasis of human prostate carcinoma in nude mice after transplantation of histologically intact tissue. Int J Cancer 52:987-990.
20. Thalmann GN, Anezinis PE, Chang SM, et al., (1994) Androgen-independent cancer progression and bone metastasis in the LNCaP model of human prostate cancer. Cancer Res 54:2577-2581.
21. Chung LWK (1994) The role of stromal-epithelial interactions in normal and malignant growth. Cancer Sur 23:33-42.
22. Wu, HS, Hsieh JT, Gleave ME, et al., (1994) Derivation of androgen-independent human LNCaP prostatic cancer cell sublines: Role of bone stromal cells. Int J Cancer 57:406-412.
23. Foster, PL (1993) Adaptive mutation. Ann Rev Microbiol 47:467-504.
24. Foulds, L (1969) Neoplastic Development, Vol 1. New York: Academic Press, pp 72-74.
25. Rubin H (1994) Experimental control of neoplastic progression in cell populations: Foulds' rules revisited. Proc Natl Acad Sci, USA 91:6619-6623.

CONCLUDING REMARKS

23

Sex Hormones and Neoplastic Transformation

Jonathan J. Li

Hormones have pervasive effects on cellular processes, since they can regulate cell growth and differentiation, metabolic activity, and the metabolism of both endogenous and exogenous substances. Moreover, hormones can affect neoplastic processes by acting either as the sole etiologic agent or in conjunction with nonhormonal chemical carcinogens or physical agents (e.g., radiation). Virtually all mammalian organs or tissues are affected, directly or indirectly, by the endocrine system. Therefore, the concept of specific target organs or tissues for hormones is perhaps only valid insofar as they are differentially sensitive to various hormones.

Of the sex hormones, estrogens clearly have been most strongly implicated in the processes of neoplastic transformation. The hallmark of all sex hormones, is their ability to elicit cell proliferation in their respective target tissues. A growing body of evidence, utilizing many species and a variety of tissues, indicates that sex hormone-induced cell proliferation plays a critical role in corresponding hormone-initiated carcinogenesis (1-7).

Because of serious inconsistencies between *in vivo*- and *in vitro*-generated data, reviewed elsewhere (8, 9), this has led us to develop the following scheme (Figure 1) based on recent studies from our laboratory (1, 2, 10-12).

This multi-stage sequence proposed for estrogen carcinogenesis in the hamster kidney is postulated to be uniquely an estrogen-driven and -dependent process acting on specific estrogen-sensitive renal cells. This process involves cell proliferation, cell damage, regenerative cell growth, and aneuploid cell induction. Chromosome imbalance and genetic instability as a consequence of these events is envisioned to play a key role in hormonally-induced cell transformation and neoplastic development. Consistent or nonrandom chromosome gains (trisomies, tetrasomies) as well as chromosome losses (monosomies) have recently been described in estrogen-treated hamster kidneys. These alterations in chromosome number may represent regions of low-level gene amplification indicating enhanced expression of oncogenes and/or growth

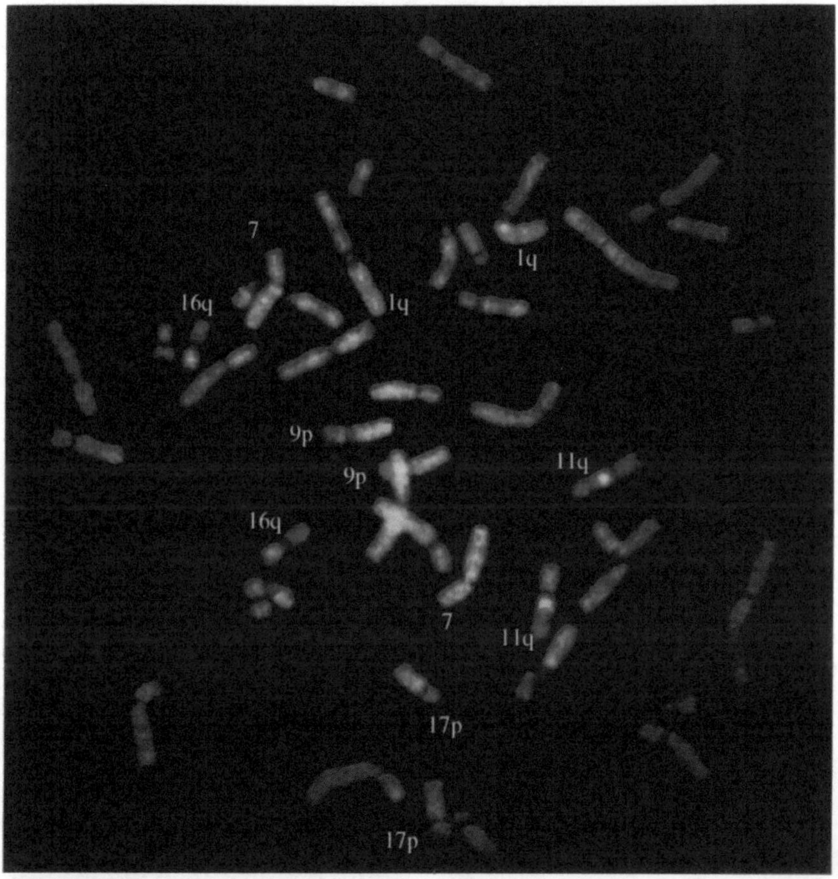

Color Plate 1. CGH analysis of human breast tumor cell line MPE600. Cell-line DNA was labeled with FITC-dUTP and hybridized, together with Texas red-dUTP-labeled normal reference DNA, to a normal metaphase spread. Note the relative gains of chromosome regions 1q, 7, and proximal 11q (green), and the relative losses of 9p, 11q distal, 16q, and 17p (red).

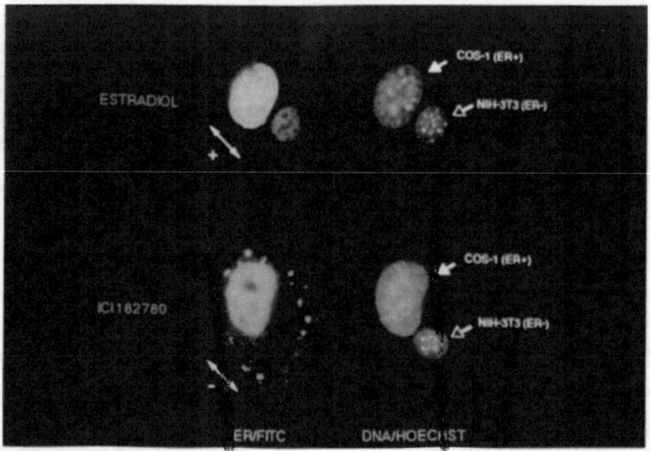

Color Plate 2. Effect of ligands on the ability of estrogen receptors to shuttle between nuclei in heterokaryons. COS-1 cells expressing the mouse estrogen receptor were fused with NIH 3T3 cells devoid of estrogen receptor and treated for three hours with 10^{-8} M 17β-estradiol or 10^{-7} M ICI 182780. The subcellular localization of the estrogen receptor was observed in cell hybrids by indirect immunofluorescence using the monoclonal antibody H222 (left panel–yellow). Nuclei from two cell types were distinguished using Hoechst 33258 dye (right panel–blue).

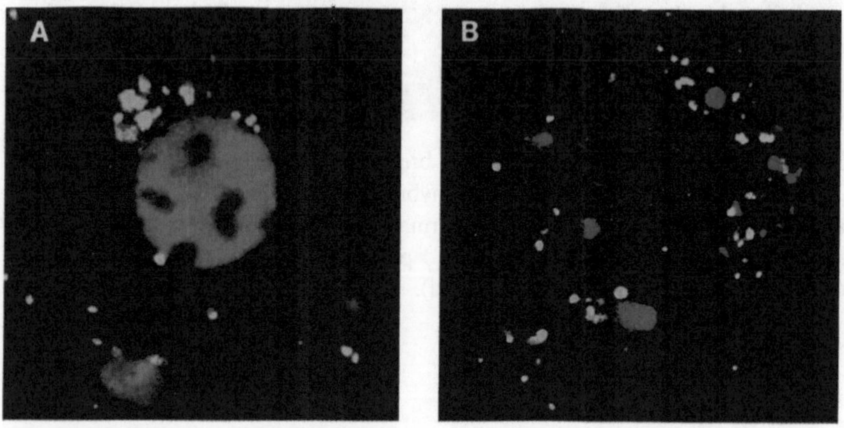

Color Plate 3. Co-immunolocalization of the estrogen receptor and cathepsin D in transiently transfected COS-1 cells. After four hours' treatment with 10^{-8} M 17β-estradiol or 10^{-7} M ICI 182780, cells were fixed and stained by indirect immunofluorescence for the estrogen receptor (red) and the lysosomal enzyme cathepsin D (yellow).

factor genes and sites of tumor suppressor functions. The consequences of such chromosomal changes may confer distinct growth advantages on early altered renal cells which progressively lead to tumor formation. In addition, chromosomal aberrations (chromatid gaps, breaks, and exchanges; chromosome breaks and exchanges; and endoreduplicated cells) seen in the hamster kidney after estrogen treatment may also contribute to estrogen-induced tumorigenic processes. The overexpression of protooncogenes (c-*myc*, c-*fos*, c-*jun*) and tumor suppressor p53 transcripts in the hamster kidney described for the first time at 5.0 and 6.0 months of estrogen treatment is consistent with our view that estrogen-driven inappropriate gene expression leads to proliferative advantages of certain proximal tubular cells (10)

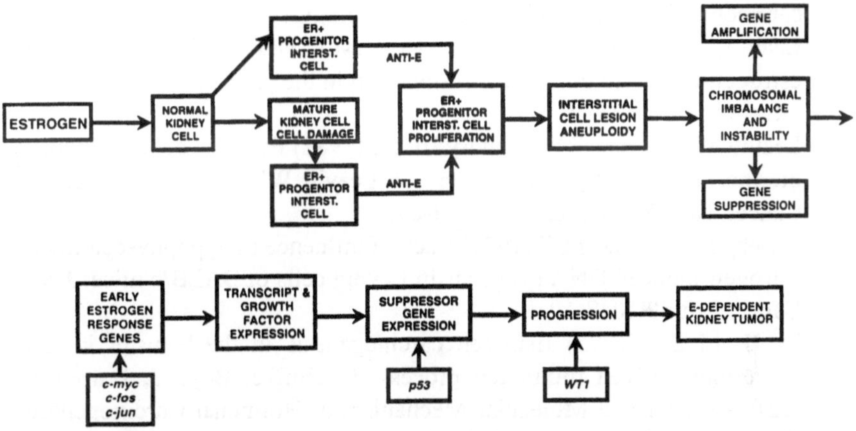

Figure 1. Multi-stage model for estrogen-induced carcinogenesis in the hamster kidney.

On this basis, the inescapable conclusion is that hormones, particularly estrogens, are nongenotoxic carcinogens. This latter term, however, is somewhat misleading since nongenotoxic carcinogens by definition ultimately effect permanent genetic changes leading to neoplastic transformation. Perhaps a more appropriate term would be "epigenotoxic," defined as referring to an agent that is not involved in direct (covalent) or indirect interactions with genetic material, but nevertheless is able to elicit heritable changes by alternative mechanisms. The ultimate result would be heritable changes in the structure or sequence of the genetic material at the level of the nucleic acids, genes, or chromosomes. It is necessary to await future studies to determine whether this estrogen-induced sequential multi-stage scheme will also be applicable to other hormonally-induced tumor model systems.

References

1. Li JJ, Gonzalez A, Banerjee S, Banerjee SK, et al., (1993) Estrogen carcinogenesis in the hamster kidney. Role of cytotoxicity and cell proliferation. Environ Health Perspect 101:259-264.
2. Li JJ, Li SA, Oberley TD, Parsons JA (1995) Carcinogenic activities of various steroidal and nonsteroidal estrogens in the hamster kidney: Relation to hormonal activity and cell proliferation. Cancer Res, in press.
3. Newsbold RR, Bullock BC, McLachlan, JA (1991) Uterus adenocarcinomas in mice following developmental treatment with estrogens: A model for hormonal carcinogenesis. Cancer Res 50:7666-7681.
4. Coe JE, Ishak KG, Ross MJ (1990) Estrogen induction of hepatocellular carcinomas in Armenian hamsters. Hepatology 11:570-577.
5. Leav I, Merk FB, Kwan PWL, Ho, SM (1989) Androgen-supported estrogen-enhanced epithelial proliferation in the prostates of intact Noble rats. Prostate 15:23-40.
6. Leavitt WW, Evans RW, Hendry III WJ (1982) Etiology of DES-induced uterine tumors in Syrian hamsters. In Leavitt WW (ed): Hormones and Cancer. New York: Plenum, pp 63-69.
7. Huseby RA, Samuels LT (1977) Lack of influence of hypophysectomy on estrogen-induced DNA synthesis in Leydig cells of BALB/c mice. J Natl Cancer Inst 58:1047-1049.
8. Li JJ, Li SA (1995) Estrogen carcinogenesis in the hamster kidney: A hormone-driven multi-step process. In Huff J, Boyd J, Barrett JC (eds): Cellular and Molecular Mechanism of Hormonal Carcinogenesis: Environmental Influences. Philadelphia: Wiley-Liss, in press.
9. Li JJ, Li SA (1990) Estrogen carcinogenesis in hamster tissues: A critical review. Endocrine Rev 11:524-531.
10. Hou X, Li JJ, Chen W, Li SA (1995) Estrogen-induced protooncogene and suppressor gene expression in the hamster kidney: Significance for estrogen carcinogenesis. Cancer Res. submitted.
11. Banerjee SK, Banerjee S, Li SA, Li JJ (1990) Cytogenetic changes in renal neoplasms during estrogen-induced renal tumorigenesis in hamsters. In Li JJ, Nandi S, Li SA (eds): Hormonal Carcinogenesis. New York: Springer-Verlag, pp 247-250.
12. Banerjee SK, Banerjee S, Li SA, Li JJ (1994) Induction of chromosome aberrations in Syrian hamster renal cortical cells by various estrogens. Mutation Res 311:191-197.

CLINICAL FORUM:

Risk and Benefits of Hormone Replacement Therapy and Oral Contraceptive Use

Introduction

Therapeutic Implications of Hormone Replacement Therapy and Oral Contraceptives

James Pickar

Although menopausal hormone replacement therapy (HRT) has been available for a considerably longer period than oral contraceptives (OC), both have been available for many years. Awareness of the benefits of preventing the long-term consequences of ovarian failure and the noncontraceptive benefits of OCs should be balanced against the risks of hormone use.

Principal benefits of menopausal HRT, in addition to the relief of menopausal symptoms, include a decrease in both fatal coronary heart disease events and the morbidity and mortality associated with osteoporotic fractures, while primary concerns include endometrial and breast cancer (1-4). These issues, along with a discussion of the use of HRT in women previously treated for breast cancer, will be presented by three of the speakers, Drs. Weiss, Samsiöe, and Eden.

Aside from the obvious benefit of reversible contraception, there are significant noncontraceptive benefits associated with OC use. The principal benefits include a reduced incidence of ovarian and endometrial cancers, of ectopic pregnancy, and of iron deficiency anemia, as well as a reduction in hospitalization for pelvic inflammatory disease. Concerns relate to breast cancer, cervical cancer, and risk of cardiovascular disease, especially in cigarette smokers (5-8). These subjects will be discussed by Drs. Grimes and Drife.

References

1. The Writing Group for the PEPI (1995) Trial: Effects of estrogen or estrogen/progestin regimens on heart disease risk factors in postmenopausal women. JAMA 273:199-208.
2. Hillard TC, Whitcroft S, Ellington MC, and Whitehead MI (1991) The long-term risks and benefits of hormone replacement therapy. J Clin Pharm and Therap 16:231-245.

3. Gorsky RD, Koplan JP, Peterson HP, Thacker SB (1994) Relative risks and benefits of long-term estrogen replacement therapy. A decision analysis. Obstet Gynecol 83:161-166.
4. Harlap S (1992) The benefits and risks of hormone replacement therapy: An epidemiologic overview. Am J Obstet Gynecol 166:1986-1992.
5. Godsland IF, Crook D, Wynn V (1992) Clinical and metabolic considerations of long-term oral contraceptive use. Am J Obstet Gynecol 166:1955-1963.
6. Kaunitz AM (1992) Oral contraceptives and gynecologic cancer: An update for the 1990's. Am J Obstet Gynecol 167:1171-1176.
7. Stergachls A (1992) Epidemiology of the non contraceptive effects of oral contraceptives. Am J Obstet Gynecol 167:1165-1170.
8. Grimes DA (1992) The safety of oral contraceptives: Epidemiologic insights from the first 30 years. Am J Obstet Gynecol 166:1950-1954.

24

Postmenopausal Estrogen-Progestin Therapy in Relation To Cancers of the Endometrium and Breast

Noel S. Weiss, Shirley A.A. Beresford, Janet L. Stanford, and Lynda F. Voigt

Abstract

Under the direction of Drs. Shirley Beresford and Janet Stanford, population-based case-control studies of endometrial and breast cancer have been completed in western Washington State during the past year. In-person interviews were obtained with women diagnosed with one of these conditions, and the responses were compared to those of a random sample of female residents of the same area.

As had been noted in a previous study of endometrial cancer in the same population, among estrogen users a supplementary progestational agent was taken more commonly by controls than cases. However, the risk of endometrial cancer was reduced to that of an estrogen non-user only if: a) progestin use occurred for 10 or more days per month; and b) such use had not been preceded by several years of unopposed estrogens. A similar proportion of breast cancer cases and controls had taken unopposed estrogens; neither was there any difference in the use of estrogens accompanied by a progestin. Long-term use of an estrogen/progestin combination was associated with, if anything, a decreased risk of breast cancer relative to hormone non-users.

The results of these studies suggest that the addition of a progestational agent to a regimen of postmenopausal estrogen therapy does not have an adverse effect on the incidence of breast cancer and affords at least partial protection against the occurrence of endometrial cancer.

Estrogen preparations are used by a great number of postmenopausal women, due both to their short-term influence on symptoms and their longer-term benefit with respect to skeletal strength and reduced risk of coronary heart disease. The potential of these hormones to predispose to the occurrence of endometrial cancer has led many physicians to prescribe a combination of estrogen plus progestin (E+P; combined hormone therapy). While this combination appears to offset the risk of endometrial cancer associated with estrogen use to at least some extent (1), the dose and/or duration of progestin that is optimal in this (and other) regard(s) is uncertain. Furthermore, the potential influence of combination therapy on cancers developing in other hormonally-sensitive organs, e.g., the breast, has only begun to be explored in epidemiologic studies (2).

Two parallel case-control studies have been performed by investigators at the Fred Hutchinson Cancer Research Center and the University of Washington to address these issues. The first, on endometrial cancer, was conducted by Shirley Beresford, Lynda Voigt, Noel Weiss, and Barbara McKnight. The second, on breast cancer, was conducted by Janet Stanford, Noel Weiss, Lynda Voigt, and Janet Daling.

Methods

Case-Control Study of Endometrial Cancer

Residents of three urban counties of western Washington State who were newly-diagnosed with an invasive epithelial carcinoma of the endometrium during 1987-90 were identified through the records of the Cancer Surveillance System, a population-based tumor registry that serves the area. Cases (n = 884) were restricted to women ages 45-74 years at diagnosis. Controls were identified through the random dialing of telephone numbers in the three counties, using the method of Waksberg (3). Women living in households contacted in this way were eligible to participate if they were 45-74 years of age and had not undergone hysterectomy.

In-person interviews were conducted with cases (n = 600, 71 percent of eligible women) and controls (n = 842). These focused on menstrual and reproductive history, and included detailed questions regarding hormonal medications taken. To facilitate recall of specific products, a pictorial display of commonly-used preparations was provided. Exposures that took place until the time of diagnosis of the cases, and a corresponding date in the controls were considered in the analysis. Interviews with cases diagnosed during 1991 are now being completed and ultimately will be included in the final analyses. Therefore, the results presented here should be viewed as preliminary.

Case-Control Study of Breast Cancer

Through the tumor registry, the study identified 50-64 year-old white female residents of King County (in which Seattle is located) diagnosed with *in-situ* or invasive breast cancer during January 1988 - June 1990. Of 660 eligible women, interviews (identical to those described earlier) were obtained from 537 (81

percent). Controls for this study were the same women as in the study of endometrial cancer, except that: 1) Only residents of King County were included.
2) Women who had undergone hysterectomy prior to interview could now be included. 3) Nonwhite women and women with a prior history of breast cancer were excluded. Altogether, 492 of the women interviewed served as control subjects.

Results

Endometrial Cancer

On average, cases were older than controls and, consistent with earlier research, more commonly were nulliparous and of greater body mass index. Adjustment was made for the effects of these variables when examining potential associations between hormone use and endometrial cancer.

Among women who used postmenopausal estrogens for at least six months, but never a supplemental progestin, there was an increased risk of endometrial cancer (Table 1). There was about a twofold elevation through five years of use, but size of the increase was considerably greater with longer durations.

Table 1. "Unopposed" Estrogen Use in Women with Endometrial Cancer and Controls[a].

	Cases	Controls	OR[b]	95% CI
No hormone use, or use <6 months	243	526		1.0
Estrogen use for:				
>0.5 but <3 years	42	66	1.6	1.1 - 2.6
≥3 years but <6 years	30	28	2.4	1.4 - 4.3
≥6 years	166	45	10.5	7.1 - 15.7

[a]Analysis restricted to women who never used a non contraceptive progestin for ≥6 months.
[b]Adjusted for age, parity, body mass index.

Overall, postmenopausal women who took estrogens exclusively in conjunction with a progestin had a risk of endometrial cancer similar to that of women who had never taken hormones (Table 2). However, in the subgroup that had taken the progestin for fewer than 10 days per month, a 2.6-fold increase in risk was present (95% CI = 1.4-4.8). No elevation in risk was seen either for "sequential" users of 10-17 days per month or for "continuous" users (defined here as women who took a progestin 18 or more days per month). Duration of use of E+P (greater or less than five years) seemed to have little bearing on the incidence of endometrial cancer (Table 3).

Table 2. Estrogen Use and Combined Hormone Therapy in Women with Endometrial Cancer and Controls.

	Cases	Controls	Crude OR	Adjusted OR[a]	95% CI
No hormone use, or HT for <6 months	243	526	1.0	1.0	
Unopposed estrogen only	244	144	3.7	4.4	3.3 - 5.9
Combined therapy only	40	93	0.9	1.2	0.8 - 1.9
Progestin <10 days	21	24	1.9	2.6	1.4 - 4.8
Progestin 10- 7 days	16	53	0.7	1.0	0.5 - 1.8
Progestin 18-30 days	2	10	0.4	0.5	0.1 - 2.2

[a]Adjusted for age, parity, body mass index.

Table 3. Combined Hormone Therapy in Women with Endometrial Cancer and Controls, by Duration of Therapy.

	Cases	Controls	Adjusted OR[a]	95% CI
No hormone use, or HT for <6 months	243	526	1.0	
Combined Therapy Only				
<10 days progestin per month				
HT ≤5 years	13	15	2.4	1.1 - 5.4
HT >5 years	8	9	2.9	1.1 - 7.7
>10 days progestin per month				
HT ≤5 years	12	48	0.7	0.4 - 1.4
HT >5 years	6	15	1.3	0.5 - 4.0

[a]Adjusted for age, parity, body mass index.

Medroxyprogesterone acetate (MPA) was the progestin used by the large majority of women in this study. Among those who used MPA for at least

10 days per month, the risk of endometrial cancer was unaffected by the particular daily dose taken (2.5 mg vs. 5-10 mg).

In the relatively small number of women who had taken estrogens alone for less than six years and then went on to take E+P, there was no observed increase in risk of endometrial cancer (Table 4). However, the use of E+P did not eliminate the excess risk associated with longer-term prior unopposed estrogen use. In such women who subsequently used a progestin for less than 10 days per month, the relative risk associated with three or fewer years of E+P was 3.8; with longer use of E+P, it was 14.2 (Table 4). In progestin users of 10 or more days per month, a three-fold elevation in risk remained that did not increase with increasing duration of E+P.

Table 4. Estrogen Use, Followed by Combined Hormone Use (E+p), in Women with Endometrial Cancer and Controls.

Duration of Unopposed Estrogen Use	Use of E + P: Years	Use of E + P: Monthly duration of P (days)	Cases	Controls	OR[a]	95% CI
<6 years	Any	Any	6	16	0.9	0.3 - 2.7
>6 years	<3	<10	6	4	4.6	1.2 - 18.1
	≥3	<10	16	5	11.2	3.9 - 31.7
	<3	≥10	6	9	2.1	0.7 - 6.3
	≥3	≥10	13	12	3.2	1.4 - 7.2

[a]Referent category comprises women who never used E + P, or who did so for <6 months. Odds ratio adjusted for age, parity, and body mass index.

Breast Cancer

The proportion of women who had taken estrogens, never accompanied by a progestin, was similar between breast cancer cases and controls (Table 5). There was no suggestion of any increase in risk associated with unopposed estrogen use of long duration; for example, the odds ratio associated with 12 or more years' use, adjusted for age, age at first full-term pregnancy, and family history of breast cancer, was 0.8 (95% CI = 0.5-1.2). Similarly, current use (as of three months prior to diagnosis) was unrelated to the risk of breast cancer (adjusted odds ratio = 0.9, 95 percent confidence interval = 0.7 - 1.3).

Women who at some time had used a progestin for one or more months to accompany their estrogen use also were at no increased risk of breast cancer

(Table 6). Most of these women had taken the progestin for less than three years, so the ability of this study to evaluate the effects of long-term use is limited. Nonetheless, there was no hint of an increasing risk with increased duration of use, and the odds ratio associated with eight or more years of E+P was 0.4 (95 % CI = 0.2-1.0).

Table 5. Unopposed Estrogen Use in Women with Breast Cancer and Controls, by Duration and Recency of Use.

Measure of estrogen use	Cases No.	Cases (%)	Controls No.	Controls (%)	OR[a]	95% C.I.
Duration						
1-3 months	17	(4.2)	13	(3.5)	1.1	0.5 - 2.4
4 mos. - 2.9 yr.	39	(9.5)	36	(9.7)	1.0	0.6 - 1.6
3 - 4.9 yr.	16	(3.9)	16	(4.3)	0.9	0.4 - 1.8
5 - 7.9 yr.	29	(7.1)	21	(5.6)	1.2	0.7 - 2.2
8 - 11.9 yr.	22	(5.4)	37	(9.9)	0.5	0.3 - 0.9
≥12 yr.	63	(15.5)	63	(16.9)	0.8	0.5 - 1.2
Years since Last Use						
Current	130	(31.7)	122	(32.7)	0.9	0.7 - 1.3
<5	12	(2.9)	17	(4.6)	0.6	0.3 - 1.3
≥5	45	(11.0)	47	(12.6)	0.8	0.5 - 1.3

[a]Odds relative to women who had never used hormones as of three months prior to diagnosis/reference date (223 cases, 187 controls); adjusted for age, age at first full-term pregnancy, and family history of breast cancer. Analysis excludes users of combined hormone therapy.

With one exception, there was no altered risk of breast cancer associated with use of E+P in any subgroup of women defined on the basis of known risk factors, e.g., family history of breast cancer. However, 12 cases and no control had undergone bilateral oophorectomy and then begun to take E+P (lower bound of the 95% CI of the OR = 2.3, relative to hormone nonusers with intact ovaries). No association was seen in women with bilateral oophorectomy who had not taken hormones or who had taken estrogens alone. Use of E+P in women with at least one intact ovary was slightly *less* common in cases than controls. Among the 12 cases who had undergone oophorectomy and then had used E+P, five had taken unopposed estrogen as well (generally for a longer period of time than E+P), but seven had not. The distribution of age at oophorectomy in these women was not atypical; five took place before 45 years, seven later in life.

Table 6. Use of Combined Hormone Therapy in Women with Breast Cancer and Controls, by Duration and Recency of Use.

	Cases No. (%)	Controls No. (%)	OR[a]	95% CI
Duration of Combined Therapy				
1-3 months	15 (4.5)	7 (2.4)	1.8	0.7 - 4.6
4 mos. - 2.9 yr.	56 (16.9)	48 (16.6)	1.0	0.6 - 1.6
3 - 4.9 yr.	17 (5.1)	20 (6.9)	0.6	0.3 - 1.3
5 - 7.9 yr.	16 (4.8)	12 (4.2)	1.0	0.5 - 2.2
≥8 yr.	8 (2.4)	15 (5.2)	0.4	0.2 - 1.0
Years since Last Use				
Current	83 (24.7)	79 (27.3)	0.9	0.6 - 1.3
<5	26 (7.7)	16 (5.5)	1.3	0.7 - 2.6
≥5	4 (1.2)	7 (2.4)	0.5	0.1 - 1.6

[a]Odds relative to hormone nonusers (223 cases, 187 controls); adjusted for age, age at first full-term pregnancy, and family history of breast cancer. Users of combined therapy may also have had one or more episodes of unopposed estrogen use.

Discussion

Endometrial Cancer

The results of the present study support the hypothesis that the use of a progestin, in addition to estrogen, is associated with a lower risk of endometrial cancer than is the use of estrogen alone, particularly if the progestin is administered for at least 10 days each month. Taken as a whole, the results of earlier studies are compatible with this hypothesis as well (1, 3-6). Two findings of the present study that will need evaluation by others are:

1) In women who had not previously taken unopposed estrogen, the similar reduction in endometrial cancer risk afforded by "sequential" regimens of 10 days per month or more and "continuous" regimens.

2) Among combined hormone users, the difference in risk between those who had and had not previously taken unopposed estrogen.

Breast Cancer

Seven prior studies have evaluated the possible relation of E+P use to the occurrence of breast cancer (2). No clear pattern of results has emerged, either for an overall association or for specific features of use (e.g., long duration of use or recent use). Our study, based on the largest number of E+P users to date, suggests no association of breast cancer incidence with any aspect of combined hormone therapy. Even for use of long duration (≥8 years), a category in which there were relatively few study subjects, the upper bound of the confidence interval of the odds ratio was only 1.0.

No prior study has addressed the possibility suggested in our data, that the small group of women without ovaries who also take E+P are at an increased risk of breast cancer. Examination of this subgroup should be a high priority in future studies of breast cancer in relation to exogenous hormones.

Acknowledgements

This publication was supported in part by grants 5 R35 CA 39779, 1 RO1 CA 47749, KO7 CA 01364, and contract NO1 CN 05230 from the NCI.

References

1. Voigt LF, Weiss NS, Chu J et al (1991) Progestogen supplementation of exogenous oestrogens and risk of endometrial cancer. Lancet 338:274-277.
2. Stanford JL and Thomas DB (1993) Exogenous progestins and breast cancer. Epidemiol Rev 15:98-107.
3. Gambrell RD, Massey FM, Castaneda TA, et al. (1979) Reduced incidence of endometrial cancer among postmenopausal women treated with progestogens. J Am Geriatrics Soc 27:389-394.
4. Persson I, Adami HO, Bergkvist L, et al. (1989) Risk of endometrial cancer after treatment with oestrogens alone or in conjunction with progestogens: Results of a prospective study. Brit Med J 298:147-151.
5. Jick SS, Walker AM, Jick H (1993) Estrogens, progesterone, and endometrial cancer. Epidemiology 4:20-24.
6. Jick SS (1993) Combined estrogen and progesterone use and endometrial cancer. Epidemiology 4:384.

25

The Use of Hormone Replacement Therapy in Women Previously Treated for Breast Cancer

John A. Eden and Barry G. Wren

Introduction

The postmenopausal woman who has had breast cancer presents the clinician with special difficulties. Most are unhappy to prescribe hormone replacement therapy (HRT) for these women for fear that HRT may stimulate their cancer. However, premenopausal women who have had early stage breast cancer are usually permitted to produce their own endogenous estrogen (1). Screening programs are detecting an increased number of early stage breast cancers and so many women may survive the tumor only to die of a cardiovascular disease or osteoporotic fracture. A number of lifestyle changes and non-sex hormone treatments are now available to reduce the risk of cardiovascular disease and osteoporotic fractures. Tamoxifen usage may also reduce the risk of heart disease and fractures but this drug is known to aggravate hot flashes and is associated with potentially serious problems including an increased risk of endometrial cancer (2). A small number of menopausal women (10-20%) will suffer severe debilitating menopausal symptoms, particularly hot flashes. For these women, life is unbearable. Quality of life is an important issue for women who have had cancer, so it is important for physicians involved in the care of such patients that treatment strategies are developed to deal with these symptoms. HRT includes a number of different sex hormone combinations of estrogen, progestins and sometimes androgens. Epidemiological studies have demonstrated small but significant differences in breast cancer risk associated with different types of estrogen (3, 4) and different combinations of progestin with estrogen (5). It is also likely that androgens will influence breast cancer risk (6, 7). The relationship between local breast estrogen metabolism and ovarian-derived estradiol (E2) remains unclear. Some breast cancer cell lines as well as breast fat and stromal cells can synthesize E2 locally from androgens (6). Breast tissue levels of E2 do not correlate well with serum levels, and local estrogen metabolism may well be more relevant than serum estrogen levels (6, 7). The effect of progestins on the breast are more complex and dependent upon the type, dosage and duration of the progestin used. In the normal luteal phase, cellular mitotic activity and apoptosis are maximal, suggesting that the effect of progesterone may be at least in part dependent upon other factors (8, 9). Studies of breast cancer cell lines in culture exposed to progestins show an initial modest increase in mitotic activity followed by a profound and continued down- regulation of breast cell proliferative activity which continues for as long as the progestin is continued (8, 10, 11). Continuous progestin reduces breast cell estrogen receptor (ER) content and induces enzymes such as E2 dehydrogenase and sulfatase which convert E2 into weaker estrogens (8). Continuous high-dose progestin is also an effective therapy for advanced breast cancer (10). There are also epidemiological data to support the hypothesis that the addition of a sequential progestin to estrogen replacement therapy is associated with a higher risk of breast cancer than either estrogen alone or continuous combined HRT (5, 12).

Thus, we postulated that, unlike sequential HRT which may increase breast cell activity, a combined continuous regimen using a moderate dose of progestin (e.g., medroxy progesterone acetate, MPA, 50 to 100 mg daily) should have an anti-mitotic action on the breast. It may be an appropriate treatment for treating menopausal women who have had breast cancer and who are having severe menopausal symptoms. We now present data on a cohort of 901 women treated for breast cancer, of whom 90 elected to take HRT to treat severe menopausal symptoms (13).

Methods

The methodology has been published elsewhere (13).

Ethics committees have been reticent to allow a formal double-blind, prospective controlled trial of HRT amongst women with breast cancer and so our group decided to use observational data and conduct a case controlled study. The study group comprised 901 subjects with surgically proven breast cancer attending 3 regional centers in southeastern Sydney, Australia. In 54 cases (6%), complete follow-up data were not available. Ninety (10%) had used estrogens after their breast cancer had been diagnosed. All had been disease free from the time of diagnosis. Seventy-two (80%) had breast-only disease. Twenty-six (29%) had been using conjugated equine estrogens, fifty-two (58%) estrone sulphate, and twelve (13%) had used other estrogens. The median estrogen dosage used, expressed as equivalent to conjugated equine estrogen was 0.625 mg daily. Eighty-two (91%) estrogen users were also taking continuous daily progestin. Forty-six (51%) used MPA, thirty-one (34%) Norethisterone (NE), five (6%) another progestin, and eight (9%) no progestin. The median progestin dosage expressed as MPA equivalent was 50 mg daily. No subject stopped therapy because of side effects.

The 90 subjects who had used estrogen since diagnosis of breast cancer, usually with a continuous moderate dose of progestin as well, were matched to two control subjects (1 to 2 ratio) who had not used hormones for menopausal symptoms since diagnosis of breast cancer. Controls were selected with outcome blinded and were matched for age at diagnosis, axillary node status, maximum tumor diameter, disease-free interval prior to starting estrogen and year of diagnosis. Results were expressed as relative risks (RR) and 95% confidence intervals.

Results

There were no deaths amongst the estrogen users (0 out of 90). Estrogen users had been taking their hormonal medication for a median of 1.5 years (range 0.3-12.0). Their median disease-free interval prior to starting estrogen was 5.0 (0-25) years. Amongst the estrogen users, only 6 out of 90 (7%) developed a recurrence, compared to 30 out of 180 (17%) of the non-user controls (RR = 0.40, 95% CI, 0.17 - 0.93).

Discussion

Haddow first successfully used synthetic estrogens to treat terminal cases of breast cancer in 1944 (14). He indicated that, although synthetic estrogens could produce mammary tumors in certain laboratory animals under specially defined conditions, the same estrogens could cause tumor regression under different conditions (14). Thus a particular hormone's effect will depend on the pathophysiological setting. There is little doubt that estrogen can induce normal and some malignant breast cells to proliferate; however, local regulation of estrogen metabolism is likely to be more relevant to breast cancer growth *in vivo* than are serum estrogen levels. We postulate that continuous moderate to high doses of progestins may be one strategy to blunt the proliferative effect of estrogen.

Confounding is always an important issue in case controlled studies. We attempted to control for factors which are known to impact on the risk of recurrence such as tumor size, stage of disease, and so on. It is possible that the estrogen users were initially more sex-steroid deficient than the controls and therefore perhaps less likely to have a recurrence. A large randomized double-blind controlled trial is needed to confirm these results. If our results are confirmed, then dose-finding studies will be needed to establish the dose of continuous progestin required to protect the breast. We would also contend that continuous combined HRT confers significant other advantages over the older sequential regimens (e.g., amenorrhea). The impact of moderate progestin dosage on cardiac risk was not addressed in this study. If continuous combined estrogen/progestin therapy does reduce the risk of breast cancer recurrence, then future debates on progestin usage and dosage may compare the risk of cardiovascular disease with breast (rather than endometrial) risk factors.

References

1. Creasman WT (1991) Estrogen replacement therapy: is previously treated cancer a contraindication? Obstet Gynecol 77:308-12.
2. Early Breast Cancer Trialists' Collaborative Group (1988) Effects of adjuvant tamoxifen and of cytotoxic therapy on mortality in early breast cancer. N Engl J Med 319:1681-1692.
3. Dupont WD, Page DL (1991) Menopausal estrogen replacement therapy and breast cancer. Arch Intern Med 151:67-72.
4. Steinberg KK Thacker SB, Smith SJ, et al. (1991) A meta-analysis of the effect of estrogen replacement therapy on the risk of breast cancer. JAMA 265:1985-1990.
5. Persson I, Yuen J, Bergkvist L , et al. (1992) Combined oestrogen-progestogen replacement and breast cancer risk. Lancet 340:1044.
6. Blankenstein MA, Szymczak J, Daroszewski J, et al. (1992) Estrogens in plasma and fatty tissue from breast cancer patients and women undergoing surgery for non-oncological reasons. Gynecol Endocrinol 6:13-17.
7. Gordon GB, Bush TL, Helzlsouer KJ, et al. (1990) Serum levels of dehydroepiandrosterone and dehydroepiandrosterone sulfate and risk of postmenopausal breast cancer. Cancer Res 50:3859-62.
8. Clarke CL, Sutherland RL (1990) Progestin regulation of cellular proliferation. Endocrine Rev 11:266-301
9. Anderson TJ, Battersby S (1989) The involvement of oestrogen in the development and function of the normal breast: histological evidence. In: Beck JS, editor. Oestrogen and the human breast. Royal Society of Edinburgh 23-32.
10. Rose C, Mouridsen M (1989) Endocrine management of advanced breast cancer. Horm Res 32: 189-97.
11. Eden JA (1992) Oestrogen and the breast - 1. Myths about oestrogen and breast cancer. Med J Aust 157:175-7.
12. Ewertz M. (1988) Influence of non-contraceptive exogenous and endogenous sex hormones on breast cancer risk in Denmark. Int J Cancer 42:832-838.
13. Eden J A, Bush T, Nand S, Wren B G (1994) The Royal Hospital for Women Breast Cancer Study - A Case controlled study of combined continuous oestrogen progestogen replacement therapy amongst women with a personal history of breast cancer. Med J Aust. Submitted.
14. Haddow A, Watkinson JM, Paterson E (1944) Influence of synthetic oestrogens upon advanced malignant disease. Brit Med J Sept 23:393-98.

26

Cardioprotection by Hormone Replacement Therapy

Göran Samsioe

Introduction

Cardiovascular disease, especially myocardial infarction, is more common in men than in women before the age of fifty. Oophorectomy in premenopausal women leads to an increase in cardiovascular morbidity and mortality (1, 2). Based on these observations, it can be inferred that female gonadal hormones may confer protection against cardiovascular disease irrespective of age (3). Experimental studies on several risk factors of cardiovascular disease carried out in both animals and humans further support this notion. These observations provide the rationale for exogenous estrogen administration in order to reduce cardiovascular disease in women.

The first studies used mainly hospital controls and were rather short term. Sample sizes were usually small. The results of these studies were conflicting, but one common feature was the large confidence intervals, suggesting methodological problems (Figure 1). With larger study groups, longer observation periods and population-based controls in combination with more rigid protocol procedures, confidence limits have been substantially narrowed, and almost all recent studies are indicative of a reduced risk of cardiovascular disease in estrogen users in comparison with the controls (Figure 1, Table 1).

Using the technique of metaanalysis, Stampfer (4) concluded that the use of estrogen is cardioprotective and that the subsequent risk of myocardial infarction is fifty percent less in estrogen users that in non-users (Figure 1).

Possible Confounders and Biases

Even after a metaanalysis, several questions remain. Observational studies do not control exposure. It is possible that women who experience climacteric symptoms severe enough to consult the medical profession and are prescribed estrogens, after duly considering potential contraindications and precautions for estrogen use, are different from women without symptoms or from women with contraindications, for instance those who carry risk factors for various diseases.

All studies agree that estrogen users are leaner that non-users, possibly for endocrine reasons. Overweight is commonly considered a risk factor for cardiovascular disease. However, recent data suggest that it is not only obesity per se but also the type of obesity. Men almost exclusively have their excessive body fat around the waist, whereas women more often show a fat distribution in limbs, buttocks and breasts. The male or android fat distribution is clearly a risk factor for cardiovascular disease, whereas the so called female or gynecoid obesity does not seem to be associated with cardiovascular morbidity or mortality to any great extent. The fact that estrogen users in epidemiological studies are leaner than non-users cannot be used as an indication of lower cardiovascular risk if not controlled for smoking or type of obesity. Several of the more recent studies are large enough to control for many of the potential confounders and biases. Such confounders comprise differences in life-style such as physical activity, smoking, socio-economic class, etc. If anything, it would seem that those women who are smokers and of low socio-economic class have more to gain from estrogen use. In other words, life-style factors do not seem to influence the results, at least not negatively (5).

CARDIOPROTECTION BY ESTROGENS
MECHANISMS OF ACTION

Metabolism

HDL-chol ↗
LDL-chol ↙
Ox-LDL ↙
Triglycerides ↙
Insulin ↙
Coagulation factors ↙
Fibrinolysis (↗)
Anticoagulation factors ↗

Vessel Wall

Endothelin(s) ↙
EDRF(NO) ↗
Thromboxane A_2 ↙
Prostacyclin (↗)
Calcium channel blocking ↗

Blood pressure ↙
Blood flow ↗
Vascular resistance ↙

Figure 1. Estrogen cardioprotection is conceivably mediated via known risk factors for cardiovascular disease. Estrogens have been described to influence a variety of markers for metabolism as well as vessel wall physiology. Only qualitative changes are marked. The importance of each change in quantitative terms vis-á-vis cardioprotection remains to be clarified. Figure represents only recognized activities of estrogens; additional important mechanisms may well exist.

Table 1. Summary of Studies of Estrogen Use in Women: Fatal and Nonfatal Cardiovascular Disease.

First author	(year)	Study design	Study size	End points	Risk Est
Rosenberg	1993	case-control	cases = 36 controls = 33	MI	0.6
Nabulsi	1993	cross-sectional	users = 1.026 non-users = 3.932	CHD	0.58
Stampfer	1991	cohort	woman years = 337 854 no of fig = 48.400	CHD	0.56
Lafferty	1985	cohort 1.100 PY	124 women	MI	0.16*
Stampfer	1985	cohort 129.000 PY	32.317 women	All CVD	0.30*
Nachtigall	1979	clinical trial 1.680 PY	168 women	MI	0.33
Hammond	1979	cohort 3.000 PY	610 women	All CVD	0.33*
Bush	1979	cohort 19.300 PY	2.270 women	CVD death	0.34*
Talbott	1977	case-control controls = 64	cases = 64	Sudden death	0.34
Rosenberg	1976	case-control controls = 6.730	cases = 336	Nonfatal MI	0.47
Henderson	1988	cohort 39.600 PY	8.807 women	MI	0.54*
Beard	1989	case-control controls = 150	cases = 86	MI and sudden death	0.55
Petitti	1987	cohort PY?	3.437 women	All CVD	0.60
Avila	1990	cohort	24.900 women	Nonfatal MI	0.70
Adam	1981	case-control controls = 151	cases = 76	Nonfatal MI	0.79
Sziko	1984	case-control controls = 39	cases = 36	Nonfatal MI	0.83
La Vecchia	1987	case-control controls = 160	cases = 603	Nonfatal MI	1.62
Wilson	1985	cohort	1.234 women	All CVD	1.76*
Finucane	1993	cohort	1.910 women	Stroke	0.66

Includes cohort and case-control studies. Nachtigall study is the only randomized clinical trial.

Furthermore, women who carry established risk factors for cardiovascular disease such as family history, hypertension or lipid perturbations seem to be even more protected than women who do not carry these risk factors (6- 8).

Recent data (9), as well as an overview of the literature (10), offer little support for the notion that estrogen given in the doses used in hormonal replacement therapy should promote thromboembolism. Rather the opposite seems to be true. Risk of thromboembolic complications would hence be higher in controls. Only a randomized trial could distinguish between a true estrogen effect and a selection bias.

In some of the observational studies it has been noted that the overall mortality is lower among estrogen users than non-users. In addition studies have reported an overall lower mortality from cancers with various origins. This observation provided the rationale for questioning the results on cardiovascular disease and osteoporotic fractures, supporting the hypothesis that estrogen users belong to a healthier population. However, recent data imply that estrogens may influence the course of cancers, and it has been reported for malignant melanoma and breast cancers that the progress of the disease is slowed-down by estrogen use (11, 12). The reduction of serum insulin and thereby the activity of several insulin dependant growth factors has been suggested, since estrogen treatment indeed impedes tumor growth in these instances. Clearly, many more data on this issue are warranted. The observed differences between current, ever and former users of estrogen could also, to some extent, be influenced by compliance (13). In several angiographic studies (7, 8, 14, 15), it has been observed that coronary stenosis is reduced by estrogen replacement therapy, most markedly for those women with severe stenosis (Table 2).

Table 2. Studies of Estrogen Use in Women and Angiographically Defined Coronary Disease.

First Author	(year)	Study design	Study size	End points	Risk est
Hong	1993	cross-sectional	18 users 72 non-users	coronary stenosis	0.13
Sullivan	1988	case-control	cases = 1.444	≥ 70% stenosis	0.44*
Gruchow	1988	cross-sectional	users = 154 non-users= 779	sever occlusion moderate occlusion	0.37* 0.59*
McFarland	1989	case-control	cases = 137	≥ 70% stenosis	0.50*

Effects of estrogens on coronary arteriosclerosis as verified by angiographic studies.

The addition of a progestogen for endometrial protection in HRT may theoretically attenuate the cardiovascular benefits by estrogens. Progestogens are generally antiestrogenic, as estrogen receptor activity is down-regulated by progestogen. Indeed, HDL cholesterol is lower in patients receiving combined therapy in comparison with those taking estrogen alone. However, LDL cholesterol may be even further reduced by addition of a progestogen (16, 17).

Observational studies on combined therapy are few (Table 3). The limited published data do not suggest any attenuation by the progestogen, be it medroxyprogestogen acetate, norethisterone or levonorgestrel (18-21). In line with these studies, past use of oral contraceptives is not associated with an increased risk of coronary heart disease (22). Rather the opposite may be true as indicated by the Finnish autopsy study in oral contraceptive users (23).

In conclusion, there is a biological plausibility that estrogens are cardioprotective. The consistency of observational studies as well as the duration-dependent effect of estrogen replacement support this view. This concept will not be altered by a controlled randomized clinical trial.

The optimal estrogen dosage is yet to be determined, as there seems to be little or no difference in cardioprotection when comparing 0.625 and 1.25 mg of conjugated equine estrogens (5).

Table 3. Observational Studies on Combined Estrogen-progestogen Preparations and Cardiovascular Disease.

First Author	(year)	Study design	Study size	End points	Risk est
Nabulsi	1993	cross- sectional	173 users	CHD	<0.58
Nachtigall	1979	clinical trial	84 pairs	MI	no increase
Thompson E alone: RR = 1.1	1989	case-control	603 cases	MI + CVD 1.206 controls	RR = 0.9 Stroke
Hunt	1990	cohort	4.544	IHD	RR = 0.3
Falkeborn	1992	cohort	23	MI, Stroke	RR = 0.5

CVdisease and combined HRT. Falkeborn study used mainly levonorgestrel-containing preparations. Hunt, et al., predominantly levonorgestrel, also norethisterone with conjugated equine estrogens, and in the remaining three studies medroxyprogesterone was the predominant estrogen co-medication.

Table 4. Differences Between Premenopausal and Postmenopausal Women in Lipid and Lipoprotein Concentrations (mean nmol/L±S.D.) Derived from Standardized Data (Triglyceride Values Are Back-converted).

	Total Cholesterol	Triglycerides	LDL Cholesterol	HDL Cholesterol	HDL_2 Cholesterol	HDL_3 Cholesterol
Premenopausal (n = 395)	4.93 ±0.76	0.65 -0.19 +0.28	2.80 ±0.32	1.81 ±0.32	0.77 ±0.27	1.03 ±0.19
Postmenopausal (n = 147)	5.60** ±1.04	0.73* -0.26 +0.41	3.56** ±0.96	1.68** ±0.34	0.58** ±0.26	1.10** ±0.15

*$p < 0.005$, ** < 0.001 compared with premenopausal women (2)

Mechanisms of Action in Lipid Metabolism

Fertile women have lower levels of serum lipids than their male counterparts of corresponding age. When a woman passes through the menopause, her serum total and LDL cholesterol as well as triglycerides are increased and HDL cholesterol is decreased (24). Longitudinal and cross-sectional data suggest that this is not an effect related merely to age but more to cessation of the endogenous hormone production (Table 4) (25). Furthermore, it seems that the decrease of HDL is confined to the HDL2 fraction, while HDL3 may, in fact, increase somewhat. The exact mechanisms for these changes in HDL and LDL are not known with certainty. For HDL, it is believed that estrogens suppress hepatic lipid activity, elevating levels of HDL2 on HDL cholesterol. LDL changes can be explained in part by its increasing rate of clearance from plasma.

Hormone replacement therapy reverses the menopausal changes in cholesterol metabolism, and it is possible that serum lipid changes may contribute to the overall cardiovascular benefit from estrogens. Using the Cox proportional hazards model, Bush, et al., (26) pointed out that serum lipid changes by estrogens, most notably the increasing HDL, could explain up to 50% of the cardioprotective effect. In subsequent analysis, the figure for HDL has been somewhat reduced and is now believed to be around 25-40%.

It is well known that arteriosclerosis and the metabolic basis for its progress are multifactorial. Disturbances in several lipid factors as well as in carbohydrate metabolism and hemostasis contribute to its development. It is virtually impossible for any study to cover all metabolic aspects and markers for arteriosclerosis, and the beneficial effect ascribed to HDL as mentioned above in fact includes all factors which co-vary with HDL cholesterol. Some of the factors that are estrogen dependent in addition to HDL include sex hormone binding globulin, factor VII, fibrinogen, and PAI-1, all of which may have an impact on the development of cardiovascular disease.

It is well documented that one of the anti-atherogenic effects by HDL is to remove cholesterol from the periphery and facilitate transport back to the liver. Excess cholesterol is subsequently excreted into the bile or reenters the cholesterol pool. To what degree HDL fulfills this purpose, also in pathological tissues, such as the arteriosclerotic plaque is less well defined (27).

Studies on nonhormonal lipid-lowering agents have repeatedly revealed almost linearity between decreased incidence of myocardial infarction and other manifestations or cardiovascular disease and lowering of LDL cholesterol or total cholesterol.

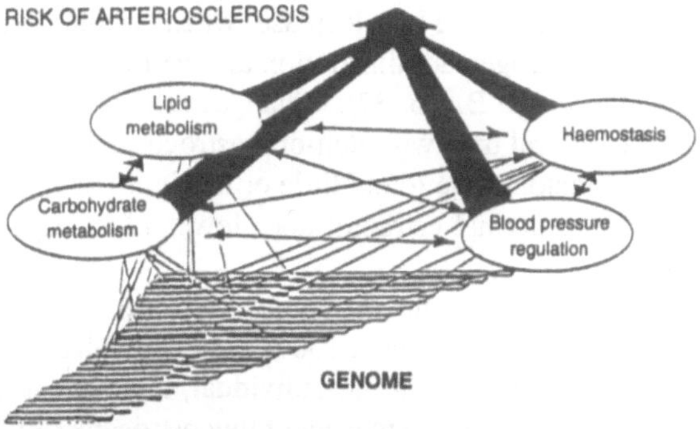

Figure 2. The metabolic syndromes sometimes referred to as Reaven's syndrome X. Figure indicates a common genome and the close interrelationships between lipid and carbohydrate metabolism as well as hemostasis and vascular function. Reprinted by permission from the front page of "Current Opinion in Lipidology," Vol. 4, 1993.

Data in men suggest that a 35% reduction of LDL cholesterol is required to achieve a 50% reduction in cardiovascular disease should LDL reductions be the sole cause of cardioprotection (28). But estrogens reduce LDL cholesterol only by some 5-10% (Table 2) (26, 29). This would seem paradoxical at first glance should lipids play an important role in cardioprotection by estrogens. However, there is the possibility that the key lipid factor is not disclosed by the routine serum lipid and lipoprotein profile.

In 1989 Steinberg and coworkers (30) summarized the evidence that a subfraction of LDL and possibly also of IDL had a much higher atherogenicity than native normal LDL. This highly atherogenic LDL is modified mainly via oxidation, but other modifications exist as well. It is an intriguing possibility that estrogens reduce formation and activity of modified LDL, which may be part of the cardioprotective effect.

Haarbo, et al. (31), showed that, despite similar serum cholesterol levels, those animals given estrogens had significantly less cholesterol in the arterial wall than those animals who were not given any hormones. There are also *in-vitro* data suggesting that estrogens may indeed reduce formation of oxidized LDL (32, 33). It is of interest to note that in animal studies the addition of progestogen doe snot seem to influence estrogen impairment of arteriosclerotic plaque formation (34, 35).

Apolipoproteins

Changes in HDL and LDL, and to some extent also in VLDL, are reflected by specific changes in apoliproprotein A1 and B100. Women have higher levels of A1 than males, and levels tend to decrease when women pass through menopause. Exogenous estrogen administration also results in increased levels of apo A1 and decreased apo B (36, 37). Changes in HDL and A1levels are much more marked with oral than with non-oral estrogens. Using transdermal estradiol, several investigators have found little or no significant effect on HDL and apo A, whereas transdermal estradiol does lower LDL levels and apo B levels (38).

Another interesting compound is lipoprotein (a) (39). The serum concentration of lipoprotein (a) is almost 100% genetically determined. There is little fluctuation throughout life in the individual, although interindividual variations are great. There is evidence to suggest that estrogens (40, 41), as well as estrogen plus progestogens (41), may in fact reduce lipoprotein (a).

Hemostasis

In addition to the effects described by Nabulsi, et al., (20) several other reports have indicated changes by estrogen on liver-derived clotting factors. Oral estrogens tend to increase fibrinogen and FVII, whereas transdermal estradiol to some extent has the reverse effect (42). Given the complexity of coagulation, anti-coagulation, fibrinolysis and anti-fibrinolysis, changes within just a few markers of the hemostatic system may be difficult to interpret in terms of clinical outcome. Observational studies do not suggest an increase of venous thromboembolism during estrogen replacement therapy. Rather the reverse seems to be true. Secondary preventive measures by estradiol yield an even lower relative risk (RR=0.5) for healthy individuals. Should estrogen promote arterial coagulation, it would be difficult to explain these differences.

Carbohydrate Metabolism

Elevated insulin concentrations are frequently found in women with coronary heart disease and are likely to be due to insulin resistance. Hyperinsulinemia may increase cardiovascular disease by directly promoting atherogenesis, and insulin propeptides may also be important in this respect. Increased insulin concentrations may adversely affect several risk factors for coronary heart disease, and it has been suggested that insulin resistance is a pivotal metabolic disturbance in a constellation of cardiovascular risk factors. There is an association between hyperinsulinemia and hypertension. Increased insulin concentrations are also associated with high triglycerides, low HDL and increased small dense LDL. Obesity is associated with insulin resistance, and it is the central or android body distribution which correlates with these metabolic disturbances. There is commonly an increase in the anti-fibrinolytic factor plasminogen activator inhibitor-1 (PAI-1) in combination with increased insulin levels. Estrogen and progesterone promote pancreatic insulin secretion, but the former reduces insulin resistance while the latter increases it. Hence, menopause reduces pancreatic insulin secretion. However, circulating insulin concentrations rise with age in postmenopausal women, due to an increase in insulin resistance and in hepatic insulin throughput (43).

Administration of estrogen to postmenopausal women improves insulin resistance. The addition of a progestogen may produce certain adverse effects on glucose and insulin metabolism. Even if differences have been observed between various progestogens in this respect, more data are urgently needed to clarify this issue.

Vessel Wall Physiology

Several reports describe vasodilatatory properties of estrogens in the uterine artery but also in the common carotid artery and the aorta. A dilatation of arteries could be explained by the fact that estrogens have been described to be calcium channel blockers (44). There is also evidence to suggest that estrogens may down-regulate one of the two principal endothelin receptors, rendering the vasoconstrictive properties of endothelins less marked (45). Yet another possibility is an influence via nitrous oxide metabolism (46). In addition, estrogens reduce peripheral resistance and may even lower blood pressure in most individuals. The lowering of blood pressure is not very marked but may be of significance from a public health aspect. However, in a small percentage (1-3%), there is an idiosyncratic reaction, and women may respond with elevated blood pressure levels. Therefore, blood pressures should be checked in women receiving hormone replacement therapy.

In conclusion, estrogens (possibly also when combined with progestogens) seem to affect lipid metabolism, hemostasis, insulin and carbohydrate metabolism as well as blood pressure. All these factors constitute the so-called metabolic syndrome, and it is postulated that estrogen may act primarily via factors that directly or indirectly influence the metabolic syndrome. In addition, several possibilities exist by which estrogens or estrogens plus progestogens may directly modulate vessel wall physiology.

Acknowledgment

Christina Mihaly is acknowledged for skillful typing of the manuscript.

References

1. Johansson BW, Kaij L, Kullander S, et al. (1975) On some effects of bilateral oophorectomy in the age range 15-30 years. Acta Obstet Gynecol Scand 54:449-461.
2. Wuest J, Dry T, Edwards J (1953) The degree of coronary atherosclerosis in bilaterally oophorectomized women. Circulation VII:801-810.
3. Kannel WV, Hjortland M, McNamara PM, et al. (1976) Menopause and risk of cardiovascular disease: The Framingham study. Ann Intern Med 85:447-552.
4. Stampfer MJ, Colditz GA, Willett WC, Manson JE, et al. (1991) Postmenopausal estrogen therapy and cardiovascular disease. N Engl J Med 325:756-762.
5. Henderson BE, Paganini-Hill A, Ross RK (1988) Estrogen replacement therapy and protection from acute myocardial infarction. Am J Obstet Gynecol 159:312-317.
6. Bush TL, Barrett-Connor E, Cowan LD, et al. (1987) Cardiovascular mortality and non-contraceptive estrogen use in women: results from the Lipid Research Clinic's Program Follow-up Study. Circulation 75:1002-1009.
7. Hong MK, Romm PA, Reagan K (1992) Effects of estrogen replacement therapy on serum lipid values and angiographically defined coronary artery disease in postmenopausal women. Am J Cardiol 69:176-178.
8. Sullivan JM, Zwaag RV, Lemp GF, et al. (1988) Postmenopausal estrogen use and coronary atherosclerosis. Ann Intern Med 108:358-363.
9. Devor M, Barrett-Connor E, Renvall M (1992) Estrogen replacement therapy and the risk of venous thrombosis. Am J Med 92:275-282.
10. Lobo R (1992) Estrogen and the risk of coagulopathy. Am J Med, 92:283-285.

11. Bergkvist L, Adami HO, Persson I, et al. (1989) Prognosis, after breast cancer diagnosed in women exposed to estrogen and estrogen-progestogen replacement therapy. Am J Epidemiol 139:221-228.
12. Brinton LA, Hoover R, Fraumeni JF Jr (1986) Menopausal oestrogens and breast cancer risk: An expanded case-control study. Brit J Cancer 54:825-32.
13. The Coronary Drug Project Research Group (1980) Influence of adherence to treatment and response of cholesteron on mortality in the Coronary Drug Project. N Engl J Med 303:1038-1041.
14. Gruchow HW, Anderson AG, Barboriak J, et al. (1988) Postmenopausal use of estrogen and occlusion of coronary arteries. Am Heart J, 115:954-963.
15. McFarland KF, Boniface ME, Hormung CA, et al. (1989) Risk factors and noncontraceptive estrogen use in women with and without coronary disease. Am Heart J 117:1209-1214.
16. Weinstein L (1987) Efficacy of a continuous estrogen-progestin regimen in the menopausal patient. Obstet Gynecol 169:929-932.
17. Crook D, Cust MP, Gangar KF, et al. (1992) Comparison of transdermal and oral estrogen/progestin replacement therapy: Effects on serum lipids and lipoproteins. Am J Obstet Gynecol 166:950.
18. Nachtigall LE, Nachtigall RH, Nachtigall RD, et al. (1979) Estrogen replacement therapy, II: A Prospective study in the relationship of carcinoma and cardiovascular and metabolic problems. Obstet Gyncol 54:74-79.
19. Thompson SG, Meade TW, Greenberg G (1989) The use of hormonal replacement therapy and the risk of stroke and myocardial infarction in women. J Epidemiol Comm Health 43:173-178.
20. Nabulsi AA, Aaron B, Folsom R, et al. (1993) Association of hormone-replacement therapy with various cardiovascular risk factors in postmenopausal women. N Engl J Med 328:1070-1075.
21. Falkeborn M, Persson I, Adami H, et al. (1992) The risk of acute myocardial infarction after oestrogen and oestrogen-progestogen replacement. Brit J Obstet Gynecol 99:821-828.
22. Stampfer MJ, Willet WC, Colditz GA, et al. (1988) A prospective study of past use of oral contraceptive agents and risk of cardiovascular disease. N Engl J Med 317:1313-1317.
23. Hirvonen E, Idänpään-Heikkilä J (1990) Cardiovascular death among women under 40 years of age using low-estrogen oral contraceptives and intrauterine devices in Finland from 1975 to 1984. Am J Obstet Gynecol 163:281-284.
24. Matthews KA, Meilahn E, Kuller LH, et al. (1989) Menopause and risk factors for coronary heart disease. N Engl J Med 321:641.

25. Stevenson JC, Crook D, Godsland IF (1993) Influence of age and menopause on serum lipids and lipoproteins in healthy women. Atherosclerosis 98:83-90.
26. Busch TL (1991) Cardioprotection by estrogens in women with and without risk factors for cardiovascular disease. 6th Int Cong on the Menopause, Abstr p. 231.
27. Gordon DG, Rifkind BM (1989) High-density lipoprotein - the clinical implications of recent studies. N Engl J Med 321:1311-1315.
28. Lipid Research Clinics Primary Coronary Prevention Program (1984) Results II. The relationship of reduction in incidence of coronary heart disease to cholesterol lowering. JAMA 251:365-374.
29. Samsioe G (1991) Lipid profiles in oestrogen users. In: The Menopause and Hormonal Replacement. Sitruk-Ware R, Utian W (eds), Marcel Dekker, Inc, New York, pp.181-200.
30. Steinberg D, Parthasarathy S, Carew TE, et al. (1989) Beyond cholesterol. Modifications of low density lipoproteins that increase its atherogenicity. N Engl J Med 320:915-920.
31. Haarbo J, Espensen PL, Stender S, et al. (1991) Estrogen monotherapy and combined estrogen-progestogen replacement therapy attenuate aortic accumulation of cholesterol in overiectomized cholesterol-fed rabbits. J Clin Invest 87:1274-1279.
32. Mazière C, Auclair M., Ronveaux M-F, et al. (1991) Estrogens inhibit copper and cell-mediated modification. Atherosclerosis 89:175-182.
33. Sack NS, Rader DJ, Cannon RO III (1994) Oestrogen and inhibition of oxidation of low-density lipoproteins in postmenopausal women. Lancet 343:269-270.
34. Clarkson TB, Adams MR, Williams JK, et al. (1990) Effects of sex steroids on the monkey cardiovascular system: Relation to changes in serum lipids and lipoproteins In: Osteoporosis, Christiansen C, Overgaard K (eds), pp 1798-1805.
35. Haarbo J, Svendsen OL, Christiansen C (1992) Progestogens do not affect aortic accumulation of cholesterol in ovariectomized cholesterol-fed rabbits. Circ Res 70:1198-1202.
36. Granfone A, Campos H, McNamara JR, et al. (1992) Effects of estrogen replacement on plasma lipoproteins and apolipoproteins in postmenopausal, dyslipidemic women. Metabolism 41:1193-1198.
37. Campos H, McNamara JR, Wilson PW, et al. (1988) Differences in low density lipoprotein subfractions and apolipoproteins in premenopausal and postmenopausal women. J Clin Endocrinol Metab 67:30-35.

38. Crook D, Cust MP, Gangar KF, et al. (1992) Comparison of transdermal and oral estrogen/progestin replacement therapy: Effects on serum lipids and lipoproteins. Am J Obstet Gynecol 166:950.
39. Hegele RA (1989) Lipoprotein (a): An emerging risk factor for atherosclerosis. Can J Cardiol 5:263-265.
40. Soma M, Fumagalli R, Paoletti R, et al. (1991) Plasma Lp(a) concentration after oestrogen and progestogen in postmenopausal women. Lancet 337:612.
41. Nabulsi AA, Aaron B, Folsom R, et al. (1993) Association of hormone-replacement therapy with various cardiovascular risk factors in postmenopausal women. N Engl J Med 328:1070-1075.
42. Lindoff C, Peterson F, Lecander I, et al. (1995) Transdermal oestrogen replacement therapy: Beneficial effects on hemostatic risk factors for cardiovascular disease. In press.
43. Stevenson JC, Proudler AJ, Walton C, Godsland IF (1994) HRT mechanisms of action: carbohydrates. Int J Ferti 39:50-55.
44. Collins P, Rosano GMC, Jiang C, et al. (1993) Cardiovascular protection by oestrogen - a calcium antogonist effect? Lancet 341:1264-1265.
45. Jiang C, Sarrel PM, Poole-Wilson PA, Collins P (1992) Acute effect of 17β-estradiol on rabbit coronary artery contractile responses to endothelin-1. Am J Physiol 263:H271-75.
46. Williams JK, Adams MR, Klopfenstein HS (1990) Estrogen modulates response of atherosclerotic coronary arteries. Circulation Res 81:1680-1687.

27

Risks and Benefits of Oral Contraceptives

David A. Grimes

Introduction

We know more today about the safety of oral contraceptives (OC) than about any other medication. Because millions of healthy women use OC for extended lengths of time, the health effects become an important public health concern. After three decades of extensive study, investigators have demonstrated that OC, when given to healthy women, protect against unwanted pregnancy and its complications, improve health, and may even prolong life (1-5).

Regrettably, women around the world do not share this reassuring assessment. A controversy surrounds OC, largely due to biased media coverage (6). A large number of women in the U.S. and in developing nations believe that OC carry serious health risks (7).

Much of this fear centers on cancer, the focus of this symposium. Indeed, the question of a potential association between OC and breast cancer is the single most important question about OC safety today. In an attempt to put the risks and benefits of OC in perspective, this chapter will briefly review several risks and benefits; comprehensive evaluations are available elsewhere (1-3,8). Although based on epidemiology, the chapter attempts to summarize epidemiological data in a fashion understandable to scientists and others who are neither clinicians nor epidemiologists.

Risks of OCs: Cancer, Infertility, and Cardiovascular Disease

Breast Cancer

Overall, use of OC has no net effect on a woman's risk of breast cancer. Over twenty case-control studies involving more than 20,000 women, as well as large cohort studies in the U.K., have shown no overall association. In addition, a large meta-analysis (9) has confirmed no association and no effect with increasing duration of use. In contrast, some studies focusing on women younger than 35 years of age have reported increased risk with increasing duration of use (10). However, the largest case-control study in the world (11) has shown a decreasing risk with increasing duration of use among older women, who are at greater risk because of their age.

Cervical Cancer

Numerous reports of this potential association appeared during the 1970's and 1980's, with conflicting results. However, in recent years, several have suggested an increased risk of cervical cancer associated with long-term use (12). "Long-term" appears to be about five years.

This association may not be causal. A number of biases can influence studies of this question. For example, women who take OC have cervical cytology screening more often than do other women, so neoplasia may be more likely to be detected should it develop. Cervical cancer is a sexually-transmitted disease; virgins do not get squamous cell cancer. However, most studies have not adequately controlled for the confounding effect of sexual behavior, such as age at first coitus and number of partners. Finally, cigarette smoking is an independent risk factor for cervical cancer and may confound this relationship.

Liver Cancer

Liver cancer is rare in developed countries. In the 1970's, epidemiologists linked long-term use of high-dose OC with benign hepatocellular adenomas. Hence, it seemed plausible that a link with cancer might exist.

Current data are conflicting. In developed countries, where incidence rates are low, a number of case-control studies have shown large increases in the relative risk among OC users. In contrast, studies done in the Third World have failed to find an association (2).

Infertility

The notion that women are infertile for a long time after OC discontinuation is widespread. Research has shown that there may be only a transient delay in the resumption of ovulation. In the Oxford-Family Planning Association cohort, a lag in fertility rates occurred in the first year after discontinuation (13). By the end of the second year, this had disappeared. More recent work at Yale University has shown that the delay in resumption of ovulation is dose-related (14). Women who had been taking higher-dose estrogen OC had about a month longer delay than did those taking lower-dose OC.

Cardiovascular Disease

After intensive study, the epidemiology is finally unfolding. Much of the risk of cardiovascular disease attributed to OC in early studies was due to the confounding effect of cigarette smoking. For healthy women who do not smoke, OC do not appear to pose an increased risk of heart disease.

For example, women in the large Royal College of General Practitioners study have had over 150 documented myocardial infarctions (15). Among nonsmokers, use of OC did not increase the risk. Among light smokers, however, the risk of a myocardial infarction increased over threefold if the woman used OC. On the other hand, among heavy smokers, this relative risk jumped to 20-fold.

In the late 1980's, the notion that progestins in OC adversely affect lipids and thus cause atherosclerosis in women received wide attention. At least four lines of indirect evidence argue against this hypothesis. First, former users of OC do not have an increased risk of heart disease. If plaque deposition were involved, the plaque should persist after discontinuation. However, in clinical studies, disease rates return to baseline, which argues against persistent atherosclerosis. Second, no dose-response relation is evident. If OC caused plaque deposition, the longer one took the OC, the greater should be the extent of disease; this is not the case. Third, an autopsy study (16) of young women who died of myocardial infarctions revealed that the majority of women who had not used the OC had atherosclerosis. In contrast, a minority of those who had used OC had atherosclerosis. Finally, elegant studies (17) with the monkey model showed that progestin-dominant OC caused adverse changes in HDL- and LDL-cholesterol. Nevertheless, the monkeys with the adverse lipids experienced significant protection against plaque formation in their coronary arteries. Hence, thrombosis related to high-dose estrogen may have increased the risk of cardiovascular disease reported in older studies.

Health Benefits: Life-Threatening and Quality-of-Life Concerns

Ovarian Cancer

Prevention of ovarian cancer is the single most important noncontraceptive health benefit of OC. Ovarian cancer is the deadliest gynecological malignancy and kills more women in the U.S. each year than do cervical and endometrial cancers combined. This stems from the fact that clinicians do not discover most cases of ovarian cancer until the cancer has spread throughout the abdomen. Each year, about 2000 women avoid this deadly cancer because they have used OC.

Twenty case-control studies and three cohort studies report a reduction in risk of about 40% overall for women who have ever used OC. A strong dose-response effect occurs: the longer a woman takes the OC, the more protection she enjoys. For example, in the Cancer and Steroid Hormone Study (18) of the Centers for Disease Control and Prevention, after ten years of use, women had an 80% reduction in the risk of this cancer. Moreover, the protection is long-lasting. The risk is lower for at least fifteen years after discontinuation.

Endometrial Cancer

As with ovarian cancer, use of OC provides strong protection against the most common gynecological malignancy in the U.S. A large number of case-control and cohort studies have confirmed this relationship. Women who have ever used OC have about a 50% reduction in their risk of developing endometrial cancer; this appears true for each of the three major histologic types. Interestingly, the protection appears strongest among women at high risk: nulliparous women have about 80% protection, and this effect lessens with increasing parity (19).

Pelvic Inflammatory Disease

Although not usually considered life-threatening, salpingitis (pelvic inflammatory disease) kills over one hundred women each year in the U.S. The costs for treating pelvic inflammatory disease and its complications, such as chronic pelvic pain, tubo-ovarian abscess, infertility, and ectopic pregnancy, run into the billions annually. In the 1970's, researchers showed that women using combination oral contraceptives (containing both an estrogen and a progestin) had only about half the risk of other women of being hospitalized for this infection (5). More recent studies confirm that contemporary low-dose OC confer similar protection. We do not know whether changes in the cervical mucus caused by the progestin, reduced retrograde menstruation, or other factors account for the protection. Nevertheless, the protection is strong and clinically important, especially for young women whose childbearing lies ahead.

Ectopic Pregnancy

Ectopic pregnancy is epidemic in the U.S. and in most developed countries. Numbers and rates of ectopic pregnancy have tripled in recent decades. This complication is the leading cause of maternal mortality in early pregnancy. If a woman does not ovulate, she cannot get pregnant in any site: uterus or fallopian tube. Even with today's low doses of steroids, ovulation almost never occurs among women taking OC (20).

Iron Deficiency Anemia

Women using OC do not have menstrual periods. Instead, they have withdrawal bleeding from the uterus when they stop OC for a week. The net effect of OC on the endometrium is progestogenic; the lining becomes thin, atrophic, and decidualized. This translates into a reduction of blood loss by about 50% (5).

Primary Dysmenorrhea

Use of OC improves or eliminates primary dysmenorrhea in most women (5). This appears to hold true for contemporary low-dose OC. Primary dysmenorrhea is that which is not due to other gynecological pathology, such as leiomyomata or endometriosis. While this disorder causes no mortality, the scope of suffering is very large. Primary dysmenorrhea is common and accounts for much absenteeism from work and school around the world.

An overproduction of prostaglandins by the uterus appears to cause both the pain and gastrointestinal complaints of primary dysmenorrhea. As noted above, the endometrium becomes atrophic in response to OC. This precludes the usual buildup of arachidonic acid (the obligate precursor of prostaglandins) in the endometrium and prevents the synthesis of prostaglandins.

Benign Breast Disease

Use of OC confers powerful protection against benign breast disease (5). The breasts are end-organs that respond to the hormonal fluxes during the normal menstrual cycle. When exposed to the steady-state hormonal milieu of OC, the breasts tend to remain quiescent. The progestin component appears to confer this protection; the stronger the progestin, the lower is the risk of benign breast disease. While no woman dies of benign breast disease, this condition causes great emotional morbidity and large expenditures of money for biopsies and other evaluations.

Functional Ovarian Cysts

Prevention of functional (physiological) ovarian cysts was one of the earliest noncontraceptive health benefits discovered for OC. This benefit remained unchallenged until a small case-series report (21) documented these cysts among women taking multiphasic OC. In response, epidemiological and clinical studies addressed the hypothesis that contemporary multiphasic OC might not offer the same protection as higher-dose, monophasic OC.

Two epidemiological studies refuted the hypothesis of an increased risk of functional cysts. However, they also noted no protective effects of these OC (22, 23). Since then, randomized controlled trials (20) of several OC types have examined ovarian function by both serum progesterone levels and by vaginal ultrasonography. These, too, have confirmed no apparent increase in the risk of functional cysts.

Osteoporosis

Estrogen replacement therapy for postmenopausal women confers strong protection against osteoporosis. Hence, some have hypothesized that use of exogenous estrogens in the reproductive years might have a similar benefit. While cross-sectional studies have yielded conflicting results, none has shown an adverse effect on bone density. In contrast, prospective studies have revealed an increase in bone density among women using OC (24). The benefit is dose-related: the longer a woman takes the OC, the more mineral deposition occurs. An important caveat is that these studies have not established protection against osteoporosis. It appears that women who have used OC may enter the menopause with denser bones than do other women; whether this will lower their risk of fractures remains unknown.

Leiomyomata

Clinical tenets hold that leiomyomata (uterine fibroids) tend to grow under the influence of exogenous estrogen. Hence, many clinicians consider leiomyomata a contraindication to use of OC. Two epidemiological studies have addressed the issue. A cohort study from the U.K. (25) found a paradoxical effect: the longer a woman took OC, the lower was her risk of leiomyomata. After ten years, the risk fell by about 30%. At present, there is no reason to withhold OC from a woman with preexisting leiomyomata.

Toxic Shock Syndrome

When toxic shock syndrome emerged in the early 1980's, investigators discovered that women using OC had a marked reduction in the risk of this disease. In a meta-analysis of five case-control studies, the risk of developing toxic shock syndrome was about 60% lower among OC users (26). This may be an indirect association, however. As discussed above, women using OC have lighter bleeding episodes and thus may be less likely to use super-absorbent tampons.

Rheumatoid Arthritis

Rheumatoid arthritis is a crippling disease that disproportionately attacks women. In the early 1970s, investigators in Europe stumbled onto an apparent protective effect of OC. Since then, others have attempted to corroborate this finding. The results were mixed. Nevertheless, a recent meta-analysis (27) has shown a modest protective effect of OC. The mechanism involved remains unknown.

Conclusions

After three decades of intensive scrutiny, a stronger case exists for cancer protection (ovarian and endometrial) than for causation (cervical and hepatic) in women. On balance, OC improve health (3) and may prolong life (4). For healthy women who do not smoke, there appears to be no increase in the risk of heart disease. Women who use OC for more than five years appear to be at increased risk of cervical neoplasia; whether this is due to the use of OC or to other factors is unclear.

Several studies suggest an increased risk of liver cancer among OC users in developed countries. Population statistics provide little support for an effect. If this effect is real, its impact should be negligible because of the rarity of this cancer. A multiple of a very tiny risk (e.g., one per million) is still a very tiny risk (several per million). Stated alternatively, the amount of disease caused by OC would be small (the attributable risk). Use of OC may cause a transient delay in the resumption of ovulation after stopping OC.

On the other hand, OC have a number of important health benefits besides contraception. Use of OC protects against four life-threatening diseases: ovarian cancer, endometrial cancer, pelvic inflammatory disease, and ectopic pregnancy. In addition, OC protect against iron-deficiency anemia, primary dysmenorrhea, and benign breast disease.

The focus of this symposium is "hormonal carcinogenesis." The title implies that hormones cause cancer. For OC, a different term has been suggested: "hormonal chemoprevention" (28). Because of the negative connotation of "chemo-" (as in chemotherapy for cancer), I propose a different neologism: "hormonal carcinoprophylaxis." Hormonal "carcinoprophylaxis" deserves the same attention by scientists as does "carcinogenesis."

References

1. Vessey MP (1990) The Jephcott Lecture, 1989: An overview of the benefits and risks of combined oral contraceptives. In Mann RD (ed): Oral contraceptives and breast cancer. Carnforth, England, The Parthenon Publishing Group, pp. 121-136.

2. Pettiti DB (1994) Safety of birth control pills. In Samuels SE, Smith MD (eds): The pill: From prescription to over the counter. Menlo Park, The Henry J. Kaiser Family Fdn, pp. 77-115.
3. Harlap S, Kost K, Forrest JD (1991) Preventing pregnancy, protecting health. A new look at birth control choices in the United States. New York, The Alan Guttmacher Institute.
4. Fortney JA, Harper JM, Potts M (1986) Oral contraceptives and life expectancy. Stud Fam Plann 17:117-125.
5. Grimes DA (1992) The safety of oral contraceptives: Epidemiologic insights from the first 30 years. Am J Obstet Gynecol 166:1950-1954.
6. Grimes DA (1990) Breast cancer, the pill and the press. In: Oral contraceptives and breast cancer, Mann RD (ed), Carnforth, England, The Parthenon Publishing Group, 309-322.
7. The Gallup Organization (1994) Women's attitudes toward oral contraceptives and other forms of birth control. Princeton, New Jersey, The Gallup Organization.
8. Prentice RL, Thomas DB (1987) On the epidemiology of oral contraceptives and disease. Adv Cancer Res 49:285-401.
9. Romieu I, Berlin JA, Colditz G (1990) Oral contraceptives and breast cancer: Review and meta-analysis. Cancer 66:2253-2263.
10. United Kingdom National Case-Control Study Group (1989) Oral contraceptive use and breast cancer risk in young women. Lancet 1:973-982.
11. Wingo PA, Lee NC, Ory HW, et al. (1991) Age-specific differences in the relationship between oral contraceptive use and breast cancer. Obstet Gynecol 78:161-170.
12. Beral V, Hannaford P, Kay C (1988) Oral contraceptive use and malignancies of the genital tract: Results from the Royal College of General Practitioners' Oral Contraception Study. Lancet 2:1331-1335.
13. Vessey MP, Smith MA, Yeates D (1986) Return of fertility after discontinuation of oral contraceptives: Influence of age and parity. Brit J Fam Plan 11:120-124.
14. Bracken MB, Hellenbrand KG, Holford TR (1990) Conception delay after oral contraceptive use: The effect of estrogen dose. Fertil Steril 53:21-27.
15. Croft P, Hannaford PC (1989) Risk factors for acute myocardial infarction in women: Evidence from the Royal College of General Practitioners' Oral Contraception Study. Brit Med J 298:165-168.
16. Engel HJ, Engel E, Lichtlen PR (1983) Coronary atherosclerosis and myocardial infarction in young women--Role of oral contraceptives. Eur Heart J 4:1-6.

17. Adams MR, Clarkson TB, Koritnik DR, Nash HA (1987) Contraceptive steroids and coronary artery atherosclerosis in cynomolgus macaques. Fertil Steril 47:1010-1018.
18. Cancer and Steroid Hormone Study of the Centers for Disease Control and the National Institute of Child Health and Human Development (1987) The reduction in risk of ovarian cancer associated with oral-contraceptive use. New Engl J Med 316:650-655.
19. Cancer and Steroid Hormone Study of the Centers for Disease Control and the National Institute of Child Health and Human Development (1987) Combination oral contraceptive use and the risk of endometrial cancer. JAMA 257:796-800.
20. Grimes DA, Godwin AJ, Rubin A et al (1994) Ovulation and follicular development associated with three low dose oral contraceptives: A randomized controlled trial. Obstet Gynecol 83:29-34.
21. Caillouette JC, Koehler AL (1987) Phasic contraceptive pills and functional ovarian cysts. Am J Obstet Gynecol 156:1538-1542.
22. Holt VL, Daling JR, McKnight B, et al. (1992) Functional ovarian cysts in relation to the use of monophasic and triphasic oral contraceptives. Obstet Gynecol 79:529-533.
23. Lanes SF, Birmann B, Walker AM, Singer S (1992) Oral contraceptive type and functional ovarian cysts. Am J Obstet Gynecol 166:956-961.
24. Recker RR, Davies M, Hinders SM, et al. (1992) Bone gain in young adult women. JAMA 268:2403-2408.
25. Ross RK, Pike MC, Vessey MP, et al. (1986) Risk factors for uterine fibroids: Reduced risk associated with oral contraceptives. Brit Med J 293:359-362.
26. Gray RH (1987) Toxic shock syndrome and oral contraception. Am J Obstet Gynecol 156:1038.
27. Spector TD, Hochberg MC (1990) The protective effect of the oral contraceptive pill on rheumatoid arthritis: An overview of the analytic epidemiological studies using meta-analysis. J Clin Epidemiol 43:1221-1230.
28. Henderson BE, Ross RK, Pike MC (1993) Hormonal chemoprevention of cancer in women. Science 259:633-638.

28

Oral Contraceptives and the Risk of Breast and Cervical Cancer

James O. Drife

Introduction

In this Symposium, we have already heard about epidemiological studies on oral contraception and breast cancer, and about biological studies on steroids and mammary carcinogenesis. I shall try to draw together the biological and epidemiological data on breast and cervical cancer. But today's workshop concentrates on issues of relevance to the practicing clinician, and I want to begin by putting our discussions into a broader perspective.

The Problem of Fertility

Although cancer is a very important issue in women's health in Western Europe and North America, other health issues also have a high priority. On a global scale, population growth is arguably our most important problem (1). In Britain, unwanted pregnancy is a subject that receives less publicity than cancer.

Abortion rates vary widely across Europe (2). The rate (measured per 1000 women aged 15-44) varies from under 8 in West Germany and the Netherlands to 180 in Russia, where most women experience several abortions during their reproductive life. In England and Sweden, the rate is about 17. This rate means that, in England, one in three of all girls leaving school will have a therapeutic abortion at some time in her life (3). A similar figure (one in four) has been calculated for the United States.

Taking another broad view, from an historical rather than a geographical perspective, the maternal mortality rate in England remained unchanged throughout history until 1935, when it suddenly began to fall dramatically (4). Nowadays, it is very safe to have a baby in Britain, Sweden, or North America but, unfortunately, that is not the case for most women in the world. Maternal mortality rates in central Africa, for example, are still the same as they were in Britain or America a hundred years ago. The result is that in central Africa one woman in 25 dies as a result of pregnancy, one girl in every class at school (5).

One way to help reduce this terrible total is to ensure that women have access to contraception (6). Unfortunately, however, European and North American concerns are often exaggerated by the news media across the globe (7).

The result is that women are showing increasing reluctance to use OCs, in spite of the fact that it is the most effective form of reversible contraception for young women. This reluctance affects women not only in developed countries but throughout the world. In a study carried out in several developing countries, educated women were asked whether OCs affect the risk of various diseases (8). Many believed that OCs increase the risk of many conditions--even those, such as uterine cancer, against which OCs actually protects (9).

Endogenous Hormones

It is worth pointing out that, for the modern woman, there is no physiological need for this relatively high level of steroid exposure.Protection from cardiovascular disease or osteoporosis can be achieved with much lower levels of estrogen, and the only purpose of progesterone secretion is to prepare the endometrium for implantation of a fertilized ovum. Otherwise, progesterone has only nuisance effects, as many women will testify, particularly those with premenstrual syndrome.

Human females ovulate every 28 days, and humans are one of the few species that are spontaneous ovulators.This could perhaps be a factor in our species' particular susceptibility to mammary cancer (10).

Exogenous Versus Endogenous Hormones

When a woman takes the combined OC, her endogenous estrogen production is suppressed. The fluctuating concentrations of ovarian estradiol are replaced by low constant levels of ethinylestradiol.The OC user substitutes one type of estrogen stimulation with another, and the question to be asked is which type of stimulation is the more harmful or beneficial (11).

Both the ovarian cycle and the contraceptive-controlled cycle have advantages and disadvantages, which have to be weighed against each other. Suppressing the natural cycle has considerable advantages, not only in terms of preventing pregnancy, as mentioned above, but also because it reduces menstrual bleeding and, more importantly, protects against ovarian and endometrial cancer.

The balance between endogenous and exogenous hormones is well illustrated by a study by Ross on uterine fibroids (12). Fibroids are benign estrogen-sensitive tumors, and are usually considered to be a contraindication to OCs. Prolonged use of combined OCs, however, leads to a reduction in the prevalence of fibroids, suggesting that--at least as far as this tissue is concerned--exogenous estrogen provides less overall stimulation than endogenous estrogen.

It is important to remember that, during the thirty-three years since OCs were introduced, their formulation has not remained constant.When originally introduced, OCs contained doses of estrogen and progestogen that would now be considered very high, and during the last thirty years there has been a steady change in both Britain (13) and the United States (14) towards the prescription of low-dose preparations. If, as I am suggesting, the balance between exogenous and endogenous steroids is important this change could be crucial.

The reduction in estrogen and progestogen dose has brought the dose ranges of the steroids in OCs close to those used in hormone replacement therapy. Although most widely-used HRT formulations use other types of estrogen, if ethinylestradiol is used to treat menopausal symptoms, the dose required is around 10-20 micrograms, which is close to the 20-35 microgram range used in OCs. The doses of progestogens are similar in both types of treatment, though the standard HRT formulations include progestogens for around 12 days, rather than the 21 days of the standard OC formulation.

Combined Oral Contraception and Cancer

The clearest effects of OCs on cancer are their protective effects against cancer of the ovary and the endometrium, which have now been demonstrated in over a dozen epidemiological studies (15, 16). These protective effects have not received nearly as much publicity as concerns about possible adverse effects on breast cancer--perhaps because breast cancer is a more common disease, or more likely because breast cancer is a more emotive disease.

Also, ovarian and endometrial cancers are of little concern to young women, as their maximum incidence is in a woman's sixties. Nevertheless, the protective effects against these cancers could be exploited by offering, say, two years of combined OC use to nonsmoking women in their forties. Theoretically, this could lead to a substantial reduction in deaths from ovarian cancer in the age group most at risk. This possible health promotion exercise is surely worth pursuing.

OCs and the Breast

Breast tissue is very sensitive to ovarian steroid hormones, particularly in the early years of reproductive life. At puberty, breast development is often the first sign of ovarian activity. The earlier this process occurs, the higher is the subsequent risk of breast cancer, demonstrating the importance of normal ovarian activity as a risk factor for breast cancer (10).

Breast Stroma

The human breast, unlike the mammary glands of other species, has a large proportion of fatty tissue surrounding the glands. At puberty, there is some growth of the glandular component, but most of the morphological development at this time is due to proliferation of the fatty tissue and stroma. Most of this stromal development takes place before the menarche and before ovulatory cycles become established. It is therefore related mainly to increasing estrogen levels, with progesterone playing little part, if any (17).

The function of the fatty tissue and stroma is unclear. It is not related to lactation, and its original function, in evolutionary terms, may well have been to act as a sexual signal (18).

The stroma of the human breast deserves study (19). The high incidence of breast cancer in our species may be related to the fact that mammary glandular tissue is invested in highly hormone-sensitive stroma. Also, women themselves strongly identify the breast with femininity: because of the psychological importance of the fatty tissue, prophylactic mastectomy (20, 21) is not a realistic option for preventing breast cancer (22), except for a few strongly-motivated women, usually from high-risk families.

Several investigators have measured breast volume during the cycle, using water displacement or other techniques (23). All studies agree that the volume increases in the second half of the normal ovarian cycle. In women taking OCs, the pattern is similar to that seen during the normal menstrual cycle. Breast volume steadily increases during OC-controlled cycles and falls during the OC-free week. This suggests that OCs and the ovarian cycle have similar effects on the fatty tissue of the breast.

Glandular Tissue

The lobules of the mammary gland are also affected by the normal menstrual cycle. Suggestions that lobular growth and degeneration occur during the cycle have not been confirmed, but Anderson (24) described an increase in mitosis and apoptosis in the second half of the menstrual cycle.If breast biopsies are incubated with tritiated thymidine, label is taken up into nuclei as cells enter the growth phase of the cell cycle (25). Counts of labeled nuclei per thousand unlabeled nuclei give the "labeling index" (LI).The LI falls during the proliferative phase of the cycle and rises sharply during the luteal phase, under the influence of progesterone.

The effects of OCs on the glandular tissue have been studied by a similar method and again show a striking similarity to those of the normal menstrual cycle. Anderson and coworkers (26) showed that the uptake of label increases in the second half of the cycle, as it does in non-users. When OC users were compared with normal women, there was a slightly greater degree of activity in users in the second week of the cycle and the fourth week of the cycle compared with non-users, but the overall pattern of activity was similar.

Of particular interest was the effect of steroid dosage on LI. There was a relationship between estrogen dosage and LI, and the total LI in women taking the lowest dose of estrogen was similar to that in the normal menstrual cycle. This might indicate that estrogen promotes the synthesis of progesterone receptors, so that ability of the breast to respond to progesterone depends on the extent to which it is primed by estrogen. Clearly the interaction between estrogen and progesterone in the breast is complex, but Anderson's work suggests that the dose of estrogen in the OC is important even in relatively low-dose OCs, and that the lowest-dose preparations stimulate the breast no more than does the normal menstrual cycle.

Epidemiological Studies

It is important to bear these biological studies in mind when considering epidemiological data. The finding that in biological studies OCs and the normal cycle have broadly similar effects on breast tissue makes it entirely unsurprising that, in epidemiological studies, no overall link has been found between OCs and breast cancer when all age groups are studied together (27, 28).

Nevertheless, the possibility has been raised in some epidemiological studies that prolonged use of OCs by women under the age of 25 may have an adverse effect on breast cancer risk (27, 29). An increased risk in this age group would make biological sense, as studies of LI show highest activity in the years immediately after puberty (30). However, there is a lack of uniformity among epidemiological studies in this age group, with some showing no increase in risk (31). These varying results could be explained by differences in OC formulations (32). The move to lower-dose formulations was later in the United States (14) than in Britain (13), but prescribing data in other countries are insufficient for us to say whether this effect explains the discrepancies between studies (31). If, however, reduction in OC dosage is important, it is likely that the recent change to low-dose formulations may reduce or eliminate any increased risk of breast cancer in young OC users.

It has been suggested that widespread OC use in young women may lead to an increase in breast cancer incidence but only after a long latent period (33). As OCs have now been in widespread use for thirty years, it is pertinent to ask how long a latent period is biologically plausible. Time may be beginning to run out for those who argue that breast cancer incidence rates are going to rise (34). Nevertheless, because these rates may reflect a long latent period after initial hormone exposure, it will be important to continue to monitor breast cancer rates for many years after the recent change to low-dose formulations (35).

Cervical Cancer

Death from cervical cancer is less common from breast cancer, but the prevention of cervical cancer is just as emotive an issue for women. Over recent years, attention has been centered on cervical screening programs, which have been the focus of almost all research on the normal cervix outside pregnancy. Very little research has been carried out on the endocrinology of the nonpregnant cervix.

The menstrual cycle has marked effects on cervical glands, with an outpouring of mucus during the preovulatory phase of the cycle. The OC replaces this with a steady state, and in OC users the cervical glands produce only a thick, sticky mucus impenetrable to sperm. This, incidentally, is probably how the OC provides some protection against bacterial pelvic inflammatory disease (9). One might therefore expect OCs to have a protective effect against cervical cancer, but this is not the conclusion of most studies.

Delgado-Rodriguez and coworkers recently published a meta-analysis of studies on cervical neoplasia (36), which showed among OC users a relative risk of 1.5 for dysplasia but a smaller relative risk (1.2) for invasive cancer. A prospective follow-up study of cervical intraepithelial neoplasia among women in Norway has shown a relative risk of 1.5 for current users and 1.4 for past users (37). It has been suggested that such findings may be explained by more frequent cytological smears being taken from OC users. It is also possible that, because of the strong influence of sexual behavior on cervical cancer risk, the apparent effect of OCs may be caused by confounding due to differences in sexual activity between OC users and non-users (38, 39). Nevertheless, the effect may be due to a direct effect of OCs and, again, it will be important to continue to monitor this disease after the move to lower-dose formulations.

The practical message for the clinician is that regular cervical smears are essential in OC users, and for the woman using OCs is that, if she or her partner have more than one partner, the risk of cervical cancer can be reduced much more effectively by using additional barrier contraception than by stopping OC use.

References

1. Smith A (1990) The population bomb has already exploded. Brit Med J 301:681-682.
2. Segal SJ, LaGuardia KD (1990) Termination of pregnancy--A global view. Bailliere's Clin Obstet Gynaecol 4:235-247.
3. Botting B (1991) Trends in abortion. Population Trends 64:19-29.
4. Loudon I (1990) Obstetrics and the general practitioner. Brit Med J 301:703-707.

5. Lettenmaier C, Liskin L, Church CA, Harris JA (1988) Mothers' lives matter: maternal health in the community. Population Reports, Series L, No 7.
6. Potts DM, Crane SF (1993) Contraceptive delivery in the developing world. Brit Med Bull 49:27-39.
7. Drife J (1993) Contraceptive problems in developed countries. Brit Med Bull 49:17-26.
8. Grubb GS (1987) Women's perceptions of the safety of the pill: A survey in eight developing countries. J Biosoc Sci 19:313-321.
9. Drife JO (1989) The benefits of combined oral contraceptives. Br J Obstet Gynaecol 96:1225-1228.
10. Henderson IC (1993) Risk factors for breast cancer development.Cancer 71:2127-2140.
11. Drife JO (1993) The benefits and risks of oral contraceptives. New York: Parthenon Publishing.
12. Ross RK, Pike MC, Vessey MP, et al. (1986) Risk factors for uterine fibroids: reduced risk associated with oral contraceptives. Brit Med J 293:359-362.
13. Thorogood M, Vessey MP (1990) Trends in use of oral contraceptives in Britain. Brit J Fam Plan 16:41-53.
14. Gerstman BB, Gross TP, Kennedy DL, et al. (1991) Trends in the content and use of oral contraceptives in the United States, 1964-88. Amer J Pub Hlth 81:90-96.
15. Cancer and Steroid Hormone Study of the Centers for Disease Control and the National Institute of Child Health and Human Development (1987) Combination oral contraceptive use and the risk of endometrial cancer. J Amer Med Assoc 257:796-800.
16. Cancer and Steroid Hormone Study of the Centers for Disease Control and the National Institute of Child Health and Human Development (1987) The reduction in risk of ovarian cancer associated with oral-contraceptive use. N Engl J Med 316:650-655.
17. Drife J (1986) Breast development in puberty. Ann NY Acad Sci 464:58-65.
18. Morris D (1977) The naked ape. St Albans: Mayflower.
19. Petrek JA, Hudgins LC, Levine B, et al. (1994) Breast cancer risk and fatty acids in the breast and abdominal tissues. J Natl Cancer Inst 86: 53-56.
20. Drife JO (1992) Are breasts redundant organs? Brit Med J 304:1060.
21. Houlihan MJ, Goldwyn RM (1991) Role of prophylactic mastectomy. In (ed) Stoll BA: Approaches to breast cancer prevention. Norwell: Kluwer, pp 135-148.
22. Drife JO (1992) Breasts and the media. Brit Med J 304:1514.
23. Milligan D, Drife JO, Short RV (1975) Changes in breast volume during normal menstrual cycle and after oral contraceptives. Brit Med J iv:494-496.

24. Anderson TJ, Ferguson DJP, Raab GM (1982) Cell turnover in the "resting" human breast: influence of contraceptive pill, age and laterality. Brit J Cancer 46:376-82.
25. Masters JRW, Scarisbrick JJ, Drife JO (1977) Cyclic variation of DNA synthesis in human breast epithelium. J Natl Cancer Inst 58:1263-1265.
26. Anderson TJ, Battersby S, King RJB, et al. (1989) Oral contraceptive use influences resting breast proliferation. Hum Pathol 20:1139-1144.
27. Schlesselman JJ (1990) Oral contraceptives and breast cancer. Amer J Obstet Gynecol 163:1279-1287.
28. Beral V, Reeves G (1993) Childbearing, oral contraceptive use, and breast cancer. Lancet 341:1102.
29. Wingo PA, Lee NC, Ory HW, et al. (1993) Age-specific differences in the relationship between oral contraceptive use and breast cancer. Cancer 71:1506-1517.
30. Drife J (1981) The effects of parity and the menstrual cycle on the normal mammary gland and their possible relationship to malignant change. MD Thesis, University of Edinburgh.
31. McPherson K, Drife JO (1986) The pill and breast cancer: why the uncertainty? Brit Med J 293:709-710.
32. Drife J (1989) The contraceptive pill and breast cancer in young women. Brit Med J 298:1269-1270.
33. McPherson K (1993) Childbearing, oral contraceptive use and breast cancer. Lancet 341:1604-1605.
34. Herbst AL, Berek JS (1993) Impact of contraception on gynecologic cancers. Amer J Obstet Gynecol 168:1980-1985.
35. Thomas DB (1993) Oral contraceptives and breast cancer. J Natl Cancer Inst 85:359-364.
36. Delgado-Rodriguez M, Sillero-Arenas M, Martin-Moreno JM, Galvez-Vargas R (1992) Oral contraceptives and cancer of the cervix uteri. Acta Obstet Gynecol Scand 71:368-376.
37. Gram IT, Macaluso M, Stalsberg H (1992) Oral contraceptive use and the incidence of cervical intraepithelial neoplasia. Amer J Obstet Gynecol 167:40-44.
38. Turnquest MA (1993) Oral contraceptive use and incidence of cervical intraepithelial neoplasia. Amer J Obstet Gynecol 168:1895.
39. Thiry L, Vokaer R, Detremmerie O, Bollen A, Hallez S (1992) Contraception, papillomavirus, and cervical cancer. Lancet 339:616.

COMMUNICATIONS
Session I. Epidemiology, Human Studies

Some Aspects of Breast Cancer Epidemiology

Leif Bergkvist

Several recent excellent reviews have been written about breast cancer epidemiology (1). The purpose of this paper is not to cover all aspects of breast cancer epidemiology, but to point out some areas of special interest.

Numerous risk factors for breast cancer have been established, and the majority of them are related to hormones in some way (Table 1). Family history of breast cancer is one of the nonhormonal factors. The risk of developing breast cancer in a woman who has a first-degree relative with breast cancer increases approximately twofold. If there is more than one first-degree relative with breast cancer, and especially if those persons were affected early in life, the risk is elevated even more. Thus there seems to be a genetic susceptibility to breast cancer, and recent research has identified specific gene abnormalities among breast cancer patients, such as overexpression of c-*erb*B-2 and defects in p53.

Table 1. Risk Factors for Breast Cancer.

Female vs. male gender	High age
Family history of breast cancer	Benign breast disease
Proliferative disorders	Atypical hyperplasia
Previous breast cancer	High age at first birth
Nulliparity	Young age at menarche
Late age at menopause	Exogenous hormones
Oral contraceptives (?)	Hormonal replacement therapy (?)
Obesity in postmenopausal women	Dietary factors (?)

It is traditionally accepted that hormonal influences start around menarche, and an early menarche is one recognized risk factor for breast cancer. However, it may well be that the hormonal effects on the breast start much earlier, perhaps *in utero*. It has been shown by Ekbom and coworkers (2) in Uppsala that daughters of mothers who developed eclampsia or preeclampsia during the last months of pregnancy have a 76% reduction in their risk of breast cancer. This is one of the largest protective effects found so far and has interesting theoretical implications. It is thus possible that the risk of developing breast cancer is determined before the child is born. It is not quite understood how this effect is mediated, but eclampsia and preeclampsia are associated with lower concentrations of pregnancy estrogens. Estrogens have a proliferative effect on breast epithelium, and intrauterine response to estrogen stimulation may have an effect on breast cancer risk throughout life. They also found a nonsignificant trend, with higher birth weight, longer birth length, and higher placental weight, toward increased risk of breast cancer. This also accords with our expectations if estrogen levels during pregnancy predict the future risk of developing breast cancer.

A couple of other studies have found associations pointing in the same direction. Pregnancy estrogens seem to be slightly higher in the first compared to the second pregnancy (3), and firstborn women have been shown to be at slightly higher risk of breast cancer than those of later birth order (4). Also, estrogens are slightly lower in young pregnant women than in older (3) and the risk of breast cancer among women born to older mothers is slightly higher (5). These associations need confirmation.

Age at first birth is related to the risk of breast cancer. Young age at first birth is regarded as having a protective effect. This is probably true overall, but later studies have shown that there might be a transient increase in the risk of breast cancer after a pregnancy. Rosner, et al. (6), in the large Nurses Health Study, found that the incidence density of breast cancer was higher for women having one child at age 20 or before, compared to nulliparous, up to the age of 55. Additional pregnancies lowered the incidence. Likewise, for single-parous women aged 20-25, the incidence density was higher than that of nulliparous women and did not fall below that of nulliparous women until after age 61. As for those with age at first birth under 20, additional pregnancies offered protection. Again, there is thus evidence that events early in life influence cancer development much later in life.

The role OC has been highly disputed. Overall, there is very little evidence of an increased risk of breast cancer, but women exposed to OCs early in life for long periods, incidence of breast cancer and mortality before age 45 seem to be slightly elevated (7).

Obesity in postmenopausal women is a risk factor for breast cancer. After menopause, the production of estrogens in the ovaries is almost zero. However, conversion of androgen takes place in the peripheral fat tissue; thus, obese postmenopausal women have higher endogenous levels of estrogens than lean women. Ovarian hormone production outweighs androgen conversion in younger women, which may explain why obese premenopausal women are not at increased risk.

Estrogens given to postmenopausal women do not increase the risk of breast cancer overall. But long-term treatment confers a slightly increased risk (8). The addition of progestogens to estrogen treatment was hoped, in concordance with the effects on the endometrium, to be protective, but there is no confirmation of that so far in the literature. On the contrary, progestogens may produce higher risks of breast cancer (9,10). On the other hand, cancers developing after estrogen use have a better prognosis than other breast cancers (11), and there is no evidence of increased breast cancer mortality among estrogen users (11).

Therefore, to conclude, there seems to be a relationship between estrogens and breast cancer throughout life, but the most profound effects of estrogens may be those exerted before life has even started.

References

1. Henderson JC (1993) Risk factors for breast cancer development. Cancer Suppl 71:2127-2140.
2. Ekbom A, Trichopoulos D, Adami HO, et al. (1992) Evidence of prenatal influences on breast cancer risk. Lancet 340:1015-1018.
3. Panagiotopoulou K, Katsouyanni K, Petrido E, et al. (1990) Maternal age, parity and pregnancy estrogens. Cancer Causes Control 1:119-124.
4. Hsieh C-C, Tzonou A, Trichopoulos D (1991) Birth order and breast cancer risk. Cancer Causes Control 2:95-98.
5. Thompson JA, Janerick DT (1990) Maternal age at birth and risk of breast cancer in daughters. Epidemiology 1:101-106.
6. Rosner B, Colditz GA, Willett WC (1994) Reproductive risk factors in a prospective study of breast cancer: The Nurses Health Study. Am J Epidemiol 139:819-835.
7. Perlman JA, Porter J, Hatz S (1994) Patterns of cancer mortality in women who used sex steroid hormones under age 60. In Li JJ, Li SA, Nandi S, Gustafsson J-A, Sekely L. (eds): Hormonal Carcinogenesis, Proceedings of the Second International Symposium. New York: Springer-Verlag, pp350-355.
8. Steinberg KK, Thacker SB, Smith SJ, et al. (1991) A meta-analysis of the effects of estrogen replacement therapy on the risk of breast cancer. JAMA 265:1985-1990.
9. Bergkvist L, Adami HO, Persson I, et al. (1989) The risk of breast cancer after estrogen and estrogen-progestin replacement. N Engl J Med 321:293-297.
10. Persson I, Yuen J, Bergkvist L, et al. (1992) Combined oestrogen-progestogen replacement and breast cancer risk. Lancet 340:1044, letter.
11. Bergkvist L, Adami HO, Persson I, et al. (1989) Prognosis after breast cancer diagnosis in women exposed to estrogen and estrogen-progestogen replacement therapy. Am J Epidemiol 130:221-228.
12. Yuen J, Persson I, Bergkvist L, et al. (1993) Hormone replacement therapy and breast cancer mortality in Swedish women: Results after adjustment for "healthy drug-user" effect. Cancer Causes Control 4:369-374.

Risk of Breast Cancer Associated with Induced Abortion

Janet R. Daling, Kathleen E. Malone, Lynda F. Voigt, Emily White, and Noel S. Weiss

Summary

Certain events of reproductive life, especially completed pregnancies, have been found to influence a woman's risk of breast cancer. Prior studies of breast cancer in relation to a history of incomplete pregnancies have not provided consistent results. Female residents of three Western Washington counties diagnosed with breast cancer to age 46 years (n = 845 and 961 control women identified through random-digit dialing) were interviewed in detail with regard to their reproductive histories, including the occurrence of induced and spontaneous abortions. Among women who had been pregnant at least once, the risk of breast cancer was 50% higher in those who had experienced an induced abortion than in those women who did not. While the size of this increased risk did not vary by the number of induced abortions or by a history of a prior completed pregnancy, it did vary according to the age at which the abortion occurred, as well as the duration of that pregnancy. Highest risks were observed when the abortion was done prior to 18 years of age (particularly, if it took place after eight weeks of gestation), or at age 30 and beyond. There was no increased risk of breast cancer associated with a prior miscarriage. The data from this and other (although not all) epidemiologic studies support the hypothesis that an induced abortion can adversely influence a woman's subsequent risk of breast cancer. This question should be re-examined in future studies, with particular attention to the potential hazards related to abortions early in life.

Introduction

Epidemiologic studies have linked reproductive and hormonal factors to breast cancer risk (1). Recently, exposures that have received attention include: alcohol use (2), smoking (3), exercise (4), body size (5), and the two exposures which are the focus of this communication, history of induced or spontaneous abortion (6). Herein, the results related to induced and spontaneous abortion from a case-control study conducted in three Western Washington Counties are presented.

Methods

The study included all women in three Western Washington Counties with a diagnosis of breast cancer from 1983 through April 1990, and who were born after 1945. This birth cohort restriction was included to insure that the women had been born recently enough so that the majority of abortions indicated would be legal. Approximately 85% of these women (n = 845) agreed to participate. Controls (n = 961) were ascertained through random digit dialing with a response rate of approximately 76%. In-person interviews, lasting one to two hours, were conducted in the respondents' homes. These analyses were adjusted for age, family history of breast cancer, religion, and age at first pregnancy or, for parous women, age at first full-term pregnancy (Table 1).

Results

Among women who had been pregnant at least once, the risk for breast cancer associated with induced abortion was 1.5 (95% CI = 1.2-1.9) (Table 1). The risk varied by age at first abortion. Women who had their first abortion prior to age 18 had the greatest risk, 2.5, followed by women whose first abortion occurred at age 30 or greater, 2.1. There was some indication that the gestational length of the pregnancy may also affect a woman's risk, with the highest risk associated with abortions occurring at 9-12 weeks of gestation. There was little difference in risk according to timing in relation to first live birth.

In order to provide women who get pregnant at a given age with the information needed to make an informed choice, we estimated the risk associated with an abortion at various ages compared to carrying that pregnancy to term. The risk was particularly high for abortions that occurred before 18 years of age that were at a gestational age greater than 8 weeks, RR = 9.0 (Table 2). A similar pattern occurred for women age 30 and over.

We performed one analysis comparing women who never had a pregnancy to women whose only pregnancy was an induced abortion. Sixty-three cases and 53 controls in this subgroup had undergone induced abortion, corresponding to an RR of 1.4 (95% CI = 0.9 - 2.2).

Table 1. Risk of *In-Situ* and Invasive Breast Cancer in Gravid Women Associated with Prior Induced Abortion.

Induced Abortion History	Cases N=689	(%)	Controls N=781	(%)	RR[a]	95% CI
Never had induced abortion	479	(69.5)	580	(74.3)	1.0	---
Induced abortion, ever	210	(30.5)	201	(25.7)	1.5	(1.2-1.9)
1 only	150	(21.8)	142	(18.2)	1.5	(1.1-2.0)
2 or more	60	(8.7)	59	(7.6)	1.6	(1.0-2.4)
Age at first abortion (years)						
< 18	20	(2.9)	15	(1.9)	2.5	(1.1-5.7)
18-19	34	(4.9)	36	(4.6)	1.7	(1.0-3.0)
20-29	115	(16.7)	123	(15.7)	1.3	(1.0-1.7)
$\geq$30	41	(6.0)	26	3.5	2.1	(1.2-3.5)
Gestational length of first aborted pregnancy						
1-8 weeks	129	(18.7)	136	(17.4)	1.4	(1.0-1.8)
9-12 weeks	64	(9.3)	48	(6.1)	1.9	(1.3-2.9)
13+ weeks	16	(2.3)	17	(2.2)	1.4	(0.7-2.8)
Unknown	1	(0.1)				
Timing of first induced abortion						
Before first birth	69	(10.0)	76	(9.7)	1.4	(1.0-2.0)
After first birth	74	(10.7)	63	(8.1)	1.5	(1.0-2.2)
Never gave birth	67	(9.7)	62	(7.9)	1.7	(1.2-2.6)

[a] Risk relative to that of women with at least one pregnancy who never had an abortion. Adjusted for age, family history of breast cancer, religion, and age at first pregnancy.

Table 2. History of Induced Abortion in Women with *In-Situ* or Invasive Breast Cancer and Controls, by Age at First Induced Abortion or Birth and Length of That Pregnancy.

Age at birth or first induced abortion (years)	Outcome of the pregnancy	Cases		Controls		RR[a] (C.I.)
		N=672[b]	(%)	N=765[b]	(%)	
< 18	Induced abortion at 1-8 weeks	5	-8.8	10	-16	1.3(0.3-7.0)
	Induced abortion at 9-24 weeks	15	-26	5	-8.2	9.0(2.0-41.2)
	Complete pregnancy	37	-65	46	-75	1
18-29[c,d]	Induced abortion at 1-8 weeks	93	-16	104	-16	1.2(0.9-1.7)
	Induced abortion at 9-24 weeks	55	-9.5	55	-8.4	1.4(0.9-2.1)
	Complete pregnancy	431	-74	498	-76	1
30+	Induced abortion at 1-8 weeks	31	-15	22	-9.2	2.1(1.2-4.0)
	Induced abortion at 9-24 weeks	10	-4.9	5	-2.1	3.3(1.1-10.2)
	Complete pregnancy	165	-80	211	-89	1

[a] Referent category is women who had a birth in the age category in question. RR adjusted for age, family history of breast cancer, religion, and age at first pregnancy.

[b] Excludes women who were never pregnant or only had a pregnancy in the indicated time frame that did not result in either an induced abortion or a birth. The sum of the numbers exceeds the total, since a woman could have had a birth in an earlier age category and so could be included in more than one category.

[c] A case of an induced abortion of unknown gestational length was excluded.

[d] Excludes women who had an induced abortion at an earlier age.

Discussion

These findings are consistent in part with animal studies and are biologically plausible (7, 8). During the first trimester of pregnancy, the breast is characterized by high mitotic ability and proliferation; only in mid- to late pregnancy does cellular differentiation predominate.

Table 3. Risk of *In-Situ* and Invasive Breast Cancer in Parous Women Associated with Prior Spontaneous Abortion.

Spontaneous abortion history	Cases N=605	(%)	Controls N=703	(%)	RR[a]	(95% CI)
Never had spontaneous abortion	432	-71.4	492	-70	1	
Spontaneous abortion, ever	173	-28.6	211	-30	0.9	(0.7-1.2)
One only	133	-22	154	-21.9	1	(0.7-1.3)
Two or more	40	-6.6	57	-8.1	0.8	(0.5-1.3)
Age at first spontaneous abortion (years)						
<18	7	(1.2)	13	(1.8)	0.7	(0.3-1.9)
18-19	26	(4.3)	19	(2.7)	1.6	(0.9-3.0)
20-29	106	(17.5)	139	(19.8)	0.9	(0.7-1.2)
≥30	34	(5.6)	40	(5.7)	0.8	(0.5-1.4)
Gestational length of first spontaneous abortion						
1-8 weeks	85	(14.1)	127	(18.1)	0.7	(0.5-1.0)
9-12 weeks	55	(9.1)	51	(7.3)	1.2	(0.8-1.9)
13+ weeks	30	(5.0)	33	(4.7)	1.1	(0.6-1.8)
Timing of first spontaneous abortion						
Before 1st birth	77	(12.7)	98	(13.9)	0.9	(0.7-1.3)
After 1st birth	96	(15.9)	113	(16.1)	1.0	(0.7-1.3)

[a] Risk relative to that of women with at least one birth who never had a spontaneous abortion. Adjusted for age, family history of breast cancer, age at first full-term pregnancy, marital status, and body mass index.

If the pregnancy is interrupted, some areas of the mammary gland are left with immature, undifferentiated cells that are susceptible to carcinogens.

We did not find the same associations with spontaneous abortion (Table 3). Among parous women, the risk associated with spontaneous abortion was 0.9 (0.4-1.2). There was indication that spontaneous abortions that occurred at age 18-19 might be associated with an elevation in risk, but this could be by chance. Women with breast cancer were more likely than controls to have had spontaneous abortion at 9-12 weeks as compared to 1-8 weeks, a pattern we found in our induced abortion analyses.

We did observe that only 14.3% of women with an induced abortion had a live birth and lactated during the five years following their abortion, compared to 46.3% of women with a spontaneous abortion. The risk for breast cancer associated with induced abortion among those women who had a live birth with lactation within five years was not elevated and was consistent with the lack of risk found with spontaneous abortion. However, when we removed these women from our analysis relating to spontaneous abortion, we still did not see an elevation in risk (RR = 1.1, 95% CI = 0.8-1.5).

The data from our study and some (although not all) other epidemiologic studies support the hypothesis that an induced abortion can adversely influence a woman's subsequent risk of breast cancer. Our data did not support an increased risk associated with a spontaneous abortion. Since these exposures are emotionally and politically charged and the results are not consistent among the studies that have addressed these issues, this question needs to be re-examined in future studies.

References

1. Kelsey JL, Gammon MD, John EM (1993) Reproductive factors and breast cancer. Epidemiol Rev 15:36-47.
2. Rosenberg L, Metzger LS, Palmer JR (1993) Alcohol consumption and risk of breast cancer: a review of the epidemiologic evidence. Epidemiol Rev 15:133-144.
3. Cigarette smoking and the risk of breast cancer (1993) Epidemiol Reviews 15:145-156.
4. Gammon MD, John EM (1993) Recent etiologic hypotheses concerning breast cancer. Epidemiol Rev 15:163-168.
5. Hunter DJ, Willett WC (1993) Diet, body size, and breast cancer. Epidemiol Rev 15:110-132.
6. Rosenberg L, Palmer JR, Kaufman DW et al (1988) Breast cancer in relation to the occurrence and time of induced and spontaneous abortion. Am J Epidemiol 127:981-989.
7. Russo J, Russo IH (1987) Biology of disease. Biological and molecular bases of mammary carcinogenesis. Lab Invest 57:112-37.
8. Russo J, Wilgus G, Russo IH (1979) Susceptibility of the mammary gland to carcinogenesis. I. Differentiation of the mammary gland as determinant of tumor incidence and type of lesion. Am J Pathol 96:721-36.

Family History of Breast Cancer as a Modifier of Other Risk Factors

Kathleen E. Malone, Janet R. Daling, Noel S. Weiss,
Barbara McKnight, Emily White, and Melissa Austin

Summary

A population-based case-control study was conducted to assess the influence of family history of breast cancer, as a surrogate for the actions of one or more susceptibility alleles, upon the effects of other breast cancer risk factors. Seven hundred thirty-four cases of primary invasive breast cancer diagnosed before age 45 were compared with 938 controls identified through random-digit dialing. The associations of selected risk factors for breast cancer were examined separately among women with no history of breast cancer in their mothers, sisters, aunts, or grandmothers (FH-), and among women with a history of breast cancer in their mothers and/or sisters (FH+). Nulliparity and induced abortion were associated with modestly increased relative risk (RR) estimates among FH- women and larger RR estimates among FH+ women. Our results suggest, at most, modest increases in risk associated with long duration and early age at first use of oral contraceptives (OCs) among FH- women. Among FH+ women, there was an increased risk associated with "ever" use of OCs for one or more years (RR = 1.7), but this was not further influenced by duration or age at first use. While these results require confirmation in other studies, they support the hypothesis that some factors preferentially influence the risk of breast cancer in women genetically predisposed to this disease.

Introduction

Past studies have consistently observed an increased risk of breast cancer among women with positive family history of breast cancer. There is strong evidence of linkage of early onset breast cancer in selected families to a marker on the long arm of chromosome 17, although it currently appears that this gene may account for only a small proportion of all breast cancers (1). Research to date suggests that the putative breast cancer gene BRCA1 is rare and highly penetrant. It has been proposed that there are also alleles with lower penetrance and greater population frequency that may involve distinct genes or different mutations in the same genes and, most importantly, may to a greater extent than the currently-sought BRCA1 locus act in conjunction with environmental exposures to influence breast cancer susceptibility (1, 2). There is evidence of associations between breast cancer incidence and certain "environmental" factors, although these associations are generally modest and appear to account for only a small portion of all breast cancers (3). We hypothesized that examination of these factors among women with a positive family history might reveal enhanced effects not otherwise seen. Specifically, the purpose of this investigation was to assess the influence of family history of breast cancer, as a surrogate for the actions of one or more susceptibility alleles, on the effects of other established or suspected risk factors for breast cancer.

Methods

In this population-based case-control study of breast cancer among white women born after 1944, all incident cases of invasive breast cancer diagnosed between January 1, 1983, and April 30, 1990, among residents of the three-county metropolitan Seattle area were ascertained through a population-based cancer registry. Controls were ascertained through random digit dialing. We interviewed 747 cases (83.2% of the eligible cases) and 961 controls (75.5% of the eligible controls). Information on factors potentially related to breast cancer risk, including reproductive, contraceptive, and menstrual histories, smoking and alcohol use, body size, general demographic characteristics, and detailed family history, was obtained similarly from cases and controls in a structured in-person interview. Thirteen interviewed cases and 23 interviewed controls who reported having been adopted were excluded.

Results

The risk of breast cancer in relation to menstrual and reproductive factors stratified by family history of breast cancer is depicted in Table 1. Earlier menarche was associated with an increased risk of breast cancer among FH- women and a reduced risk of breast cancer in FH+ women. Nulliparity and past induced abortion were both associated with a somewhat increased risk of breast cancer in FH- women and a more substantially increased risk of breast cancer among FH+ women. There was no apparent association with prior miscarriage.

Table 1. Risk of Breast Cancer in Relation to Menstrual and Reproductive Factors Stratified by Family History of Breast Cancer.

Characteristic	*NO FAMILY HISTORY*				*1ST-DEGREE FAMILY HISTORY*			
	Case n=461	**Cont. n=747**	**OR**	**95% CI**	**Case n=119**	**Cont. n=49**	**OR**	**95% CI**
Age at menarche (years)[1]								
> 12	210	424	1.0	REF.	60	17	1.0	REF.
≤ 12	250	322	1.5	1.2- 1.9	59	32	0.5	0.3- 1.1
History of miscarriage[2]								
No miscarriage	266	425	1.0	REF.	69	29	1.0	REF.
Yes miscarriage	107	183	0.9	0.7- 1.2	28	13	0.9	0.4- 2.0
History of induced abortion[2]								
No Ind. Abortion	260	442	1.0	REF.	71	36	1.0	REF.
Yes Ind Abortion	113	166	1.3	1.0- 1.8	26	6	2.6	1.0- 6.9
Parity status								
Parous	330	553	1.0	REF.	83	39	1.0	REF.
Nulliparous	131	194	1.4	1.1- 1.9	36	10	2.0	0.9- 4.4
Number of live births								
None	132	198	1.0	REF.	36	10	1.0	REF.
1	90	150	0.8	0.5- 1.1	20	7	0.7	0.2- 2.2
2	142	269	0.6	0.5- 0.9	41	19	0.5	0.2- 1.3
3+	97	130	0.9	0.6- 1.2	22	13	0.4	0.1- 1.1
Age at first live birth								
None	132	98	1.0	REF.	36	10	1.0	REF.
< 20	63	110	0.7	0.4- 1.0	16	9	0.4	0.1- 1.2
20 - 24	146	221	0.8	0.6- 1.1	30	16	0.5	0.2- 1.2
> 24	120	218	0.7	0.5- 0.9	37	14	0.6	0.3- 1.6

[1] Difference in the odds ratios between FH- women and FH+ women was significant, $p < 0.05$.

[2] Among gravid women only.

Table 2. Risk of Breast Cancer in Relation to OC Use Stratified by Family History of Breast Cancer

Characteristic	*No Family History*				*1st-Defree Family History*			
	Case n=461	Cont. n=747	OR	95% CI	Case n=119	Cont. n=49	OR	95% CI
Ever used OCs for > one year								
Never or < 1 year	109	181	1.0	REF.	20	13	1.0	REF.
≥ 1 year use	352	566	1.0	0.8- 1.4	99	36	1.7	0.8- 3.9
Lifetime duration of OC use (years) [1]								
Never or < 1 year	109	181	1.0	REF.	20	13	1.0	REF.
1-2 years	86	165	0.9	0.6- 1.2	27	10	1.7	0.6- 4.7
3-5 years	95	167	1.0	0.7- 1.3	31	12	1.7	0.6- 4.4
6-9 years	104	160	1.1	0.8- 1.5	26	9	1.8	0.6- 5.0
> 9 years	67	74	1.4	0.9- 2.1	15	5	1.9	0.5- 6.3
Age at first use (years) [1]								
Never or < 1 year	109	181	1.0	REF.	20	13	1.0	REF.
> 22	41	76	0.8	0.5- 1.3	16	5	2.0	0.6- 6.8
20-22	117	190	0.9	0.7- 1.3	33	12	1.5	0.6- 4.0
17-19	157	253	1.1	0.8- 1.5	39	14	1.8	0.7- 4.6
< 17	37	47	1.6	1.0- 2.6	11	5	1.8	0.5- 6.5
Time since most recent OC use (years)								
Never or < 1 year	109	181	1.0	REF.	20	13	1.0	REF.
Current user	41	66	1.5	0.9- 2.4	7	2	3.6	0.6-20.6
< 10 years	146	248	1.1	0.8- 1.5	40	10	2.8	1.1- 7.7
≥ 10 years	165	252	0.9	0.7- 1.2	52	24	1.2	0.5- 2.8
Time since first OC use (years) [2]								
Never or < 1 year	109	181	1.0	REF.	20	13	1.0	REF.
≤ 15 years	123	260	1.0	0.7- 1.4	40	7	4.5	1.6-13.3
> 15 years	229	306	1.1	0.8- 1.4	59	29	1.1	0.5- 2.6

[1] Linear trend of the odds ratios significant among FH- women, $p < 0.05$.

[2] Difference in the odds ratios between FH- women and FH+ women was significant, $p < 0.05$. Linear trend of the odds ratios significant among FH+ women, $p < 0.05$.

The risk of breast cancer in relation to OC use stratified by family history of breast cancer is shown in Table 2. Ever use of OC for at least one year was in either group. Among FH- women, there was little suggestion of altered risk in relation to various indices of body size (except perhaps a reduced risk of breast cancer in relation to maximum body size [data not shown]). Some association between breast cancer risk and both heavier body size early in life (age 10 and age 18) and leaner body size later in life (reference age minus one year and maximum body size) was seen among FH+ women. unrelated to breast cancer risk among FH- women, but was associated with a 1.7-fold excess risk among FH+ women. Among FH- women, there was a modest increased risk of breast cancer associated with long duration of use, early age at first use, and current use. Among FH+ women, the general pattern of risk in relation to various indices of OC use suggests that low levels of exposure to OC may be sufficient to increase breast cancer risk and that such risk may not further increase with age at first use or increased duration of use. An increased risk of breast cancer observed in relation to earlier age at first use of alcohol and increased lifetime weekly average number of alcoholic drinks was confined to FH- women (data not shown).

Conclusions

Our results suggest that there are some suspected or established risk factors for breast cancer that may vary in effect according to family history of breast cancer. Continuing advances in our knowledge about genes that contribute to breast cancer risk will eventually allow examination of how environmental and behavioral factors may influence the expression of such genes. The identification of risk factors which can affect the penetrance of a susceptibility allele would offer insight into breast cancer etiology, and any that can be modified offer as well the possibility of interventions to reduce the incidence of breast cancer in genetically susceptible women.

References

1. Easton DF, Bishop DT, Ford D, et al. (1993) Genetic linkage analysis in familial breast cancer ovarian cancer: Results from 214 families. Am J Hum Gen 52:678-701.
2. Andrieu N, Clavel F, Demenais F (1989) Familial susceptibility to breast cancer: A complex inheritance. Int J Cancer 44:415-418.
3. Adami H-O, Adams G, Boyle P, et al. (1990) Chapter 2: Breast cancer etiology. Int J Cancer Suppl 5:22-39.

Oral Contraceptive Use and Reproductive Risk Factors for Breast Cancer: A Comparison of Results among Black and White Women

Julie R. Palmer, Lynn Rosenberg, and Samuel Shapiro

Summary

Studies of breast cancer have yielded conflicting results on the effects of age at first term birth, parity, and oral contraceptive (OC) use. We assessed these relations separately in black and in white US women aged 25-59 years in a hospital-based case-control study; 524 black cases were compared with 1021 black controls, and 3540 white cases were compared with 4488 white controls. In both racial groups, the risk of breast cancer increased with increasing age at first birth. Multiparity was associated with an increased risk before age 35 in both groups, and a reduced risk in black women aged 35-59 and in white women aged 35-44. OC use was associated with an increased risk among black women under age 45 and among white women under age 35.

Introduction

Breast cancer risk is influenced by reproductive factors, but the exact relation and the mechanisms for the effects have not been established. Many epidemiologic studies have indicated a reduced risk of breast cancer associated with early age at first full-term pregnancy and with multiparity (1). However, the relationship with parity may be more complex. Some studies indicate an increased risk in the few years immediately following a birth, and some suggest a crossover effect with age, with multiparity associated with an increased risk at young ages and a reduced risk at older ages (2, 3).

The relation of OC use to breast cancer risk is not clear (4). While studies of breast cancer in women over age 45 have indicated no association, long-term OC use has been associated with an increased risk of breast cancer in young women in a number of recent studies.

We had the opportunity to evaluate the relation of these factors to breast cancer risk in two racial groups, US white and black women, in data collected in the same study. The few data available on breast cancer risk factors in black women suggest that age at first birth, nulliparity, and multiparity have a similar relation to that observed in most studies of white women (5-7). However, results within different age groups have not been reported, and almost no information is available on OC use. A comparison of results in black and white women is of particular interest, because both reproductive patterns and breast cancer incidence differ between black and white US women. Below age 40, incidence is higher in blacks than in whites, whereas above age 40 the reverse holds (6).

Methods

The data were drawn from our ongoing hospital-based Case-Control Surveillance Study, in which hospital patients with cancer or a variety of other conditions are interviewed regarding their medical history, reproductive history, smoking and other behavioral factors, and lifetime contraceptive history. Ninety-five percent of patients approached have agreed to participate in the study. Women aged 25-59 who had been interviewed from 1977 through 1992 in hospitals in four eastern US cities were included in this analysis.

Cases had been admitted to the hospital for a primary breast cancer and had no previous cancer. Controls were drawn from among the pool of interviewed women hospitalized for nonmalignant conditions judged to be unrelated to OC use or childbearing; up to four controls per case were selected by frequency matching on race, five-year age group, and city. Included were 524 black breast cancer cases and 1021 black controls, and 3540 white cases and 4488 white controls. In each analysis, multiple logistic regression was used to control for age, geographic region, interview year, years of education, and the other two of the three reproductive variables under study.

Results

Consistent with US demographic data, patterns of childbearing were strikingly different in the two groups. Among all controls, 15% of black women and 25% of white women were nulliparous; among parous controls, 54% of black and 19% of white women had given birth before age 20, and 23% of black and 14% of white women had at least five births. A similar difference between black and white women was observed in each of the three age groups.

Among both black and white women and at all ages, late age at first full-term pregnancy was associated with an increased risk of breast cancer: the relative risk estimates for a first birth at age 30 or later compared with first birth before age 20 were 2.3 (95% CI 0.8-6.2) and 2.4 (95% CI 1.0-5.6) among black women aged 25-44 and 45-59, respectively, and were 2.1 (95% CI 1.4-3.1) and 2.9 (95% CI 2.0-4.1) among younger and older white women, respectively.

The relation of number of births to risk of breast cancer varied according to age in both black and white women (Table 1). Before age 35, multiparity was associated with an increased risk; beyond that age, high parity relative to uniparity was associated with a reduced risk, except in the oldest group of white women, among whom there was an increased risk in the intermediate parity categories.

Table 1. Parity in Relation to Breast Cancer Risk.

Number	Black women			White women		
of births	Cases	Controls	MVRR[a]	Cases	Controls	MVRR[a]
Age <35						
1	10	52	1.0[b]	45	144	1.0†
2	16	56	4.0(1.2-12.9)	103	177	2.6(1.6-4.3)
3	8	26	3.3(0.9-12.9)	38	91	2.1(1.1-4.0)
4+	4	24	2.9(0.5-15.2)	7	34	1.7(0.6-5.0)
Age 35-44						
1	34	53	1.0[b] 158	154	1.0†	
2	48	95	0.8(0.4-1.5)	378	387	1.3(0.9-1.7)
3	25	53	0.9(0.4-1.9)	205	276	1.3(0.9-2.0)
4	20	44	0.9(0.4-2.0)	63	154	0.9(0.6-1.5)
5+	14	67	0.6(0.2-1.3)	25	151	0.5(0.3-1.0)
Age 45-59						
1	50	46	1.0[b] 227	202	1.0†	
2	59	61	0.8(0.5-1.5)	651	393	1.7(1.3-2.2)
3	45	74	0.6(0.3-1.1)	445	374	1.5(1.1-2.0)
4	35	59	0.6(0.3-1.2)	197	275	1.0(0.7-1.4)
5+	63	116	0.7 (0.4-1.2)	141	276	1.0(0.7-1.4)

[a] Multivariate relative risk estimate (95% confidence interval).

[b] Reference category.

The prevalence of OC use was similar among black and white women. Use was associated with an increased risk of breast cancer among black women under age 45 and among white women under age 35 (Table 2). Among white women, risk increased with increasing duration of use; among black women, the relative risk estimate was greatest for 3-4 years of use. Analyses of OC use according to the timing of use (recency, use at a young age, use before a first birth) did not appear to explain the observed associations.

Table 2. OC Use in Relation to Breast Cancer Risk.

Duration of OC use (years)	White women			Black women		
	Cases	Controls	MVRR[a]	Cases	Controls	MVRR[a]
Age <35						
<1	20	111	1.0[b]	134	464	1.0 [b]
1-2	8	27	1.4 (0.5-3.8)	59	149	1.6 (1.1-2.3)
3-4	8	22	2.2 (0.7-6.3)	43	101	1.7 (1.1-2.7)
5-9	13	35	1.7 (0.7-4.2)	60	138	1.7 (1.1-2.6)
10+	3	14	1.3 (0.3-5.6)	22	34	2.6 (1.4-4.8)
Age 35-44						
<1	80	249	1.0 [b]	618	943	1.0 [b]
1-2	24	53	1.4 (0.8-2.6)	171	194	0.9(0.7-1.2)
3-4	19	20	3.3 (1.6-7.0)	104	130	1.0 0.7-1.3)
5-9	29	31	2.4 (1.3-4.5)	131	192	0.9(0.6-1.2)
10+	15	20	1.6 (0.8-3.5)	49	90	0.6(0.4-0.8)
Age 45-59						
<1	258	373	1.0 [b]	1687	1662	1.0 [b]
1-2	15	22	1.1 (0.5-2.4)	132	131	1.0(0.8-1.4)
3-4	8	7	1.9 (0.6-5.9)	65	54	1.0 0.7-1.5)
5-9	14	12	2.0 (0.8-4.7)	128	78	1.8 1.3-2.5)
10+	5	12	0.5 (0.2-1.5)	64	70	0.9(0.6-1.3)

[a] Multivariate relative risk estimate (95% confidence interval).
[b] Reference category.

Discussion

This study of US white and black women provides further evidence for a complex effect of pregnancy on the risk of breast cancer. The findings of an increased risk with later age at first birth and a crossover effect of parity with age lend support to the idea that pregnancy has both a beneficial effect on risk due to the terminal differentiation of breast stem cells that occurs with a term pregnancy, and an adverse effect due to the proliferative effect of the increased hormone levels present during pregnancy. The results concerning OC use confirm recent studies in other populations which indicate a positive association with breast cancer at young ages. The factors studied may contribute to differences in breast cancer incidence between black and white women.

References

1. Kelsey JL, Gammon MD, John EM (1993) Reproductive factors and breast cancer. Epidemiol Rev 15:36-47.
2. Janerich DT, Hoff MB (1982) Evidence for a crossover in breast cancer risk factors. Am J Epidemiol 116:737-42.
3. Bruzzi P, Negri E, La Vecchia C, et al. (1988) Short term increase in risk of breast cancer after full term pregnancy. Brit Med J 297:1096-8.
4. Malone KE, Daling JR, Weiss NS (1993) Oral contraceptives and breast cancer risk. Epidemiol Rev 15:80-98.
5. Austin H, Cole P, Wynder E (1979) Breast cancer in black American women. Int J Cancer 24:541-4.
6. Gray GE, Henderson BE, Pike MC (1980) Changing ratio of breast cancer incidence rates with age of black females compared with white females in the United States. J Natl Cancer Inst 64:461-3.
7. Schatzkin A, Palmer JR, Rosenberg L, et al. (1987) Risk factors for breast cancer in black women. J Natl Cancer Inst 78:213-17.

Patterns of Cancer Mortality in Women Who Had Used Sex Steroid Hormones under Age 60

Jeffrey A. Perlman and Jane Porter

Summary

Studies of oral contraceptive (OC) use show its health benefits outweigh its risks, but controversy continues concerning the balance of lifetime effects, particularly for cancer. We examine proportional mortality and cause-specific risks for OC users in the National Mortality Follow-Back Survey, representing 121,000 USA female deaths in 1986. Among OC users, standardized proportional mortality ratios (SPMRs) for heart disease were slightly lower than for non-users. There was a shift to death by the malignant diseases for OC users. For all-neoplasms, the SPMR was higher by 41% in OC users. However, for all nonbreast malignancies, the SPMR was only 24% higher than that of non-users. A nested case-control study demonstrated increased breast cancer mortality risk (RR = 1.9, $p<.05$) limited to women diagnosed before age 45 who took OCs beginning before age 25. Breast cancer mortality over age 45 was not related to past OC use. Other reproductive cancers were a rare cause of death. Thus, patterns of mortality were not greatly affected by OC use, except for rare premature breast cancer deaths arising in early-age OC users.

Introduction

The first cohort studies of OC users were launched in the late 1960's to characterize its morbidity and mortality risks (1). Recent correspondences from the Royal College of General Practitioners' Cohort Study came in 1983 and 1988, when its original cardiovascular (CV) mortality concerns were minimized and interest shifted to the risk of premature breast cancer in young women (2). The most recent study from the Oxford Family Planning Association in 1989 indicated that, after 20 years of follow-up, patterns of increases in CV and breast cancer mortality appeared to be offset by decreases in ovarian cancer mortality (3). Although these findings were statistically insignificant, this has given rise to the notion that OC users may have slight increases or decreases in the lifetime probability of reproductive system cancers and that OCs might affect life expectancy if women were followed out further in time.

We reviewed data from the National Mortality Follow-Back Survey to: 1) Distinguish differences in patterns of mortality (proportional mortality) for OC users vs. non-users from a national survey of death certificates representing all USA female deaths in 1986. 2) Examine the possible etiologic role of OC in major causes of death. 3) Isolate contingencies of OC use contributing to mortality, as in incidence-based case-control studies of OC use (4).

Methods

The Survey

Decedents were identified from a near 1-in-10 sampling of the National Center for Health Statistics' Current Mortality Sample. Questionnaires to family members included 2 questions on how old the decedent was when she first used OCs and "how long" she used them. General mortality and reproductive risk factors were also obtained this way. More than 8495 women were sampled. Deaths before age 45 and ischemic heart disease deaths were oversampled.

Proportional Mortality Studies

Analyses were first restricted to 2614 of 8495 women under age 60, since OCs were not used in cohorts born before 1925. Cause-specific proportional mortality ratios (PMRs) were calculated as the ratio of cause-specific deaths to all deaths among 621 known OC users and 1614 known never-users. Age-adjusted, cause-specific standardized proportional mortality ratios (SPMRs) were calculated using as denominator the expected number of deaths in each disease category, as provided by the OCMAP Program from the University of Pittsburgh. The SPMR rate ratio for OC users to non-users was calculated for each ICD-9 category. The SPMR estimates the disease-specific Relative Standardized Mortality Ratio. SPMRs greater than 1.00 indicate that the observed number of events in the survey exceeds expected numbers based on national mortality rates.

Case-Control Studies

Nested case-control studies of ischemic heart disease and breast cancer were undertaken, comparing prevalence of OC use in these "case" categories to the prevalence of use among all other decedents. The RR of breast cancer and IHD mortality were estimated for various contingencies of OC use. Separate case-control analyses were undertaken for women under and over the age of 45. Multiple logistic regression analyses were performed. Odds ratio estimates and 95% confidence intervals were calculated. Adjusted odds ratios were generated from logistic regression analyses incorporating both OC use and risk factor variables. The variables available from the informant questionnaire for consideration in the multivariate models included standard CV risk factors (which also serve as general mortality risk factors), hysterectomy status, menopausal status, marital status, number of live births, education, and family income.

Results

There were 2037 deaths in women under age 60 with known OC use status in the NMFBS. Related to a beneficial risk factor pattern, the proportion of CV deaths was somewhat lower in the OC user group than in the non-user group (20% vs. 30%). This was offset by a relatively higher proportion of malignant neoplasms among OC users (34% vs. 27% of all deaths). The % of OC use ranged from 11% to 48% for 10-year cohorts born between 1925 and 1961.

In Table 1, Selected SPMRs are presented for users and non-users. All of the SPMRs for CV diseases are artificially elevated due to the survey oversampling of ischemic heart disease, but less so for OC users. The latter point is reflected in a reduced rate ratio of 0.91. The greatest disparity between SPMRs that also included an elevated SPMR for OC users, was for breast cancer. The SPMR for breast cancer exceeded 1.00 and was 1.71 times as high in OC users as in non-users. There were only three endometrial cancer cases. The remaining SPMRs and ratios were unremarkable. However, there did appear to be a general tendency of the rate ratios to exceed 1.0 for the malignant diseases.

The results of the case-control study of ischemic heart disease death, before or after age 45, showed no association with OC use. In Table 2, the case-control study of women who died of breast cancer before age 45, there was increased risk associated with ever-use of OCs. There were no significantly increased odds ratios among breast cases occurring over the age of 45. Among premenopausal women, the risk of breast cancer appeared greatest for first use before age 25, short duration of use (< 5 years), and 20 or more years since first use.

Discussion

Of the three large world cohort studies, only the Royal College of General Practitioners' Study has ever demonstrated an increased overall mortality in OC

Table 1. Selected Proportional Mortality Ratios (SPMRs) Standardized By Age and Other Common Disease or Mortality Risk Factors.

Cause of Death (ICDA-9 Code)		SPMRs USED OCs YES	NO	RR#
CV Diseases (390-8,402,404,410-429)	1.54[a]	1.69[a]	0.91	
Ischemic Heart Dis. (410-414)		2.60[a]	2.59[a]	1.00
Cerebrovascular Disease (430-438)		0.75	0.74	1.00
Malignant Neoplasms (140-208)		1.03	0.73	1.41
All Neoplasms Except Breast Cancer (140-173,176-208)		0.90	0.72	1.24
Breast Cancer (174-175)		1.11	0.65	1.71
Endometrial Cancer+ (182)		0.90	0.53	1.71
Cervical Cancer (180)		0.87	0.74	1.19
Ovarian Cancer (183)		0.78	0.72	1.09
Melanoma (172)		1.12	1.00	1.12
Accidents (800-949, 960-999)		0.81	0.77	1.05
Suicide (950-959)		0.88	0.85	1.04

\# Rate Ratio adjusted by risk factors as applicable. + Only three cases.

[a] Rate Ratio affected by survey oversampling of IHD.

Table 2. Relationship Between OC Use and Breast Cancer Mortality by Age at Death and OC Use Contingency (Adjusted).

OC Use Contingency	< 45 Years		45-59 Years	
	OR[a]	95% CI[b]	OR[c]	95% CI
Never-use	1.0		1.0	
Ever-use	1.9	(1.2, 3.2)	1.6	(0.9, 2.9)
Age at first use				
< 25 years	2.2	(1.3, 3,7)	1.8	(0.5, 6.4)
25+ years	1.1	(0.5, 2.7)	1.7	(0.9, 3.2)
Unknown	2.1	(0.9, 5.3)	0.7	(0.1, 5.7)
Duration of Use				
<5 years	2.6	(1.5, 4.4)	2.1	(0.9, 4.8)
5+ years	1.3	(0.7, 2.4)	1.3	(0.6, 2.9)
Unknown	2.1	(0.7, 6.4)	1.3	(0.2, 10.0)
Time since first use				
< 15 years	1.6	(0.7, 3.3)		
15-19 years	1.8	(0.9, 3.4)	1.5	(0.6, 3.8)[d]
20+ years	2.6	(1.3, 5.4)	1.7	(0.8, 3.6)
Unknown	2.2	(0.9, 5.3)	0.7	(0.1, 5.0)

a Odds ratio estimates controlled for age, race, income, menopause, and ever smoking.
b Confidence interval.
c Odds ratio estimates controlled for age, race, income, and ever smoking.
d Estimate is for less than 20 years since first use.

users, attributed largely to CV deaths in older reproductive-age women, many of whom smoked. There has been no support for mortality effects in more recently published cohort studies, including the Nurses' Cohort and the Puget Sound Health Cooperative Cohort, or general vital statistics studies (5, 6).

This vital statistics study was based on a national survey of deceased persons. The survey had multiple surveillance purposes and was not specifically designed to study OC. The study had the advantage of being based on a representative sampling of all deaths in the USA in 1986. OC use rates in the study cohorts appear to be lower than the expected national rates of use, reflecting a degree of under-ascertainment or, conceivably, the inverse of the highly-touted healthy contraceptive-user effect. Although the PMR study suggested a net reduction in CV risk, the SPMR and case-control studies suggest that CV disease is not affected by OC use.

In this study, a sample deficit of heart diseases resulted in a shift to slight excesses in deaths from the malignant diseases. The latter was related almost exclusively to excesses in premenopausal breast cancer, a diagnosis accounting for only 3700 of 965,000 female deaths in the USA. Due to the rarity of endometrial and ovarian cancer death through age 60, there was no evidence that their reductions provided any of the anticipated offsets. Breast cancer mortality for the majority of ever-users is not associated with OC use. Taken with the total lack of effect on heart disease, overall patterns of mortality do not appear to be greatly affected by OC use, with the exception of only onset breast cancer.

In Table 2, the short-term as opposed to long-term OC duration-of-use effect on premature breast cancer mortality defies biological explanation. However, it could be consistent with a transient, OC-related breast cancer risk, similar to that seen in pregnancy (7). Tumors arising during pregnancy are known to have poor prognosis and survival. The data herein may simply suggest that tumors arising in the presence of high progestational agent levels may be associated with decreased survivability. This would be consistent with the results of a study in Sweden showing an increased risk of breast cancer in early-age OC users whose tumors demonstrated Her-2/neu amplification, a poor prognostic indicator, as well as poor survival independent of adjuvant treatment given (8).

References

1. Stergachis A (1992) Epidemiology of the noncontraceptive effects of oral contraceptives. Am J Obstet Gyn 167:1165-1170.
2. Kay CR, Hannaford PC (1988). Breast cancer and the pill: A further report from the Royal College of General Practitioners' Oral Contraception Study. Br J Cancer 58:675-680.
3. Vessey MP, Villiard-Mackintosh I, McPherson K, Yeates D (1989). Mortality among oral contraceptive users: 20 year follow-up of women in a cohort study. Brit Med J 200:1487-1491.
4. UK National Case-Control Study Group (1989). Oral contraceptive use and breast cancer risk in young women. Lancet 8645:973-982.
5. Porter J, Jick H, Walker A (1987). Mortality among oral contraceptive users. Obstet Gynecol 70: 29-32.
6. Colditz GA (1994). Oral contraceptive use and mortality during 12 years of follow-up: The Nurses' Health Study. Ann Int Med 120: 821-826.
7. Williams I, Jones L, Vessey MP, McPherson K (1990). Short term increase in risk of breast cancer with full term pregnancy. Brit Med J 300:578-579.
8. Olsson H, Borg A, Ferno M, Ranstaum J, Sigurdsson H (1991). Her-2/neu and INT 2 proto-oncogene amplifications in malignant breast tumors in relation to reproductive factors and exposure to exogenous hormones. J Natl Cancer Inst 83:1483-1487.

Progestogen Use and Risk of Breast Cancer

Geneviève Plu-Bureau, Monique G. Lê, Jean-Christophe Thalabard, Régine Sitruk-Ware, and Pierre Mauvais-Jarvis

Summary

The role of progestins in the etiology of breast cancer (BC) remains unclear. Controversial results have been reported regarding the effects of progestins on the breast epithelial tissue and in epidemiological studies. In the late seventies, progestins were known to counteract the effects of estrogens, both at the target level and at the hypothalamopituitary level. We hypothesized that chronic administration of progestins, particularly 19-nortestosterone derivatives, at levels which exert an antigonadotropic effect, could reduce the risk of breast cancer. To investigate the role of chronic progestin use on the risk of breast cancer, a cohort study including 1150 premenopausal French women (aged 20-50) diagnosed with benign breast disease was carried out over a 10-year period. The analysis was performed using the Poisson regression and Cox proportional hazards models. The use of 19-nortestosterone derivatives was found to be significantly associated with a lower risk of breast cancer than that in untreated women (RR 0.48, 95% CI 0.25-0.90). In addition, there was a significant linear relationship between the duration of use and the decrease in the breast cancer risk (p for trend = 0.02). Our results suggest that progestin use was not associated with increased breast cancer risk in this population of women with benign breast disease. Furthermore, they suggest that some types of progestins might have a beneficial effect on the risk of breast cancer.

Introduction

Over the past twenty years, OC use has generated a large number of epidemiological studies about chronic progestin use. These have provided conflicting results about the effects of the progestin component on breast cancer risk, ranging from protective to deleterious effects (1, 2). Changes in of both OC formulation and dosage, the exposed population, and the increase in the duration of follow-up studies may partially explain some of the discrepancies. Similarly, the interpretation of the biological effects of progestins on normal breast epithelial tissue seems to be evolving (3, 4).

The class of progestin compounds assessed included various molecules with different binding capacities to the androgen receptor, as well as different metabolism. In this respect, 19-nortestosterone derivatives may be viewed as distinct agents.

In the late 1970's, it was hypothesized that progestins were able to counteract the carcinogenic effect of estrogens on the breast epithelial cells by: 1) Antagonizing the estrogen effect at the target level. 2) Altering the hypothalamopituitary axis and thus reducing the ovarian secretions. A cohort study of premenopausal women with a diagnosis of benign breast disease was designed to address the hypothesis that chronic administration of progestins, at doses exerting an antigonadotropic effect, could reduce the risk of breast cancer.

Material and Methods

The cohort and methodology have already been described (5). Briefly, 1150 women were recruited over the period 1976-1979 in two breast clinics located in the Paris area. Inclusion criteria were 20-50 years old, premenopausal, without previous history of breast cancer, with a diagnosis of benign breast disease. The initial and follow-up questionnaires were recorded by specially-trained physicians qualified in breast pathology and included, in addition to all relevant medical data, doses and durations of progestin treatments, types of progestin used (19-nortestosterone derivatives versus all other progestins), and all other medications. No estrogen was associated with these progestin treatments.

The analysis was performed using the Poisson regression analysis and a Cox proportional hazards model, taking into account the main risk factors for breast cancer.

Results

Subjects' mean age at time of inclusion was 38.5 years. The cumulative follow-up of the cohort reached 12,462 person-years. Sixty-seven percent of the women have used progestins. The proportion of ever-users was higher in women with high socioeconomic status, familial history of breast cancer, and early age at menarche. During this period, 44 breast cancers were diagnosed. The relative risk of breast cancer according to the duration of use is shown in Figure 1.

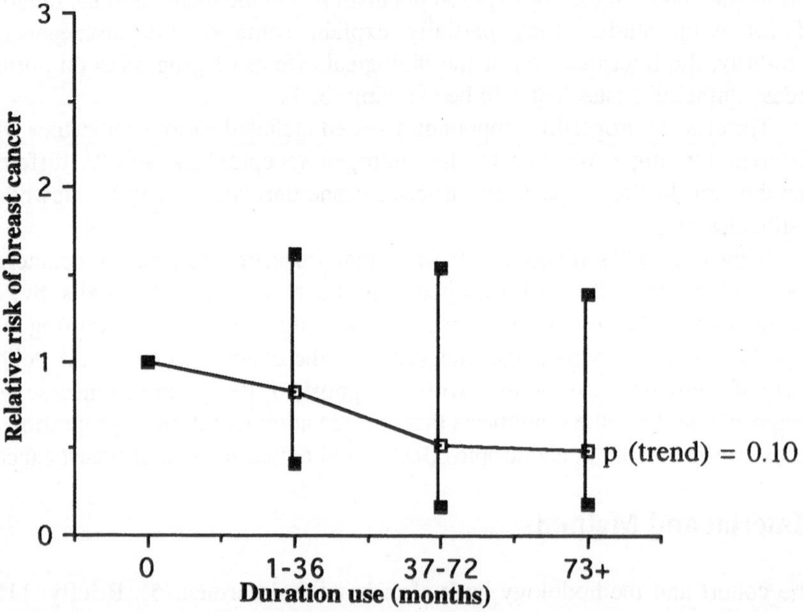

Figure 1. Relative risk of breast cancer associated with the cumulative duration of all categories of progestin treatment, calculated with an adjusted Cox model.

No significant difference in breast cancer risk could be found between progestin users and non-users. However, when progestins were categorized, the relative risk of breast cancer was significantly lower in 19-nortestosterone users than in non-users. In addition, a significant linear trend was observed according to the duration of use (Figure 2). The other types of progestins were not significantly associated with a reduction in the relative risk compared with non-users.

Conclusion

Our results suggest that, in this population of premenopausal women diagnosed with benign breast disease, the chronic administration of 19-nortestosterone derivatives may have a protective effect against breast cancer. Hence, these results support the assumption that different types of progestins may have different effects on the mammary gland, and could partially explain the controversial reports relating progestin use and breast cancer risk.

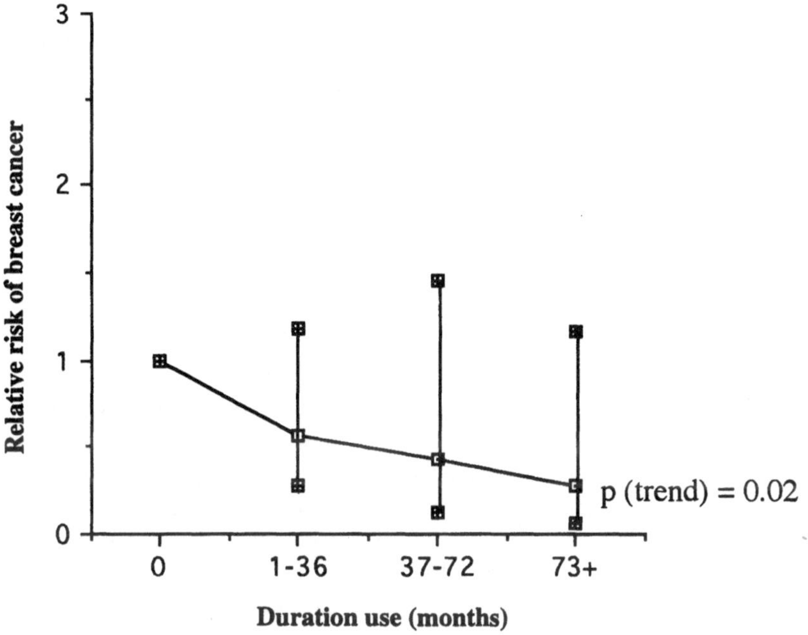

Figure 2. Relative risk of breast cancer associated with the cumulative duration of 19-nortestosterone derivatives, calculated with an adjusted Cox model.

References

1. Staffa JA, Newschaffer CJ, Jones JK, Miller V (1992) Progestins and breast cancer: An Epidemiologic Review. Fertil & Steril 3:473- 491.
2. Stanford JL, Thomas DB (1993) Exogenous progestins and breast cancer. Epidemiol Rev 15:98-107.
3. Ferguson DJP, Anderson TJ (1981) Morphological evaluation of cell turnover in relation to the menstrual cycle in the "resting" human breast. Brit J Cancer 44:177181.
4. Mauvais-Jarvis P, Kuttenn F, Gompel A (1986) Antiestrogen action of progesterone in breast tissue. Breast Cancer Res Treat 8:179-187.
5. Plu-Bureau G, Le MG, Sitruk-Ware R, et al. (1994) Progestogen use and decreased risk of breast cancer in a cohort study of premenopausal women with benign breast disease. Brit J Cancer, in press.

Five-Year Survival of Women with Breast Cancer According to Prior Use of Oral Contraceptives

Phyllis A. Wingo, Harland Austin, Andrzej Kosinski, Howard Ory, James J. Schlesselman, William Eley, and Bert Peterson

Summary

More than 80% of US women have used oral contraceptives (OCs) at some time during their lives. Between the ages of 15 and 54, the breast is the leading site for cancer among women, and breast cancer is the leading cause of death from cancer. We examined the relationship between the survival of 4197 women diagnosed with primary breast cancer between December 1, 1980, and December 31, 1982, and their prior use of OCs. We used interview data from the Cancer and Steroid Hormone Study and survival information from the Surveillance, Epidemiology, and End Results program. In general, although the differences were not statistically significant, Kaplan-Meier estimates suggested that survival was slightly better for women who had used OCs than for never-users. Duration of OC use, age at first use, and exclusive use of specific OC formulations were not related to survival. However, survival appeared to increase with increasing time between diagnosis and first or last OC use. Women with at least 20 years latency or 15 years recency had an 83% probability of surviving five years (95% CL: 78, 88), whereas women with the shortest intervals had a 77% probability of surviving five years (95% CL: 72, 83). Women who had never used OCs had survival rates consistent with the survival experience for all users combined. OC use did not appear to have an adverse effect on the long-term risk of death from breast cancer.

Introduction

Breast cancer accounts for 32% of all new cancers diagnosed and 18% of all cancer deaths among USA women (1). In 1994, 182,000 women are expected to be newly diagnosed with breast cancer and 46,000 are expected to die from the disease. More than 80% of women in the USA have used OCs at some time in their lives (2), and approximately 19% of all women aged 15-44 years were currently using OCs as their method of contraception in 1988 (3). Thus, any possible relationships between this common method of birth control and the risk of developing or dying from breast cancer should be carefully considered. Moolgavkar and colleagues proposed a two-stage model for carcinogenesis that appeared to be applicable to the epidemiology of breast cancer and suggested that hormones probably acted as promoters (4). As promoters, hormones might stimulate growth in existing but undetected breast tumors. Better survival after a diagnosis of breast cancer among women who have used OCs than among nonusers may be attributable to the development of less aggressive tumors, or poor survival among users may be due to the development of more aggressive tumors (5). Alternatively, bias may explain differences in survival between users and nonusers. If users were screened more effectively than nonusers, then tumors might be detected at earlier stages of disease and survival would appear to be better for users than nonusers as a result of detection bias (5). If young women, perceived to be at low risk of malignant breast disease by themselves and by their clinicians because of their age, received minimal screening, then tumors may be detected at later stages of disease and survival may be poorer, regardless of OC user status.

Methods

We used a retrospective cohort design to examine the relationship between prior use of OCs and survival among women who were diagnosed with breast cancer during the Cancer and Steroid Hormone (CASH) Study. In general, information on risk factors for survival was derived from the CASH interview data, and survival, stage, and histology data were derived from the SEER program.

CASH was a large population-based case-control study primarily designed to assess the relationship between OC use and the risk of breast, ovarian, and endometrial cancer (6). The Centers for Disease Control and Prevention (CDC) in Atlanta, Georgia, coordinated data collection in eight geographic locations in the US: the metropolitan areas of Atlanta, Detroit, San Francisco, and Seattle; the states of Connecticut, Iowa, and New Mexico; and the four urban counties of Utah.

Cases were women aged 20-54 years who resided in these locations and who were diagnosed with histologically confirmed primary breast cancer between

December 1, 1980, and December 31, 1982 (6). Of the 4742 women with completed interviews, 11.4% were excluded from this analysis: 0.7% reported that they were uncertain whether they had ever used OCs, 1.9% failed to match with the SEER records, 2.4% had sites other than the breast as the first primary site; 1.7% had a year of birth in the SEER record that did not match with the exact year of birth in the CASH record, and 4.7% had *in-situ* disease. A total of 4197 women were available for the survival analyses.

We used Kaplan-Meier estimates of the probability of survival and logrank statistics to examine differences in survival among specific subgroups of women. Cox regression methods were used to estimate hazard rate ratios (HRR) and to adjust simultaneously for the variables considered *a priori* as potential confounders (7).

Results

Among the cohort, 941 women (87.2%) had breast cancer listed as the cause of death, 85 women (7.9%) had causes of death other than breast cancer, and death certificates were unavailable for 53 women (4.9%). Follow-up for women with deaths due to causes other than breast cancer and women whose death certificates could not be located was censored at date of death.

Duration of OC use, age at first use of OCs, and exclusive use of specific OC formulations were not associated with the risk of dying from breast cancer (data not shown). The relationships between survival and times since first and last use of OCs were more difficult to interpret (Table 1). Women with at least 20 years latency or 15 years recency had a risk of dying from breast cancer reduced an approximately 20%. The Kaplan-Meier estimates of survival (data not shown) increased with increasing time since first or last OC use. Women who never used OCs had survival rates that were consistent with the experience of all users combined. Cox models (Table 1) and stratification by historic stage of disease did not alter these conclusions.

Discussion

Because power is important in interpreting negative findings, we estimated the power of the present analysis. The power was high,at least 80% for hazard rate ratios smaller than .80 or larger than 1.25. Examination of all upper limits of the confidence bounds for the hazard rate ratios also provided reassurance regarding the magnitude of the relationship between the risk of dying from breast cancer and the prior use of OCs. For analyses that included all women, 1.5 was the largest limit among all measures of OC use, and most upper limits were 1.2 or less. Thus, OC use is probably not associated with more than a 20% increased risk of dying from breast cancer.

Although this study was large, we must consider the possible limitations of our analysis. First, this study only included women who completed an interview for the CASH Study; it did not include women who died before interview (1%), who were too ill to be interviewed (4%), whose physicians refused consent for interview (3%), or who were not interviewed for other reasons (6%). This study also did not include women who were eligible for enrollment for CASH but were not ascertained. Presumably, most of the women who died, or who were too ill to be interviewed, or whose physicians refused consent to contact their patients were women with more advanced disease.

Table 1. 5-Year Risk of Death from Breast Cancer By Time since Using OCs.

Months Latency	**Patients**	**Cox's Proportional Hazards Model HRR[&]**	**(95% CL)**
Never	1636	1.00	Referent
< 121	284	0.92	(0.69, 1.2)
121-180	703	0.91	(0.74, 1.1)
181-240	969	0.91	(0.75, 1.1)
241+	231	0.81	(0.59, 1.1)
Months Recency			
Never	1636	1.00	Referent
< 25	234	0.97	(0.72, 1.3)
25- 60	316	0.87	(0.65, 1.1)
61-120	677	1.00	(0.83, 1.2)
121-180	619	0.85	(0.68, 1.1)
181+	341	0.78	(0.59, 1.0)

[&] Adjusted for age at diagnosis, body mass index, and education.

A previous analysis of the CASH data supports this opinion (6). Although women who were not interviewed were similar in age and race and were only slightly different by marital status from the women who completed interviews, they were slightly more likely to have late stage disease (8% vs. 5%). Because of the general lack of an association between OC use and breast cancer survival, it seems unlikely that very poor survival could be related to OC use. If most of the difference between women interviewed and not interviewed was related to stage of disease, then the separate analyses by stage should have minimized bias in the estimates of the effects of OC use.

We did not have data about key factors that predict for better survival, i.e., estrogen and progesterone receptor status, flow cytometry results, oncogenes, growth factors, or tumor grade. Although many studies have examined the relationships between these factors and survival, little information about the effects of hormones on these factors exists.

These analyses suggest that OC use does not increase the risk of death from breast cancer. In particular, OC use does not adversely affect breast cancer survival whether assessed by duration of use before diagnosis, age at first use, use of specific OC formulations, or time between first or last use and diagnosis. Examination of these relationships by historic stage of disease did not produce different conclusions. These data are reassuring for OC users and provide no reasons for changing OC-prescribing practice.

References

1. Boring CC, Squires TS, Tong T (1993) Cancer statistics. CA 43:7-26.
2. Institute of Medicine (IOM) Committee on the Relationship Between Oral contraceptives and Breast Cancer (1991) Oral Contraceptives and Breast cancer. Washington, DC: National Academy Press.
3. Lee ET (1992) Statistical Methods for Survival Data Analysis. New York: John Wiley & Sons.
4. Moolgavkar SH, Day NE, Stevens RG (1980) Two-stage model for carcinogenesis: Epidemiology of breast cancer in females. J Natl Cancer Inst 65:559-569.
5. Mosher WD, Pratt WF (1990) Contraceptive use in the United States, 1973-1988. Advance data from vital and health statistics, No. 182. Hyattsville, MD: National Center for Health Statistics.
6. Thomas DB (1991) In Institute of Medicine (IOM), Committee on the Relationship Between Oral Contraceptives and Breast Cancer: Oral Contraceptives and Breast cancer. Washington, DC: National Academy Press.
7. Wingo PA, Ory HW, Layde PM, Lee NC (1988) The evaluation of the data collection process for a multicenter, population-based, case-control design. Am J Epidemiol 128:206-217.

Plasma Prolactin and IGF-1 Levels in Young, Healthy, Nulliparous Women in Relation to Low Dose Oral Contraceptive Use

Helena C. B. Jernström and Lars Håkan Olsson

Summary

Plasma prolactin (p-PROL) and IGF-1 were studied in relation to low-dose combined oral contraceptive (OC) use in young, healthy, nulliparous women in order to find out whether modern OCs create any temporary or permanent hormonal changes in this group of women, which could be important for development of malignant tumors. Paired blood samples for plasma prolactin and IGF-1 were drawn on cycle days 5–10 and 18–23 from each woman ($n = 43$). There was no significant difference between present, former, and never-users in p-Prol levels. Time since awakening was strongly negatively associated with p-Prol levels in both cycle phases ($p = 0.035$ and $p = 0.001$, respectively) when present users were compared with non-users. Present OC use was significantly negatively associated with absolute levels of p-IGF-1 during cycle days 18–23 compared with never-users ($p = 0.0013$), after adjusting for age. The difference in IGF-1 levels between the two menstrual cycle phases in each woman (ΔIGF-1) was significantly negatively correlated with present OC use, compared with non-users ($p = 0.0002$). No significant difference in ΔIGF-1 was seen between former and never-users.

Introduction

Estrogens increase the level of Prol (1), growth hormone (GH) and its peripheral tissue mediator, insulin-like growth factor-1 (2). These hormones are also influenced by aging and pregnancies (3-6). High-dose OCs augment Prol levels (7), while an increase in GH levels with a subsequent decrease in IGF-1 levels has been reported after oral estrogen administration to postmenopausal women (2). In the present study, we wanted to investigate whether there are any temporary or permanent hormonal changes after low-dose OC intake on p-PRol and IGF-1 levels, which could be important for future tumor development. Paired blood samples were drawn on cycle days 5-10 and 18-23 from 43 young, healthy, nulliparous women aged 19-25, who had never been pregnant. In three women, the second blood sample was drawn after cycle day 23 to adjust for a long menstrual cycle.

For statistical analyses, regression analyses and analyses of covariance were used.

Results

The women were divided into present, former (previous OC use for at least one month), and never-users of OCs. Seventeen women were present, 16 former, and 10 never-users. The women were interviewed and height, weight, hip, waist, and breast size were measured by the same person (Table 1).

Table 1. Physical Characteristics of the Women Surveyed.

	Present Users N=17 median (range)		Former Users N=16 median (range)		Never Users N=10 median (range)	
Age (years)	20	(19-25)	21	(19-25)	21	(9-24)
Menarche	14	(11-16)	13	(11-15)	13.5	(12-17)
Physiological menstrual cycle length (days)	28	(21-56)	28.5	(26-31)	29.5	(25-40)
BMI	21.1	(18.1-28.7)	20.5	(17.6-24.8)	20.0	(18.2-24.2)
Height (cm)	170	(161.5-178.5)	165.5	(152-174)	164.5	(158.5-180.5)
Weight (kg)	60	(52.2-82.8)	53.3	(46.2-70.8)	52.5	(48.4-74.8)

There were no significant differences in p-Prol levels in either cycle phase among present, former, and never users; neither was there any significant difference between women in the follicular and luteal phases (Figure 1). A strong negative association was found between time since awakening and time when the blood sample was drawn, when present users were compared with nonusers, both during cycle days 5–10 (p = 0.035) and cycle days 18–23 (p = 0.001). An early menarche was significantly associated with higher p-Prol levels in the luteal phase (p = 0.014), after adjusting for time since awakening.

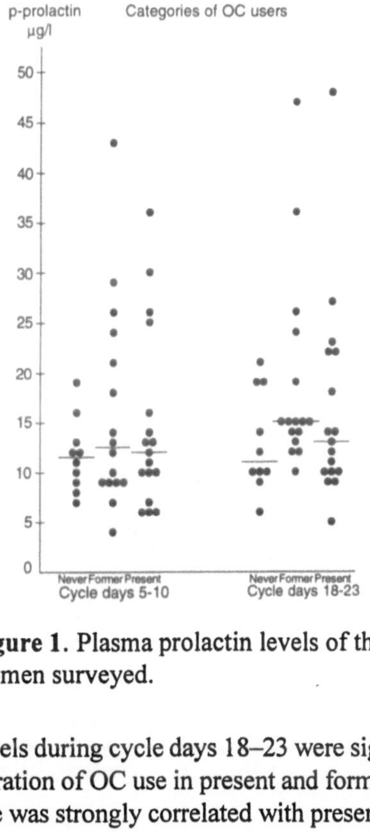

Figure 1. Plasma prolactin levels of the women surveyed.

Present OC use was significantly negatively associated with absolute levels of p-IGF-1 during cycle days 18–23 compared with never-users (p = 0.0013), after adjusting for age. The difference in IGF-1 levels between the two menstrual cycle phases in each woman (ΔIGF-1) was significantly negatively correlated with present OC use, compared with non-use (p = 0.0002). No significant difference in ΔIGF-1 was seen between former and never-users (Figures 2-4). A significant correlation between estradiol and IGF-1 levels was found during cycle days 18–23 (r=0.30; p=0.05). P-IGF-1 levels during cycle days 18–23 were significantly negatively associated with total duration of OC use in present and former users (p = 0.006). Total duration of OC use was strongly correlated with present OC use (r = 0.49; p = 0.003). There was no significant effect from total duration among the group of former users on p-IGF-1 levels during cycle days 18–23. It is therefore difficult to separate the effects of present OC use from total duration on p-IGF-1 levels.

Only three women were presently smoking; therefore, no conclusions on the effects of smoking on p-Prol or p-IGF-1 could be drawn from this material. None of the constitutional factors (height, weight, hip/waist-ratio, or breast size) were significantly related to p-Prol or p-IGF-1 in either cycle phase. P-Prol and p-IGF-1 levels were not significantly related to each other.

Discussion

In this homogenous group of young, healthy, nulliparous women, who had never used any hormonal or psychotropic medication except combined OCs that could influence p-Prol levels, low dose OCs did not significantly change p-Prol, in contrast to a significant suppression of p-IGF-1 levels from present OC use. The finding of no change in p-Prol levels is in line with our former study (8).

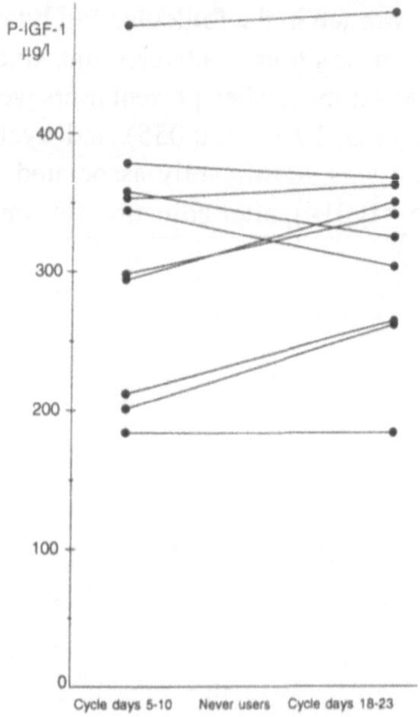

Figure 2. Levels of p-IGF-1 in never users of OCs.

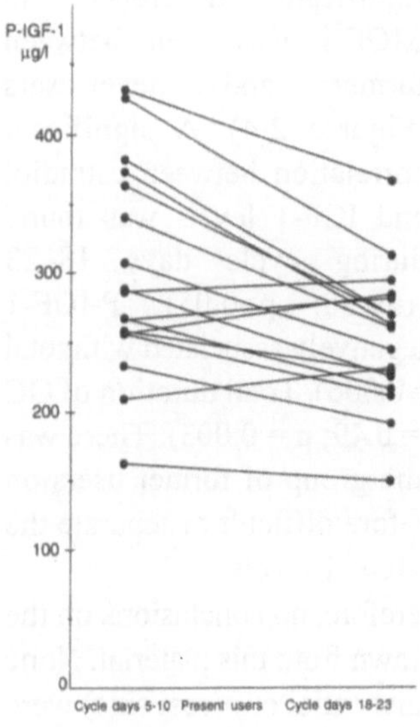

Figure 3. Levels of pIGF-1 in present OC users.

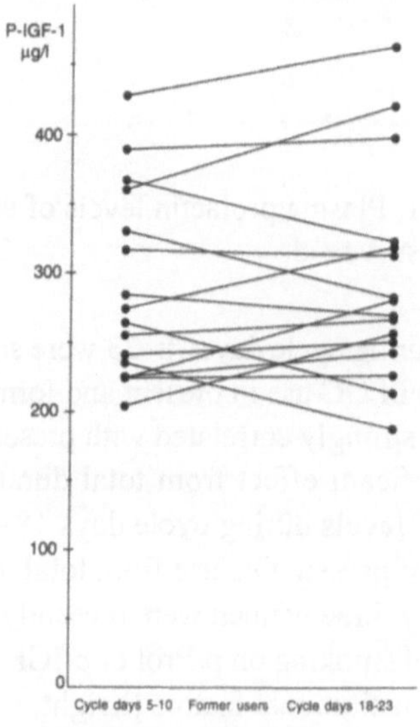

Figure 4. Levels of pIGF-1 in former OC users

Since this was an extension of our previous study (9), the cyclic suppression of IGF-1 levels in healthy, nulliparous present OC users has not been described earlier. Karlsson, et al. (10), found a change in GH secretion pattern during low-dose OC use, which, together with our finding, could indicate that there is a dissociation between GH and IGF-1 levels in present OC users.

Acknowledgment

Supported from grants from the Swedish Cancer Society; Gustav V's Jubileumsfond; Berta Kamprad's Foundation; the Medical Faculty, University of Lund; and the Hospital Foundation, University Hospital of Lund.

References

1. Nagasawa H (1989) Prolactin and estrogen in mammary tumorigenesis. In Levine AS (ed): Etiology of Cancer in Man. Dordrecht: Kluwer Acad. Pub.
2. Bellantoni MF, Harman SM, Cho DE, Blackman MR (1991) Effects of Progestin-opposed transdermal estrogen administration on growth hormone and insulin-like growth factor-1 in postmenopausal women of different ages. J Clin Endocrinol Metab 72:172–178.
3. Yu MC, Gerkins R, Henderson BE, et al. (1981) Elevated levels of prolactin in nulliparous women. Brit J Cancer 43:826–831.
4. Musey VC, Collins DC, Musey PI, et al. (1987) Long term effects of a first pregnancy on the secretion of prolactin. N Eng J Med 316:229–234.
5. Ho KY, Evans WS, Blizzard RM, et al. (1987) Effect of sex and age on the 24–hour profile of growth hormone secretion in man: Importance of endogenous estradiol concentrations. J Clin Endocrinol Metab 64:51–58.
6. Lang I, Schernthaner G, Pietschmann P, et al. (1987) Effects of sex and age on growth hormone response to growth hormone–Releasing hormone in healthy individuals. J Clin Endocrinol Metab 65:535–540.
7. Luciano AA, Sherman BM, Chapler FK, Hauser, et al. (1985) Hyperprolactinemia and contraception: A prospective study. Obstet Gynecol 65:506–510.
8. Jernström H, Knutsson M, Taskila P, Olsson H (1992) Plasma prolactin in relation to menstrual cycle phase, oral contraceptive use, arousal time and smoking habits. Contraception 46:543–548.
9. Jernström H, Olsson H (1994) Suppression of plasma insulin-like growth factor-1 levels in healthy, nulliparous, young women using low dose oral contraceptives. Gynecol Obstet Invest, in press.
10. Karlsson R, Edén S, von Schoultz B (1990) Altered growth hormone secretion during oral contraception. Gynecol Obstet Invest 30:234–238.

Effect of a Low-Fat Diet on Estrogen and Bile Acid Metabolism in Normal Human Subjects

Charles P. Martucci, Daniel W. Sepkovic, H. Leon Bradlow, Daniel G. Miller, and G. Stephen Tint

Summary

The urinary estrogen metabolites, 2-hydroxyestrone (2-OHE_1) and 16α-hydroxyestrone (16α-OHE_1), as well as serum testosterone (T), estradiol (E_2), and bile acids were measured in fasting normal subjects ($n = 12$) before and three months after the adoption of a low-fat diet (initial fat content *ca.* 40% reduced to *ca.* 20%). The serum bile acids measured were lithocholic (LCA), deoxycholic (DCA), cholic (CA), chenodeoxycholic (CDCA), and ursodeoxycholic acids (UDCA). No significant differences were found in the absolute or relative ratios for 2-OHE_1, 16α-OHE_1, E_2, or T. The composition of the serum bile acids was changed; comparison of the percentage of each plasma bile acid at baseline with that after three months on the diet indicated an increase in CDCA (14.9% to 24.4%; $p < 0.001$) and a reduction in UDCA (17.9% to 13.2%; $p < 0.01$). A marked change in the ratio of CDCA/UDCA was also observed; the ratios of the three-month values compared with baseline were consistently increased by the diet (mean 2.72; SEM $\pm$ 0.71). Only modest or no differences were observed with the ratios for the other bile acids. Because of the product-precursor relationship between CDCA and UDCA, these data can be interpreted as an indication of a dietary change in bile acid metabolism which resulted in inhibition of the oxidation at the 7-hydroxy group of CDCA and epimerization to UDCA.

Introduction

Epidemiological studies have associated the consumption of Western-type diets, which are typically high in fat, with the development of breast (1), prostate (2), and colon cancer (3). The mechanism by which dietary fat is involved in the etiology of these cancers is unknown. Alteration of estrogen and bile acid metabolism by diet has been reported, but remains controversial. Studies of estrogen metabolism in women (4) and chimpanzees (5) before and after the adoption of a modified diet indicate an increase in 16α-hydroxylated estrogens and a decrease in 2-hydroxylated estrogens with an increase in dietary fat. However, many other dietary components appear to affect sex hormone metabolism, and no clear understanding of the role of dietary fat has been achieved (6). A similar difficulty has been encountered with the effect of diet on bile acid metabolism (7). In an effort to examine the effect of dietary fat on estrogen and bile acid metabolism, we carried out a study in which normal men and women (n=12) changed their diet from an average of 40% fat to an average of 20% fat for three months. Diets were accessed from food records which were evaluated for macro- and micronutrient content; in general, the reduction in dietary fat content was achieved by replacement with carbohydrate. At baseline and three months after initiation of the diet, urine and blood samples were collected. Measurements were made of urinary 2-hydroxyestrone (2-OHE_1) and 16α-hydroxyestrone (16α-OHE_1) (8), as well as serum estradiol (E_2), testosterone (T), and the following bile acids: lithocholic (LCA), 7-ketolithocholic (7KLCA), deoxycholic (DCA), chenodeoxycholic (CDCA), cholic (CA), and ursodeoxycholic (UDCA) (9).

Results

No significant differences were found in the serum E_2 (baseline 65.2 ± 22.6pg/ml vs. three-month 79.4 ± 31 pg/ml) or in serum T. Similarly, the urinary values for 2-OHE_1, 16α-OHE_1 and the ratio for the 2-OHE_1/16α-OHE_1 (Figure 1) did not show any significant differences. No significant differences were found in the individual or total amounts of bile acids.

A comparison of the percentage of individual bile acids at baseline versus three months on the low-fat diet (Figure 2) indicated a statistically significant increase in CDCA (14.9 ± 5.0% to 24.4 ± 18.6%, $p < 0.001$) and a reduction in UDCA (17.9 ± 7.6% to 13.2 ± 6.6%, $p < 0.01$). In addition, a marked increase in the ratio of CDCA/UDCA for each subject was consistently observed (Figure 3). The mean of the ratios of the three-month values for CDCA/UDCA compared with the mean of the baseline values was 2.72 ± 0.71 SEM. Only modest or no differences in the ratios for the other bile acids were observed. In two of the subjects, 7KLCA was measurable at baseline (0.24 and 0.18 μg/ml), but became unmeasurable after three months on the diet. In the other 10 subjects, 7KLCA was not measurable before or after the diet period.

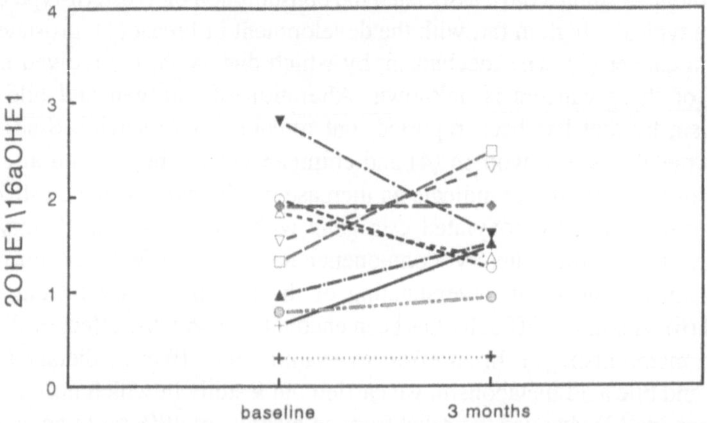

Figure 1. Ratios of 2-hydroxyestrone/16α-hydroxyestrone at baseline and after three months on a low-fat diet.

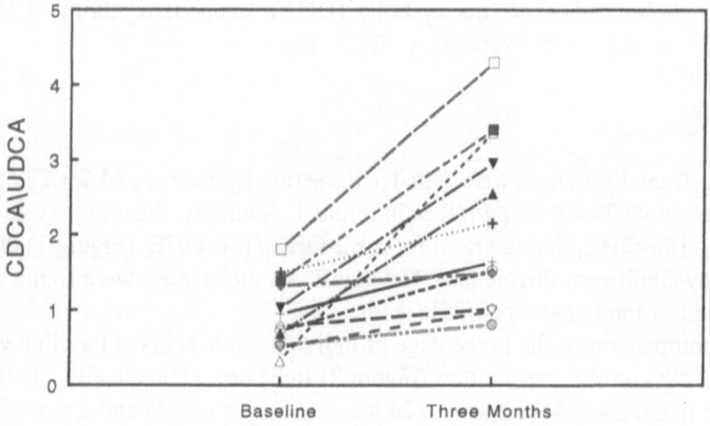

Figure 2. Composition of serum bile acids as a percentage of total bile acids at baseline and after three months on a low-fat diet.

Discussion

Following adoption of a low-fat diet, no change in estrogen metabolism via the 2-hydroxylation and 16α-hydroxylation pathways was observed in our subjects. Presumably, the numerous dietary factors which can influence these pathways (10) may have obscured any effect fat may have had on 2-OHE_1 and 16α-OHE_1. The observed changes in serum CDCA and UDCA can be interpreted as the result of a reduction in the activity of 7α-dehydrogenase affecting CDCA, brought about by the low-fat diet.

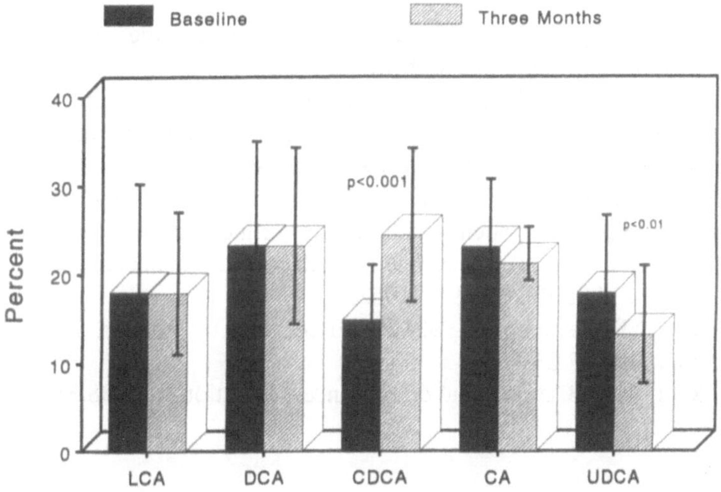

Figure 3. Ratios of chenodeoxycholic/ursodeoxycholic at baseline and after three months on a low-fat diet.

The inhibition of this reaction would result in reduced formation of 7KLCA, which is considered an obligatory intermediate in the epimerization of CDCA to UDCA. It is of interest that, in the two subjects with measurable 7KLCA, it fell to undetectable levels following the three months on the low-fat diet. As opposed to the primary bile acids (CA and CDCA), secondary bile acids appear to be capable of initiating and/or promoting carcinogenesis (11). The benefit of a low-fat diet in reducing the risk of cancer may reside in a decreased metabolism of primary bile acids (especially CDCA) to secondary bile acids.

References

1. Kelsey, JL (1991) The epidemiology of breast cancer. CA-A Cancer J Clin 41:146-165.

2. Carter BS, Carter HB, Isaacs JT (1990) Epidemiologic evidence regarding predisposing factors to prostate cancer. Prostate 16:187-197.
3. Willett W (1989) The search for the causes of breast and colon cancer. Nature 338:389-394.
4. Longcope C, Gorbach S, Goldin B, et al. (1987) The effect of a low fat diet on estrogen metabolism. J Clin Endocrinol Metab 64:1246-1250.
5. Musey PI, Collins DC, Bradlow HL, et al. (1987) Effect of diet on oxidation of 17B-estradiol *in vivo*. J Clin Endocrinol Metab 65:92-95.
6. Adlercreutz H (1990) Western diet and western diseases: Some hormonal and biochemical mechanisms and associations. Scand J Clin Lab Invest 50:3-23.
7. Van Munster IP, Nagengast FM (1991) The influence of dietary fibre on bile acid metabolism. Eur J Cancer Prev 1:35-44.
8. Klug TL, Bradlow HL, Sepkovic DW (1994) Monoclonal antibody based enzyme-immonuassay for simultaneous quantitation of 2- and 16α-hydroxyestrone in urine. Steriods, in press.
9. Batta AK, Arora R, Salen G, et al. (1989) Characterization of serum and urinary bile acids in patients with primary billiary cirrhosis by gas-liquid chromatography-mass spectrometry: Effect of ursodeoxycholic acid treatment. J Lipid Res 30:1953-1962.
10. Martucci CP, Fishman J (1993) P 450 enzymes of estrogen metabolism. Pharmac Ther 57:237-257.
11. Hill MJ (1990) Bile flow and colon cancer. Mutation Res 238:313-320.

Expression of GSTπ in Breast Cancer and Its Relationship to the Expression of Estrogen and Progesterone Receptor

Chun-Hai Li, Jiang Zhou, Shan-Chun Guo, Yuan-Ji Xu, and Yue-Tang Zhao

Summary

We examined the expression of the glutathione S-transferase π (GSTπ), estrogen (ER), and progesterone receptor (PR) genes in breast tumor samples and in the corresponding adjacent normal tissues from 120 patients. We also analyzed DNA amplification and RNA expression of GSTπ in fresh breast tumor tissues and in adjacent normal tissue from 32 patients. Immunohistochemical staining revealed that 78% of the breast cancer samples and 68% of the adjacent normal tissues were positive for GSTπ. GSTπ expression was inversely correlated with ER and PR expression. Nucleic acid hybridization with dot-blot analysis showed that three out of 32 breast cancer patients (9.4%) had GSTπ gene amplification, and 9 out of 32 cases (28.1%) showed GSTπ mRNA overexpression. These results suggest that the increased expression of GSTπ in breast cancer could occur at the level of DNA and mRNA, but mainly occurs at the post-transcriptional level.

Introduction

Glutathione S-transferases (GSTs E.C. 2.5.1.18) are a family of multifunctional proteins which catalyze conjugation of an electrophilic substrate with GSH. These also function as intracellular binding proteins for a variety of lipophilic compounds. GSTs, especially the π form (GSTπ), have been associated with the development of some tumors and are considered as markers for various malignant tumors (1, 2). A negative correlation between GSTπ expression and the expression of estrogen (ER) and progesterone receptor (PR) has been reported (3, 4). We therefore analyzed the expression of GSTπ, ER, and PR in breast tumor samples and in their respective adjacent normal tissues.

Results

Immunohistochemical staining revealed that 25%, 28%, and 25% of the tumor samples from 120 patients exhibited weak, moderate, and strong GSTπ positivity, respectively, for an overall total of 78% positivity; 68% of the respective adjacent normal tissues showed GSTπ-positive staining. In the tumor samples, the GSTπ expression was mainly localized in the ductus epithelial cells. In the normal samples, the staining was seen at the ducts. Analysis of the relationships between GSTπ expression and different histologic locations exhibited no correlation. The positive rates were 81%, 64%, 88%, and 60% for invasive tumors, invasive lobular cancer, mucous cancer, and marrow-like cancer, respectively. GSTπ-positive rates of ER-positive (ER+) and -negative (ER-) samples were 67.2% and 90.6%, respectively; whereas GSTπ-positive rates in PR-positive (PR+) and -negative (PR-) samples were 68.2% and 88.6%, respectively. The inverse correlation of GSTπ with ER and PR was statistically significant ($p < 0.05$).

DNA hybridization using dot-blot analysis with biotin-labeled GSTπ cDNA probe revealed that, in three out of 32 cases (9.4%), the copies of GSTπ gene were increased more than 2.0-fold compared with the corresponding adjacent normal tissue. RNA hybridization, using the same probe, showed an increased RNA expression in nine out of 32 cases (28.1%), including those three with gene amplification.

Discussion

An inverse relationship between GSTπ expression and ER and PR expression has been reported in breast cancer (3, 4). In this study, we have observed a similar correlation. However, the extent of the inverse correlation in our samples was actually less pronounced than in previous reports from other countries (3, 4). We cannot exclude the possibility that this difference may indicate that the etiology of breast cancer in China differs from that in other countries. In spite of this inverse relationship, GSTπ-positive rates in (ER-) and (PR-) breast cancer were still relatively high, which indicates that this gene could still be used as a marker for these types of breast cancer in China.

The mechanism involved in the increased expression of GSTπ in breast cancer is unclear. Morrow, et al. (5), found that, in (ER-) breast cancer cell lines, GSTπ transcription was unchanged, although the mRNA was highly stable and the stability was not different from that of (ER+) cell lines (5). They suggested that, while the mRNA stability might contribute significantly to the high expression of GSTπ in (ER-) breast cancer, the differential expression of GSTπ in (ER+) and (ER-) cells might be governed by other post-transcriptional processes. In this study, we found a low rate (9.7%) of DNA amplification and a moderate rate (28.1%) of mRNA overexpression, but a very high rate (90.6%) of increase of GSTπ gene protein. These indicate that the increased expression of GSTπ in breast cancer may occur at the DNA and mRNA level, or more likely at the post-transcriptional level.

References

1. Beckett JG, Hayes JD (1993) Glutathione S-transferases: Biomedical applications. Adv Clin Chem 30:281-380.
2. Tsuchida S, Sato K (1992) Glutathione transferases and cancer. Crit. Rev. Biochem Mol Biol 27:337-384.
3. Gilbert L, Elwood LJ, Merino M, et al. (1993) A pilot study of pi-class glutathione S-transferase expression in breast cancer: Correlation with estrogen receptor expression and prognosis in node-negative breast cancer. J Clin Oncol 11:49-58.
4. Peters WH, Roelofs HM, van Putten WL, et al. (1993) Response to adjuvant chemotherapy in primary breast cancer: No correlation with expression of glutathione S-transferases. Brit J Cancer. 68:86-92.
5. Morrow CS, Chiu J, Cowan KH (1992) Posttranscriptional control of glutathione S-transferase pi gene expression in human breast cancer cells. J Biol Chem 267:10544-10550.

p53 and Neu (c-*erb*B-2) Overexpression in Relation to Risk Factors for Breast Cancer

Karin van der Kooy, Matti A. Rookus, Hans L. Peterse, and Flora E. van Leeuwen

Summary

Different types of breast cancer, with and without protein overexpression of neu (c-*erb*B-2) or p53, were studied to determine whether there was an association with known risk factors for breast cancer. Neu and p53 protein expressions were evaluated by immunohistochemistry in primary breast tumor samples from a series of patients aged 20-54 years, who participated in a case-control study. Neu overexpression was shown in 117 out of 597 tumors (19.6%) and p53 overexpression in 142 out of 528 tumors (26.9%). Diagnosis at a young age (≤35 years) was found to be associated with p53 and neu overexpression. Age at first full-term pregnancy (≥29 years) was more strongly related to neu[+] than to neu[-] tumors. Other risk factors were not significantly associated for breast tumors categorized according to neu or p53 protein expression.

Introduction

Many risk factors for breast cancer have been recognized; however, only about 30% of breast cancer cases can be accounted for by known risk factors (1). Therefore, the question arises whether breast cancer is a single disease with one etiology. It may be important to examine whether different types of breast cancer can be recognized that bear a stronger relation to specific risk factors.

Oncogenes and tumor suppressor genes play a significant role in carcinogenesis. Mutations or other aberrations in these genes can reveal clues about the etiology and molecular pathogenesis of multiple cancers. Amplification of the neu oncogene and mutations of the tumor suppressor gene p53 have been found in 10-30% and 20-50% of breast cancers, respectively (2). Our purpose was to investigate whether known risk factors might be strongly associated with overexpression of neu or p53 protein.

Methods

Cases and controls had participated in a population-based case-control study. Information on risk factors was obtained during home interviews. Population controls (n = 918) were used as part of our control group, because overexpression of neu and p53 are generally not seen in nontumorous breast tissue.

Immunohistochemistry was used to determine p53 and neu overexpression in paraffin-embedded, formalin-fixed tumor tissue sections of 528 and 597 primary invasive breast cancer cases, respectively. An immunoperoxidase staining technique was used with monoclonal antibodies, 3B5 (NKI, The Netherlands) for neu (3), and DO7 (DAKO) for p53 (4). For neu expression, staining of tumor cell membranes was scored as neu-positive (neu[+]), and no membrane staining as neu-negative (neu[-]). Dark nuclear staining of more than 10% of the tumor cells was scored as p53-positive (p53[+]), weak nuclear staining or nuclear staining between 1 and 10 % of tumor cells as p53-intermediate (p53[±]), and no staining as p53-negative (p53[-]).

Unconditional multivariate logistic regression analysis was used to estimate relative risks separately for neu[+], neu[-], p53[+] and p53[-] breast cancers. Polytomous logistic regression was used to compare risk factor associations between tumors with and without overexpression.

Results

Neu overexpression was found in 117 tumors (19.6%). p53 overexpression, defined as dark nuclear staining, was found in 142 breast cancer cases (26.9%), while weak and scattered staining was found in 143 cases (27.1%). Relative risks (RRs) of this last group of cases are not shown. The comparison between the p53[+] group versus the p53[-] group is presented here.

Most risk factors studied (biopsy for benign breast disease, family history of breast cancer, age at menarche, body mass index, parity, number of children, age at first full-term pregnancy, and duration of breastfeeding) were not more strongly related to neu[+] than to neu[-], or more strongly related to p53[+] than to p53[-]. However, a higher frequency of neu and p53 overexpression was found in the tumor samples from women younger than 35 years compared with older women (Table 1).

Furthermore, women aged 29 years or older at first full-term pregnancy (FFTP) had a significantly increased risk of breast tumors with neu protein overexpression compared with women who had their first birth before aged 22; however, the association between FFTP and neu[-] tumors was much weaker (Table 2). Prolonged breast-feeding was found to protect against tumors without neu overexpression. However, this association did not differ significantly from the association between breast-feeding and neu[+] tumors. Opposite findings were observed for the associations between breast-feeding and p53[+] and [-] tumors.

Table 1. The Frequency of Breast Tumors with Overexpression of neu[-] or p53 [-] Protein in a Specific Age Range.

Age (yr)	neu				p53			
	[+]		[-]		[+]		[-]	
	n	%	n	%	n	%	n	%
≤ 35	28	23.9	59	12.3	33	23.2	28	11.5
36 - 40	29	22.2	118	24.6	29	20.4	68	28.0
41 - 45	53	30.8	186	38.8	53	37.3	91	37.4
46 - 54	27	23.1	117	24.4	27	19.0	56	23.0
		χ^2 - *p=0.015*				χ^2 - *p=0.014*		

Having breast-fed for 24 weeks or longer decreased the risk of p53[+] tumors, but this association was not significantly different from the association between breast-feeding and p53[-] tumors.

Conclusions

In this study, risk factor patterns were examined for breast cancer categorized according to neu and p53 overexpression of the tumor samples. A significantly inverse association between age and both neu and p53 protein overexpression was found. In addition, increasing age at FFTP was more strongly associated with neu[+] than with neu[-] tumors. For the other risk factors studied, only small, insignificant differences in risk factor associations were found between neu[+] versus neu[-], and p53[+] versus p53[-] tumors. Other, more controversial risk factors, such as oral contraceptive (OC) use and diet, are still under study. The few weak differences in associations found need to be further examined to confirm the associations and to explore feasible mechanisms.

It should be noted that the present study may have some methodological limitations. We examined only one tissue slide per tumor. This may not be representative because of the heterogeneity of tumor tissue. Also, variation in quality and fixation of tumor tissue may have affected the immunohistochemistry results (5). In addition, the quantitation of immunohistochemical staining is troublesome (6). Another problem of a more fundamental character is that overexpression of p53 protein may not be due only to mutations, but also to stabilization of the p53 protein by other cell constituents (7). In addition, immunohistochemistry may also underestimate p53 gene aberrations in case of deletion of the total p53 gene or of mutations resulting in a stop codon (6).

Table 2. Risk of neu[+] versus neu[-] and p53[+] versus p53[-] Breast Cancers by Selective Risk Factors.

Variable	Controls n	neu[+] cases n	RR[1]	95%CI[1]	neu[-] cases n	RR[1]	95%CI[1]	*p*[2]
Age at first full-term pregnancy								
≤ 22 (years)	208	39	1.0	-	96	1.0	-	-
23 - 25	289	35	1.4	0.8-2.6	131	1.0	0.7-1.4	0.34
26 - 28	194	18	1.1	0.5-2.2	107	1.3	0.9-1.9	0.77
≥ 29	110	25	3.6	1.7-7.3	61	1.3	0.9-2.1	0.02
trend[3]				*$p < 0.01$*			*$p = 0.21$*	*0.06*
Breast-feeding								
0 (weeks)	187	24	1.0	-	99	1.0	-	-
1 - 6	186	26	1.2	0.6-2.2	121	1.2	0.9-1.8	0.68
7 - 24	216	19	0.7	0.4-1.4	96	0.8	0.5-1.1	0.68
> 24	212	29	1.2	0.6-2.3	79	0.7	0.5-1.1	0.18
trend[3]				*$p = 0.90$*			*$p = 0.02$*	*0.14*
Breast-feeding								
0 (weeks)	187	31	1.0	-	43	1.0	-	-
1 - 6	186	36	1.2	0.7-2.0	55	1.3	0.8-2.0	0.70
7 - 24	216	30	0.8	0.5-1.5	51	1.0	0.6-1.7	0.59
> 24	212	20	0.6	0.3-1.0	53	1.2	0.8-2.0	0.06
trend[3]				*$p = 0.09$*			*$p = 0.51$*	*0.37*

1 RR and 95% 95%CI adjusted for age (continuous), family history of breast cancer, biopsy for benign breast disease, education, smoking, age at menarche, parity, duration of breast-feeding, number of children, age at first full-term pregnancy, and OC use.

2 Resulting from polytomous logistic regression.

3 Test for trend, using continuous variables.

Particularly for p53, it is necessary to investigate the relationship between variations in protein overexpression and mutational spectra of the p53 gene. It is not clear whether weak and scattered p53 protein overexpression, which was found in 27% of our breast tumors, is caused by mutations of the p53 gene (8). Finally, p53 mutations as well as neu amplification might be late steps in the progression of advanced tumors and thus, may not be related to early, initiation, or promotion events. This should be the subject of further study.

References

1. Kelsey JL. (1993) Breast cancer epidemiology: Summary and future directions. Epidemiol Rev 15:256-63.
2. Van de Vijver MJ (1993) Molecular genetic changes in human breast cancer. Adv Cancer Res 61:25-56.
3. Van de Vijver, et al. (1988) Neu-protein overexpression in breast cancer. N Engl J Med 319:1239-45.
4. Cattoretti G, et al. (1993) Antigen unmasking on formalin-fixed paraffin-embedded tissue sections. J Path 171:83-98.
5. Fisher CJ, et al. (1994) Problems with p53 immunohistochemical staining. Brit J Cancer 69:26-31.
6. Hall PA, Lane DP (1994) p53 in tumour pathology. J Path 172:1-4.
7. Pietenpol JA, Vogelstein B (1993) No room at the p53 inn. Nature 365:17-8.
8. Jacquemier J, et al. (1994) P53 immunohistochemical analysis in breast cancer with four monoclonal antibodies. Brit J Cancer 69:846-52.

Cost Effectiveness of Hormone Replacement Therapy in French Postmenopausal Women

Geneviève Plu-Bureau, Dominique Bureau, and
Jean-Christophe Thalabard

Summary

The beneficial effects of hormone replacement therapy (HRT) in preventing osteoporosis and, reportedly, ischemic heart disease gives support to its extension to symptom-free women. The purpose of our study is a reevaluation of the risks, benefits, and costs of HRT primary prevention in the French context. In addition to the methodology of benefit analysis, we used two methodological tools, Markovian cohort simulations and meta-analyses. We took into account potential long- term effects, and particularly possible increase in breast cancer risk. The life expectancy gain was evaluated to 0.16 year for a 10-year HRT duration, and the discounted cost/year gained from HRT is FF 143,000 (US $28,600). Our study led to an intermediate cost effectiveness ratio as compared with other accepted health care interventions. However, prevention might be cost effectiveness in specific subpopulations. The study underlines the lack of data about the net effect of combined HRT and the persistence of the effect of HRT after discontinuation.

Introduction

Hormone replacement therapy (HRT) was initially prescribed in postmenopausal women suffering from functional symptoms such as hot flushes, irritable mood, and genito-urinary disorders, that usually follow the cessation of the ovarian estrogen secretion. HRT has been recognized now in the general population of menopausal women for primary prevention of osteoporosis and, reportedly, ischemic heart diseases.

The purpose of the study was to evaluate the risks, benefits and costs associated with HRT prescription in symptom-free nonhysterectomized French women. The french context is characterized; 1) by specific HRT and health care costs; 2) a slightly lower risk of coronary heart disease in postmenopausal women as compared to that of other westernized countries; and finally, 3) the choice of a combined estrogen-progestin treatment to suppress an increased risk of endometrial cancer associated with unopposed therapies.

Methods

The basic methodology of the study is a cost effectiveness analysis. As we focus on symptom-free women, the evaluation has mainly to balance the gain in life expectancy in HRT treated versus untreated women, with the cost of combined HRT implementation, and net of expected health care economic savings due to the risk modifications. As the corresponding gains and costs occur on different time horizons, the cost effectiveness ratio discounts these net changes.

In addition to the methodology of analysis, we used two methodological tools, like the Markov cohort simulation approach and the meta-analysis methodology (1). In the untreated cohort, the incidence and mortality rates were calibrated using the French registry data. In the treated cohort, estimations of the changes in incidence strictly relied upon meta-analyses of published epidemiological studies (Table 1). Furthermore, as the majority of the studies suggested a protective effect of HRT on the cardiovascular system limited to the period of exposure, this hypothesis was adopted in the model.

Results

A ten-year HRT was associated with a 0.16 year improvement in expected life expectancy. When the HRT was extended to 15 years, taking into account the possible increase in breast cancer risk lowered the total gain in life expectancy. To this result, the contributions of the different changes in incidence rates to this result are detailed in Figure 1.

Discussion

As the gain in life expectancy seems rather limited, our results contrast with the conclusions of many recent reports. However, our data are in accordance with the study by the Oxford Hormone Therapy Group (2), which concluded that HRT in hysterectomized women or in women with severe functional symptoms was highly advantageous, while its prescription in asymptomatic British women was economically less favorable. Analytical and epidemiological methods used to evaluate the cost effectiveness ratios differed between the studies. The Markovian approach, previously used in this context (3), appeared particularly hypotheses of the simulations between conflicting or selected epidemiological results.

Table 1. Relative Risks Hypotheses according to the HRT implicated pathologies.

Risks	Combined Relative Risk	Confidence Interval (95%)	Hypothesis
Coronary heart disease	0.69	(0.60 - 0.79)	Current users
Hip fracture	0.53	(0.44 - 0.64)	Duration of use > 5 years
Breast cancer	1.23	(1.07 - 1.42)	Duration of use > 12 years

Although cost-effectiveness analysis appears to be a very efficient tool to summarize the net impact of a preventive therapy with various targets , we have to admit our limited knowledge regarding various assumptions necessary to validate the consequences of HRT, i.e., the different effects of a combined treatment, the impact of HRT on specific risk subpopulations, the duration of the residual effect of HRT, and its dependance on the exposure duration.

Figure 1. Decomposition of the 15-year HRT simulation.

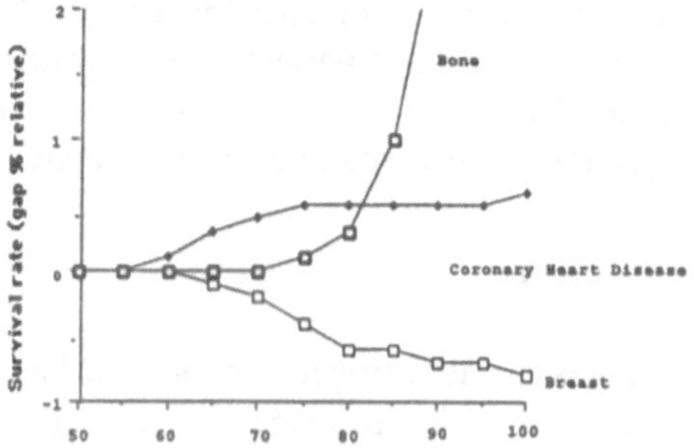

References

1. Bureau D, Plu-Bureau G, Thalabard JC (1993) L'évaluation des traitements médicaux: Méthodes et enjeux. Economie et Statistiques 6:65-75.
2. Daly E, Roche M, Barlow D, et al. (1992) An analysis of benefits, risks and costs. Brit Med Bull 48:368-400.
3. Tosteson AN, Rosenthal DI, Melton J, et al. (1990) Cost effectiveness of screening preimenopausal white women for osteoporosis: Bone densitometry and hormone replacement therapy. Ann Int Med 113: 594-604.

Tamoxifen Treatment Does Not Induce Endometrial Hyperplasia in Postmenopausal Women

Philippe Touraine, Pierre A. Driguez, I. Cartier, H. Yaneva, F. Kuttenn, and Pierre Mauvais-Jarvis

Summary

Tamoxifen-induced histological changes in the endometrium of postmenopausal breast cancer patients were determined in 40 women who had undergone hysteroscopy examinations. Endometrial samples were obtained from 15 of these women for histological examination. With hysteroscopy, the endometrium showed very specific changes combining diffuse epithelial atrophy and round protuberances due to cystic glands. Histological examination confirmed the atrophy of the luminal and glandular epithelia, coexisting with large, dilated cystic glands. The stroma was dense and rich in collagen fibers associated with edematous areas. Polyps were present in 17 patients. The endometrial thickening observed in ultrasonography has often been interpreted as "endometrial hyperplasia," whereas the endometrium of postmenopausal patients treated with tamoxifen showed "cystic glandular atrophy." The cystic glands account for the thickened appearance observed in ultrasonography.

Introduction

Tamoxifen is an anti-estrogen currently used as an adjuvant treatment in patients with breast cancer (1). However, several adverse side effects have been reported, especially the induction of "endometrial hyperplasia" (2), and an increased risk of endometrial carcinoma has been observed (3). Indeed, like many antihormones, tamoxifen has both antagonistic and agonistic effects depending on the species, the type of tissue analyzed, the dose used, and the presence or absence of estrogen secretion. Tamoxifen has an agonist estrogenic effect on the endometrium, in contrast to its anti-estrogenic effect on the mammary gland. Herein, we report the results of hysteroscopy in 40 tamoxifen-treated patients.

All had endometrial atrophy, which was associated in 50% of the cases with the presence of multiple, large, cystic glands.

Results

Hysteroscopy showed the same pattern of endometrial atrophy in all 40 patients. The endometrium was smooth, white, and brilliant. In addition, in 20 of these patients, there were multiple scattered protuberances. Twenty-six polyps (diameter > 15 mm) were noted in 17 patients. Whenever a biopsy could be performed, the atrophy of the luminal and glandular epithelium was confirmed. The luminal epithelium was very thin and unstratified. The glands were cystic with a flattened epithelium and little or no secretion. Mitoses were rare. In some cases, the endometrium was more than 10 mm thick due to the cystic transformation of the endometrium. The most intriguing result was the presence of these large, dilated cystic glands with an inactive epithelium (Figure 1).

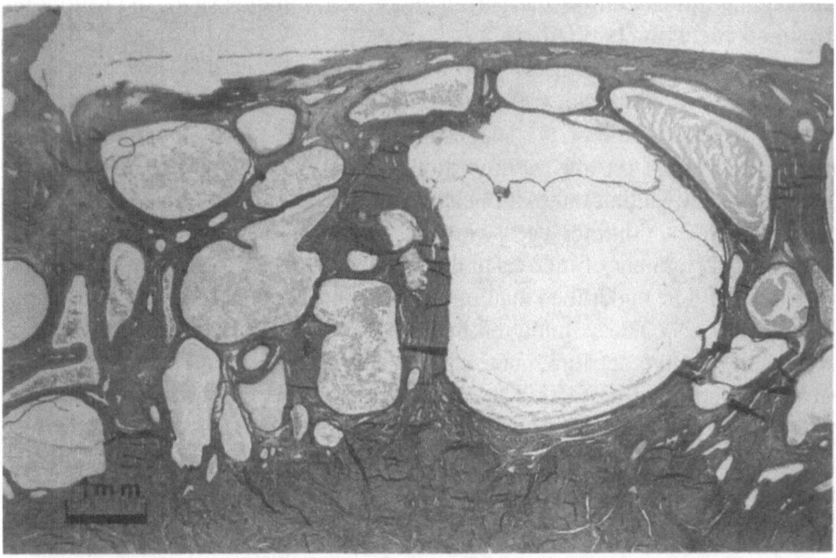

Figure 1. Histologic appearance of the endometrium. Dilated cystic glands are spread throughout the whole depth of the endometrium. No hyperplastic epithelium is was visible.

The appearance of the endometrium on tamoxifen can be described as "cystic glandular atrophy" rather than "endometrial hyperplasia." In contrast, in epithelial atrophy, the stroma was always dense and rich in collagen fibers, associated with edematous areas. Blood vessels were numerous and of variable size.

Discussion

We observed endometrial atrophy in all postmenopausal, tamoxifen-treated patients. Fifty percent of the patients had areas of atrophy with areas of cystic, dilated glands, in a "Swiss cheese pattern." The large majority of patients on tamoxifen treatment show endometrial atrophy associated with cystic glands. This observation differs from a previous report (4) based on ultrasound examination of the "endometrial hyperplasia" (5). The presence of large, dilated cystic glands (Figure 1) may be responsible for the thickened appearance of the mucosa on ultrasound examination. The edematous stroma may also contribute to this observation. Some authors have suggested that tamoxifen has a direct effect on the myometrium (6). Finally, the "endometrial hyperplasia" observed on ultrasound may be reinforced by the presence of polyps which on ultrasound detection cannot always be clearly distinguished from the endometrium. Therefore, it seems that hysteroscopy is a more accurate maneuver than transvaginal ultrasound for measuring endometrial thickness for the follow-up of patients receiving tamoxifen.

The effect of tamoxifen on the endometrium is still not very well understood. In the genital tract, tamoxifen is largely considered an estrogen agonist. Several studies have suggested that, in the uterus, tamoxifen has a proliferative effect on the endometrium *in vivo* (7) and *in vitro* (8). Animal studies have shown that tamoxifen causes an increase in the total weight of the uterus. However, this increased weight is due to tissue edema rather than an increase in the number of mitoses in the endometrium (9). Regarding the mitotic index, it should be underlined that, in the endometrial epithelium, anti-estrogens induce a much weaker stimulation than estrogens, whereas in the endometrial stroma and the myometrium, anti-estrogen stimulation is marked and as strong as estrogen stimulation (10). The simultaneous presence of atrophic mucosa, cystic glands, and a dense, edematous stroma could be due to a dissociated effect of tamoxifen, which would vary according to the endometrial area. It is likely that tamoxifen has little or no estrogenic action on the epithelium, since we observed only marked atrophy of the luminal epithelium with rare mitoses. It is more difficult to understand the presence of large dystrophic cystic glands. It may exert weak estrogenic stimulation on glandular epithelium secretion, thus increasing the volume of the fluid inside the glands and therefore their size. This contrasts with the absence of any stimulatory effect on the growth of the epithelium, which remains thin and atrophic. Finally, the thick and edematous aspect of the stroma could be due to an estrogenic effect of tamoxifen on this area of the endometrium.

In conclusion, tamoxifen on the human endometrium seems to involve more than the accepted description of a simple estrogenic effect inducing hyperplasia.

References

1. Early Breast Cancer Trialists' Collaborative Group. Systemic treatment of early breast cancer by hormonal, cytotoxic or immune therapy. Lancet (1992) 339:1-15, 71-85.
2. Neven P, De Muylder X, Van Belle Y, et al. (1990) Hysteroscopic follow-up during tamoxifen treatment. Eur J Obstet Gynecol 35:235-238.
3. Fornander T, Cedermark B, Mattsson A, et al. (1989) Adjuvant tamoxifen in early breast cancer: Occurrence of new primary cancers. Lancet 1:117-119.
4. Cohen I, Rosen DJD, Shapira J, et al. (1994) Endometrial changes with tamoxifen: Comparison between tamoxifen-treated and non-treated asymptomatic, postmenopausal breast cancer patients. Gynecol Oncol 52:185-190.
5. Lahti E, Blanco G, Kauppila A, et al. (1993) Endometrial changes in postmenopausal breast cancer patients receiving tamoxifen. Obstet Gynecol 81:660-664.
6. Anteby E, Yagel S, Zacut D, et al. (1992) False sonographic appearance of endometrial neoplasia in postmenopausal women treated with tamoxifen. Lancet 340:433-434.
7. Gottardis MM, Robinson SP, Satyaswaroop PG, Jordan VC (1988) Contrasting actions of tamoxifen on endometrial and breast tumor growth in the athymic mouse. Cancer Res 48:812-815.
8. Anzai Y, Holinka CF, Kuramoto H, Gurpide E (1989) Stimulatory effects of 4-hydroxytamoxifen on proliferation of human endometrial adenocarcinoma cells (Ishikawa line). Cancer Res 49:2362-2365.
9. Dix CJ, Jordan VC (1980) Subcellular effects of monohydroxytamoxifen in the rat uterus: Steroid receptors and mitosis. J Endocr 85:393-404.
10. Mukku VR, Kirkland JL, Hardy M, Stancel GM (1981) Stimulatory and inhibitory effects of estrogen and antiestrogen on uterine cell division. Endocrinology 109:1005-1010.

Uterine Leiomyomas: A Model to Study Hormonal Regulation of Growth Related Genes

Agneta Blanck, Inger Gustafsson, Katarina Englund, Peter Sjöblom, and Bo Lindblom

Summary

The mRNA expression of the progesterone receptor (PR) and of insulin-like growth factor I (IGF-I) was measured in leiomyomas and in myometrium from women with and without preoperative treatment with an agonist to gonadotropin-releasing hormone (GnRHa) for three months. Both the PR and IGF-I were down-regulated in myomas from GnRHa-treated women, whereas the decrease in myometrium was significant only for IGF-I.

Introduction

Leiomyomas are the most common uterine tumors and constitute a major health problem in women in their fourth or fifth decades of life. Although benign myomas are a common cause of bleeding disturbances and/or cause of infertility. Myoma growth is dependent on gonadal hormones, they stop growing and usually decrease in size after menopause. In fertile women, their size can be reduced by treatment with an agonist to gonadotropin-releasing hormone (GnRHa), which is known to decrease the serum levels of both estrogen and progesterone (1). Also, treatment with antiprogestins has been shown to induce regression of myomas (2). The mechanisms behind the relative growth advantage of myomas compared with normal myometrium have not been investigated in detail. A hormonal influence is clear, and some investigations have supported the hypothesis that both hormone receptors, as well as growth factors and their receptors might be of importance for the differences in growth regulation (3, 4). The present study was designed to investigate mRNA-expression of the progesterone receptor (PR) and of insulin-like growth factor I (IGF-I) in myomas, and normal myometrium from women with or without preoperative GnRHa treatment.

Materials and Methods

Myomas and myometrial tissue were collected at hysterectomy or enucleation of myomas from women preoperatively treated with a GnRHa (Goserelin; Zoladex[R]) for three months, and from women without such pretreatment. The myomas were enucleated either by endoscopic surgery, or at laparotomy. The possible locations of myomas are presented in Figure 1. Based on information about the menstrual cycle pattern and histopathological examination of the endometrium, the tissue from women not receiving GnRHa was subclassified according to the phase of the menstrual cycle (proliferative or secretory). Total nucleic acids (TNA) were prepared, and mRNA levels of the PR and IGF-I were measured by solution hybridization (5) technique. Complementary RNA (cRNA) probes, labeled with [^{35}S]-UTP were synthesized using a riboprobe system (Promega, Falkenberg, Sweden). An 800-bp fragment of the gene coding for the PR, and a 775-bp stretch of the IGF-I gene were used as templates for probe synthesis.

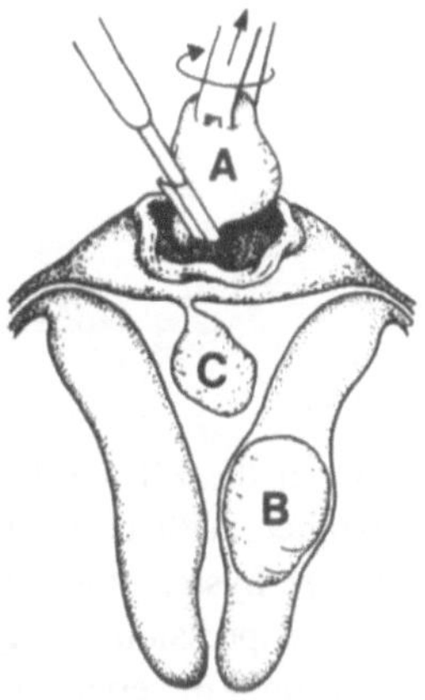

Figure 1. Uterus with myomas.
A. Subserosal
B. Intramural
C. Submucosal

Results

Myomas from GnRH a pretreated individuals exhibited a lower mRNA expression of both the PR and IGF-I genes, whereas a significant decrease in the myometrium was seen only for IGF-I (Figure 2). IGF-I expression in myomas from women not receiving GnRHa was significantly higher than in the corresponding myometrium. Preliminary data indicate that expression of the PR in normal myometrium is dependent on alterations in hormonal status during the menstrual cycle, whereas no such dependency seemed to be present for the IGF-I gene (data not shown). The average mRNA level of the PR in myometrium during the secretory phase was only 37% of the level during the proliferative phase (n = 7 and n = 3, respectively).

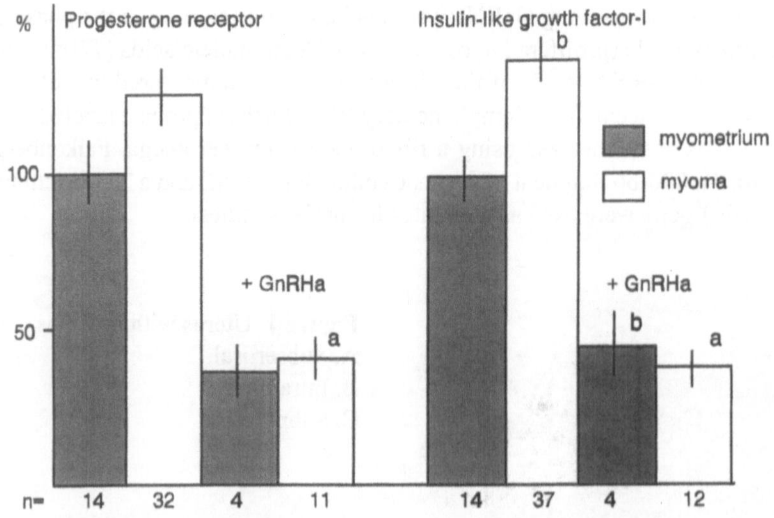

Figure 2. mRNA expression (cpm/μg TNA) of the PR and the IGF-I genes in normal myometrium and myomas from women with or without preoperative treatment with GnRHa for three months before hysterectomy or enucleation of myomas. Data are shown as percent of the expression in myometrium from untreated women (average ± SEM). A significantly different from myomas from untreated women ($p < 0.05$; Wilcoxon rank sum test). [b]Significantly different from normal myometrium from untreated women.

Discussion

The present study demonstrates a down-regulation of IGF-I mRNA expression in both myomas and normal myometrium following GnRHa therapy. These findings are in agreement with the findings by Giudice, et al. (4), who reported markedly decreased IGF-I mRNA levels in leiomyomas from women pretreated with GnRHa. Slightly higher mRNA levels of IGF-I were observed in myomas than in myometrium from women not receiving preoperative GnRHa. This difference might contribute to a comparative growth advantage of myoma cells compared with the surrounding myometrium.

A hormonal influence was demonstrated for the PR, with decreased expression in myomas from GnRHa-treated patients and a variability of expression in normal myometrium related to the phase in the menstrual cycle. No significant differences between expression of the PR in myometrium and myomas were observed. This is somewhat surprising in view of the reports by Brandon, et al. (6), demonstrating higher levels of PR expression at both the mRNA and protein levels. Quantification of the receptor levels by enzyme immunoassay is in progress.

These findings represent a step in characterizing genes of possible importance for the endocrine regulation of myoma growth. Further studies on gene expression of hormone receptors, protooncogenes, and growth factors and their receptors will serve as a basis to clarify mechanisms whereby steroid hormones regulate growth of benign as well as malignant uterine tumors.

References

1. Friedman AJ, Barbieri RL, Benacerraf B, Schiff I (1987) Treatment of leiomyomata with intranasal or subcutaneous leuprolide, a gonadotropin-releasing hormone agonist. Fertil Steril 48:560-564.
2. Alvarez Murphy A, Kettel LM, Morales AJ, et al. (1993) Regression of uterine leiomyomata in response to the antiprogesterone RU 486. J Clin Endocrinol Metab 76:513-517.
3. Koutsilieris M (1992) Pathophysiology of human leiomyomata. Biochem Cell Biol 70:273-278.
4. Giudice LC, Irwin JC, Dsupin BA, et al. (1993) Insulinlike growth factor (IGF), IGF binding protein (IGFBP), and IGF receptor gene expression and IGFBP synthesis in human uterine leiomyomata. Human Reprod 8:1796-1806.
5. Durnam DM, Palmiter RP (1983) A practical approach for quantifying specific mRNA by solution hybridization. Anal Biochem 131:385-393.
6. Brandon DD, Bethea CL, Strawn EY, et al. (1993) Progesterone receptor messenger ribonucleic acid and protein are overexpressed in human uterine leiomyomas. Am J Obstet Gynecol 169:78-85.

Transforming Growth Factor ß1 Facilitates Cell Proliferation and Invasion in MAT-LyLu Prostatic Cancer Cells

Ching-Jey G. Chang, YunFu Hu, Yasuro Sugimoto, WilliamChang, Samuel Kulp, Robert W. Brueggemeier, and Young C. Lin

Summary

Murine MAT-LyLu cell line is a fast-growing, androgen-independent, highly metastatic and transplantable prostatic carcinoma cell line. Within six hours of treatment under serum-free culture conditions, TGF ß1 is more mitogenic than bovine serum albumin (BSA, 10 ng/ml, control) for MAT-LyLu cells by recruiting more MAT-LyLu cells from the G_0/G_1 phase into the S-phase of the cell cycle. TGF ß1 also facilitates the invasion process within 24 hours of treatment as indicated by the *in-vitro* invasion assay using conditioned media from prostatic stromal cells as an attractant. Our results indicate that TGF ß1 represents a potent mitogen and a differentiation inducer for MAT-LyLu cells.

Introduction

The prostatic epithelium, from which prostatic carcinoma is derived, is under the constant surveillance of various types of systemic and local factors which influence cellular proliferation, differentiation, and biological behavior. The murine MAT-LyLu cell line (a subline of R3327 line), derived from prostatic glandular epithelium (1), is a fast-growing, androgen-independent, locally invasive, highly metastatic, and transplantable prostatic cancer cell line (2). Because of its rapid metastatic capability to regional lymph nodes and lung, the MAT-LyLu cell line provides one of the few models to investigate human prostatic malignancies. Routine *in-vitro* propagation of MAT-LyLu cells requires supplementation with 10% fetal bovine serum (FBS) (3), and depletion of FBS leads to cell arrest within 2 hours (unpublished data). It has been suggested that TGF ß1 is secreted from MAT-LyLu cells (5), and this expression is evidently facilitated by TGF ß1 itself, apparently acting as an autocrine factor. The expression level of TGF ß1 under TGF ß1 treatment was comparable to that under treatment with 10% FBS (unpublished data). The current study was aimed at investigating the functional impact of TGF ß1 on MAT-LyLu cells, with emphasis on cell proliferation and invasion.

Methods

The methods employed in these studies included: thymidine incorporation to determine cell proliferation rate; flow cytometry to monitor the cell cycle; and an *in-vitro* invasion assay, to monitor the invasiveness of MAT-LyLu cells after selected treatments. The results were analyzed using Student t-test. A significance level of 0.05 was chosen empirically.

Results

As determined by the [^{3}H]-thymidine incorporation, TGF ß1 is more mitogenic in MAT-LyLu cells than in control samples (Figure 1).

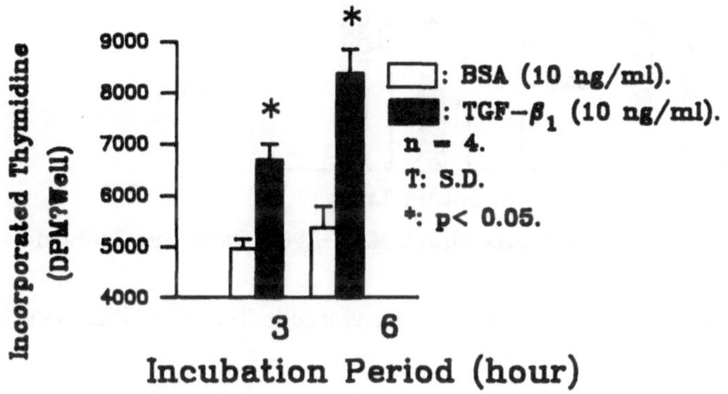

Figure 1. Thymidine incorporation of MAT-LyLu cells. TGF ß1 is more mitogenic than BSA for MAT-LyLu cells as determined by thymidine incorporation method.

TGF ß1 recruits more Mat-LyLu cells into S-Phase than control, as shown in Figure 2.

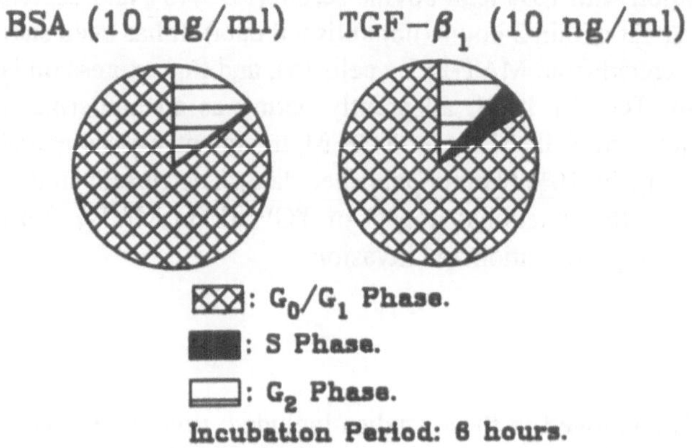

Figure 2. Distribution of MAT-LyLu cells in cell cycle. Serum-free culture conditions, TGF ß1, 10 ng/ml, recruits more MAT-LyLu cells from the G_0/G_1 phase into the S-phase of the cell cycle within six hr of treatment than BSA, 10 ng/ml.

TGF ß1 facilitates a higher degree of invasiveness than control samples, as depicted in Figure 3.

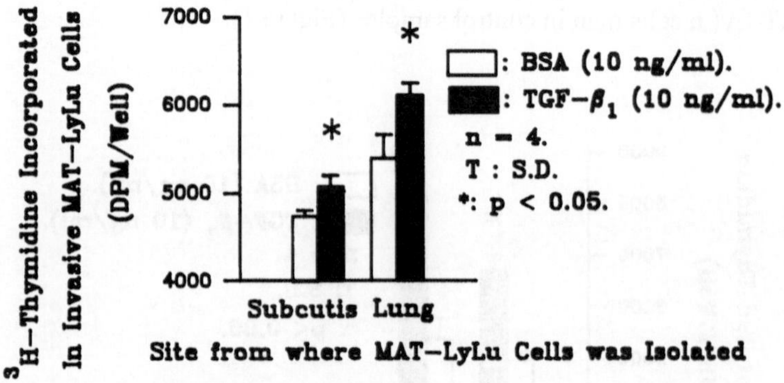

Figure 3. *In-vitro* invasiveness of MAT-LyLu cells. TGF ß1 facilitates the invasion process within 24 hs of treatment.

Discussion and Conclusions

TGF ß1 is a 25-kDa dimeric polypeptide, and is referred to as a prototypical epithelial cell inhibitor. Our results indicate that, for MAT-LyLu prostatic cancer cells, TGF ß1 represents a potent mitogen and differentiation inducer which leads to a highly invasive nature preceding the metastatic process. We found that stromal cells from prostate and lung inhibit MAT-LyLu cell proliferation, while only FSH and LH at physiological concentrations could enhance MAT-LyLu cell replication in serum-free culture conditions. Evans, et al. (7), claimed that an autocrine motility factor (AMF), of molecular weight less than 30 kDa, was secreted by Dunning R-3327 AT-2.1 tumor cells (a closely related subline) and stimulated motility of AT-2.1 cells. Based on our results, which indicate chemotactic and chemokinetic characteristics for TGF $ß_1$, this factor may very likely be TGF ß1. Our results indicate that TGF ß1 represents not only a potent mitogen but also a differentiation inducer for MAT-LyLu prostatic cancer cells.

Acknowledgment

This work was supported by grants from the National Institute of Health, P30CA16058 and DK45916.

References

1. Feitz WFJ, Debruyne FMJ, Vooijs GP, et al. (1986) Intermediate filament proteins as tissue specific markers in normal and malignant urological tissues. J Urol 136:922-931.
2. Isaacs JT, Yu GW, Coffey DS (1981) The characterization of a newly identified, highly metastatic variety of Dunning R 3327 rat prostatic adenocarcinoma system: The MAT LyLu tumor. Invest Urol 19:20-23.
3. Isaacs JT, Isaacs WB, Feitz WFJ, Scheres J (1986) Establishment and characterization of seven Dunning rat prostatic cancer cell lines and their use in developing methods for predicting metastatic abilities of prostatic cancers. Prostate 9:261-281.
4. Steiner MS and Barrack ER (1990) Expression of transforming growth factors and epidermal growth factor in normal and malignant rat prostate. J Urol 143:240A.
5. Krishan A (1975) Rapid flow cytofluorometric analysis of mammalian cell cycle by propidium iodide staining. J Cell Biol 66:188-193.
6. Albini A, Iwamoto Y, Kleinman HK, et al. (1987) A rapid in vitro assay for quantitating the invasive potential of tumor cells. Cancer Res 47:3229-3245.
7. Evans CP, Walsh DS, Kohn EC (1991) An autocrine motility factor secreted by the Dunning R-3327 rat prostatic adenocarcinoma cell subtype AT2.1. Int J Cancer 49:109-113.

Quantitative Effects of Antiandrogen Therapy on High-Grade Prostatic Intraepithelial Neoplasia in Radical Prostatectomy Specimens

Fernando U. Garcia, Kevin L. Bashman, and Mark S. Austenfeld

Summary

High-grade prostatic intraepithelial neoplasia (HGPIN) is considered to be the premalignant lesion of prostatic adenocarcinoma. This study was performed to evaluate the quantitative effects of antiandrogen (AA) therapy on HGPIN. Glass slides from totally embedded radical prostatectomy specimens from ten patients treated with leuprolidine and/or flutamide prior to surgery were examined for area of HGPIN, and compared with those of ten patients who had received no therapy. Patients from each group had comparable serum PSA levels, age, and histologic tumor grades. Only HGPIN located one half low-power field or more from the tumor were measured. The total area of HGPIN from each case was then expressed as the percentage of the total area of prostate examined. Our results showed that, in the treated group ($n = 10$), the mean area involved by HGPIN was $0.034 \pm 0.028\%$ of the total area examined, and $0.087 \pm 0.061\%$, $p < 0.038$, in the untreated group ($n = 10$). This retrospective study suggests that HGPIN may be reversed with AA therapy.

Introduction

High-grade prostatic intraepithelial neoplasia (HGPIN) is a proliferative lesion considered to be the premalignant phase of prostatic adenocarcinoma (1, 2). HGPIN is found in the peripheral zone, at the border of the neoplasm (3). The histological features of HGPIN have been described elsewhere (1, 2). Briefly, they include cellular crowding and stratification, cellular and nuclear pleomorphism, hyperchromasia, and prominent nucleoli. The cytologic features of HGPIN are indistinguishable from invasive cancer of the prostate; however, in specimens showing HGPIN, there is preservation of at least part of the basal cell layer in the acini structures (4). Although the clinical behavior of this lesion has not been determined, progression to carcinoma *in situ* and invasive carcinoma is believed to be part of its natural history.

In the past, AA therapy was used as palliative for high-stage prostatic cancer (5). Recently, however, it has been used to downstage clinical T2/3 prostatic cancer because AA therapy has been shown to reduce tumor volume (6). The changes induced by AA therapy on prostatic adenocarcinoma and normal prostate glands have been the subject of recent studies (7, 8). However, the morphological changes induced by AA therapy on HGPIN have not been fully characterized. In a recent study of 23 patients with prostate cancer receiving AA therapy (8), HGPIN was reported in only 13% of the patients treated (three cases). This finding suggests a marked reduction in the frequency of this lesion, since HGPIN is usually present in about 73% of untreated patients with prostatic adenocarcinoma (9). Herein ,we examined the presence of HGPIN in totally embedded prostate specimens obtained after radical prostatectomy from twelve prostatic adenocarcinoma patients receiving AA therapy.

Materials and Methods

Prostate specimens from twelve patients with confirmed prostatic adenocarcinoma underwent preoperative androgen blockade treatment with leuprolide and/or flutamide. Prostate specimens from twelve other patients, confirmed prostatic adenocarcinoma but receiving no preoperative treatment, were used as controls. Preoperative tumor biopsies and PSA levels were obtained for all the patients studied.

Measurements of the total area of the prostate, the carcinoma, and the HGPIN areas were scored using a microscope mounted with a camera. The areas containing the HGPIN were outlined as described by Bostwick (1).

Selected paraffin blocks from six treated patients and from six control patients were stained by an immunoperoxidase method (10) with a monoclonal mouse antibody 34βE12 (DAKO) and stained by a peroxidase technique.

The data were analyzed using standard t test, and simple regression was determined by statistical software from Statworks, Cricket Software, Inc.

Results

Radical prostatectomy specimens from androgen blockade treated and control groups were reviewed. The average weight of the prostates in the AA treated group was 46.4 ± 5.4 g. The average weight of the prostates in the control group was 42.5 ± 4.8 g. The Gleason scores were compared with those of the diagnostic needle biopsies. Overall, there were no significant changes in Gleason score for the neoplasm between the biopsy and the prostatectomy for each group (Tables 1 and 2). However, the correlation between the Gleason score given at biopsy and the Gleason score given from the prostatectomy specimen was significantly different from the control group (r = 0.63) and that of the AA-treated group (r = 0.36).

Prostatectomy specimens in both groups were divided into stage A+B and stage C+D and analyzed for the proportion of prostate area occupied by the carcinoma. The mean % of tumor area for stage A+B was 3.38 ± 0.15 (n = 9)

Table 1. Patients Receiving Androgen Blockade Treatment. RRP (retropubic radical prostatectomy). The Gleason score indicates the primary pattern followed by the secondary pattern and the sum. Note the different AA regimen therapies. The staging protocol used is the modified Whitmore and Jewett. The stage presented is the pathological stage after prostatectomy. The stage D patient is a D1 with focal regional lymph node metastases.

PATIENT	AGE	BIOPSY GLEASON SCORE	RRP GLEASON SCORE	ANDROGEN BLOCKADE TREATMENT	%HGPIN OF TOTAL PROSTATE AREA	% TUMOR OF TOTAL PROSTATE AREA	STAGE
1	60 y	2,4=6	2,2=4	Flutamide 250 mg tid/2 months Leuprolide 7.5	0.0	.153	A
2	72 y	3,5=8	3,4=7	Flutamide 250 mg tid/1 month	0.0192	8	B
3	76 y	3,3=6	3,3=6	Flutamide 250 mg tid/2 months Leuprolide 7.5 mg/month/2 months	0.0234	1.003	B
4	68 y	2,2=4	2,3=5	Leuprolide 7.5 mg/ 1 month	0.0360	5.645	B
5	57 y	3,2=5	3,4=7	Leuprolide 7.5 mg/ 1 month	0.0050	5.9	B
6	51 y	2,3=5	3,4=7	Flutamide 250 mg tid/2 months Leuprolide 7.5 mg/month/2 months	0.0142	8	B
7	78 y	2,2=4	2,4=6	Leuprolide 7.5 mg/month/6 months	0.0	.070	B
8	60 y	2,2=4	2,4=6	Leuprolide 7.5 mg/month/6 months	0.0029	.328	B
9	64 y	3,3=6	3,3=6	Leuprolide 7.5 mg/month/3 months	0.0115	1.019	C
10	53 y	3,4=7	4,3=7	Flutamide 250 mg tid/1 month	0.0850	19	C
11	62 y	3,4=7	3,4=7	Flutamide 250 mg tid/2 months	0.0702	16	D
12	68 y	3,3=6	3,4=7	Flutamide 250 mg tid/2 months Leuprolide 7.5 mg/month/2 months	0.0636	.788	D

Table 2. Controls. Similarly, the stage denotes the pathological stage after prostatectomy.

PATIENT	AGE	BIOPSY GLEASON SCORE	RAP GLEASON SCORE	%HGPIN OF TOTAL PROSTATE AREA	% TUMOR OF TOTAL PROSTATE AREA	STAGE
1	52 y	2,4=4	2,2=4	0.0778	.38	A
2	67 y	2,2=4	2,4=6	0.1237	.25	A
3	51 y	2,2=4	2,3=5	0.173	6.50	B
4	70 y	2,2=4	2,3=5	0.026	2.75	B
5	63 y	2,2=4	2,3=5	0.0453	1.57	B
6	75 y	2,2=4	3,2=5	0.2085	2.90	B
7	50 y	3,3=6	3,3=6	0.1486	5.20	B
8	71 y	3,4=7	3,4=7	0.0498	.68	B
9	66 y	5,3=8	3,4=7	0.0412	5.42	B
10	59 y	2,3=5	3 4=7	0 0543	4 12	B
11	70 y	4,4=8	2,4=6	0.0438	19.90	C
12	59 y	2,3=5	4,3=7	0.0358	18.89	D

and 2.90 ± 0.71 (n = 10), p ≤ 0.765, in the treated and control groups, respectively. The mean % of tumor area for stage C+D was 11.93 ± 5.64 (n = 3) and 19.3 ± 0.51 (n = 2), p≤ 0.38, in the treated and control groups, respectively. Therefore, no significant differences were found in the amount of tumor present within the prostatectomy specimens in either the treated or control groups, regardless of the stage.

The portion of total prostate area occupied by HGPIN for all stages expressed as mean % in the treated and control groups respectively was 0.028 ± 0.009 (n = 12), p ≤ 0.008 (Figure 1). When only stage A+B were compared for the proportion of HGPIN in the treated and control groups respectively, the mean % was 0.012 ± 0.004 (n = 9) and 0.095 ± 0.020 (n = 10), p ≤ 0.001.

Histologically, the prostates from patients with androgen blockade showed an increase in atrophic glands with flattened epithelium and a relative increase in stroma. These glands were preferentially seen in the peripheral zone in areas surrounding the prostatic adenocarcinoma. These atrophic glands exhibited a discontinuous basal cell layer when stained with 34βE12, suggesting that they were the end result of androgen blockade on HGPIN. In addition, several glands showed one or two layers of epithelial cells with nuclear atypia. These glands were seen to represent HGPIN with involutionary changes.

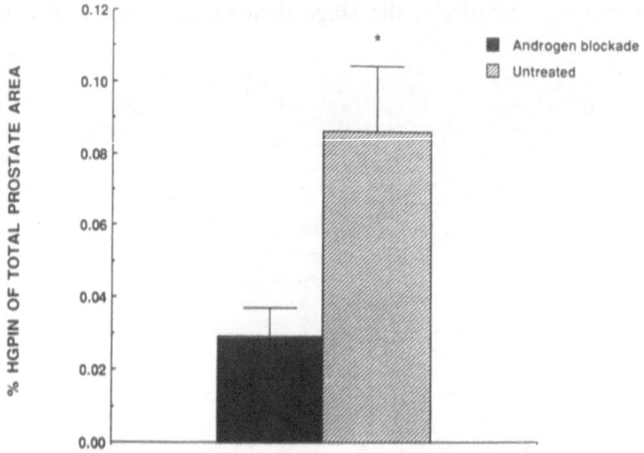

Figure 1. The proportion of total prostate area occupied by HGPIN for all stages expressed as mean percent in the androgen-blockade and control groups was 0.028 ± 0.009 (n = 12) and 0.086 ± 0.018 (n = 12), respectively . *p ≤ 0.008. This may indicate that, like prostatic adenocarcinoma, HGPIN is androgen sensitive.

Discussion

The present retrospective study is the first description of changes attributed to AA therapy on HGPIN. Androgen deprivation is known to reduce and degenerate malignant and benign prostate glands (11). By using morphometric analysis, we have shown that HGPIN is markedly reduced in quantity after AA therapy. This finding suggests that, like prostatic adenocarcinoma (12), HGPIN is androgen dependent, which is consistent with the hypothesis that it is the premalignant lesion of prostatic adenocarcinoma. In our study, patients receiving AA therapy varied with regard to AA dosage given. This may partially account for the presence of comparable size tumors when compared to the control group. In addition, this relatively low AA therapy may explain why there were no significant reductions in the size of the prostate glands when compared to controls. However, the fact that there were marked differences in the amount of HGPIN suggests that this lesion may be more sensitive to androgen deprivation than prostatic adenocarcinoma, suggesting the presence of increased activity or number of androgen receptors.

The differences found in the correlation of the Gleason scoring between the biopsy and the prostatectomy may be due to several factors. First, the distinctive changes of androgen deprivation in tumors have recently been described.

Second, the Gleason grading system, which uses architectural features, does not account for glands that are breaking apart and showing involutionary changes.

Prostates that received AA therapy had increased numbers of atrophic glands. In addition, we found glands that had both HGPIN and atrophic epithelium. These glands, stained with 34βE12, showed a discontinuous basal cell layer, suggesting that they contain HGPIN lesions with involutionary changes produced by AA therapy.

This first report, demonstrating a marked reduction in the amount of HGPIN after AA therapy, suggests that HGPIN is reversible and that AA therapy might play a role in the treatment of patients at high risk of developing prostate cancer.

References

1. Bostwick, DG (1989) Prostatic intraepithelial neoplasia (PIN). Urology 34:16-20.
2. McNeal JE, Bostwick DG (1986) Intraductal dysplasia: a premalignant lesion of the prostate. Hum Path 17:64-69.
3. Bostwick, DG, Brawer MK (1987) Prostatic intra-epithelial neoplasia and early invasion in prostate cancer. Cancer 56:788-792.
4. Bostwick, DG, Amin MB, Dundore P, et al. (1993) Architectural patterns of high-grade prostatic intraepithelial neoplasia. Hum Path 24:298-302.
5. The Leuprolide Study Group. Leuprolide versus diethylstilbestrol for metastatic prostate cancer (1984) N Eng J Med 311:1281-1283.
6. Soloway, MS: Newer methods of hormonal therapy for prostate cancer. Urology 24:30, 1984.
7. Murphy, WM, Soloway, MS, Barrows, GH (1991) Pathological changes associated with androgen depravation therapy for prostate cancer. Cancer 68:821.
8. Tetu, B, Srigley, JR, Boivin, JC, et al. (1991) Effect of combination endocrine therapy LHRH agonist and flutamide) on normal prostate and prostatic adenocarcinoma. Am J Surg Path 15:111-115..
9. Brawer, MK. Prostatic intraepithelial neoplasia (1992) Hum Path. 23:242-250.
10. Gown, AM, Vogel AM (1982) Monoclonal antibodies to intermediate filaments proteins of human cells: unique and cross-reacting antibodies. J Cell Biol 95:414-416.
11. Levine AC, Kirschenbaum A, Kaplan P, et al. (1989) Serum prostate-antigen levels in patients with benign prostatic hypertrophy treated with leuprolide. Urology 34:10-14.
12. Huggins C, Hodges CV (1941) Studies on prostatic cancer: I. The effect of castration, of estrogen and of androgen injection on serum phosphatase in metastatic carcinoma of the prostate. Cancer Res 1:293-297.

Alterations in Circulating Levels of Androgens and PSA During Treatment with Finasteride in Men at High Risk for Prostate Cancer

Frank Z. Stanczyk, Eila C. Skinner, Susan Mertes, Maryann F. Spahn, Rogerio A. Lobo, and Ronald K. Ross

Summary

Circulating levels of both 5α-dihydrotestosterone (DHT) and prostate specific antigen (PSA) are of interest with regard to the pathogenesis and progression of cancer of the prostate. However, little is known about the relationships among PSA, DHT, as well as precursors and products of DHT. Recently, we have initiated a one-year study in which circulating levels of PSA and androgens are measured during treatment with finasteride, a 5α-reductase inhibitor, and under conditions of no treatment (controls) in men at high risk for prostate cancer. Our preliminary data show that PSA levels decrease slowly with finasteride treatment, in contrast to the rapid drop in levels of DHT and its metabolite, 5α-androstane-3α, 17ß-diol glucuronide. However, the fall in PSA levels correlates highly with inhibition of 5α-reductase activity after three months of treatment.

Introduction

In the adult male, the potent androgen 5α-dihydrotestosterone (DHT), formed primarily from testosterone via the enzyme, 5α-reductase, appears to play an important role in the development of the normal prostate and the pathogenesis of both benign and malignant diseases of the gland (1). Men with 5α-reductase deficiency have small prostates compared with their unaffected male siblings. Also, DHT is essential for the maintenance of benign prostatic hyperplasia (BPH).

A competitive inhibitor of 5α-reductase activity, namely, finasteride, (Proscar; Merck, Sharpe & Dohme, Rahway, New Jersey), has, recently, been used to treat men with BPH. Studies have shown that this drug causes a chronic decrease in intraprostatic DHT, accompanied by a progressive decrease in prostatic volume and an improvement in clinical symptoms (2).

We have recently initiated a pilot chemoprevention trial of finasteride to examine potential biological markers of carcinogenicity in men at high risk for prostate cancer. Herein, we present preliminary data from one segment of that study, on the effect of finasteride on circulating levels of PSA, DHT, and precursors and metabolites of DHT, in men at high risk for prostate cancer. An additional objective was to assess whether correlations exist between PSA and ratios of DHT precursors to either DHT or its products.

Methods

A total of 19 men, who were at high risk for prostate cancer because of their age (60-85 years) and serum PSA level (>4 ng/ml), but who had no evidence of adenocarcinoma on biopsies of the prostate, were studied. The men were assigned randomly to either treatment group ($n = 12$) or the control group ($n = 7$). The treatment group received one tablet of finasteride daily for six months. Blood specimens were obtained prior to treatment and after one, three, and six months of treatment. The serum from each sample was removed and stored at -20°C.

PSA, testosterone (T), DHT, 5α-androstane-3a and 17ß-diol glucuronide (3α-diol G) were measured in the serum by specific immunoassays. PSA was measured by enzyme-linked immunoabsorbent assay (Hybritech, San Diego, California). Tand DHT were quantified by RIA, and 3α-diol G by a direct RIA kit (Diagnostic Systems Laboratories, Webster, Texas). Data were analyzed by Student's t test and Pearson's correlation coefficient.

Results

Serum levels of PSA, T, DHT, and 3α-diol G measured in the treatment and control groups are depicted in Figure 1. Neither PSA nor androgen levels changed significantly in the control group during the six-month study period. In the treatment group, PSA levels were only 15% less than baseline after one month, but dropped significantly to 40% ($p = 0.01$) and 51% ($p = 0.001$) at three and six months post-treatment, respectively. In contrast, DHT and 3α-diol G levels fell dramatically (80% and 70%, respectively) one month post-treatment and remained low thereafter. T levels, however, increased moderately (22-28%).

Significant negative correlations were found between PSA levels and the T/DHT ratios after six months of finasteride treatment (r = -0.61; p = 0.001), and between PSA levels and T/3α-diol G ratios after three (r = -0.52; p = 0.004) and six months (r = -0.56; p = 0.002), respectively.

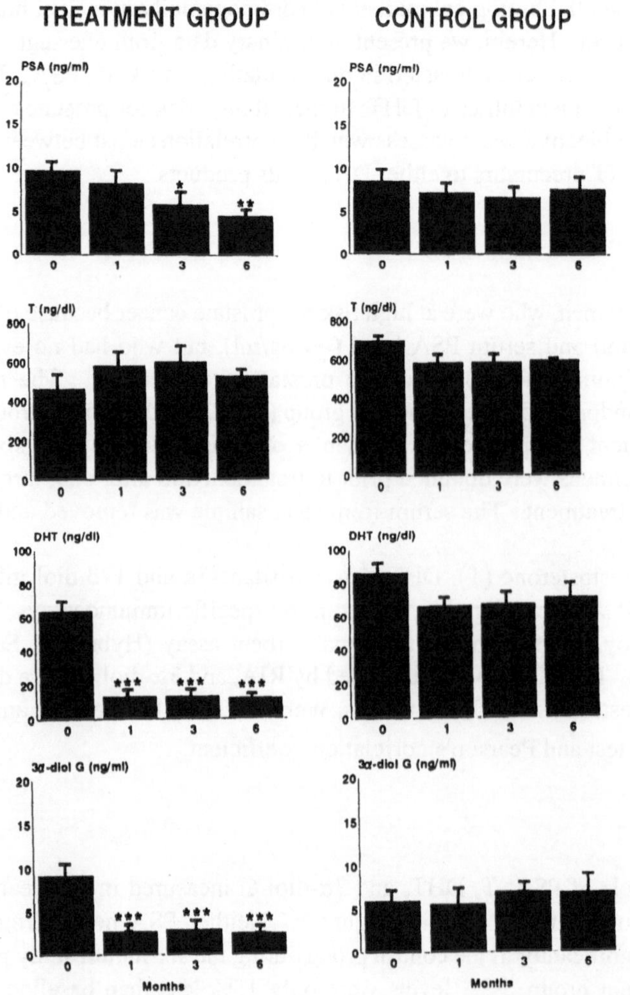

Figure 1. Serum levels (mean ± standard error) of PSA, T, DHT, and 5α-androstane-3a, 17ß-diol glucuronide (3α-diol G) in men at high risk for prostate cancer, either treated daily with finasteride for 6 months (treatment group) or not treated (control group). * p = 0.01; ** p = 0.001; *** p = <0.001, s ignificantly different between treatment and control groups

Discussion

Little is known about the relationship between PSA and 5α-reductase activity. Previous studies have shown that prolonged finasteride treatment of men with BPH causes marked reduction in prostatic volume, accompanied by a large decrease in circulating PSA levels and significant reductions in serum and intraprostatic concentrations of DHT and/or its metabolites (2-3). However, correlations between PSA levels and 5α-reductase activity have not been determined after prolonged treatment with finasteride. Our preliminary data in the present study show that, although PSA levels declined relatively slowly with finasteride treatment, the fall in PSA levels correlates highly with inhibition of 5α-reductase activity at three months of treatment. The PSA levels were correlated with either the T/DHT or T/3α-diol G ratios in the treatment group.

Despite the fact that 3α-diol G is a distal metabolite of testosterone, it is not surprising that the T/3α-diol G ratios correlated highly with PSA levels. After DHT is formed from testosterone via 5α-reductase, it can be converted to a number of different metabolites. One important route of DHT metabolism involves reduction of DHT by action of 3α-hydroxysteroid oxidoreductase, forming 3α-diol, which can then be conjugated via glucuronyl transferase to give 3α-diol G (4). The formation of 3α-diol G from DHT has been shown to be an important pathway of DHT metabolism in peripheral tissues of both men and women (4, 5).

It is possible that the ratios of other precursors/products of DHT may also correlate with the decrease in PSA levels. We are currently determining serum levels of androstenedione, a DHT precursor, as well as DHT sulfate and androsterone glucuronide, both of which are DHT metabolites, to gain a better understanding of the relationship between 5α-reductase activity and PSA.

References

1. Tenover JS (1991) Prostates, pates, and pimples. In: Strauss III JF (ed) Clinics in Endocrinology and Metabolism, Philadelphia, W.B. Saunders Co., Volume 20, pp 893-909.
2. Gormley GJ, Stoner E, Braskewitz RC (1992) The effect of finasteride in men with benign prostatic hyperplasia. NEJ Med 327:1185-1191.
3. Rittmaster RS (1994) Finasteride. NEJ Med 330:120-124.
4. Horton RA (1990) Testicular steroid transport, metabolism, and effects. In: Becker KL (ed) Principles and Practice of Endocrinology and Metabolism. Philadelphia, JB Lippincott Co, pp 937-941.
5. Lobo RA (1990) Hirsutism, alopecia and acne. In: Becker KL (ed) Principles and Practice of Endocrinology and Metabolism. Philadelphia, JB Lippincott Co, pp 834-838.

COMMUNICATIONS
Session II. Cell & Molecular Biology, Metabolism

Effect of Estradiol on MCF-7 Human Breast Cancer Cells: Ultrastructural Studies

Xue-Min Zhang and De-Hui Chen

Summary

The effect of estradiol on MCF-7 breast cancer cells was studied using transmission and scanning electron microscope, as well as an immunogold labeling method. After eight days of estradiol (10nM) treatment, an increase in the number and length of microvilli on the cell surface of the MCF-7 was observed, as well as numerous blebs on the cell surface. These observations suggest that estrogen exposure may facilitate tumor cell adhesion and movement, thus strengthening the ability of breast tumor cells to invade and metastasize. The cell cytoplasm contained developed organelles. Using a colloidal gold probe, epidermal growth factor receptors (EGFR) were localized. The membrane surface, including microvilli, was covered with electron-dense gold particles. A continuing course of internalization and transfer of EGFR labeled with colloidal gold was observed. Our results suggest that the autocrine secretion of transforming growth factor binding to EGFR may be a possible mechanism involved in MCF-7 cell growth, as well as in the growth-stimulating effects of estradiol.

Introduction

It is known that breast cancer patients whose tumors are estrogen receptor (ER) positive usually experience longer survival rates than those whose tumors are ER-negative. These patients also respond to endocrine therapy (1-3). The mechanism(s) behind these phenomena, however, are not well understood. In order to probe whether estrogen is related to these phenomena, we studied the effect of estradiol on the ultrastructure of MCF-7 cells, a human breast cancer cell line, as well as the presence of epidermal growth factor receptor (EGFR) in these cells.

Results

After eight days of estradiol treatment, the cell surface of MCF-7 cells was covered with a dense network of long microvilli and numerous blebs; while the surface of untreated control cells remained smooth, with short and rare microvilli and infrequent blebs. Most of the cells treated with estradiol showed well-developed Golgi complexes and rough endoplasmic reticula, as well as numerous large mitochondria characterized by clear cristae. Under fluorescence microscope, estradiol-treated MCF-7 cells stained positively for EGFR, while untreated cells did not. The EGFR in estradiol-treated cells was further localized using an immuno-gold labeling technique. Under transmission electron microscope (TEM), the membrane surface, including microvilli, of most MCF-7 cells was covered with electron-dense gold particles labeled with antibody against EGFR (Figure 1a). Pinosome formation was observed in the membrane surface, usually containing inward gold-labeled EGFR. A continuing course of internalization of the gold particles was observed from the cell surface toward the nucleus (Figure 1a-c). The surface of the control cells did not exhibit appended gold particles. The ultrastructure of labeled cells was well preserved, with many intermediate filaments in the cytoplasm.

Discussion

Our study demonstrates that estradiol treatment of MCF cells gives rise to an increased number and length of microvilli, as well as an increased number of blebs in the cell surface. It is possible that the increased microvilli might promote adhesion and movement of the tumor cells, thus strengthening the invasiveness required for tumor metastasis. The abundance of microvilli expands to the cell surface, improving their nutrition and proliferation capacity. The well-developed cytoplasmic organelles indicate that the cells may be active in many functional respects (4).

Transforming growth factor α (TGF α) is known to be secreted by MCF-7 cells via an autocrine mechanism, and estradiol treatment has been reported to stimulate this process (5, 6). TGF α exerts its function through binding to EGFR (5, 6). Therefore, the localization of EGFR in the membrane surface of estradiol-treated MCF-7 cells suggests that one mechanism by which estradiol stimulates MCF-7 cell growth might be stimulation of autocrine secretion of TGF α and increasing EGFR on the cell membrane. More intriguing was the continuing course of internalization of the gold particles, from the cell surface toward the nucleus. From a morphologic aspect, this event may represent an unfamiliar pathway by which EGFR-related growth factors target nuclear DNA to exert their functions.

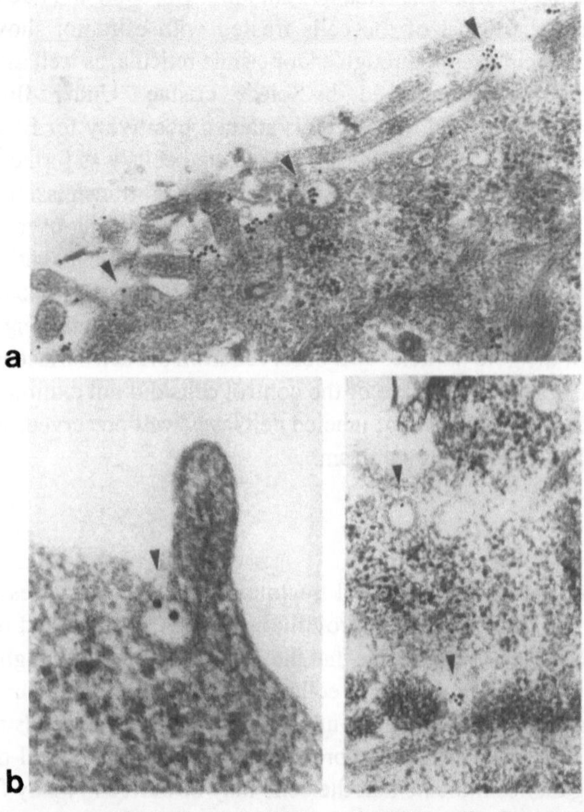

Figure 1. Immuno-gold labeling of EGFR on MCF-7 cells. Suspension cultures of MCF-7 cells were treated with 10 nM estradiol for eight days. For EGFR detection, the cells were incubated at room temperature with a monoclonal antibody against EGFR, followed by incubation with a second antibody coupled to gold particles. The cells were fixed in glutaraldehyde and osmium tetroxide, then embedded for TEM by routine procedures. a) Gold particles present on the surface membrane (× 43,200). b) Formation of a pinosome (× 82,800). c) Pinosomes with gold particles inside moving toward the nucleus (× 55,200).

In this study, the antibodies were directly added to cell suspensions, and the incubation was performed before embedding the cells for TEM, to keep the viability of the cells intact while EGFR was labeled. This maneuver ensured well-preserved cell ultrastructure while the interaction between the membrane receptors and the antibodies labeled with immuno-gold was observed.

References

1. Oza AM, Tannock IF (1994) Clinical relevance of breast cancer biology. Hematol Oncol Clin North Am 8:1-14.
2. Klijn JG, Berns EM, Bontenbal M, Foekens J (1993) Cell biological factors associated with the response of breast cancer to systemic treatment. Cancer Treat Rev 19:45-63.
3. Donegan WL (1992) Prognostic factors. Stage and receptor status in breast cancer. Cancer 70:1755-64.
4. Vic P, Vignon F, Derocq D, Rochefort H (1982). Effect of estradiol on the ultrastructure of the MCF7 human breast cancer cells in culture. Cancer Res 42:667-673.
5. Lippman ME, Dickson RB, Gelmann EP, et al. (1988) Growth regulatory peptide production by human breast carcinoma cells. J Steroid Biochem 30:53-61.
6. Kasid A, Lippman ME (1987) Estrogen and oncogene mediated growth regulation of human breast cancer cells. J Steroid Biochem 27:465-70.

Estradiol and Tamoxifen Induce TGF-β1 Gene Expression In Cultured Normal Human Breast Stromal Cells

YunFu Hu, Ching-Jey G. Chang, Yasuro Sugimoto,
Robert W. Brueggemeier, William B. Farrar, and Young C. Lin

Summary

We have previously demonstrated that 17β-estradiol (E_2) inhibits the growth of human breast cancer cells co-cultured with normal human breast stromal cells (1). In the present study, the effects of E_2 and the anti-estrogen tamoxifen (TAM) on expression of the growth-inhibitory TGF-ß1 gene in cultured normal human breast stromal cells was investigated. A highly-enriched population of human breast stromal cells was isolated from breast tissues of human patients and cultured to confluency. The cultured cells were then treated for 24 h with E_2 (20 nM) and/or TAM (1 mM) in a chemically-defined, serum- and phenol red-free culture medium. Expression of the TGF-ß1 gene was examined with semiquantitative polymerase chain reaction techniques following reverse transcription of total RNA from the cultured cells. Our results indicate that the TGF-ß1 gene was expressed in human breast stromal cells treated with either E_2 or TAM, but not in untreated controls or those treated with a combination of both E_2 and TAM. Expression of the co-amplified housekeeping glucose-6-phosphate dehydrogenase gene was detected at constant levels regardless of treatment. Since TAM possesses intrinsic estrogenic activities, our data strongly suggest that TAM induces TGF-ß1 gene expression by acting upon estrogen receptor (ER) as a partial estrogen agonist. In the presence of estrogen, TAM acts as an anti-estrogen. Thus, regulation of TGF-ß1 production in breast stromal cells appears to take place at the transcriptional level as a result of ER-mediated gene activation by full (e.g., E_2) or partial (e.g., TAM) estrogen agonists.

Introduction

The most important endocrine factor identified so far for the etiology of human breast cancer is estrogen. Estrogen has been shown to act on specific ER proteins influencing expression of genes instrumental in the development, growth, and metastasis of human breast cancer (2). Typically, endocrine-responsive breast cancer cells depend upon estrogenic stimulation for their growth. As a result, TAM, a nonsteroidal anti-estrogen, has been extensively studied as an adjuvant therapy for human breast cancer patients (3). The tumoristatic action of TAM is attributed primarily to blockade of estrogen-stimulated growth in cancer cells. TAM has also been shown to inhibit the growth of breast cancer cells by inducing the autocrine production of TGF-ß1 (4), a member of the multigene family, each consisting of a 25-kd polypeptide homodimer with potent multifunctional regulatory activities (5). In fact, a favorable response to adjuvant TAM has been observed in patients with breast cancer containing undetectable levels of ER (3). Such a beneficial effect of TAM has been subsequently associated with production of the growth-inhibitory TGF-ß1 in stromal cells (6, 7).

However, based on *in-situ* hybridization studies in the mouse embryo (8-10), the authors claimed that breast epithelium is the source of the extracellular TGF-ß1 immunoreactivity observed in breast stroma. Thus, it remains unclear whether cultured breast stromal cells could produce TGF-ß1 in response to TAM treatment. In addition, mechanisms associated with TAM induction of TGF-ß1 gene expression have not been determined. Accordingly, the objective of the present study was to examine the expression of the TGF-ß1 gene in cultured stromal cells from normal human breast tissues in response to estrogen and/or anti-estrogen treatments.

Results

As shown in Figure 1, expression of the TGF-ß1 gene in cultured normal human breast stromal cells was detected at a very low level in the untreated controls. Treatment of the cultured human breast stromal cells with either E_2 (20 nM) or TAM (1 mM) dramatically increased expression of the TGF-ß1 gene. However, a combination of E_2 and TAM had no effect on TGF-ß1 gene expression.

Discussion

Breast cancer represents the most common neoplasm in women. An intricate interplay of stromal and epithelial cells has been implicated in the development of breast epithelial malignancy (11). Breast stroma has been shown to play an important paracrine role influencing breast cancer cell function and growth (1).

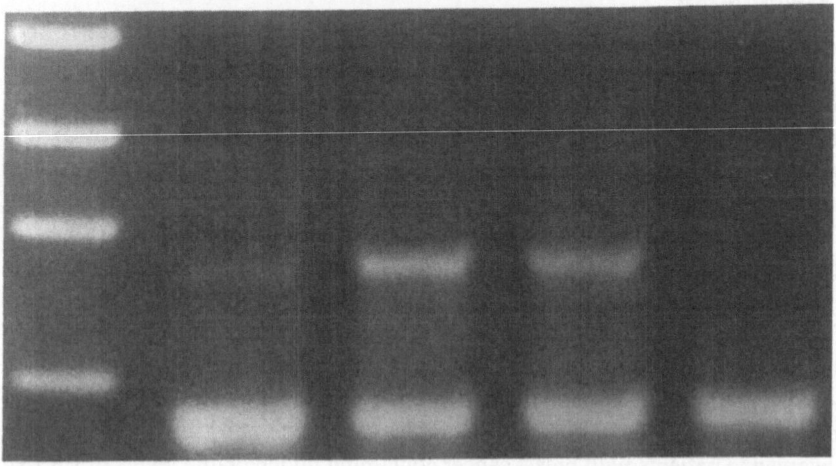

Figure 1. Expression of TGF-ß1 gene (top bands) in confluent cultures of normal breast stromal cells treated with: vehicle only (Lane 2), 20 nM E_2 (Lane 3), 1 mM TAM (Lane 4), a combination of 20 nM E_2 and 1 mM TAM (Lane 5). Phi X RF DNA/Hae III fragments were used as DNA MW markers (Lane 1).

Breast stroma has also been implicated as a potential paracrine mediator of therapeutic agents for breast cancer patients. In fact, it is now generally believed that the anti-estrogen TAM provides beneficial effects for human patients with ER-negative breast cancer by stimulating stromal production of TGF-ß1 (7), which has been shown to inhibit the growth of cells derived from various epithelial tissues, including human breast cancer cell lines (4, 5). In agreement with previous *in-vitro* studies using human fetal fibroblasts (6), our results clearly indicate that TAM treatment can induce expression of the TGF-ß1 gene in cultured normal human breast stromal cells.

Mechanisms associated with TAM induction of TGF-ß1 gene expression are largely unknown. Previous studies have shown that production of receptor-reactive TGF-ß1 was increased by anti-estrogens in the MCF-7 breast cancer cell line via an ER-mediated mechanism at the post-transcriptional level (4). Results from the current study indicate that estrogen can also induce expression of the TGF-ß1 gene in normal human breast stromal cells. Since TAM possesses intrinsic estrogenic activities, our data strongly suggest that TAM induces TGF-ß1 gene expression by acting upon ER as a partial estrogen agonist. In the presence of estrogen, TAM acts as an anti-estrogen. Our experiments support this concept. Iit is reasonable to speculate that regulation of TGF-ß1 production in stromal cells is at the transcriptional level as a result of ER-mediated gene activation by full (e.g., E_2) or partial (e.g., TAM) estrogen agonists.

Acknowledgment

Supported by NIH grants CA58003, 2P30 CA16058, and DK45916.

References

1. Hu YF, Chang CJG, Brueggemeier RW, Lin YC (1993) Gossypol inhibits basal and estrogen-stimulated DNA synthesis in human breast carcinoma cells. Life Sci 53:433-438.
2. Lippman ME, Dickson RB (1989) Mechanisms of growth control in normal and malignant breast epithelium. J Steroid Biochem 34:107-121.
3. Early Breast Cancer Trialists' Collaborative Group (1992) Lancet 339:1-15, 71-85.
4. Knabbe C, Lippman ME, Wakefield LM, et al. (1987) Evidence that transforming growth factor β is a hormonally-regulated negative growth factor in human breast cancer cells. Cell 48:417-428.
5. Roberts AB, Anzano MA, Wakefield LM, et al. (1985) Type beta transforming growth factor: A bifunctional regulator of cellular growth. Proc Natl Acad Sci USA 82:119-123.
6. Colletta AA, Wakefield LM, Howell FV, et al. (1990) Anti-oestrogens induce the secretion of active transforming growth factor beta from human fetal fibroblasts. Brit J Cancer 62:405-409.
7. Butta A, MacLennan K, Flanders KC, et al. (1992) Induction of transforming growth factor beta 1 in human breast cancer *in vivo* following tamoxifen treatment. Cancer Res 52:4261-4264.
8. Lehnert SA and Akhurst RJ (1988) Embryonic expression pattern of TGF beta type-1 RNA suggests both paracrine and autocrine mechanisms of action. Development 104:263-273.
9. Millan FA, Denhez F, Kondaiah P (1991) Embryonic gene expression patterns of TGF beta 1, beta 2 and beta 3 suggest different developmental functions *in vivo*. Development 111:131-144.
10. McCune BK, Mullin BR, Flanders KC, et al. (1992) Localization of transforming growth factor-beta isotypes in lesions of the human breast. Hum Path 23:13-20.
11. Dickson RB (1992) Regulation of tumor-host interactions in breast cancer. J Steroid Biochem Molec Biol 41:389-400.

Estrogen and Anti-Estrogen Regulation of Amplified *erb*B2 Gene Expression in Human Breast Cancer Cells

Anni M. Wärri, Jorma J. Isola, and Pirkko L. Härkönen

Summary

Amplification and overexpression of the *erb*B2 oncogene have been shown to lead to aggressive growth of breast cancer cells and poor patient prognosis. We have studied estrogen and anti-estrogen regulation of *erb*B2 expression in BT-474 human breast cancer cells, which have an amplified *erb*B2 gene and contain functional estrogen receptors. Both estrogens and the anti-estrogen toremifene (TOR) caused a transient suppression of *erb*B2 mRNA (4.0-fold) and protein (1.5-fold) expression over a five-day period. In addition to the transient regulation of *erb*B2, TOR also acted as an estrogen-like agonist on cell growth, which differs from growth suppression in most "truly" estrogen-sensitive cell lines. The results indicate that estrogen receptor-positive breast cancer cells containing *erb*B2 gene amplification are not responsive to anti-estrogen therapy.

Introduction

The *erb*B2/*neu* oncogene product p185 is a transmembrane glycoprotein which belongs to a family of structurally-related growth factor receptors, including EGFR, *erb*B3/HER-3 (1) and *erb*B4/HER-4 (2). The function and regulation of the normal *erb*B2 gene are not well known. This gene is widely expressed in the epithelial cells of human fetal and adult tissues, including the mammary gland (3, 4).

Oncogenic activation of *erb*B2 in human tumors appears to occur through amplification and/or overexpression of the gene; no point mutations or other aberrations have been found (5). Both amplification and overexpression of *erb*B2 (detected in ~ 20-25% of human breast cancers) have been shown to correlate with aggressive growth of the cells and poor prognosis of breast cancer patients (6-12).

Estrogen has previously been shown to down-regulate expression of *erb*B2 oncogene in human breast cancer cells, which contain a normal, unamplified *erb*B2 gene (13). It is not known, however, whether an amplified *erb*B2 gene is hormonally regulated. We therefore studied whether estrogen and anti-estrogen can regulate expression of reamplified *erb*B2 gene, as well as their effect on cell proliferation of BT-474 human breast cancer cells.

Results

Estrogen Stimulation of BT-474 Cell Growth and pS2 mRNA Expression

Estradiol (E_2) enhanced the growth rate of BT-474 cells. In the presence of 10 nM E_2, cells had a doubling time of 2 days, whereas the doubling time without estrogen was 9 days. TOR had a small agonistic effect at a concentration of 1 μmol/l, while at higher concentration (7.5 μmol/l), it did not exert either a stimulatory or an inhibitory action (in the absence of E_2). Expression of the 0.6-kb mRNA encoding pS2 protein was analyzed as an additional marker for functional estrogen receptors and estrogen action. The pS2 mRNA was stimulated by 10 nM E_2 and down-regulated by E_2 withdrawal and TOR addition, also observed in ZR-75-1 cells (Figure 1).

Estrogen and Anti-Estrogen Effects on Expression of *erb*B2 Protein in BT-474 Cells

The expression of *erb*B2 protein in BT-474 cells was examined to determine whether expression of an amplified *erb*B2 was regulated by E_2 and anti-estrogens. The cells cultured in steroid-free (E-) medium were analyzed for the membrane-bound cellular *erb*B2 protein by using a sandwich-type enzyme immunoassay and a specific polyclonal antibody against the cytoplasmic domain of the *erb*B2 protein (Triton Diagnostics, Alameda, CA). Addition of 10 nM E_2 for 48 h decreased the *erb*B2 protein concentration by 30% in BT-474 cells (Figure 2). Addition of TOR to the culture medium at low or high concentration for 48 h resulted in different effects: 1.0 -5.0 μM concentrations of TOR were agonistic (i.e., suppressed protein levels to approximately the same extent as E_2), while 10.0 μM TOR decreased the protein level only slightly compared with the levels observed in control medium (93% of control, Figure 2). A similar result (89% of control) was obtained with 7.5 μM tamoxifen (data not shown).

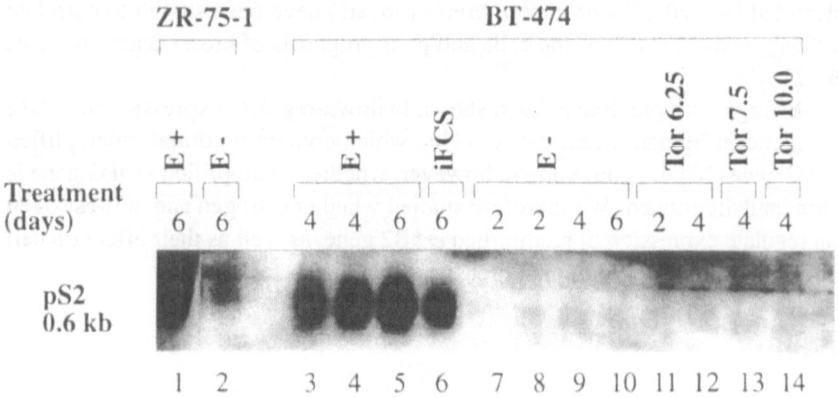

Figure 1. Expression of 0.6-kb pS2 mRNA in BT-474 cells (Lanes 3-14). Northern blot analysis of total RNA (5 µg/lane). pS2 expression was stimulated by the addition of 10 nM E_2 in cells cultured with 10% stripped serum (for 4 and 6 days, respectively, "E +" Lanes 3-5) or in full serum (iFCS, for 6 days, Lane 6) and down-regulated by E_2-withdrawal ("E -" Lanes 7-10) and TOR addition ("TOR" Lanes 11-14). pS2 expression in ZR-75-1 cells is shown for comparison in the presence (Lane 1) and absence (Lane 2) of E_2.

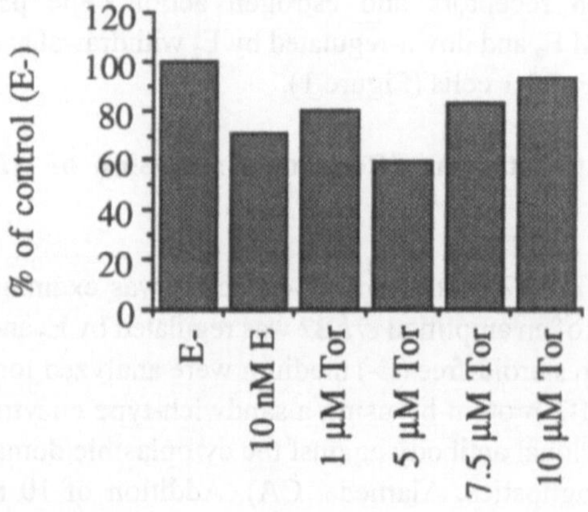

Figure 2. Expression of *erb*B2 protein in BT-474 cell lysates and soluble *erb*B2 (ECD) in culture media after a 48-h incubation. Effect of 10 nM E_2 and TOR (1-10 µM). One million cells were plated in 5 ml of steroid-depleted (E-) medium (day 0). Forty-eight hrs after addition of the test media (day 3), cells and culture media were collected and analyzed for cellular *erb*B2 protein (Triton Diagnostics, Alameda, CA, USA). The columns are expressed as % of control. Each column represents the mean of two parallel cultures.

Time course studies using 7.5 μM TOR revealed a gradual decrease in expression of *erb*B2 protein during the first 72-96 h of TOR treatment, whereafter the level increased again (Table 1).

Table 1. Time Course of the Cellular *erb*B2 Protein Content in BT-474 Cells.

Treatment	Time (h)			
	48	72	96	120
TOR (7.5 μM)	89.5 ± 5.3	70.0 ± 4.2	55.0 ± 1.4	80.3 ± 3.7
E_2 (10 nM)	70.3 ± 4.0	n.d.	n.d.	90.3 ± 7.5

BT-474 cells grown in the presence of 7.5 μM TOR or 10 nM E_2 expressed as % of control (cells grown in estrogen-deprived E-medium for 120 h). Duplicate determinations were done from 2-3 parallel samples at each time point. (n.d.= not determined)

Transient Suppression of mRNA Expression of Amplified erbB2 Gene by E_2

Figure 3 shows the 4.8-kb *erb*B2 mRNA expression levels in BT-474 cells cultured in steroid-deprived medium (E-) after treatment with 10 nM E_2 (E +), 7.5 μM TOR, or tamoxifen (TAM) *in vitro*. The level of full-length 4.8-kb *erb*B2 mRNA expression in BT-474 cells was about 30 times higher than that observed in ZR-75-1 cells, based on densitometric analysis (Figure 3). Expression of *erb*B2 mRNA was initially stimulated by E_2 withdrawal and suppressed by E_2 addition (48-h white columns), but after a 144-h culture time, expression levels were almost the same in the absence or presence of E_2 at about 60% of the initial E-level (black columns).

Expression levels after 48-h TOR and tamoxifen treatments were 60% and 50% of that of E-, respectively. After a 144-h culture time, 70% and 90% of the initial E- level were observed (after TOR and tamoxifen treatment, respectively, Figure 3). Thus, the *erb*B2 mRNA expression levels after anti-estrogen treatments were not as high as the initial E and E_2 levels, but were higher than in the presence of E_2, at both the beginning (48 h) and the end (144 h) of the experiment.

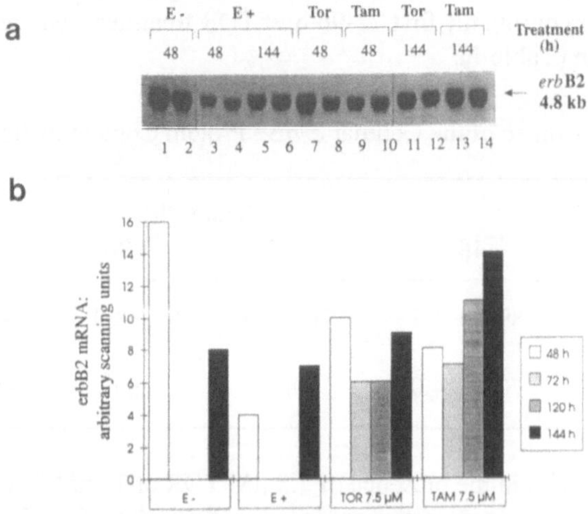

Figure 3. Time course of changes in 4.8-kb *erb*B2 mRNA expression in BT-474 cells upon addition of 10 nM E_2 or 7.5 μM TOR or tamoxifen for 48 and 144 h. (A) Northern blot analysis of total RNA, 5 μg/lane. (B) Quantitation of the *erb*B2 mRNA levels expressed as arbitrary scanning units corrected for RNA loading. Each RNA sample was prepared from 2-3 parallel dishes. The cultures and Northern blot analyses were repeated twice.

Discussion

Our results indicate that expression in BT-474 cells of amplified *erb*B2 gene is transiently down-regulated by estrogen and the anti-estrogen TOR. The suppression of *erb*B2 mRNA and protein was released when cultures were continued, and expression of both mRNA and protein finally became independent of the presence of estrogen and anti-estrogen. Conversely, the stimulation of cell growth by estrogen continued.

In BT-474 cells, the anti-estrogen TOR unexpectedly behaved as an estrogen-like agonist on both *erb*B2 expression and cell growth, and did not exhibit the antagonistic effects seen in most "truly" estrogen-sensitive cell lines. Thus, the *erb*B2 amplification in this cell line may lead to cell growth that cannot be suppressed by anti-estrogens despite the fact that growth can be induced by estrogen.

References

1. Lemoine NR, Barnes DM, Hollywood DP, et al. (1992) Expression of the *erb*B3 gene product in breast cancer. Brit J Cancer 66:1116-1121.

2. Plowman GD, Culouscou J-M, Whitney GS, et al. (1993) Ligand-specific activation of HER-4/p180^{erbB4}, a fourth member of the epidermal growth factor receptor family. Proc Natl Acad Sci USA 90:1746-1750.
3. Borg Å (1992) Gene Aberrations in Human Breast Cancer. Academic Dissertation, Wallin & Dalholm Boktryckeri AB, Lund, Sweden.
4. Prigent SA, Lemoine NR (1992) The type 1 (EGFR-related) family of growth factor receptors and their ligands. Progr in Growth Factor Res 4:1-24.
5. Lemoine NR, Staddon S, Dickson C, et al. (1990) Absence of activating transmembrane mutations in the c-*erb*B-2 proto-oncogene in human breast cancer. Oncogene 5:237-239.
6. Slamon DJ, Clark GM, Wong SG, et al. (1987) Human breast cancer; correlation of relapse and survival with amplification of the HER-2/neu oncogene. Science 235:177-182.
7. Slamon DJ, Godolphin W, Jones LA, et al. (1989) Studies of the HER-2/neu proto-oncogene in human breast and ovarian cancer. Science 244:707-712.
8. Wright C, Angus B, Nicholson S, et al. (1989) Expression of c-*erb*B-2 oncoprotein: A prognostic indicator in human breast cancer. Cancer Res 49:2087-2090.
9. Paik S, Hazan R, Fisher ER, et al. (1990) Pathologic findings from the national surgical adjuvant breast and bowel project: Prognostic significance of *erb*B-2 protein overexpression in primary breast cancer. J Clin Oncol 8:103-112.
10. Borg Å, Baldetorp B, Fernö M, et al. (1991) ERBB2 amplification in breast cancer with a high rate of proliferation. Oncogene 6:137-143.
11. Gullick WJ, Love SB, Wright C, et al. (1991) c-*erb*B-2 protein overexpression in breast cancer is a risk factor in patients with involved and uninvolved lymph nodes. Brit J Cancer 63:434-438.
12. Kallioniemi O.-P, Holli K, Visakorpi T, et al. (1991) Association of c-*erb*B-2 protein overexpression with high rate of cell proliferation, increased risk of visceral metastasis and poor long-term survival in breast cancer. Int J Cancer 49:650-655.
13. Wärri AM, Laine AM, Majasuo KE, et al. (1991) Estrogen suppression of *erb*B2 expression is associated with increased growth rate of ZR-75-1 human breast cancer cells *in vitro* and in nude mice. Int J Cancer 49:616-623.

Evaluation of Conformational Changes in hER-HBD by Pharmacological Dissection of Hormone Dissociation Rates in a *Homogeneous* Hormone-Binding Assay

Johan Häggblad, Bo Carlsson, and Jean-Pierre Raynaud

Summary

Hormone dissociation rates can be used as a surrogate parameter to monitor changes in conformational states, as the hormone binding affinity is directly related to the dissociation rate and is dependent on receptor conformation. High-resolution studies of dissociation rates are tedious because the bound radioligand has to be separated from unbound. In contrast, *homogeneous* radioligand binding assays are rapid and have high temporal resolution.The *homogeneous* assay format allows rapid mixing, followed by continuous measurement of bound radioactivity. We describe a *homogeneous* hormone-binding assay for human estrogen-receptor hormone-binding domain (hER-HBD) produced in a yeast expression system. The binding assay was used for quantitation of dissociation rates for [^{3}H]-estradiol (E_2) by addition of different estrogens and anti-estrogens. Our results revealed that different compounds caused significantly different monophasic dissociation rates (expressed as min) for [^{3}H]-E_2: DES (0.046 min^{-1}) > >TAM (0.040) >E3 (0.032) ≈ E_2 (0.029). However, a closer examination of the concentration dependence showed that at higher concentrations certain nonsteroidal compounds induced biexponential dissociation curves. We propose that, depending on the concentration and identity of the chaser used, different conformational states of the hER-HBD are induced. These different states have different affinities for the hormone.

Introduction

Most studies that deal with drug interaction, assessed by binding competition with a radiolabeled hormone, quantitate effects at equilibrium conditions. However, as life is hardly an equilibrium endpoint, kinetic parameters may be better indicators for monitoring biological events. This study employs new methodology, allowing a high throughput of time-resolved binding experiments.

Due to its importance as a drug target, the estrogen receptor was chosen as a model receptor for the assay development. The dissociation rate constant of steroidal and nonsteroidal estrogens and anti-estrogens was determined by competitive binding. New characteristics of their action were observed.

Methods

Recombinant human estrogen-receptor hormone-binding domain (hER-HBD), produced in yeast (1), was employed in all experiments. Bound [^{3}H]-estradiol was measured with a *homogeneous* assay employing a scintillating microtitration plate as a platform for detection of radioactivity. Thus, the reaction vial, rather than scintillation proximity beads or liquid scintillators, is the means of detection. Radioactivity was measured in a multidetector scintillation counter designed for the microtitration plate format. Plates and instrument were obtained from Wallac Oy, Turku, Finland. The hER-HBD was allowed to attach to the scintillating plastic of the plates. After washing excess receptor from the wells, the plates were ready to use for experimentation, i.e., dissociation-rate determinations in the presence of common binding inhibitors of [^{3}H]-estradiol. All experiments were performed in phosphate-buffered saline at room temperature. Data were evaluated with linear and nonlinear exponential decay models.

Results

In previous experiments, the methodology was validated for the hER-HBD (2) in equilibrium- and kinetic-binding assays employing [^{3}H]-estradiol as radioligand. The present results clearly show that different compounds influence the dissociation rates of [^{3}H]-estradiol (Figure 1). Some compounds induce monoexponential dissociation rates, while others induce biexponential dissociation rates. Among the compounds that induce biexponential dissociation rates, there is a shift from biexponential rates to monoexponential when the concentration of chaser is decreased.

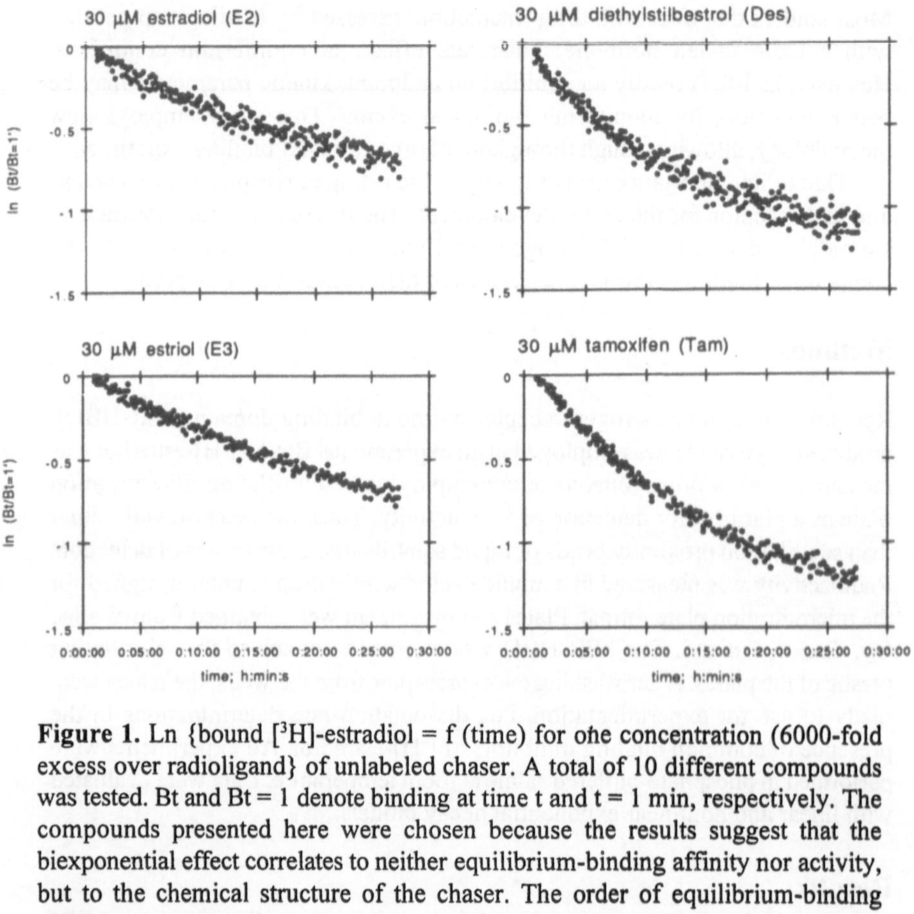

Figure 1. Ln {bound [^{3}H]-estradiol = f (time) for one concentration (6000-fold excess over radioligand} of unlabeled chaser. A total of 10 different compounds was tested. Bt and Bt = 1 denote binding at time t and t = 1 min, respectively. The compounds presented here were chosen because the results suggest that the biexponential effect correlates to neither equilibrium-binding affinity nor activity, but to the chemical structure of the chaser. The order of equilibrium-binding affinities for this series of compounds is the activity within parentheses: DES (agonist) > E_2 (agonist) > Tam (partial agonist/antagonist) > E_3 (partial agonist).

Discussion and Conclusions

This report describes a novel way of studying ligand-receptor interactions. Effects unrelated to equilibrium-binding affinity and pharmacological activity but related to chemical structure were observed.

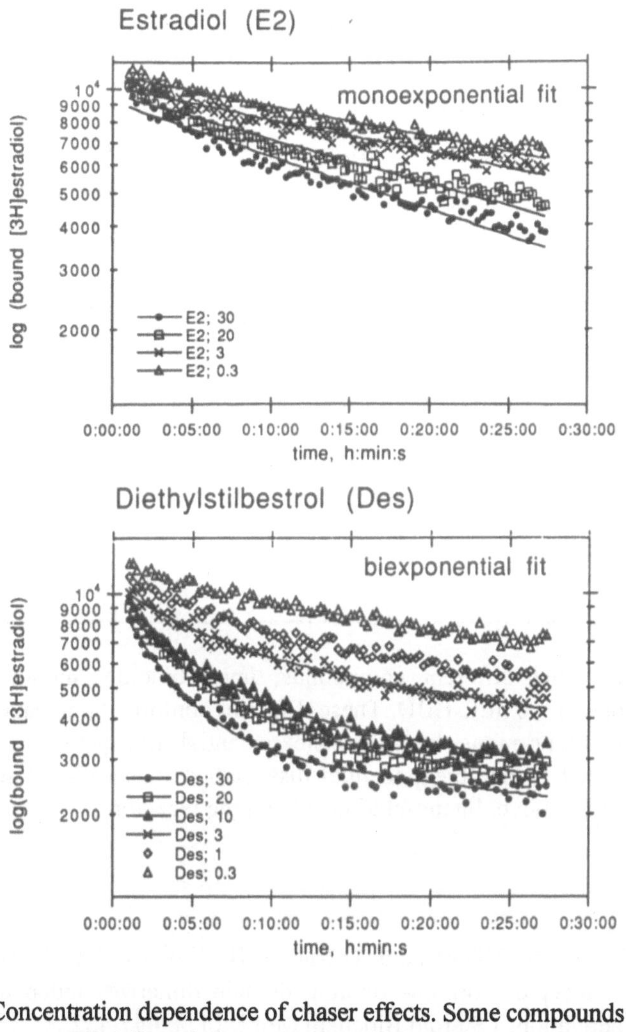

Figure 2. Concentration dependence of chaser effects. Some compounds fitted best with monoexponential dissociation rates (steroids), while other compounds fitted with biexponential rates at high concentrations, although at lower concentrations, the latter turned toward monoexponential rates (nonsteroids). Estradiol fitted equally well with biexponential and monoexponential dissociation at the highest concentration used (30 μM). In general, the higher the tendency to induce biexponential dissociation, the more "fast" sites were present. Concentrations of chaser used were 0.3-30 μM.[^{3}H]-estradiol concentration was 5 nM. Due to legibility considerations, not all estradiol concentrations are shown in this figure.

Table 1. Summary of Effects on [^{3}H]-Estradiol Dissociation Rates at High Concentrations of Chaser. Compounds termed borderline fitted equally well with biexponential dissociation rates and monoexponential dissociation rates, at high concentrations (≥20 μM). The two structural families (steroids, nonsteroids) are indicated.

Monoexponential dissociation rate	Borderline	Biexponential Dissociation rate
Estriol	Estradiol	
ICI 164384	Moxestrol	
	Estrone	
	Nafoxidin	Diethylstilbestrol
	Raloxifen	Tamoxifen
		4-OH-Tamoxifen

We conclude that different compounds, upon binding, induce different conformations in the hER-HBD. These different conformations have different affinities for the hormone, hence the biexponential dissociation of hormone. Compounds that induce biexponential dissociation rates may do so through interaction with a novel ligand-binding site on the receptor.

References

1. Salomonsson M, Häggblad J, O'Malley B, Sitbon G (1994) The human estrogen receptor hormone binding domain dimerizes independently of ligand activation. J Steroid Biochem Mol Biol 48:447-452
2. Häggblad J, Carlsson B, Kivelä P, Siitari H (1994) Scintillating microtitration plates as platform for determination of [^{3}H]-estradiol binding constants for the hER-HBD. Biotechniques, in press.

Estradiol Promotion of Size and Number of Hepatic Enzyme-Altered Foci in Rats in the Absence of Cell Proliferation

Nancy Bower, Louis Mylecraine, Adam P. Stein,
Kenneth R. Reuhl, and Michael A. Gallo

Summary

High-dose (95 µg/kg/day) 17β-estradiol (E_2) administered as a promoter in a two-stage model of hepatocarcinogenesis significantly ($p < 0.05$) increases the number, size (mean volume and mean surface area) and volume density of enzyme-altered foci (EAF) larger than 0.003 mm^2 in the livers of ovariectomized F344 rats when compared with diethylnitrosamine-initiated (150 mg/kg) controls. Changes observed at 9.5 µg E_2/kg/day, which mimics serum E2 levels of intact animals, were limited to a qualitative increase in the number of EAF $\leq$ 0.003 mm^2. At termination (days 39/40), the mean cumulative bromodeoxyuridine (BrdU) labeling indices for the E2-treated groups did not differ from the mean labeling index for initiated controls. E_2's failure to increase the cumulative BrdU-labeling index indicates that generalized hepatic growth stimulation (proliferation as measured by BrdU-labeling index) does not completely explain the increase in number and/or size of EAF observed. It is hypothesized that an alteration of the cell-cycle kinetics of initiated hepatic EAF may contribute to the promotion of hepatocarcinogenesis by E_2.

Introduction

Epidemiological studies conducted in the 1970s and 1980s have shown a link between prolonged use of oral contraceptives (OC) and liver neoplasms in women (1-3). Although estrogens are not genotoxic in classical assays, synthetic steroidal estrogens, such as ethinylestradioland 17-α ethinylestradiol are effective promoters of preneoplastic, enzyme-altered foci and hepatocellular carcinoma in the rat two-stage model of hepatocarcinogenesis (4, 5). In addition, ovarian hormones such as E_2 enhance the tumor-promoting activity of xenobiotics such as 2,3,7,8-tetrachlorodibenzo-*p*-dioxin (TCDD) in the rat liver (6). However, the promotion of experimental hepatocarcinogenesis by naturally-occurring estrogens such as E_2 has received little study. Our study used a two-stage model of hepatocarcinogenesis in ovariectomized female Fisher 344 rats to evaluate the ability of E_2 to promote experimental hepatocarcinogenesis. The parameters measured included liver weight, number and size of enzyme-altered foci, cell proliferation, and serum E2 levels. The rats received a single intraperitoneal initiating dose of diethylnitrosamine (DEN) (150 mg/kg) and were fed a choline- and methionine-deficient diet for 6 weeks, commencing 72 h after initiation. E_2, administered as subcutaneously-implanted, timed-release pellets, was given as a low dose (approximately 9.5 μg/kg/day) for 3 weeks preinitiation and 6 weeks postinitiation (to simulate the intact animal) or as a high dose (approximately 95 μg/kg/day) for 6 weeks postinitiation.

Table 1. Number of GST-P+ Hepatic Foci at Study Termination (Days 39/40).

GROUP*	SMALL FOCI/LIVER ($< 0.003\ mm^2$)	LARGE FOCI/LIVER ($> 0.003\ mm^2$)
E_2(L)+DEN+E_2(L)	190.5**[a,b,d]	17.0[a]
	133-248	5-37
CON+DEN+E_2(H)	87.7	59.7[a,b,c,d]
	59-102	17-143
CON+SAL	2.0	0.8
	0-5	0-2
CON+DEN	53.0	18.7[a]
	35-67	11-27
E_2(H)+CON	13.0	0.3
	1-21	0-1

*Pretreatment+ Initiation + Postinitiation treatment

**Mean Range n = 3 - 8

[a] = Significantly different from uninitiated control (CON + SAL)

[b] = Significantly different from initiated control (CON + DEN)

[c] = Significantly different from E_2(L)-treated animals [E_2(L) + DEN +E_2(L)]

[d] = Significantly different from E_2(H)-treated control animals [E_2(H) + CON]

Neuman-Keuhl's multiple comparisons test used for all statistical comparisons, $p<0.05$

Abbreviations used in Tables 1-3: DEN = Diethylnitrosamine

E_2(L) = Low-dose E_2 (~9.5 μg/kg/day) CON = Placebo control

E_2(H) = High-dose E_2 (~95.0 μg/kg/day) SAL = 0.9% Sodium Chloride Injection, USP

Results and Discussion

Enzyme-altered foci were identified by the phenotypic marker glutathione-S-transferase, placental type (GST-P+), which is considered a sensitive immunohistochemical marker for hepatocarcinogenesis in rodents (7), and were quantified using computer-enhanced image analysis (*Presage®*, Advanced Imaging Concepts, Inc., Princeton, NJ). E_2 effectively promoted the formation of enzyme-altered foci in ovariectomized rats. Compared with the initiated control (CON + DEN), administration of low doses of E_2 before and after initiation significantly increased the number of small (< 0.003 mm^2) hepatic GST-P+ foci present, but not the number or size of the large foci present. Administration of high doses of E_2 after initiation significantly increased both the number of large (> 0.003 mm^2) foci present, and the volume of liver occupied by these foci (Tables 1 and 2).

Table 2. Size and Volume Density of GST-P+ Hepatic Macrofoci at Study Termination (Days 39/40).

GROUP	MEAN SURFACE AREA (mm^2 x10^{-3})	MEAN FOCUS VOLUME (mm^3 x 10^{-3})	VOLUME DENSITY (No./mm^3 x 10^{-3})
E_2(L)+DEN+E_2(L)	105.2*[a,b] 74-175	251.5[a] 11-578	1.13[a] 0.3-2.0
CON+DEN+E_2(H)	206.5[a,b,c] 103-332	29841.0[a,b,c] 164-117819	8.70[a,b,c] 1.2-23.2
CON+SAL	18.0 0-36	0.1 0	0.04 0
CON+DEN	69.4[a] 26-149	180.4[a] 1-754	1.08[a] 0.2-2.6

*Mean Range n = 6 - 8

[a] = Significantly different from uninitiated control (CON + SAL)

[b] = Significantly different from initiated control (CON + DEN)

[c] = Significantly different from E_2(L)-treated animals [E_2(L) + DEN +E_2(L)]

Neuman-Keuhl's multiple comparisons test used for all statistical comparisons, $p < 0.05$.

Additionally, high doses of E_2 promoted the clonal expansion of these foci, as indicated by the significant (3.0- and 165.0-fold, respectively) increases in mean focus surface area and volume (Table 2).

Pre- and/or post-initiation administration of E_2 to DEN-initiated, ovariectomized rats stimulated a hypertrophic/hyperplastic response in the liver, as indicated by a significant increase in relative liver weight compared with both initiated (CON + DEN) and uninitiated (CON + SAL) controls (Table 3). Serum E2 levels in the animals given the low dose of E_2 were in the range of physiological values (1.2-80 pg/mL) reported for rats during normal reproductive cycles (8). In rats administered the high dose of E_2, mean serum E2 levels averaged approximately two times greater than the maximum physiological value reported for rats during normal reproductive cycles (Table 3).

Table 3. Comparison of Mean Serum E_2 Values, Relative Liver Weights, and Cumulative Liver Labeling Indices (%) at Study Termination (Days 39/40).

DOSE GROUP	Serum Estradiol (pg/mL)	Relative Liver Weight (g/100g bwt)	Labeling Index[a]	
			Percent	Fold Increase over Control
6 E_2(L) + DEN + E_2(L)	44.4[b,c] 35-55	4.766[d,e,g] 4.35-5.49	34.03[d,g] 33.4-34.7	2.49
CON + DEN + E_2(H)	156.0 72-223	5.526[d,e,f,g] 5.06-6.66	26.56 22.3-30.8	1.95
CON + SAL	ND	2.502 2.35-2.76	13.65* 7.4-20.7	--
CON + DEN	ND	3.731[d,g] 3.04-4.47	29.43 24.1-34.3	2.15
E_2(H) + CON	11.2 ND-22	3.639[d] 2.87-4.28	8.40 8.36-8.44	--

[a] Labeling Index = No. Labeled Nuclei 4000 Nuclei Counted (1000 Nuclei Counted/Lobe X 4 Lobes)
[b] Mean Range
[c] Normal rat serum estradiol levels = 1.2 - 80 pg/mL
[d] = Significantly different from uninitiated control (CON + SAL)
[e] = Significantly different from initiated control (CON + DEN)
[f] = Significantly different from E_2(L)-treated animals [E_2(L) + DEN + E_2(L)]
[g] = Significantly different from E_2(H)-treated control animals [E_2(H) + CON]
Neuman-Keuhl's multiple comparisons test used for all statistical comparisons, $p<0.05$.
* Combined data for Days 1 and 39/40.
ND = Not detectable (below detection limit of 10 pg/mL)

BrdU pellets were implanted scduring the period of E2 treatment. The labeling index (LI) was determined as the number of labeled nuclei/4000 nuclei counted. Although an increase in the LI was observed in both groups treated with E_2, the increase was similar to the CON + DEN control group. Because the LIs in the E_2-treated groups and the initiated control group are similar, promotion of EAF in rats by E_2 cannot be explained by cell proliferation alone. Preliminary data from our lab suggest effects on cell-cycle kinetics as a possible mechanism in the promotion of experimental hepatocarcinogenesis by E_2.

Conclusion

Our study has shown that, like synthetic estrogens, E_2 effectively promotes the formation of enzyme-altered foci in rats. This promotion cannot be explained by cell proliferation alone, and preliminary results indicate that it may result from the effects of E_2 on cell-cycle kinetics.

References

1. Huggins GR, Zucker PK (1987) Oral contraceptives and neoplasia: 1987 update. Fert and Steril 47:733-761.
2. Palmer JR, Rosenberg L, Kaufman DW, et al. (1989) Oral contraceptive use and liver cancer. Am J Epidemiol 130:878-882.
3. Henderson BE, Preston-Martin S, Edmondson HA, et al. (1983) Hepatocellular carcinoma and oral contraceptives. Brit J Cancer 48:437-440.
4. Yager J, Roebuck BD, Paluszcyk TL, Memoli VA (1986) Effect of ethynyl estradiol and tamoxifen on liver DNA turnover and new synthesis and appearance of gamma glutamyl transpeptidase-positive foci in female rats. Carcinogenesis 7:2007-2014.
5. Campen DB, Sloop TC, Maronpot RR, Lucier GW (1987) Continued development of hepatic γ-glutamyltranspeptidase-positive foci upon withdrawal of 17 α-ethynyl estradiol in diethylnitrosamine-initiated rats. Cancer Res 47:228-2333.
6. Tritscher AM, Clark GC, Lin, FH, Lucier GW (1992) Ovarian hormones enhance the carcinogenic activity of 2,3,7,8-TCDD in rats. In Li JJ, Nandi S, Li SA (eds): Hormonal Carcinogenesis. New York: Springer-Verlag, pp 326-329.
7. Satoh K, Hatayama I, Tateoka N, et al. (1989) Transient induction of single GST-P positive hepatocytes by DEN. Carcinogenesis 10:2107-2111.
8. Günzel P, Putz B, Lehmann M, Hasan SH, et al. (1989) Steroid toxicology and the "pill": Comparative aspects of experimental test systems and the human. In Dayan AD and Paine AJ (eds): Advances in Applied Toxicology. New York: Taylor and Francis, Ltd, pp 19-49.

Expression of Liver-Enriched Transcription Factors During Progression in the Resistant-Hepatocyte Model

Joshua DeZhong Liao, Per Flodby, Agneta Blanck,
Kleanthis G. Xanthopoulos, and Inger Porsch-Hällström

Summary

The expression of the CCAAT/enhancer binding proteins (C/EBP) α and β, and hepatocyte nuclear factors (HNF) 1 and 4 were measured in Wistar rats treated according to the resistant-hepatocyte (RH) model. A decreased nuclear transcription and mRNA content of C/EBPα was observed in early nodules, persistent nodules, and hepatocellular carcinomas, compared with their respective control or surrounding livers. A lower level of C/EBPα protein was observed in persistent nodules. HNF-1 and HNF-4 were decreased at the protein level but not at the mRNA or at the transcriptional levels in persistent nodules. No changes were observed in C/EBPβ.

Introduction

Persistent nodules, occurring in rat liver after treatment with carcinogens, display a cascade of phenotypic changes, including several features rendering nodules resistant to hepatotoxic substances (1). Development of the "resistance phenotype" in experimental hepatocarcinogenesis has been suggested as a key event in the clonal proliferation of initiated cells and their subsequent development into hepatocyte nodules and cancer (1). In general, the levels of several phase II enzymes and the multidrug-resistance protein are markedly increased, while the levels of most cytochrome P-450 (CYP) enzymes are decreased (1). Recently, the regulation of the genes involved in this process has begun to unravel, but in most aspects, the regulatory events leading to the resistance phenotype remain unclear.

Liver-enriched transcription factors, including the hepatocyte nuclear factors (HNF) and the CCAAT/enhancer binding protein (C/EBP) family, are involved in the positive control of genes expressed predominantly in the liver, such as albumin, and in the negative control of α-fetoprotein (2). A changed expression of these genes has been observed in hepatic nodules. Furthermore, binding sites for liver-specific transcription factors have been identified in the regulatory regions of genes involved in the resistance phenotype, such as the placental form of glutathione S-transferases (GST-P) and GST Ya (3), as well as many CYP forms (4, 5). Therefore, an altered control of liver-specific transcription factors may contribute to the development of the resistance phenotype of liver nodules. This study was designed to investigate whether an altered expression of these factors may be associated with liver carcinogenesis.

Results

After eight months of initiation in the RH model, C/EBPα mRNA expression decreased in nodules, and in hepatomas after 11-14 months, when compared with the corresponding surrounding tissues (Figure 1). The transcription in nuclei prepared from nodules and hepatomas isolated 11 and 14 months after initiation was also decreased. These changes were observed after two weeks of 2-acetylaminofluorene (2-AAF) treatment in the early nodules treatment, compared with livers of rats receiving only 2-AAF and partial hepatectomy (PH) (Figure 1).

There were no significant changes in mRNA expression or transcription of C/EBPβ, HNF-1, and HNF-4 (data not shown). No sex differences were observed in the expression of these genes. Western blot analysis revealed decreased protein levels of C/EBPα, HNF-1, and HNF-4 in nuclear extracts from persistent nodules (Figure 2). No differences were apparent in the content of C/EBPβ protein (data not shown).

Discussion

Liver-enriched transcription factors play a pivotal role in induction and maintenance of the differentiated state of the hepatocyte. This study demonstrates decreased expression of C/EBPα at the level of transcription, mRNA, and protein. These changes were observed shortly after promotion, indicating that C/EBPα down-regulation is an early event in liver carcinogenesis in this model. C/EBPα is regarded as a negative growth regulator and its transcription is down-regulated during the prereplicative phase of liver regeneration (2, 8). HNF-4 serves as a common regulator for the liver-specific transcriptional activation of many genes of the CYP2 family (4, 5), and an HNF-1-like element has been localized in the promoter of the rat CYP2E1 gene (9).

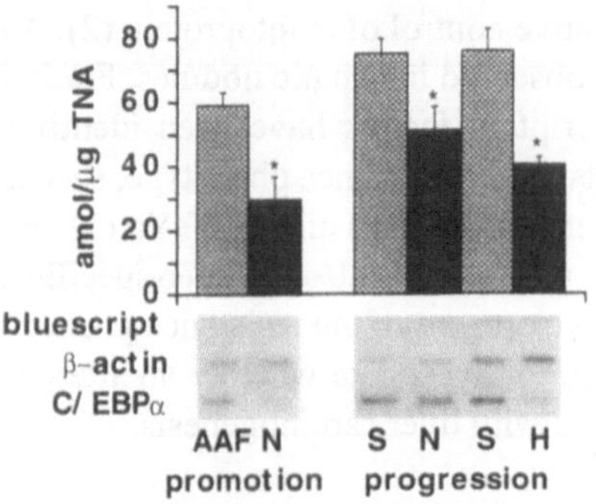

Figure 1. C/EBPα mRNA expression (upper panel) and nuclear transcription (lower panel). Male Wistar rats initiated with DEN and promoted with 2-AAF + PH according to the RH model (6). Early male nodules (N) isolated 2 weeks after 2-AAF treatment and liver of male rats receiving only 2-AAF/PH during promotion, individual persistent nodules (N), hepatomas (H), and respective surrounding livers (S) from male rats during progression were analyzed. *Significantly lower than the 2-AAF-treated controls or the respective surrounding liver ($p < 0.05$, t test).

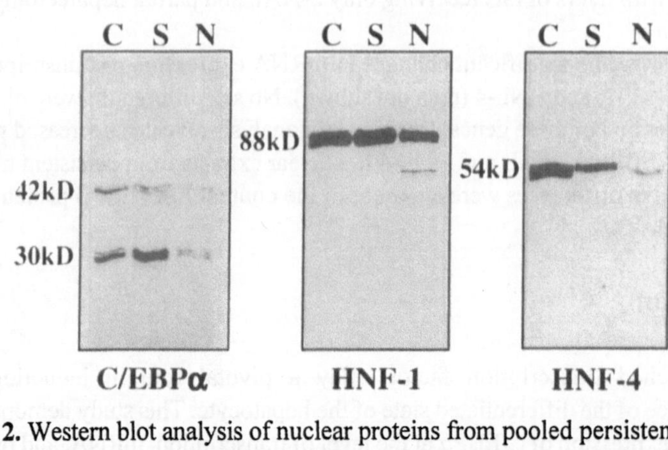

Figure 2. Western blot analysis of nuclear proteins from pooled persistent nodules (N), their respective surrounding livers (S), and age-matched control rat liver (C), performed as previously described (8), with 20 μg nuclear protein per well.

Therefore, the early down-regulation in liver carcinogenesis supports that C/EBPα expression is incompatible with cell proliferation in target organs. It is possible that the reduction of the C/EBPα level contributes to a higher proliferative advantage of the pre- and neoplastic lesions by preventing hepatocytes from returning to their normal quiescent state.

Whereas the C/EBPα gene shows a reduced transcription, the decrease in the expression of the HNF-1 and -4 genes seems to be post-transcriptionally regulated. It has been suggested that additionally, the GST-Ya promoter contains putative binding sites for HNF-1 and -4 (3), and C/EBPα has to transactivate the xenobiotic response element in the regulatory regions of this gene (10). Therefore, the possibility cannot be excluded that the decreased expression of C/EBPα and HNF-1 and -4 might be involved in development of the resistance phenotype in preneoplastic liver nodules, and could have impact on the regulation of several genes during liver carcinogenesis.

References

1. Farber E, Sarma DSR (1987) Hepatocarcinogenesis: A dynamic cellular perspective. Lab Invest 56:4-22.
2. De Simone V, Cortese R (1992) Transcription factors and liver-specific genes. Biochim Biophys Acta 1132:119-126.
3. Daniel V (1993) Glutathione S-Transferase: Gene structure and regulation of expression. Crit Rev Biochem Mol Biol 28:173-207.
4. Venepally R, Chen D, Kemper B (1992) Transcriptional regulatory elements for basal expression of cytochrome P450IIC genes. J Biol Chem 267:17333-17338.
5. Chen D, Lepar G, Kemper B (1993) A transcriptional regulatory element common to a large family of hepatic cytochrome P450 genes is a functional binding site of the orphan receptor HNF-4. J Biol Chem 269:5420-5427.
6. Solt D, Farber E (1976) New principle for the analysis of chemical carcinogenesis. Nature 263:702-703.
7. Blanck A, Porsch Hällström I, Svensson D, et al. (1993) Increased expression of the female-predominant cytochrome P450 2C12 in liver nodules from male Wistar rats. Carcinogenesis 14:755-759.
8. Flodby P, Antonson P, Barlow C, et al. (1993) Differential expression of three C/EBP isoforms, HNF-1 and HNF-4 after partial hepatectomy in rats. Exptl Cell Res 208:248-256.
9. Ueno T, Gonzalez FJ (1990) Transcriptional control of the rat hepatic CYP2E1 gene. Mol Cell Biol 10:4495-4505.
10. Pimental R, Liang B, Yee G, et al. (1993) Dioxin receptor and C/EBP regulate the function of the glutathione S-transferase Ya gene xenobiotic response element. Mol Cell Biol 13:4365-4373.

Nuclear Protein Binding to the AP-1 and CRE Sites During Sex-Differentiated Promotion of Rat Liver Carcinogenesis

Lena C.E. Ohlson and Inger Porsch Hällström

Summary

The binding of liver nuclear proteins to the AP-1 and CRE sites was studied. Decreased binding to both sites was observed in liver extracts from 2-AAF-promoted and unpromoted male and female rats initiated with DEN. Overall, binding to the AP-1 and CRE sites was lower in the extracts from females than in those from the males.

Introduction

Promotion of rat liver carcinogenesis in the resistant-hepatocyte (RH) (1) model is markedly sex differentiated (male > female) (2). During the week after partial hepatectomy (PH) and during 2-acetylaminofluorene (2-AAF) promotion, when the growth hormone (GH)-dependent difference in outgrowth of preneoplastic foci becomes manifest, we have observed increased mRNA levels of c-*jun*, *jun*B, and liver regeneration factor (LRF-1)(3). The protein products of these genes are implicated in growth control by binding to AP-1 recognition sites, but also have affinity to cAMP-responsive elements (CREs). To investigate whether the observed increases in mRNA levels are reflected by increased binding to these recognition sequences, the binding capacity of nuclear extracts from rats treated in the RH model was studied with gel mobility shift analysis.

Materials and Methods

Male and female Wistar rats were treated according to the RH model (diethylnitrosamine 200 mg/kg i.p., 0.02 % dietary 2-AAF for two weeks with a PH in the middle of the feeding period) (1). Total livers from males and females were isolated at PH, 1-7 days and three weeks after PH. Nuclear proteins were prepared according to Hattori, et al. (4), and gel mobility shift analysis was performed on 4% BIS-acrylamide gels. [^{32}P]-end labeled oligonucleotides covering the recognition sites and flanking regions from the collagenase (AP-1), and choriogonadotropin (CRE) genes were used as probes. The sense strands of the annealed oligonucleotides were as follows (lowercase letters denote nucleotides present in double-stranded oligonucleotides after fill-in reaction):

AP-1: ctagAAGCATGAGTCAGACACctag

αCRE: ctagCGAGAAATTGACGTCATGGTAActag

Results

The interaction of rat liver nuclear proteins with the CRE sequence gave rise to five specific complexes. Supershifts were produced by the addition of c-*jun*, *jun*-D, LRF-1 and CREB antibodies. As shown in Figure 1, decreased binding to the site was observed one week after PH with nuclear protein extracts from DEN-initiated animals with or without 2-AAF treatment compared with extracts prepared at PH. Band 1, containing LRF-1 and c-*jun* proteins, is reduced in extracts taken 7 days after PH from females and males in both groups. Band 4, which contains unknown proteins, seemed to decrease only in female and male extracts in the DEN+2-AAF group. Overall, lower binding was observed in extracts from females than from males in this group.

Six complexes were formed with the AP-1 oligonucleotide, which were competed with excess of unlabeled competitor. Supershifts were produced by the addition of antibodies toward c-*jun*, *jun*-B, *jun*-D, and c-*fos*. LRF-1 antibodies resulted in loss of one complex.

As shown in Figure 2, decreased binding was also noted at this site one week after PH in 2-AAF-promoted and unpromoted DEN-initiated animals. Band 1, containing complexes between c-*fos* and the *jun* proteins (c-, -B and -D), decreased in extracts taken 7 days after PH from both sexes in the DEN+2-AAF-treated and the DEN-initiated non-2-AAF-treated group.

Band 4, not yet identified, and band 5 containing LRF-1, were almost abolished in extracts taken 7 days after PH from both females and males in the DEN+2-AAF-treated group, and were decreased in extracts taken 7 days after PH from initiated non-2-AAF-treated females and males. The total binding was lower in liver extracts from females compared with males. When the binding capacity of nuclear extracts prepared from initiated males and females at 1, 3, and 7 days and three weeks after PH was studied, a gradually decreased binding was observed with time; the complex formation was almost abolished at three weeks after PH.

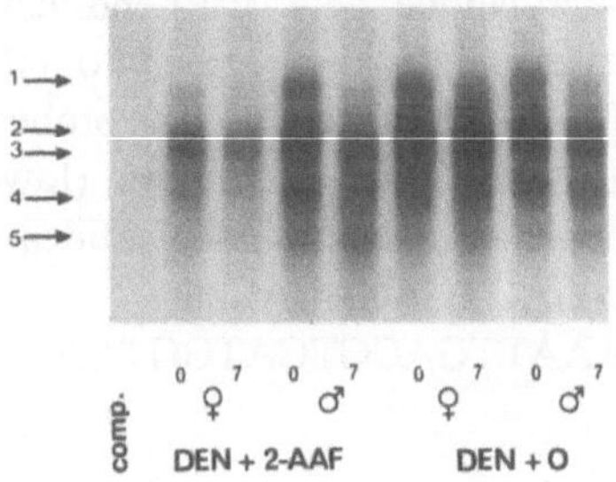

Figure 1. Binding to the CRE site in nuclear protein extracts prepared from DEN+2-AAF-treated male and female rat livers, compared with DEN-initiated male and female rats not treated with 2-AAF, at PH (0) and 7 days after PH in the RH model. Competitor oligonucleotide (unlabeled CRE oligo in 200.0-fold excess) added to extract from female PH (0) DEN+2-AAF. Complexes representing specific interactions with the CRE oligonucleotide are indicated to the left.

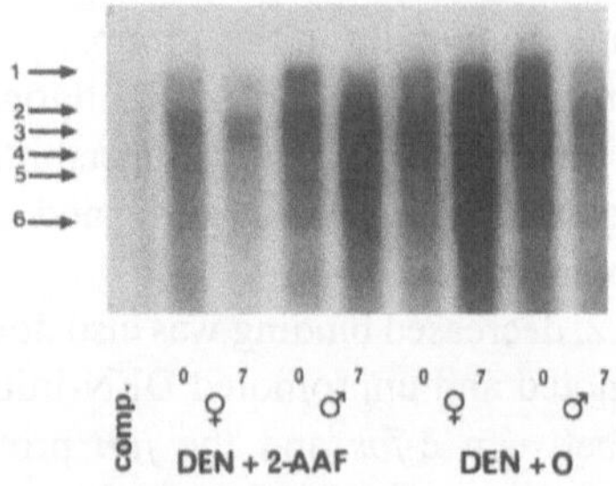

Figure 2. Binding to the AP-1 site in nuclear protein extracts prepared from DEN+2-AAF-treated male and female rat livers, compared with DEN-initiated male and female rats not treated with 2-AAF, at PH (0) and 7 days after PH in the RH model. Experimental protocol as described in figure 1.

Also, in early male nodules isolated three weeks after PH, binding to the AP-1 site was extremely low.

Discussion

The decreased binding to the AP-1 site, during promotion in the RH model and in isolated early nodules, suggests that this site is not important for growth stimulation in early preneoplastic lesions. The decreased binding to the AP-1 and CRE sites in spite of increased mRNA expression of several genes, which products bind to these elements, could be due to modification at the translational or post-translational level, the latter including phosphorylation or nuclear translocation. It could also be due to higher affinity of the proteins for binding to other AP-1 sites not involved in growth signaling during 2-AAF treatment. It is worth noting that AP-1 and TRE-like sequences reside in both promoter and enhancer motifs involved in the response to xenobiotics of GST-Ya, GST-P, and the quinone reductase genes (5-7).

Acknowledgments

This work has been supported by the NIEHS, NIH (1RO1 CA 57925-01), and the Swedish Cancer Society.

References

1. Solt D, Farber E (1976) New principle for the analysis of chemical carcinogenesis. Nature 262:701-703.
2. Blanck A, Hansson T, Eriksson LC, et al. (1987) Growth hormone modifies the growth rate of enzyme-altered hepatic foci in male rats treated according to the resistant hepatocyte model. Carcinogenesis 8:1585-1588.
3. Liao D, Blanck A, Gustafsson J-Å, Porsch Hällström I (1994) Expression of the c-*jun*, LRF-1, *jun*-B and *ets*-2 genes during promotion and progression of rat liver carcinogenesis in the resistant hepatocyte model. In prep.
4. Hattori M, Tugores A, Veloz L, et al. (1990) A simplified method for the preparation of transcriptionally active liver nuclear extracts. DNA and Cell Biol 9:777-781.
5. Daniel V (1993) Glutathione S-transferase: Gene structure and regulation of expression. Biochem Mol Biol 28(3):173-207.
6. Morimura S, Okuda A, Sakai M, et al. (1992) Analysis of glutathione transferase P gene regulation with liver cells in primary culture. Cell Growth & Diff 3:685-691.
7. Bergelson S, Pinkus R, Daniel V (1994) Induction of AP-1 (*fos/jun*) by chemical agents mediates activation of glutathione S-transferase and quinone reductase gene expression. Oncogene 9:565-571.

Induction of Endometrial Cancer by Tamoxifen in the Rat

Eero T. E. Mäntylä, Stefan H. Karlsson, and Lauri S. Nieminen

Summary

Induction of preneoplastic and neoplastic changes by two anti-estrogenic drugs was studied in female rats. Groups of female rats were treated with tamoxifen citrate (TAM) (11.3 or 45 mg/kg) or toremifene citrate (TOR) (3, 12, or 48 mg/kg) by daily oral gavage to rats. Also, a group of animals was treated for 20-52 weeks followed by a 12-13 recovery period. The animals were killed 13-52 weeks after treatment, including control rats. The uteri of all the rats were examined macroscopically and histopathologically. No pathological uterine changes were detected in the control group, the three TOR groups, or the low-dose TAM group. However, in the high-dose TAM group, 10 cases (10/104; 9.6%) of squamous cell metaplasia were detected. In three (3/104; 2.9%) of the metaplasias, a focal dysplastic change was observed. In two (2/104; 1.9%) of these, focal invasive squamous cell carcinomas were verified. The present data on endometrial cancer induced by TAM provides a link between human and animal data and suggests that a similar mechanism of cancer induction may be at work in both species.

Introduction

TAM is a valuable anti-estrogenic anticancer drug used in adjuvant and primary therapy of breast cancer for over a decade. Recently, prophylactic trials in high-risk groups have been initiated (1). The adverse effects of TAM are mild, but recent data suggest that TAM therapy can cause ocular toxicity (2) and secondary endometrial cancers (3-5). In the Stockholm adjuvant trial, the incidence of endometrial cancer was 1.4% (13/941) (3), but these cancers can be more aggressive with considerable mortality (4). Also, TAM is a genotoxic hepatocarcinogen in the rat (6, 7) and induces hepatic DNA adducts (7). Herein, we examined whether long-term TAM treatment causes endometrial cancer in the rat. The new anti-estrogenic drug, TOR, was used as a reference compound. TOR is chemically closely related to TAM. Its clinical efficacy is similar to TAM's, but it induces neither malignant tumors nor adduct formation in rat liver (6, 7).

Methods

TAM citrate and TOR citrate (Z isomers, at least 99% pure) were used as test chemicals. Female Sprague-Dawley rats were treated by daily oral gavage. Equimolar dose levels of TAM (11.3 or 45 mg/kg) and TOR (3, 12, or 48 mg/kg). The animals were killed after 13, 20, 26, or 52 weeks of treatment. Recovery groups were also included after 20, 26, or 52 weeks. Following sacrifice, uteri were removed. After fixation in formalin, samples were cut from 1-3 macroscopically normal sites on the right uterine horn. All macroscopic abnormalities were recorded. Paraffin sections were stained with hematoxylin and eosin, and examined by light microscopy. The histological study was based on a critical classification of the endometrial changes, i.e., only metaplasia with keratinization was evaluated as true squamous cell metaplasia. The classification of carcinomas was based on criteria from human pathology of the cervix uteri.

Results

No squamous cell metaplasia was observed in the control group (0/109), the low-dose TAM group (0/25), or in the three TOR groups (0/38, 0/62 or 0/64 animals, in increasing dose-level order). However, squamous cell metaplasia of the endometrial epithelium was observed in 10/104 animals from the high-dose TAM group. The earliest finding was made after 13 weeks of treatment. Focal dysplastic changes in the squamous cell epithelium were observed in three cases. In addition, two of these cases showed focal squamous cell carcinoma. Of the carcinomas, one case was found after 20 weeks of treatment with 12 weeks of recovery, and another after 26 weeks of treatment with 13 weeks of recovery. Both cases displayed a diffuse pattern of densely packed small atypical cells with strongly basophilic nuclei and typical large squamous carcinoma cells in the subepithelial tissue. Signs of invasion through the myometrium, lymphocytic infiltration, and epithelial necrosis were also observed in both cases (Table 1).

Discussion

Results indicate that long-term oral treatment with a high dose of TAM causes well-differentiated keratinized endometrial squamous cell metaplasia in the rat. Spontaneous squamous cell metaplasia of the endometrium is rare in rats, except on superficial and glandular epithelium of some endometrial polyps.

Table 1. Cumulative frequency of the pathological endometrial changes in control and TOR citrate- (TOR) or TAM citrate-treated (TAM) rats.

Treatment Daily Dose	Squamous Cell Metaplasia	Dysplasia	Squamous Cell and Carcinoma
Control	0[a]/109[b] (0%)	0/109 (0%)	0/109 (0%)
TOR 48 mg/kg	0/64 (0%)	0/64 (0%)	0/64 (0%)
TAM 45 mg/kg	10/104 (9.6%)	3/104 (2.9%)	2/104 (1.9%)

[a] Number of observed cases. [b] Number of animals studied.

Diethylstilbestrol (DES) induces squamous metaplasia in mice exposed transplacentally (8) or neonatally (9). In these cases, the squamous metaplasia has been poorly differentiated, with focal stratification but no keratinization (8, 9). The effect of DES may occur directly on the developing uterine epithelium, and these lesions may respond later in development to estradiol, resulting in squamous metaplasia (9). Thus these lesions may result from abnormal hormonal stimuli (8). This finding is supported by the fact that squamous cell carcinoma is rarely found in DES-treated rats (10).

However, in chemically-induced endometrial squamous cell carcinomas, well-differentiated keratinized squamous metaplasia is the earliest epithelial preneoplastic change. In most of these studies, the carcinogen was directly applied to the uterine wall (11). Spontaneous squamous cell carcinoma is also very rare in rats (11). In a study by Curtis, et al., (12), uterine squamous cell carcinoma occurred only in animals older than 16 months with a frequency of 0.2%. Spontaneous squamous cell carcinomas do not seem to be hormone-dependent. On the other hand, when a carcinogenic chemical is applied to the uterine wall, squamous cell carcinoma is more frequent and is the most common malignant change (10). These findings suggest that the endometrial cancer in TAM-treated rats may be hormone independent, indicating a possible nonhormonal effect of TAM on the endometrium. The fact that TOR, with hormonal effects equal to TAM's (13), induced neither endometrial squamous cell metaplasia nor squamous cell carcinoma supports this idea. In our study, the two highest doses decreased uterine weights about 50%, indicating a similar hormonal effect on this target organ.

The present results indicate that, in addition to the liver, the endometrium may be a target tissue for TAM genotoxicity. At equal TAM concentrations in the rat, induction of uterine cancers occurred later, and the frequency of the neoplasias (1.9%) was lower than that for liver carcinomas (30-80%) (6, 7).

In conclusion, our results suggest that TAM can induce preneoplastic changes and squamous cell carcinomas in the rat endometrium. TOR does not.

The mechanism of TAM-induced endometrial carcinogenesis may be nonhormonal and may be related to direct genotoxicity of the parent compounds or its metabolites.

References

1. Powles TJ, Tillyer CR, Jones AL, et al. (1990) Prevention of breast cancer with tamoxifen. An update on the Royal Marsden Hospital pilot programme. Eur J Cancer 26:680-684.
2. Pavlidis NA, Petris C, Briassoulis E, et al. (1992) Clear evidence that long-term, low-dose tamoxifen treatment can induce ocular toxicity. Cancer 69:2961-2964.
3. Fornander T, Rutqvist LE, Cedermark B, et al. (1989) Adjuvant tamoxifen in early breast cancer: Occurrence of new primary cancers. Lancet 1:117-120.
4. Magriples U, Naftolin F, Schwartz PE, Carcangiu ML (1993) High-grade endometrial carcinoma in tamoxifen-treated breast cancer patients. J Clin Oncol 11:485-490.
5. vanLeeuwen FE, Benraadt J, Coebergh JWW, et al. (1994) Risk of endometrial cancer after tamoxifen treatment of breast cancer. Lancet 343:448-452.
6. Hirsimäki P, Hirsimäki Y, Nieminen L, Payne BJ (1993) Tamoxifen induces hepatocellular carcinoma in rat liver: A 1-year study with two antiestrogens. Arch Toxicol 67:49-54.
7. Hard GC, Iatropoulos MJ, Jordan K, et al. (1993) Major difference in the hepatocarcinogenicity and DNA adduct forming ability between toremifene and tamoxifen in female Crl:CD(BR) rats. Cancer Res 53:4534-4541
8. Walker BE (1983) Uterine tumors in old female mice exposed prenatally to diethylstilbestrol. J Natl Cancer Inst 70:477-484.
9. Ostrander PL, Mills KT, Bern HA (1985) Long-term responses of the mouse uterus to neonatal diethylstilbestrol treatment and to later sex hormone exposure. J Natl Cancer Inst 74:121-135.
10. Baba N, von Haam E (1967) Experimental carcinoma of the endometrium. Adenocarcinoma in rabbits and squamous cell carcinoma in rats and mice. Prog Exptl Tumor Res 9:192-260.
11. Goodman DG, Hildebrandt PK (1987) Squamous Cell Carcinoma, Endometrium/Cervix, Rat. In Jones,TC, Mohr, U, Hunt, RD (eds): Monographs on Pathology of Laboratory Animals: Genital System. Berlin: Springer-Verlag, pp 82-83.
12. Curtis MR, Bullock FD, Dunning WF (1931) A statistical study of the occurrence of spontaneous tumors in a large colony of rats. Am J Cancer 15:67-121.
13. Kendall ME, Rose DP (1992) The effects of DES, tamoxifen, and toremifene on estrogen-inducible hepatic proteins and estrogen receptor proteins in female rats. Toxicol Appl Pharmacol 114:127-131.

Mechanisms of Anti-Estrogen and Retinoid Inhibition of Breast Cancer Cell Proliferation

Colin K.W. Watts, N. Wilcken, A. Warlters, E.A. Musgrove, and R.L. Sutherland

Summary

In breast cancer, anti-estrogens and retinoids may be chemopreventive because they have antiproliferative activities in the normal breast epithelium similar to those elicited in breast cancer cells. Using estrogen-responsive breast cancer cell lines, we investigated the mechanisms by which these compounds control cell cycle progression. Anti-estrogens and retinoids block cells in the G_1 phase of the cell cycle, hence reducing the S-phase population. However, the effects of retinoids are delayed by about 16 h. Northern blot analysis showed that anti-estrogens rapidly down-regulate c-*myc* mRNA expression, followed by a decrease in cyclin D1 mRNA levels, an effect reversible by the addition of estradiol. Inhibition of entry into S-phase began 9-12 h after treatment; therefore, inhibition of c-*myc* and cyclin D1 expression may not be merely a consequence of changes in cell cycle progression. In contrast, retinoid treatment of breast cancer cells did not affect expression of either gene. However, both retinoids and anti-estrogens increased the proportion of the underphosphorylated, growth inhibitory form of the retinoblastoma protein (Rb), with a time course comparable to changes in decreased entry into S-phase. These data implicate Rb in the mechanism of inhibition of cell cycle progression by both anti-estrogens and retinoids.

Introduction

Anti-estrogens and retinoids have chemopreventive actions in animal models of breast cancer. The mechanisms are unknown, but these compounds may have antiproliferative activities in normal mammary epithelial cells similar to those anti-estrogens and retinoids control cell-cycle progression, using the MCF-7 and T-47D estrogen-responsive human breast cancer cell lines treated with retinoic acid (RA), steroidal pure anti-estrogens (ICI 164384 and ICI 182780) or nonsteroidal anti-estrogen (4-hydroxytamoxifen, OHT). The protooncogene c-*myc*, cyclins, and their associated cyclin-dependent kinases are potential key targets for steroid hormone/steroid antagonism and growth factor regulation of breast epithelial cell proliferation (1-4). Herein, the effects of retinoids and anti-estrogens on these genes are the focus of the present study.

Results

Both RA and anti-estrogens inhibit MCF-7 and T-47D breast cancer cell proliferation, which results from arrest within the G_1-phase of the cell cycle, leading to a fall in the proportion of cells synthesizing DNA (% S-phase, Figure 1). The effect of RA was delayed by about 16 h compared with that of anti-estrogens, although, by 32 h, both inhibited entry into S-phase to a similar maximal extent.

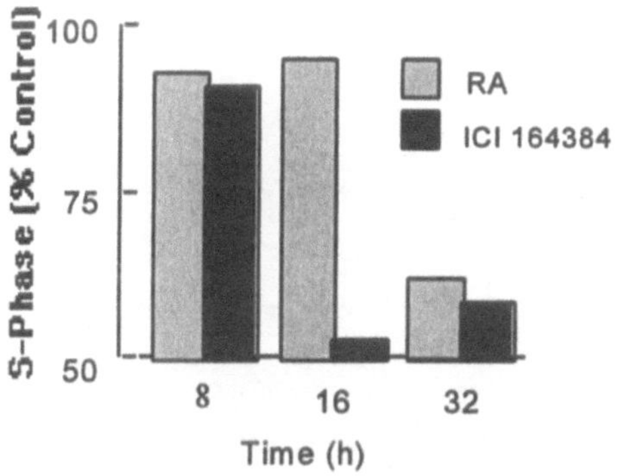

Figure 1. T-47D cells growing exponentially in medium containing 5% fetal calf serum were treated with 1 μM RA or 100 nM ICI 164384. These concentrations produced equivalent degrees of growth inhibition. DNA flow cytometry was used to determine % S-phase.

Anti-estrogens ICI 182780 and ICI 164384 markedly and rapidly reduced levels of c-*myc* protein, several hours before significant reductions were seen in the % S-phase (Table 1). In contrast, RA had no significant effect on c-*myc* protein.

Anti-estrogens rapidly down-regulated cyclin D1 mRNA, following effects on c-*myc* but preceding changes in % S-phase (Figure 2). The decreases were fully reversed by estradiol, suggesting that the effects of anti-estrogens are estrogen receptor-mediated. Changes in total cyclin D1 protein were of lesser magnitude (not shown). In contrast, RA had no significant effects on cyclin D1 magnitude (not shown). In contrast, RA had no significant effects on cyclin D1 mRNA or protein levels. Cyclins D2, D3, and E, as well as cdk4 protein levels, were not significantly altered by anti-estrogens or RA (not shown).

Table 1. Effects of Anti-Estrogens and RA on Cell Cycle-Regulatory Genes in Breast Cancer Cells.

Treatment	c-*myc*[a]	cyclin D1[b]	Rb[c]	cdk2[d]	cyclin A[a]
Anti-estrogen	Ø	Ø	Ø	Ø	Ø
Retinoic acid	NS	NS	Ø	ND	ND

[a] Protein expression. [b] mRNA level. [c] Phosphorylation. [d] Kinase activity.
Ø Down-regulation. NS no significant change. ND not determined.

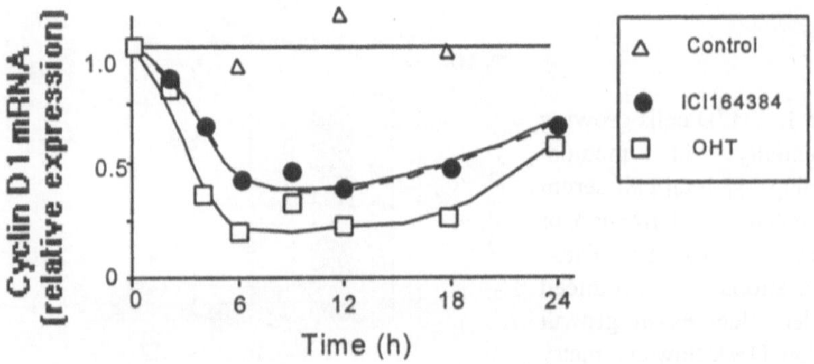

Figure 2. Total RNA extracted from MCF-7 cells treated with ICI 164384 or OHT. Northern blots probed with cyclin D1 cDNA. Quantitation by video densitometry.

Treatment with anti-estrogen or RA resulted in a time-dependent decrease in retinoblastoma-protein (Rb) phosphorylation, with a corresponding increase in the growth-inhibitory, hypophosphorylated form of Rb from about 6-8 h. For anti-estrogens, this immediately preceded alterations in % S-phase. The effect on Rb was more marked with anti-estrogen than RA, and after 18- 24 h of anti-estrogen treatment, total Rb protein decreased.

To study the mechanisms responsible for changes in Rb phosphorylation, cdk2 activity was measured using a histone H1 substrate. It decreased in response to anti-estrogen at 18-24 h, corresponding to the late changes seen in Rb phosphorylation. Since Cdk2, cyclin E protein, and mRNA levels were unaffected, alterations in cdk2 activity may be explained by the corresponding decrease in levels of cyclin A. However, the decrease in cdk2 activity is more likely to be a consequence than the cause of the inhibition of cell-cycle progression, as anti-estrogens had earlier inhibitory effects on entry into S phase.

Discussion

In breast cancer cells, anti-estrogens and RA achieve the same degree of cell-cycle arrest, but the effects of RA are delayed. Anti-estrogens rapidly decreased c-*myc* and cyclin D1 expression, while RA had no effect, suggesting different mechanisms of action.

Anti-estrogens and RA decrease the degree of Rb phosphorylation, which may at least partially account for the inhibition of entry into S phase. The mechanisms responsible for early decreases in Rb phosphorylation are unknown, but are likely to involve changes in cdk activity. For anti-estrogens, this could involve alterations in the activity of cyclin D1/cdk4 complexes.

Acknowledgments

Supported by grants from the Nat. Health & Med. Res. Council, the New South Wales State Cancer Council, MLC-Life Ltd., and the Leo & Jenny Leukaemia and Cancer Foundation. Elizabeth A. Musgrove is an MLC-Life Research Fellow.

References

1. Buckley MF, Sweeney KJE, Hamilton JA, et al. (1993) Expression and amplification of cyclin genes in human breast cancer. Oncogene 8:2127-2133.
2. Musgrove EA, Hamilton JA, Lee CSL, et al. (1993) Growth factor, steroid and steroid antagonist regulation of cyclin gene expression associated with changes in T- 47D human breast cancer cell cycle progression. Mol Cell Biol 13:3577-3587.
3. Watts CKW, Sweeney KJE, Warlters A (1994) Anti-estrogen regulation of cell cycle progression and cyclin D1 gene expression in MCF-7 human breast cancer cells. Breast Cancer Res Treat, in press.
4. Musgrove EA, Lee CSL, Buckley MF, Sutherland RL (1994) Cyclin D1 induction in breast cancer cells shortens G_1 and is sufficient for cells arrested in G_1 to complete the cell cycle. Proc Natl Acad Sci USA, in press.

Bisphenol-A Disturbs Microtubule Assembly and Induces Micronuclei *In Vitro*

Erika Pfeiffer, Brigitte Rosenberg, and Manfred Metzler

Summary

The environmental estrogen bisphenol-A (BP-A), a monomer of polycarbonate and other plastics, resembles diethylstilbestrol (DES) in its chemical structure. Herein, we report that BP-A and several of its analogs inhibit assembly of microtubules under cell-free conditions with activities similar to that of DES. In Chinese hamster V79 cells, BP-A disrupts the cytoplasmic microtubule complex and induces micronuclei containing whole chromosomes. Our study suggests that BP-A and other bisphenols have aneuploidogenic potential.

Introduction

Bisphenol-A (BP-A, Figure 1) and structurally related bisphenols are used on a large scale for the manufacture of polycarbonate and other plastics, as antioxidants and tanning agents, and for the synthesis of dyes. BP-A has recently been identified as an environmental estrogen. Its chemical structure resembles that of the stilbene estrogen diethylstilbestrol (DES, Figure 1).

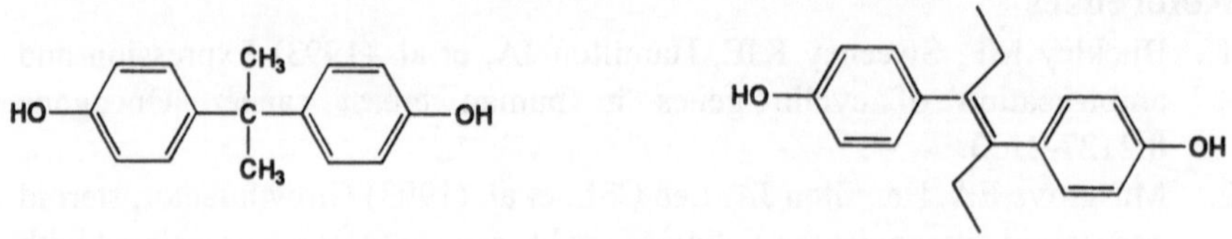

Figure 1. Chemical structures of BP-A (left) and E-DES (right).

DES has previously been reported to have colchicine (COL)-like effects on microtubules (MT) under cell-free conditions and to disrupt the mitotic spindle in various mammalian cells (1), thereby inducing aneuploidy and micronuclei (MN). Moreover, the aneuploidogenic activity of DES has been related to its well-documented carcinogenicity (2). Because of the structural similarity between DES and BP-A, we have now tested BP-A and several of its analogs for their potential to interfere with the formation of MT in a cell-free assay. BP-A was also assayed for disruption of the cytoplasmic microtubule complex (CMTC) and MN induction in cultured Chinese hamster V79 cells.

Materials and Methods

The MT polymerization assay was carried out with MT proteins from bovine brain. The test compound, (50-200 μM) dissolved in dimethylsulfoxide (DMSO) was added to 0.5 ml (final volume) assembly buffer (3) containing 10 μM MT proteins. After 20 min at 37°C, MT assembly was started by adding 0.5 mM guanosine triphosphate, and the increase in turbidity was measured at 350 nm for 30 min. The control incubation containing all components except the test compound was used as reference (100% assembly). Depolymerization was carried out at 4°C to confirm MT formation and detect aggregation of MT proteins.

For the MN assay, V79 cells were plated onto sterile glass slides and incubated overnight. The medium was replaced and the test compounds added in DMSO at various concentrations. After 6 h, the cultures were rinsed with phosphate-buffered saline (PBS) and fresh medium was supplied. At various time points after treatment, the cells were fixed in methanol at -20°C for at least 1 h. Methanol-fixed coverslips were kept in acetone at -20°C for 10 min, then washed with PBS at room temperature and incubated with CREST serum for 1 h at 37°C. After PBS washing, slides were incubated with FITC-labeled antihuman IgG for 45 min at 37°C. The coverslips were again washed and mounted in antifade solution containing 1 μg/ml DAPI and 1 μg/ml propidium iodide.

To assay the disruptive effects on the CMTC, V79 cells were incubated with the test compounds for 3 h and subsequently fixed as described above. Tubulin was then stained using a monoclonal anti-a-tubulin antibody as primary and a Cy3-labeled antimouse immunoglobulin as secondary antibody.

Results

Inhibition of Microtubule Assembly Under Cell-Free Conditions.

When the compounds listed in Figure 2 were tested for their inhibitory effect on the *in-vitro* polymerization of MT proteins from bovine brain, marked differences were noted. BP-A and its alkyl-fluorinated analog were clearly positive with an activity similar to that of E-DES. A BP-A derivative containing one methoxy activity similar to that of E-DES. A BP-A derivative containing one methoxy group in each ring in the ortho-position to the hydroxy group also proved to be positive, with about the same activity as BP-A. In contrast, the bisphenol without alkyl substitute (named BP-0) or the corresponding sulfone, sulfide and ketone were negative or only marginally positive.

Disruption of Cytoplasmic Microtubules and Induction of Micronuclei.

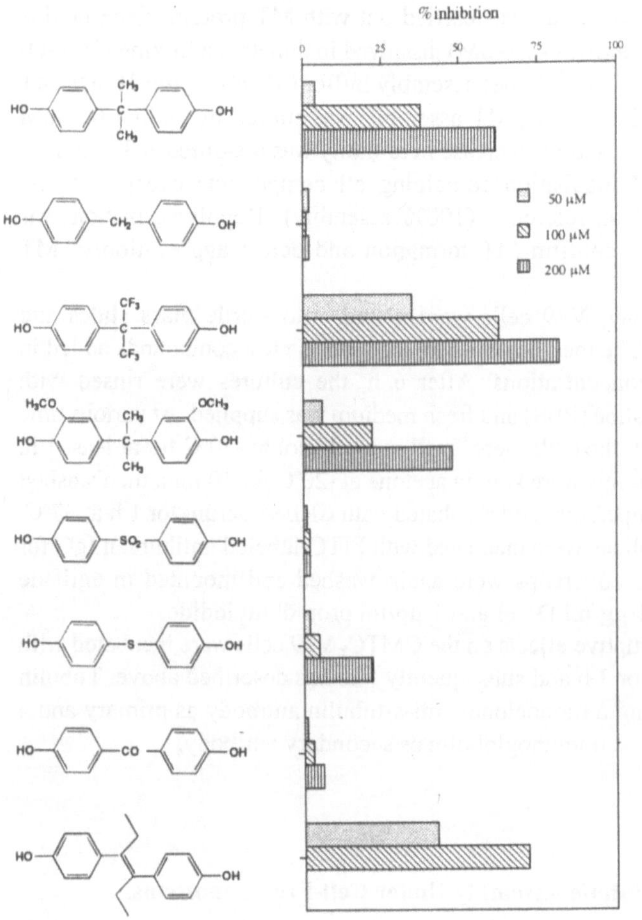

Figure 2. Inhibition of MT assembly by various bisphenols *in vitro.*

The MT-disrupting compounds COL and DES have been reported to lead to marked alterations of the CMTC and to induction of MN in various mammalian cells. MN caused by disturbance of the mitotic spindle contain whole chromosomes or chromatides; therefore, it stains with CREST antibodies, which are directed against kinetochore proteins. When V79 cells were exposed to concentrations of 50, 100, 150, and 200 μM BP-A and BP-0 for 3 h and the CMTC stained with anti-a-tubulin antibodies, a clear and concentration-dependent

CMTC (data not shown). Induction of MN in V79 cells was studied at various time points after treating the cells for 6 h with 100 and 200 μM concentrations of BP-A and BP-0. BP-A, but not BP-0, caused a clear increase in micronucleated cells; staining with CREST antibodies showed that BP-A induced only kinetochore-positive MN (Figure 3).

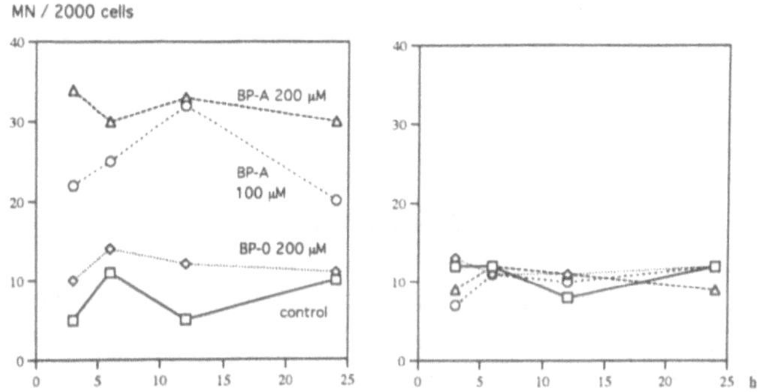

Figure 3. Time dependency of the incidence of CREST-positive (left) and CREST-negative (right) MN in V79 cells after treatment with BP-A and BP-0.

Conclusions

Our study clearly indicates that BP-A and two of its analogs inhibit MT assembly. This colchicine-like effect, which is not shown by all bisphenols, caused a disrupted CMTC and elevated incidence of MN with whole chromosomes in V79 cells. Therefore, BP-A is a potential aneuploidogenic compound. In view of its similarity with DES, it is now important to clarify whether BP-A can neoplastically transform cells and whether it may be a carcinogenic hazard.

References

1. Hartley-Asp B, Deinum J, Wallin M (1985) Diethylstilbestrol induces metaphase arrest and inhibits microtubule assembly. Mutation Res 143:231-235.
2. Tsutsui T, Maizumi H, McLachlan JA, Barrett JC (1983) Aneuploidy induction and cell transformation by diethylstilbestrol: A possible chromosomal mechanism in carcinogenesis. Cancer Res 43:3814-3821.
3. Epe B, Harttig U, Stopper H, Metzler M (1990) Covalent binding of reactive estrogen metabolites to microtubular protein as a possible mechanism of aneuploidy induction and neoplastic cell transformation. Environ Health Perspect 88:123-127.

Natural Estrogens Induce Modulation of Microtubules in Chinese Hamster V79 Cells in Culture

Yoshihiro Sato and Eriko Aizu-Yokota

Summary

The effect of estrogens on the modulation of cytoplasmic microtubules and on growth inhibition in Chinese hamster V79 cells are described herein. We used 28 compounds, including estradiol, estrone, their metabolites, and some chemically-derived compounds. The cytoplasmic microtubule networks were observed by the indirect immunofluorescence method using anti-b-tubulin antibody. The effective concentration of estradiol required for induction of microtubule disruption in 50% of V79 cells (EC_{50}) was 10 mM; however, 2-methoxyestradiol showed the strongest activity (EC_{50}: 2 mM). The 3-alkyl ethers (C1 to C3) of estradiol showed values similar to that of estradiol. There was a correlation between the EC_{50} value and the growth inhibitory activity of the compounds studied. Taxol, at 1 mM concentration, protected the microtubule disruption induced by 50 mM estradiol. Cycloheximide and actinomycin D did not inhibit the effect of estradiol, suggesting that the microtubule disruptive effect of estradiol is not associated with newly synthesized proteins and mRNAs.

Introduction

We have reported that cytoplasmic microtubules are disrupted by diethylstilbestrol (1) and estradiol (2) in Chinese hamster V79 cells, and that estradiol is capable of disrupting microtubules in estrogen receptor-positive (MCF-7 cells) and -negative (MDA-MB-231) human breast cancer cell lines, indicating that the estradiol-induced microtubule disruption is not estrogen-receptor mediated (3). Herein, we examined the microtubule-disruptive activities of 28 compounds, including 17β-estradiol (E_2), estrone, their metabolites, and some chemically-derived compounds in Chinese hamster V79 cells.

Results

We observed the cellular microtubule network by an indirect immunofluorescence method using anti-b-tubulin antibody. Some estrogens used were supplied by Dr. Toshio Mambara, Tohoku University. Figure 1A shows the normal microtubule network in control cells. One hr after E_2 treatment (50 mM), the cells were rounded, losing normal microtubule networks (Figure 1B). A quantitative assessment of the microtubule disruption induced by different compounds is shown in Table 1. The values represent the effective concentration of a compound required for the induction of microtubule disruption in 50% of the cells (EC_{50}). The EC_{50} for E_2 was 10 mM. Of the compounds studied, 2-methoxyestradiol (EC_{50}: 2 mM), showed the strongest activity. It is noteworthy that the EC_{50} for estrone was >100 mM, however, its catechol derivatives showed great ability to modulate microtubule disruption. Interestingly, 17α-E_2 showed almost the same activity (EC_{50}: 9 mM) as E_2. Furthermore, the 3-alkyl ethers of E_2 showed almost the same level of activity as E_2. The EC_{50} of methyl, ethyl, and propyl ether were 9, 9, and 10 mM, respectively.

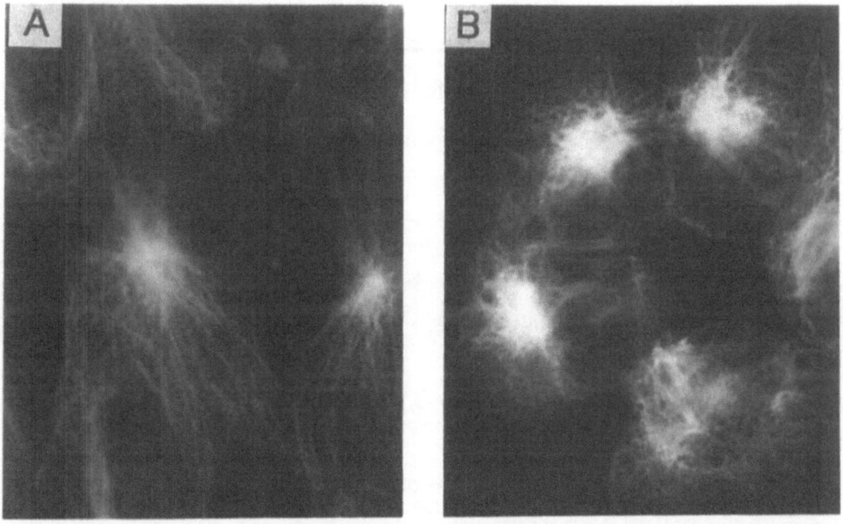

Figure 1. Effect of estradiol on the cellular microtubule networks of V79 cells. Cells were incubated with 0.1 % of DMSO (A) or 50 mM E_2 for 1h (B). The cells were then stained by indirect immunofluorescence using anti-b-tubulin antibody. All magnifications x 1000.

Table 1. Structure-Activity Relationships of Estrogens and Their Catechol Derivatives on Microtubule Disruption in V79 Cells.

Compound	EC_{50} (mM)[a]
17β-Estradiol	10
17α-Estradiol	9
Estrone	>100
2-Hydroxyestradiol	14
2-Methoxyestradiol	2
2-Hydroxyestradiol 3-methyl ether	17
4-Hydroxyestradiol	25
4-Methoxyestradiol	27
2-Hydroxyestrone	50
2-Methoxyestrone	30
4-Hydroxyestrone	40
4-Methoxyestrone	70
4-Hydroxyestrone 3-methyl ether	>100

[a] Compound concentration required for disruption of microtubule networks in 50% of the cells after 1 h treatment.

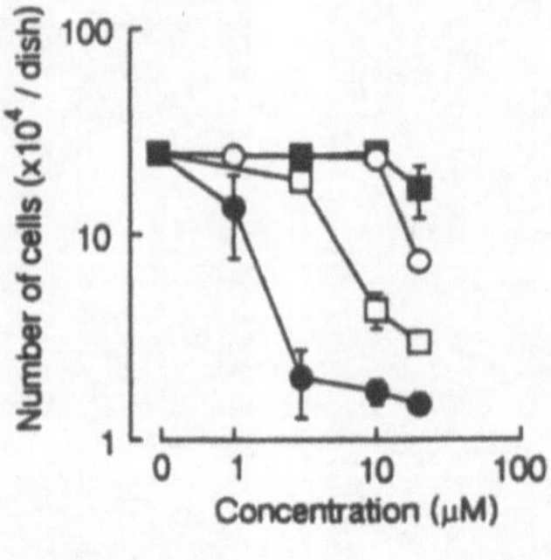

Figure 2. Effect of catechol derivatives of E_2 on growth of V79 cells. Cells were incubated for 2 days with various concentrations of E_2 (○), 2-MeOE_2 (●), 2-OHE_2 (□), or 2-OHE_2 3-Me (■). Number of viable cells was counted by the trypan blue exclusion test.

In relation to the above results, it is interesting that 3-methyl-3-deoxyestradiol also showed a similar activity (EC_{50}: 13 mM) to that of E_2. 6-Hydroxy derivatives of estradiol were less active, and some sulfates showed >100 mM for their EC_{50} values.

The cell-growth studies indicated that 2-methoxyestradiol was the most potent inhibitor of cell growth. Some degree of correlation was found between the microtubule disruption level of a particular compound and its growth inhibitory activity. Further, in the presence of 1 mM taxol, estradiol-induced microtubule disruption was protected. Cycloheximide and actinomycin D did not inhibit the microtubule disruption induced by E_2.

Discussion

We have demonstrated that natural estrogens and their chemical derivatives disrupt the cellular microtubule network of Chinese hamster V79 cells. This estrogenic effect is independent of its binding to the estrogen receptor. Cycloheximide and actinomycin D did not block the estradiol-induced microtubule disruption, indicating that this effect does not require newly synthesized proteins and mRNA. Of the tested compounds, 2-methoxyestradiol was the most potent estrogen capable of disrupting the microtubule network and growth of V79 cells. Recently, 2-methoxyestradiol was reported to inhibit angiogenesis and suppress tumor growth (5), suggesting that endogenous estrogens may exhibit tumor suppressor activity.

References

1. Sakakibara Y, Saito I, Ichinoseki K, et al. (1991) Effects of diethylstilbestrol and its methyl ethers on aneuploidy induction and microtubule distribution in Chinese hamster V79 cells. Mutation Res 263:269-276.
2. Sato Y, Sakakibara Y, Oda T, et al. (1992) Effect of estradiol and ethynylestradiol on microtubule distribution in Chinese hamster V79 cells. Chem Pharm Bull 40:182-184.
3. Aizu-Yokota E, Ichinoseki K, Sato Y (1995) Microtubule disruption induced by estradiol in estrogen receptor-positive and -negative human breast cancer cell lines. Carcinogenesis, in press.
4. Schiff PB, Fant J, Horwitz SB (1979) Promotion of microtubule assembly *in vitro* by taxol. Nature 277:665-667.
5. Fotsis T, Zhang Y, Pepper MS, et al. (1994) The endogenous oestrogen metabolite 2-methoxyoestradiol inhibits angiogenesis and suppresses tumor growth. Nature 368:237-239.

Induction of Micronucleation, Spindle Disturbances, and Mitotic Arrest in Human Chorionic Villi Cells by 17ß-Estradiol, Diethylstilbestrol, and Coumestrol

Maik Schuler, Katrin Huber, Heinrich Zankl, and Manfred Metzler

Summary

17β-estradiol (E_2) and diethylstilbestrol (DES) are known to induce micronuclei (MN), aneuploidy, and mitotic arrest in different cell systems. These parameters, however, have not been studied in any cell system after treatment with the phytoestrogen coumestrol (COUM). We now report that E_2, DES, and COUM induce MN in human chorionic villi cells. Whereas DES clearly inhibited the cytoplasmic microtubule complex and the mitotic spindle, E_2 exhibited only a slight effect, and COUM had no effect when tested with anti-a-tubulin antibodies. MN were characterized by immunofluorescence staining using antikinetochore antibodies (CREST-staining), and by fluorescence *in-situ* hybridization with the centromere-specific DNA probe p82H. Nearly 80% of the DES-induced MN showed positive signals by either technique, whereas COUM-induced MN were negative for both color reactions. Of the E_2-induced MN, only 45% were CREST-positive, but 70% gave positive hybridization signals.

Introduction

Diethylstilbestrol (DES) and 17β-estradiol (E_2) cause metaphase arrest, anaphase abnormalities, micronucleation, and aneuploidy in different cell systems (1-3). Coumestrol (COUM), however, has not been tested with respect to its cytological effects thus far. Negative results in the Salmonella/mammalian microsome assay (4-6) have suggested that all three estrogens lack mutagenic properties.

Several reports (7, 8) have indicated that DES inhibits *in-vitro* polymerization of tubulin, while E_2 (2) and COUM (data not shown) do not.

In Don cells, Wheeler, et al. (9), showed that DES inhibited spindle assembly and disassembled the cytoplasmic microtubule complex (CMTC), while E_2 arrested mitosis, allowing spindle assembly, but had no definite effect on the CMTC. On the other hand, Sato, et al. (2), could detect disruption of the normal microtubule network with both substances in V79 cells.

Using diploid human chorionic villi cells as test material, we compared the effects of DES, E_2, and COUM on induction of mitotic arrest, tubulin-disrupting activities, and micronucleation. Micronuclei (MN) were analyzed by CREST-immunofluorescence staining and *in-situ* hybridization with the centromere-specific probe p82H, which allows us to test whether the MN contain whole chromosomes or acentric fragments

Results

Mitotic Index

Table 1 depicts the mitotic indices of HCV cells treated with the three estrogens for 10 h. DES and E_2 caused a significant increase in mitotic index at concentrations higher than 50 μM. The phytoestrogen COUM did not induce metaphase arrest in this cell system.

Table 1. Effects of DES, E_2, and COUM on the Mitotic Index of HCV Cells[a]

	Control	10 μM	25 μM	50 μM	75 μM	100 μM	125 μM
DES	12	17	17	71	81	59	51
E_2	4	8	9	18	14	6	10
COUM	9	9	11	12	8	3	n.d.

[a] The data are expressed as the mitotic index of HCV cells/1000 cells after 10 h of treatment

Microtubule Behavior

Under the influence of colchicine (COL) or colcemid, cells entering mitosis are unable to assemble spindle microtubules, and are subsequently blocked at metaphase. Our immunofluorescence experiments using an anti-a-tubulin antibody showed striking similarities between DES- and colcemid-treated microtubules, whereas E_2 and COUM allowed spindle formation with clearly present equatorial plates.

Micronucleus Induction

All three estrogens induced a significant increase in MN in human chorionic cells. The most potent inductor of MN was COUM, with maximum induction frequencies 12 and 24 h post treatment. E_2 displayed a maximum effect after 3 h, and DES, 6 and 12 h post treatment (Figure 1). Typical aneugens like COL lead to maximum induction frequencies after 12 h. Typical clastogens like 4-nitroquinolineoxide displayed their maximum effect 24 h post treatment in our system.

MN were characterized with respect to the presence of whole chromatids or chromosomes at different concentrations, and to times after removal of the compound when maximum induction of MN occurred (Figure 1). For kinetochore detection by CREST-staining, we used the method described by Eastmond, et al.(10). To prove the centromeric region in MN, we utilized *in-situ* hybridization with the centromere-specific DNA probe p82H (11). MN containing whole chromosomes should show positive signals with both methods.

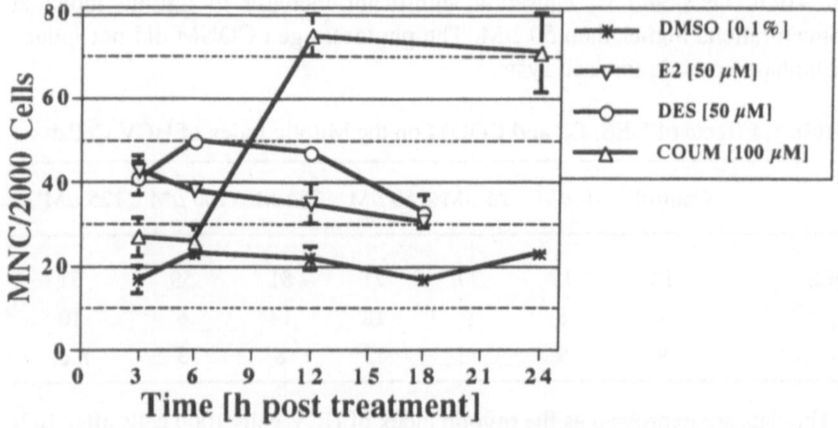

Figure 1. Time course of MN induction by DES, E_2, and COUM in HCV cells. Cultures were treated for 6 h with the various estrogens. After removal of the compounds, cells were fixed with methanol at different times post treatment. All data points represent the mean ± standard deviation from three experiments.

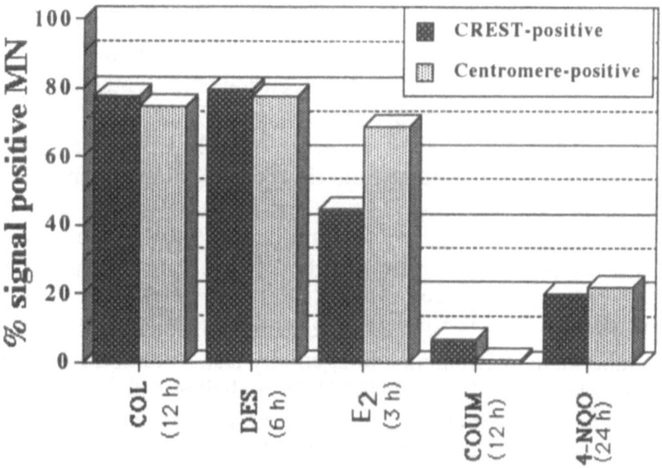

Figure 2. Proportions of CREST- and centromere-positive MN obtained with different compounds. To compare the frequencies of signal-positive MN, induction frequencies were corrected against controls. The data represent the frequencies induced by the compound.

Conclusions

Whereas DES and E_2 were potent mitotic arrestors, COUM did not cause an increase in mitotic index. DES, like COL, disrupted the formation of a bipolar spindle, and partially dismantled the CMTC. E_2 allowed some degree of spindle assembly with no apparent effect on the CMTC. COUM did not affect these structures. Our results are in agreement with previous reports in the cell-free microtubule polymerization assay. DES- and COL-induced MN showed concordant results with the immunofluorescent staining and the *in-situ* hybridization. Most of the MN contained whole chromosomes. E_2 led to different results with the two methods. CREST-staining gave only 45% positive MN, whereas, after *in-situ* hybridization, most MN were positive as in DES- and COL-treated cells. COUM induced MN that were negative for both color reactions.

We conclude that DES leads to chromosomal nondisjunction by binding to tubulin dimers and preventing polymerization into microtubules, similar to the action of COL and nocodazole. MN induction by E_2 may result from interference with processes other than tubulin assembly. The differences in frequencies of CREST- and centromere-positive MN could be explained by interaction with certain kinetochore proteins, so that MN containing whole chromosomes appear CREST-negative. Another explanation might be extensive breakage in the heterochromatin region of the chromosomes. COUM seems to have clastogenic properties and is the most potent MN inductor in human chorionic villi cells.

References

1. Wheeler WJ, Cherry LM, Downs T, Hsu TC (1986) Mitotic inhibition and aneuploidy induction by naturally occuring and synthetic estrogens in Chinese hamster cells *in vitro*. Mutation Res 171:31-41.
2. Sato Y, Sakakibara Y, Oda T, et al. (1992) Effect of estradiol and ethinylestradiol on microtubule distribution in Chinese hamster V79 cells. Chem Pharm Bull. 40:182-184.
3. Tsutsui T, Suzuki N, Maizumi H, Barrett JC (1990) Aneuploidy induction in human fibroblasts: Comparison with results in Syrian hamster fibroblasts. Mutation Res 240:241-249.
4. Lang R, Redmann U (1979) Non-mutagenicity of some sex hormones in the Ames Salmonella/microsome mutagenicity test. Mutation Res 67:361-365.
5. Glatt HR, Metzler M, Oesch F (1979) Diethylstilbestrol and 11 derivates. A mutagenicity study with Salmonella typhimurium. Mutation Res 67:113-121.
6. Bartholemew R, Ryan D (1980) Lack of mutagenicity of some phytoestrogens in the Salmonella/microsome assay. Mutation Res 87:317-321.
7. Sato Y, Murai T, Tsumuraya M, Kodama M (1984) Disruptive effect of diethylstilbestrol on microtubules. GANN 75:1046-1048.
8. Sharp DC, Parry JM (1985) Diethylstilbestrol: The binding and effects of diethylstilbestrol upon polymerisation and depolymerisation of purified microtubule protein *in vitro*. Carcinogenesis 6:865-871.
9. Wheeler WJ, Hsu TC, Tousson A, Brinkley BR (1987) Mitotic inhibition and chromosome displacement induced by estradiol in Chinese hamster cells. Cell Motil Cytoskeleton 7:235-247.
10. Eastmond DA, Tucker JD (1989) Kinetochore localization in micronucleated cytokinesis-blocked Chinese hamster ovary cells: A new and rapid assay for identifying aneuploidy-inducing agents. Mutation Res 224:517-525.
11. Becker P, Scherthan H, Zankl H (1990) The use of a centromere-specific DNA probe (p82H) in nonisotopic *in situ* hybridization for classification of micronuclei. Genes, Chrom Cancer 2:59-62.
12. Pfeiffer E, Metzler M (1992) Effects of steroidal and stilbene estrogens and their peroxidative metabolites on microtubular proteins, In Li JJ, Nandi S, Li SA (eds): Hormonal Carcinogenesis. New York: Springer-Verlag, pp 313-317.
13. Sakakibara Y, Hasegawa K, Oda T, et al. (1990) Effects of synthetic estrogens, (R,R)-(+), (S,S)-(-)-,dl- and meso-hexestrol stereoisomers on microtubule assembly. Biochem Pharmacol 39(1):167-172.
14. Sato Y, Murai T, Oda T, et al. (1987) Inhibition of microtubule polymerization by synthetic estrogens: Formation of a ribbon structure. J Biochem 101(5):1247-1252.

Increased Nuclear IGF-1 Receptor Level, Coupled with Attenuation in DNA Repair, Plays an Important Role in Estrogen-Induced Carcinogenesis

Deodutta T.E. Roy, Chiao-wen Chen, and Zhi-Jie Yan

Summary

We have previously shown an increase in the level of plasma membrane IGF-1 receptor (IGF-1R) in DES-induced kidney tumors. In this study, we demonstrate that DES treatment of Syrian hamsters increases the level of renal nIGF-1R by 3.0- to 4.0-fold over control levels, while the transcripts of DNA polymerase β, a repair enzyme, decreased by 8.0- to 10.0-fold in DES-induced kidney tumors compared with control levels. These findings suggest that increased mitogenic signals, through enhanced level of nIGF-1R, coupled with attenuation in the DNA repair system may play an important role in the induction of DES-induced carcinogenesis.

Introduction

Diethylstilbestrol (DES), a synthetic estrogen, is carcinogenic to both humans and animals (1). In Syrian hamsters, DES treatment for 7-8 months produces 90-100% tumor incidence, specifically in the kidneys (1). The mechanism of DES-induced carcinogenesis is not clear. Estrogen treatment has been shown to produce genetic instability in the kidney, the site of the carcinogenic effect (2-4). We have previously shown increased levels of plasma membrane IGF-1R in DES-induced kidney tumors (5). IGF-I, through IGF-1 receptor, is known to produce mitogenic effects (6). Increased mitogenicity is considered to fix genetic instability or to increase the probability of infidelity in DNA repair, which may allow genetic instability to occur (7). We propose that increased mitogenic signal(s), coupled with alteration(s) in the DNA repair system induced by DES exposure, may provide the necessary environment for the induction of carcinogenesis. In order to provide evidence to confirm this concept, we have examined the level of nuclear IGF-IR, an indicator of mitogenic signals, and the expression of DNA polymerase β, an indicator of DNA excision repair, in the kidneys of untreated and DES-treated animals. Our findings revealed that DES treatment led to an increased level of nuclear IGF-IR, and a decrease in the DNA repair system in the kidney, the target organ of carcinogenesis.

Methods

Male Syrian hamsters were treated with a carcinogenic dose of DES (5). Highly pure nuclei were prepared (4). Total RNA from age-matched normal control animals and tumor kidney tissues was isolated by using an RNA isolation kit (Biotecx Laboratories, Inc.). The binding of [^{125}I]-IGF-1 to nuclear protein and cross-linking assay were carried out as described previously (5). For the studies involving cDNA synthesis and PCR amplification, a pair of primers based on the rat DNA polymerase β mRNA sequence (8) was designed to amplify a 306-bp specific cDNA product. A pair of primers for β-actin mRNA sequence yielding a 245-bp product was used as an internal standard. For quantitation, 2 μCi of [α-^{32}P]-dCTP was added to each reaction mixture, consisting of 1 μg RNA, 2.5 μM oligo d(T)$_{16}$ primer, 10 mM Tris-HCl, pH 8.3, 50 mM KCl, 5 mM $MgCl_2$, 250 μM dNTPs, 50 U reverse transcriptase and 20 U of RNAse inhibitor. After 30 min at 42°C, cDNA was diluted with water. PCR was carried out with cDNA derived from 50 ng RNA, 1 U AmpliTaq polymerase, and reaction buffers supplied by Cetus. PCR cycle temperature condition consisted of 1 min denaturation at 94°C, 1 min primer annealing ° at 55 C, and 40 sec extension/synthesis at 72°C. After electrophoresis and autoradiography, the band corresponding to each specific PCR product was excised from the gel and the radioactivity counted by Cerenkov counting. Changes in the level of DNA polymerase β transcript were further confirmed by slot-blot and Northern blot hybridization.

Results

Nuclei from kidneys of untreated and DES-treated hamsters were incubated with [^{125}I]-IGF-1 in either the absence or presence of increasing concentrations of unlabeled IGF-1. The binding data showed that DES treatment for 8 days increased the level of nIGF-1R by 3.0- to 4.0-fold in the kidney nuclei compared with control untreated samples (Figure 1). This finding was further confirmed by affinity-labeling experiments using Scatchard plot analysis. The transformation affinity and the number of receptors for IGF-1 were altered by DES exposure compared with the control values (Figure 1).

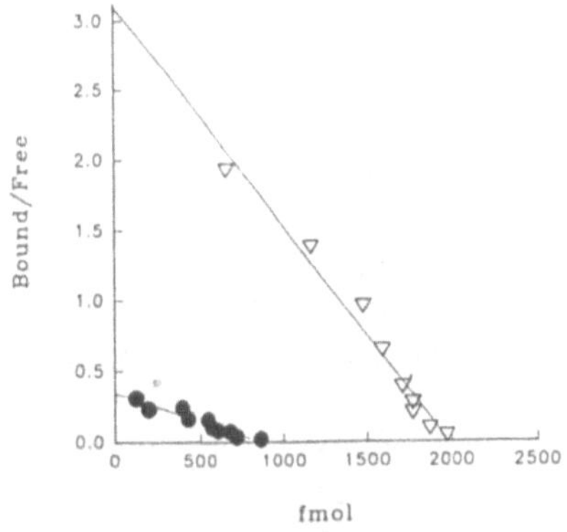

Figure 1. Comparison of [^{125}I]-IGF-1 binding activity between normal (●) and DES-treated (∇) kidney nuclei. Binding of [^{125}I]-IGF-1 to nuclear proteins was performed in either the absence or presence of increasing concentrations of IGF-1 (5). The data were transformed into Scatchard plot.

RT-PCR revealed that, when compared with control samples, the level of transcripts of DNA polymerase β was drastically lower in kidney tumor samples (Figure 2). The decrease in the level of mRNA of DNA polymerase β in tumor samples was further confirmed by slot-blot and Northern blot hybridization.

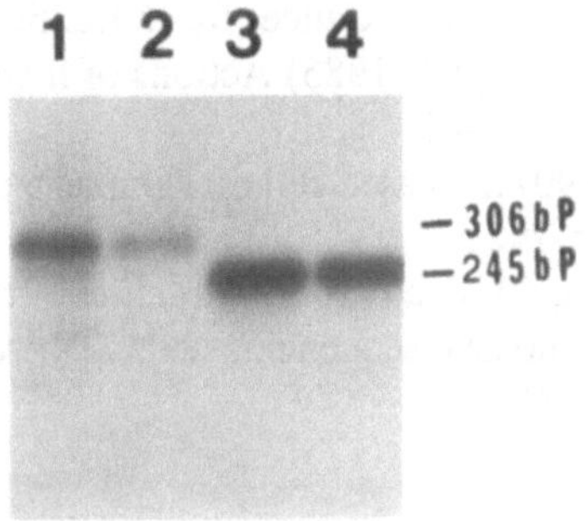

Figure 2. Autoradiography of RT-PCR products separated on 2% agarose gel. Lanes 1 & 2: 306-bp product of DNA pol β mRNA from normal and kidney tumor, respectively. Lanes 3 & 4: 245-bp product of β-actin mRNA from normal and tumor tissues, respectively.

Discussion

The attenuation in the level of DNA polymerase β observed in these studies may lead to a decrease in DNA damage-repair capacity that may allow genetic damage or genetic instability to accumulate. The increase in the nIGF-1R level observed after DES treatment may provide stimuli to increase mitogenicity, which may allow the fixation of genetic instability. Thus, enhanced cell division, presumably through increase in nIGF-1R, coupled with attenuation in DNA repair may play an important role in DES-induced carcinogenesis.

Acknowledgment

This study was supported by the NIH grant CA52584.

References

1. IARC Monogr Eval Carcinog Risk Chem Human: Sex hormone (1979) 21:139-362.
2. Banerjee SK, Banerjee S, Li SA, Li JJ (1992) Cytogenetic changes in renal neoplasms and during estrogen-induced renal tumorigenesis in hamsters. In Li JJ, Nandi S, Li SA (eds): Hormonal Carcinogenesis. New York: Springer-Verlag, pp 247-250.
3. Roy D, Floyd R, Liehr JG (1991) Elevated 8-hydroxyguanosine levels in DNA of diethylstilbestrol-treated Syrian hamsters: Covalent DNA damage by free radicals generated by redox cycling of diethylstilbestrol. Cancer Res 51:3882-3885.
4. Palangat M, Roy D (1993) Inhibition of an organelle nuclear transcription by diethylstilbestrol through modulation of phosphorylation. Proc Am Assoc Cancer Res 34:530.
5. Narayan S, Roy D (1993) Insulin like growth factor 1 receptors are increased in estrogen-induced kidney tumors. Cancer Res 53:2256-2259.
6. Forsech ER, Schmid C, Schwander J, Zapf J (1985) Actions of insulin-like growth factors. An Rev Physiol 47:443-467.
7. Martin S, Pike M, Ross R, et al. (1990) Increased cell division as a cause of human cancer. Cancer Res 50:7415-7421.
8. Zmudzka BZ, Sengupta D, Matsukage A, et al. (1986) Structure of rat DNA polymerase β revealed by partial amino acid sequencing and cDNA cloning. Proc Natl Acad Sci, USA, 83:5106-5110.

Pattern of Reproductive Aging in Female Rats Can Affect Mammary Tumor Incidence

J. Charles Eldridge, Lawrence T. Wetzel, Merrill O. Tisdel, and James T. Stevens

Summary

Reproductive senescence in female rats is due primarily to a developing neuroendocrine failure. Aging Sprague-Dawley (SD) animals demonstrate episodes of persistent vaginal estrus (PVE), while aging Fischer 344 (F-344) rats have episodes of pseudopregnancy. In the present studies, the ratio of serum estrogen to progesterone rose during PVE episodes in mature SD rats. The ratio was also higher than in F-344 rats at the same ages. In addition, SD rats demonstrated a much higher incidence of spontaneous mammary tumors (MT) than did F-344 rats. Among SD female rats fed 400 ppm atrazine, a broad-spectrum herbicide, for 2 years, there was an earlier appearance of PVE episodes, higher estrogen-progesterone ratios, and an earlier appearance of MT, compared to age-matched SD controls. Final tumor incidence, at 24 months, was not increased over control incidence. Indices of reproductive aging were unaffected in F-344 rats fed 400 ppm atrazine, and MT incidence remained low through 24 months. F-344 rats can develop a high rate of MT in response to chemical carcinogens, but did not do so in response to atrazine feeding. Results suggest that the spontaneous MT rate of aging female rats is influenced by exposure to the animals' own reproductive steroids during middle age, and that the pattern of aging and alteration of the endocrine milieu can correspondingly affect the timing of MT development in senescence.

Introduction

The albino rat is perhaps the most-used experimental subject for a variety of reasons that certainly include a presumption that the model is an effective reasons that certainly include a presumption that the model is an effective surrogate for physiologic and pathologic characteristics of the human species. However, different rat strains can display varied characteristics, particularly with advancing age, that confound a decision regarding choice of the most appropriate surrogate. Such is the case with rodent models of mammary cancer. It is well known that females of some rat strains exhibit a very high incidence of spontaneous mammary tumors during the second year of life (e.g., Sprague-Dawley, SD) (1), whereas other strains do not (e.g., Fischer 344, F-344) (2).

The influence of hormones in the development of rodent mammary tumors has also been well documented. Estrogens and/or prolactin appear to promote tumor growth, while removal and/or inhibition of either hormone results in tumor regression (3, 4). Because the levels of activity of these hormones also undergo substantial changes with advancing age in female rats, it seems reasonable to suggest that mammary tumor incidence may be functionally related to the nature of strain-related aging. However, tests of this hypothesis have not been frequently reported.

In the SD rat, the hallmark of estrous cycle change is a persistent vaginal estrus (PVE) (5), due essentially to neuroendocrine deficiencies in stimulating the preovulatory gonadotropin surge (6, 7). Ovarian follicles persist and continue to secrete estrogens, which in turn enhance prolactin release (8, 9). The aging F-344 rat continues to ovulate with less difficulty but exhibits prolonged corpora luteal activity and increased progesterone, rather than estrogen, secretion (10).

Results from concomitant long-term feeding studies in SD and F-344 female rats administered atrazine, a broad-spectrum herbicide that affects normal estrous cycling but is not mutagenic or estrogenic, permitted us to test whether treatment-related changes in mammary tumor incidence follow the nature of reproductive endocrine changes.

Results

When examined at several intervals up through 18 months of age, parameters of reproductive endocrine activity in control SD and F-344 rats were initially close but soon became extremely divergent. As expected, the percentage of life days in estrus rose substantially in SD rats by 9 months of age. In contrast, F-344 rats demonstrated a reduction of estrous days through 18 months of age.

The ratios of plasma estradiol (E_2) to progesterone (P_4) also increased substantially in aging SD rats, primarily due to increased E_2. The E_2/P_4 ratios declined in aging F-344 rats, principally because of rising P_4 titers.

Mammary tumor incidence increased to nearly 30% of all SD rats by 18 months of age, while it remained nearly zero for F-344 rats through the same age.

Compared to SD controls, the percent days in estrus rose even faster in SD animals continuously fed 400 ppm atrazine beginning at 2 months of age. The E_2/P_4 ratio also moved higher, due to an even higher plasma E_2 level, and the mammary tumor incidence began to rise at an earlier time during the study. Mammary tumor incidence was not different at 24 months (data not shown). Among F-344 rats, no significant alterations of these measures were produced, and no increase in tumors was observed. It was determined that both strains consumed identical amounts of atrazine, adjusted for body weight, and also that 400 ppm exceeded the maximum tolerated dose (MTD) for each strain .

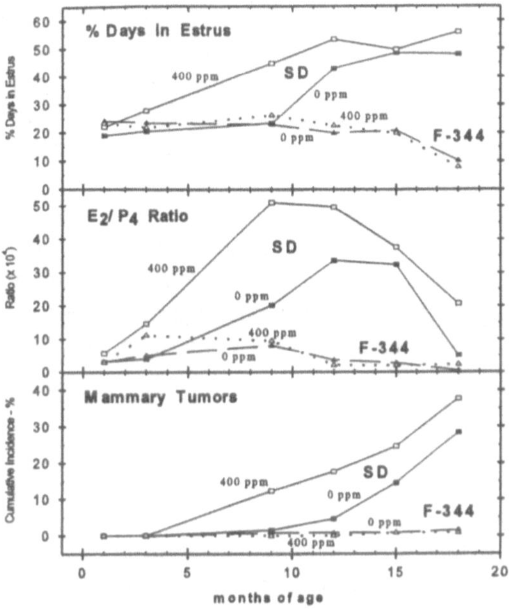

Figure 1. Parameters of reproductive endocrine activity in control SD and F-344 rats. % days in estrus, E_2/P_4 ratio measured from 1-18 months of age.

Discussion

Although the spontaneous rate is low, Fischer 344 rats can develop mammary tumors when stimulated by chemical carcinogens such as DMBA (4). The low spontaneous rate observed in the present studies was likely due to the relatively higher P_4 secretion in this strain. This suggests that the hormonal environment of aging strongly controls the rate of mammary tumor formation in the F-344 strain.

As the the percent days in estrus of the SD animals increased, a higher relative estrogen balance was also created, and an increase in mammary tumors followed. The contrast with events in F-344 rats leads to a prediction that alterations of the endocrine environment to promote estrogen secretion will similarly foster mammary tumor growth. The appropriate environment could be created by disrupting the ovulatory cycle in a manner that causes arrested ovarian follicular growth and persistent estrogen secretion. When estrous cycling of SD rats was further disrupted by administration of the atrazine, the E_2/P_4 ratio rose even faster, and the curve of rising mammary tumor incidence shifted further to the left. The sequence of effects did not occur in atrazine-treated F-344 rats. This suggests that incidence of mammary tumors can be enhanced by manipulation of the estrous cycle.

Because reproductive aging in humans (menopause) is an estrogen-deficient environment resulting from reduced ovarian rather than neuroendocrine activity, it is concluded that the SD rat model is a poor surrogate for human mammary tumor incidence. The rat tumors appear to be dependent upon a particular estrous cycle manipulation and endocrine environment that does occur in humans.

References

1. McMartin DN, Sahota PS, Gunson DE, et al. (1992) Neoplasms and related proliferative lesions in control Sprague-Dawley rats from carcinogenicity studies. Toxicol Pathol 20:212-224.
2. Solleveld HA, Haseman JK, McConnell EE (1984) Natural history of body weight gain, survival and neoplasia in the F-344 rat. J Natl Cancer Inst 72:929-940.
3. Welsch CW (1985) Host factors affecting the growth of carcinogen-induced mammary carcinomas. Cancer Res 43:3415-3443.
4. Russo J, Gusterson BA, Rogers AE, et al. (1990) Comparative study of human and rat mammary tumorigenesis. Lab Invest 62:244-278.
5. Mandl AM (1961) Cyclical changes in the vaginal smears of senile nulliparous and multiparous rats. J Endocrin 22:257-268.
6. Wise PM (1987) The role of the hypothalamus in aging of the female reproductive system. J Steroid Biochem 27:713-719.
7. Nelson JF, Bergman M, Karelus K, Felicio LS (1987) Aging and the hypothalamo-pituitary-ovarian axis: Hormonal influences and cellular mechanisms. J Steroid Biochem 27:699-705.
8. Finch CE (1978) Reproductive senescence in rodents: Factors in the decline of fertility and loss of regular estrous cycles. In EL Schneider (ed): The Aging Reproductive System. New York: Raven Press, pp 159-192.
9. Huang HH, Steger RW, Bruni JF, Meites J (1987) Patterns of sex steroid and gonadotropin secretion in aging female rats. Endocrinology 103:1855-1859.
10. Estes KS, Simpkins JW, Kalra SP (1982) Normal LHRH neuronal function and hyperprolactinemia in old pseudopregnant F-344 rats. Neurobiol Aging 3:247-252.

Antiproliferative Activity of Luteolin, a Naturally-Occurring Edible Plant Flavone, Against Estrone-Induced Cell Proliferation in the Mammary Gland of Noble Rats

Michael B. Holland and Deodutta T.E. Roy

Summary

In this study, we present evidence that estrone treatment leads to an imbalance in mammary gland cell proliferation in the Noble rat, which can be blocked by the naturally-occurring plant flavone, luteolin. Estrone treatment drastically increased cell proliferation, while luteolin treatment significantly inhibited estrone-induced cell proliferation. Increased cell proliferation, by providing the necessary milieu for promoting genetic instability, may play a key role in estrone-induced mammary cancer in the Noble rat. The antiproliferative action of luteolin suggests that it may play a protective role against estrone-induced mammary carcinogenesis in this species.

Introduction

Recent epidemiological observations have revealed that increased levels of estrogens are associated with high risk of mammary cancer (1-3). Both natural and environmental estrogens have been shown to induce mammary tumors in various animal species (4). The mechanism of mammary cancer induction is not clear. In the present study, we have examined the possibility of uncontrolled cell growth in this target organ by estrone exposure in Noble rats. This model of estrogen-induced mammary carcinogenesis has been used because: (a) estrone exposure in female Noble rats induces 80-90% mammary tumors after 10-11 months (5, 6); (b) estrone acts, evidently, as both an initiator and a promoting agent; and (c) the growth of the tumor is estrogen dependent (5, 6). Our findings demonstrated that estrone treatment leads to an imbalance in cell proliferation in the Noble rat mammary gland, which can be blocked by the naturally-occurring plant flavone, luteolin.

Methods

Female Noble rats (NCI) at 50 days of age received a sc pellet of estrone alone (11 mg) (5, 6) or ip injections of estrone + luteolin (1 mg, 2 x daily). On the eleventh day post-estrone implant, all rats were sacrificed and the superior abdominal mammary gland excised. Tissues were fixed in 10% buffered formalin for 3 hours and dehydrated in alcohols before storing in xylene. The glands were then routinely processed, paraffin embedded, and sectioned to 5 μm.

Cell kinetics was measured by immunohistochemical analysis for reactivity to antiproliferating cell nuclear antigen (PCNA) monoclonal antibody PC10 clone (Signet) (7). Immunohistochemical staining was performed using a protocol modified from Hall, et al. (8). Positive staining was scored as either S-phase (nuclei with uniform, brown-dark brown to black nuclear staining without cytoplasmic staining) or G_1-phase (patchy to uniform, light brown nuclear staining, without cytoplasmic staining) as determined by the method of Foley, et al. (9).

The number of S-phase or G_1-phase nuclei was scored, divided by the total number of nuclei counted, and expressed as a percentage. The total number of nuclei in a fixed area of the gland was determined and multiplied by the fraction of immunopositive cells to obtain the total number of proliferating, S-, or G_1-phase nuclei/mm^2 in a uniform area of the gland. Only epithelial cells of the mammary ductolobular structures were counted. Myoepithelial, fibroblast, or adipocyte nuclei were excluded. For each animal, the sum of all structures was counted in one gland. At least 4 structures per gland were counted and >2000 nuclei were assessed for immunoreactivity per animal. Examination of gland differentiation revealed that the majority of the structures (>62% ± 4% in controls, and 100% in treated) were some form of lobule, with a small percentage (10% ± 2%) of terminal end buds (TEBs) found only in control females.

Results

Estrone treatment of female Noble rats induced a significant increase in total proliferation of mammary gland epithelial cells as compared with controls

(8.8-fold, $p < 0.05$) (Table 1). Luteolin treatment inhibited estrone-induced total cell proliferation by 92% ($p < 0.01$), but had little effect on control glands (control, 51 ± 30 nuclei/mm^2 vs. luteolin, 61 ± 35). Estrone treatment also increased the absolute fraction of proliferating nuclei (> 1.5-fold), and luteolin cotreatment reduced the estrone effect to control levels (estrone + luteolin, 9.4 ± 3.9% vs. control, 13.4 ± 5.2 %).

Table 1. Effect of Luteolin on Estrone-Induced Mammary Gland Cell Proliferation.

	Nuclei/mm^2	S%	S/mm^2	G_1%	G1/mm^2
Control	51 ± 30	0.25 ± 0.12	0.98 ± 0.37	12.25 ± 5.77	50.9 ± 29.85
Estrone	448 ± 127	2.19 ± 0.61	46.72 ± 18.92	20.3 ± 2.77	441.7 ± 164.7
Luteolin	61 ± 35	0.26 ± 0.21	1.02 ± 0.68	13.22 ± 5.25	60.75 ± 35.9
Estrone + Luteolin	190 ± 94	0.67 ± 0.11	12.62 ± 6.24	10.3 ± 2.95	224.4 ± 89.41

Estrone treatment increased the S-phase fraction almost 9.0-fold, whereas the fraction of nuclei in G_1 phase increased about 2.0-fold. Also, estrone treatment increased the total number of S-phase nuclei/mm^2 (48.0-fold), and G_1-phase nuclei/mm^2 (8.8-fold), as compared with controls. Luteolin intervention led to a decrease in the percent of S-phase fraction (3.7-fold) as compared with estrone treatment, whereas G_1-phase fraction was reduced 2.0-fold. G_1-phase nuclei/mm^2 decreased 2.0-fold with luteolin treatment as compared with estrone treatment. S-phase and G_1-phase fractions, as well as total nuclei/mm^2, in luteolin alone were equivalent with control.

Discussion

Our studies demonstrate for the first time that estrone exposure induces significant alterations in proliferation, as well as overall cell growth and differentiation in the mammary gland of female Noble rats. Luteolin blocked the effects of estrone on cell proliferation.

Cell division is necessary for conversion of modifications in DNA or other single-stranded DNA damage to gaps or mutations. Cell division also allows for mitotic recombination (e.g., nondisjunction, gene conversion), which results in more profound changes than those of a single mutation (11). Our findings suggest that the probable window of initiation may be between days 7 and 14, as defined by the period between the rise in proliferation rate, day 11, but before day 14, when differentiation has removed the majority of the ductolobular epithelial cell mass from the pool of susceptible tissue (unpublished data). The ductolobular terminal ducts have been postulated as the site of tumor initiation in human mammary cancer (10, 11). The undifferentiated equivalent structures in the rat (TEB and TD) have been likewise implicated as the site of cancer initiation (10). As these structures differentiate into lobules, they may be refractile to carcinogen and spontaneous initiation. Given the known cellular targets of initiation and the observed timing of the process, the elucidation of cytogenetic factors relevant to the human pathology may now be readily addressed in this estrogen-induced mammary cancer model.

References

1. Pike M, Spicer D (1992) Endogenous estrogen and progesterone as the major determinants of breast cancer risk: Prospects for control by natural and technological means. In Li JJ, Nandi S, Li SA (eds): Hormonal Carcinogenesis. New York: Springer-Verlag, pp 209-216.
2. Poliner F (1993) A holistic approach to breast cancer research. Environ Health Perspect 101:116-120.
3. Marshall E (1993) Search for a Killer: Focus shifts from fat to hormones. Science 259:518-520.
4. IARC (1979) Monographs on the Evaluation of the Carcinogenic Risk of Chemicals to Humans: Sex hormones. IARC Monographs 21:173-221.
5. Cutts JH, Noble RL (1964) Estrone-induced mammary tumors in the rat. I. Induction and behaviour of tumors. Cancer Res 24:1116-1123.
6. Cutts JH (1964) Estrone-induced mammary tumors in the rat. II. Effects of alterations in the hormonal environment on tumor induction, behavior, and growth. Cancer Res 24:1124-1130.
7. Garcia R, Coltrera M, Gown A (1989) Analysis of proliferative grade using anti-PCNA/cyclin monoclonal antibodies in fixed, embedded tissues: Comparison with flow cytometric analysis. Am J Path 134:733-739.
8. Hall PA, Levison DA, Woods AL, et al. (1990) Proliferating cell nuclear antigen (PCNA) immunolocolization in paraffin sections: An index of cell proliferation with evidence of deregulated expression in some neoplasms. J Path 162:285-294.
9. Foley J, Ton T, Maronpot R, et al. (1994) Proliferating Cell Nuclear Antigen (PCNA): Comparison to tritiated thymidine (^{3}H-TdR) as a marker of proliferating hepatocytes in rats. Environ Health Perspect, in press.
10. Russo J, Russo IH (1987) Biological and molecular basis of mammary carcinogenesis. Lab Invest 57:112-137.
11. Martin S, Pike M, Ross R, et al. (1990) Increased cell division as a cause of human cancer. Cancer Res 50:7415-7421.

Sex Hormone-Induced Prostatic Carcinogenesis in Noble Rats Involves Genetic Damage and Cell Proliferation

Shuk-mei Ho

Summary

Noble (NBL) rats are uniquely susceptible to sex hormone-induced carcinogenesis of the dorsolateral prostate (DLP) (1, 2). We have previously demonstrated the induction of dysplasia, an early neoplastic lesion, selectively in the DLPs of all NBL rats treated with testosterone (T), 17β-estradiol (E_2), or both (T + E_2), for 16 weeks (3, 4). In this study, rats were treated with T, E_2, or T + E_2, while controls were left untreated for 16 weeks. Using a modified alkaline elution assay procedure (5), we detected augmented frequency of nuclear DNA strand breaks in the DLPs of T + E_2-treated rats, while those found in the DLPs of T-treated and E_2-treated animals remained unchanged from untreated control values. Dramatic increases in total cell number (200%), cell density (100%), and wet weight (65%) were also observed in the DLPs of T + E_2-treated animals. Treatment of rats with T and E_2, either separately or jointly, did not increase nuclear DNA strand breakage in rat VPs. A mild growth response was observed in the DLPs of T-treated animals, while E_2-treatment caused atrophy in both prostatic lobes. These results indicate the dual hormone treatment exerts tissue-specific genotoxicity and mitogenicity in rat DLP, which may exert early oncogenic action in this prostatic lobe.

Introduction

We and others have established that Noble (NBL) rats are uniquely susceptible to sex hormone-induced carcinogenesis of the dorsolateral prostate (DLP). Long-term (over 52 weeks) treatment of NBL rats with T and an estrogen induces a high incidence of prostatic adenocarcinomas selectively in the dorsolateral lobe (1-4). As in prostatic carcinoma of the human gland (6), development of carcinoma in rat DLP is preceded by the appearance of an intraepithelial lesion, termed dysplasia, which is considered the putative predecessor of the carcinoma (3, 4). Following 16 weeks of $T + E_2$ treatment of NBL rats, a 100% incidence of dysplasia was found in the DLPs, while treatment of rats for the same length of time with T or E_2 alone did not produce these precancerous lesions in this lobe (3, 4). These findings indicate that dual hormone (androgen plus estrogen) treatment exerts potent oncogenic action in rat DLP. The aim of this study is to uncover the mechanisms by which the combined androgenic-estrogenic action causes tumor development in the rat prostate.

Results

Using a modified alkaline elution assay (5), relative frequencies of single-strand DNA breaks in the ventral prostates (VPs) and the DLPs of rats treated with T, E_2, or $T + E_2$, or left untreated for 16 weeks were estimated (Figure 1). A significantly higher frequency of DNA strand breaks, as indicated by a more negative slope of the elution curve, was observed in the DLPs of $T + E_2$-treated rats when compared with those found in untreated controls or animals treated with T or E_2 alone. Conversely, none of the hormonal treatments tested induced any change in the frequency of DNA strand breakage in rat VP (data not shown).

Direct cell counting was used to determine the number of cells in cell suspensions obtained by complete dissociation of prostatic lobes in a calcium-free buffer. Total number of cells in a prostatic lobe was determined from cell counts in multiple aliquots and the volume of the dissociated cell suspension (Table 1). Cell densities were calculated as the number of cells per gram tissue. $T + E_2$ treatment of NBL rats for 16 weeks increased the total cell number in the DLP by 200% and the cell density of the lobe by 100%. In contrast, T treatment of rats for the same period of time altered neither total cell number nor cell density of the DLPs, while E_2 treatment reduced total cell number. In the VPs, neither $T + E_2$ nor T treatments changed total cell number and cell density from the values found in untreated controls, while E_2 did induce reduction in total cell number. DLP wet weight showed a 63% increase following $T + E_2$ treatment, a 30% increase following T treatment, and a 57% decrease after E_2 treatment for 16 weeks. VP wet weight exhibited only a 14% increase following $T + E_2$ treatment, no change after T treatment, and a 67% reduction after E_2 treatment for 16 weeks.

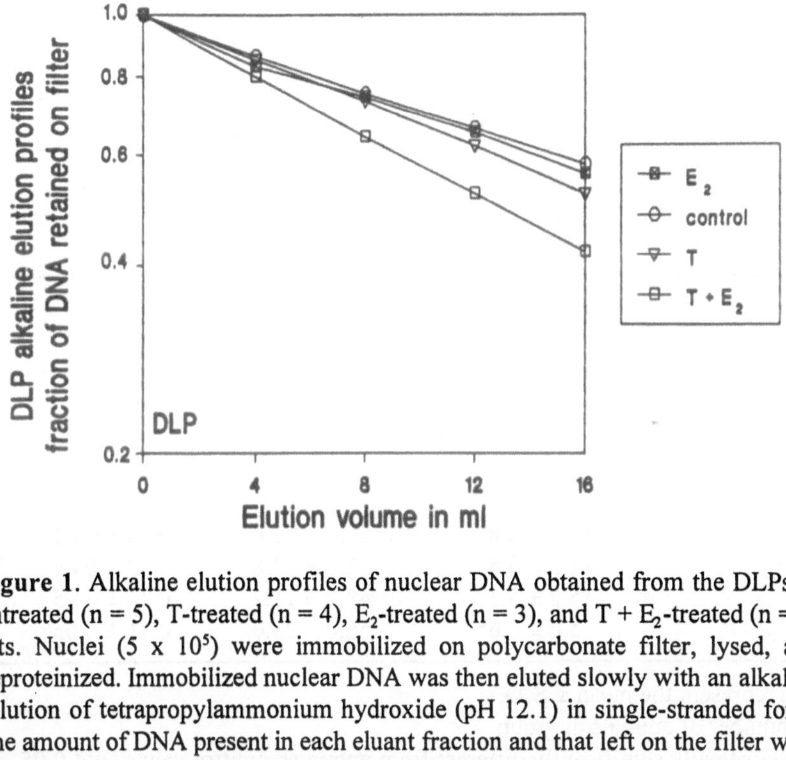

Figure 1. Alkaline elution profiles of nuclear DNA obtained from the DLPs of untreated (n = 5), T-treated (n = 4), E_2-treated (n = 3), and T + E_2-treated (n = 5) rats. Nuclei (5 x 10^5) were immobilized on polycarbonate filter, lysed, and deproteinized. Immobilized nuclear DNA was then eluted slowly with an alkaline solution of tetrapropylammonium hydroxide (pH 12.1) in single-stranded form. The amount of DNA present in each eluant fraction and that left on the filter were determined by a fluorometric assay (5), and the data were used to obtain the % DNA retained in the filter. The percentage of DNA retained in the filter was plotted against the elution volume to obtain an elution curve. DNA breakage reduces the size of the DNA strands and generates elution curves with more negative slopes. Each elution profile represents composite data from elution curves of individual nuclei samples prepared from individual glands. n = number of animals in each group.

Discussion

Data from this study demonstrate that treatment of NBL rats with T + E_2 for 16 weeks induces a profound growth response and an increase in DNA strand breaks selectively in the DLPs, but not in the VPs, of the treated rats. Similar changes were not observed in the DLPs of T- or E_2-treated rats, suggesting that combined androgenic-estrogenic action is needed to induce these changes. We postulate that the combined oncogenic action of T + E_2 in rat DLP is mediated by hormone-induced, combined mitogenicity-genotoxicity in this prostatic tissue. Recent studies suggest that the most oncogenic environment is one with the combined occurrence of genetic damage and persistent cell proliferation (7, 8).

Table 1. Changes in Total Cell Number, Cell Density, and Gland Wet Weight in DLPs and VPs of NBL Rats Treated with T + E_2, T, or E_2 for 16 Weeks.

Treatment	n	Total cell number (million)	Gland Cell Density (million/g)	Gland Wet Weight (g)
DLP				
Untreated	5	14.4 ± 3.6	38.4 ± 3.2	0.36 ± 0.03
T + E_2	5	41.0 ± 4.2*	75.0 ± 15.6*	0.59 ± 0.06*
T	4	22.7 ± 4.8	48.5 ± 5.4	0.47 ± 0.12
E_2	3	7.6 ± 1.8*	50.7 ± 11.0	0.16 ± 0.06*
VP				
Untreated	5	13.0 ± 4.8	36.4 ± 12.0	0.34 ± 0.01
T + E_2	5	11.3 ± 3.8	31.0 ± 2.8	0.39 ± 0.06
T	4	15.2 ± 5.3	43.7 ± 11.2	0.34 ± 0.05
E_2	3	5.6 ± 0.6*	56.0 ± 23.0	0.11 ± 0.05*

Data represent the mean ± S.D.
n = number of animals in group
* = statistically significantly different from untreated control values, $p < 0.05$.

Most premutagenic DNA lesions are transient, since they are readily eliminated by DNA repair and cell death. However, if augmented cell proliferation occurs simultaneously with genetic damage, it dramatically increases the probability of converting primary genetic lesions into mutations. Additionally, sustained cell proliferation induces a state of genetic instability while concurrently creating the promotional environment for clonal establishment and expansion. Epidemiological data indicate a higher ratio of free estrogen to free androgen in the circulation of elderly men (9). It remains to be discovered whether the observed age-related perturbation of sex hormone milieu in men could induce genetic damage and cell proliferation and, therefore, exert oncogenic action in the human gland, as now suggested by our Noble rat model.

Acknowledgments

Supported in part by NIH grants CA-60923 and CA-15776.

References

1. Noble RL (1982) Prostate carcinoma of the Nb rat in relation to hormones. Int Rev Exp Pathol 23:113-159.
2. Drago JR (1984) The induction of Nb rat prostatic carcinomas. Antican Res 4:255-256.
3. Leav I, Ho S-M, Ofner P, et al. (1988) Biochemical alterations in sex hormone-induced hyperplasia and dysplasia of the dorsolateral prostates of Noble rats. J Natl Cancer Inst 80:1045-1053.
4. Ho S-M, Yu M, Leav I, Viccione T (1992) The conjoint actions of androgens and estrogens in the induction of proliferative lesions in the rat prostate. In Li JJ, Nandi S, Li SA (eds): Hormonal Carcinogenesis. New York: Springer-Verlag, pp 18-25.
5. Ho S-M, Roy D (1994) Sex hormone-induced DNA damage and lipid peroxidation in the dorsolateral prostates of Noble rats. Cancer Let, in press.
6. McNea JE, Bostwick DG (1986) Intraductal dysplasia: A premalignant lesion of the prostate. Hum Pathol 17:64-71.
7. Ames BN and Gold LS (1990) Chemical carcinogenesis: Too many rodent carcinogens. Proc Natl Acad Sci USA 87:7772-7776.
8. Preston-Martin S, Pike MC, Ross RK, et al. (1990) Increased cell division as a cause of human cancer. Cancer Res 50:7415-7421.
9. Zumoff B, Levin JJ, Strain GW, et al. (1982) Abnormal levels of plasma hormones in men with prostate cancer: Evidence toward a "two disease" theory. Prostate 3:579-588.

Catechol Estrogen Analogs as Probes of Estrogen Carcinogenesis

Robert W. Brueggemeier, Carl J. Lovely, Nancy E. Gilbert, YunFu Hu, and Young C. Lin

Summary

The roles of estrogens in biochemical events leading to tumor formation remain to be fully elucidated. Induction of renal tumors in the estrogen-treated Syrian hamster offers a unique model to study. Catechol estrogen analogs have recently been prepared to examine the effects of A-ring substituents on estrogen metabolism and on estrogen-induced responses. Methylenedioxy- and 2-hydroxyalkyl-estrogens contain oxygen atoms available for protein interactions (receptor, enzyme); however, these analogs are not susceptible to oxidative reactions, such as redox cycling. 2-Substituted estrogens were recently prepared from 2-formylestradiol (2-CHO-E_2), synthesized via ortholithiation and formylation. Reduction of 2-CHO-E_2 provided 2-hydroxymethylestradiol (2-HME_2). The compounds were examined for their ability to inhibit estrogen metabolism. Several of the catechol estrogen analogs inhibited microsomal estrogen 2-hydroxylase activity, with apparent Ki's ranging from approximately 2 to 40 μM. These synthetic estrogen analogs provide new approaches to probing the relationships of estrogen structure with biological and tumorigenic effects.

Introduction

Estrogens play a pivotal role in approximately 50% of breast cancer cases (1). Understanding the functions of E_2 and its metabolites in estrogen-induced tumorigenesis, as well as in tumor cell growth, will provide additional approaches to prevention and treatment of breast cancer. Induction of renal tumors in estrogen-treated Syrian hamster offers a unique *in-vivo* model for examining the

effects of estrogens in tumorigenesis, while human breast cancer cells in culture provide *in-vitro* models for studying effects of estrogens on tumor growth.

Catechol estrogen analogs were designed to examine the effects of A-ring substituents on estrogen metabolism. Methylenedioxy- and 2-OH-alkyl-estrogens contain oxygen atoms available for protein interactions; however, these analogs are not susceptible to oxidative reactions such as redox cycling. Replacement of the OH group with the bioisosteric amino group provides 2-aminoalkyl analogs for further elaboration of structure-activity relationships. Various 2-substituted catechol estrogens are being examined for their inhibition of 2- and 4-hydroxylases,

Chemistry

The 2,3-methylenedioxyestradiol derivative **1** was prepared by the reaction of 2-hydroxy-estrone with dibromoethane and Adogen 464, followed by reduction to the 17-alcohol. The 3,4-methylenedioxyestrogen analogs were prepared from 4-hydroxyestrone following the same procedure. For 2-substituted estradiol homologs, we required a route that was flexible enough to provide access to all the desired targets. Previously, 2-substituted aminoestrogen derivatives were prepared from the intermediate 2-(N,N-dimethylamino)methylestradiol, formed by reaction of E_2 with dimethylamine and aqueous formaldehyde (2). However, this route was low yielding. A survey of recent literature revealed that formyl estradiol **4** can be prepared via ortholithiation and formylation of a protected estradiol derivative **2** (3). 2-Substituted estrogens were prepared from 2-formylestradiol (2-CHO-E_2, **4**), as shown in Figure 1. Deprotonation of **2** with *s*-BuLi, followed by the addition of freshly distilled DMF, gave **3** exclusively in 86% yield. None of the regioisomeric 4-substituted compound was isolated. Deprotection with dilute HCl gave the desired 2-CHO-E_2. Reduction of 2-CHO-E_2 with LiAlH4 provided 2-hydroxymethylestradiol (2-HME2, **5**), and further synthetic elaborations yielded additional analogs (4).

Biochemistry

The compounds were examined for their ability to inhibit estrogen metabolism. Initial enzyme inhibition studies examined the ability of the synthetic estrogens to block liver microsomal estrogen 2-hydroxylase activity using a specific radiometric assay under initial velocity conditions (5). Additional inhibition studies are performed in cells using an HPLC assay for estrogen metabolites (6). A few of the catechol estrogen analogs competitively inhibited estrogen 2-hydroxylase, with apparent Ki's ranging from 2 to 40 μ M (Table 1).

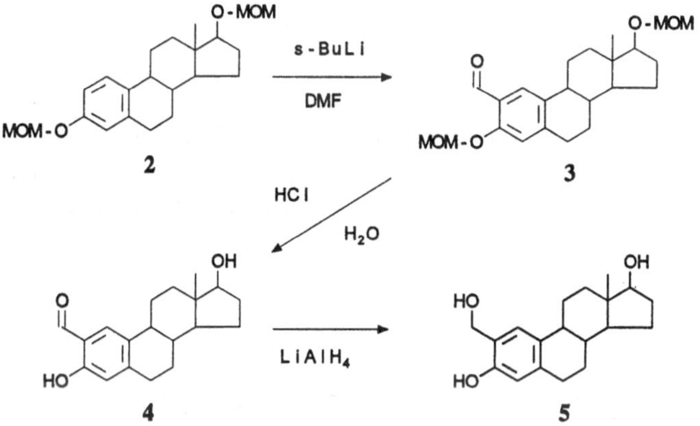

Figure 1

Conclusions

The methylenedioxy- and 2-nitrogen-substituted estrogens are competitive inhibitors of estrogen 2-hydroxylase activity; however, their affinity for the enzyme, as reflected in the apparent Ki values, is lower than the affinity for the substrate E_2 or 2-haloestrogens (5). Larger chemical substitutions at the 2 position resulted in lowered affinities and higher apparent Ki values. A change

in electronic or hydrogen-bonding nature of the substituent, such as replacement with a hydroxyl or amine, had less effect on the inhibitory activity than steric characteristics. In the methylenedioxyestrogen series, the analogs with substituents at the 3,4-position showed greater inhibition of the enzyme than their 2,3-isomers. Finally, inhibition by 2-hydroxymethylestradiol was different from the competitive inhibition exhibited by the other homologs. This compound, 2-HME$_2$, had initially demonstrated mixed inhibition at low concentrations. Further biochemical investigations are necessary to elucidate the nature of the interaction of 2-HME$_2$ with the enzyme system and the type of inhibition. In summary, these synthetic cathechol estrogen homologs provide new approaches for probing the relationships between estrogen structure and biological and tumorigenic effects.

Table 1. Inhibition of Microsomal Estrogen 2-hydroxylase Activity by Various Catechol Estrogen Homologs. Apparent Km for E_2 ranged from 2.9 to 5.6 μM.

Compound	Apparent Ki	S.E.	Type of Inhibition
2,3-methylenedioxyestradiol	20.2 μM	3.1	competitive
2,3-methylenedioxyestrone	21.3 μM	3.7	competitive
3,4-methylenedioxyestradiol	8.0 μM	1.6	competitive
3,4-methylenedioxyestrone	9.3 μM	2.0	competitive
2-(dimethylamino)methylestradiol	29.2 μM	4.5	competitive
2-cyanomethylestradiol	16.8 μM	4.2	competitive
2-aminoethylestradiol	25.7 μM	10.0	competitive
2-hydroxymethylestradiol	35.8 μM	6.8	mixed

Acknowledgments

This research is supported by NIH grants R01-CA58003 (RWB) and P30 CA16058 (CCC).

References

1. Harris JR, Lippman ME, Veronesi U, Willett W (1992) Breast cancer. NEJ Med 327:319-328, 390-398, 473-480.
2. Stevens JM (1988) PhD Dissertation, The Ohio State University.
3. Pert DJ, Ridley DD (1989) Formylation of oestrogens. Aust J Chem 42:405-419.
4. Lovely CJ, Brueggemeier RW (1994) Synthesis of 2-substituted hydroxyalkyl and aminoalkyl estradiols. Tetrahedron Lett 35:8735-8738.
5. Brueggemeier RW, Kimball JG (1983) Kinetics of inhibition of estrogen 2-hydroxylases by haloestrogens. Steroids 42:93-103.
6. Brueggemeier RW, Tseng K, Katlic NE, et al. (1990) Estrogen metabolism in primary kidney cell cultures from Syrian hamsters. J Steroid Biochem 36:325-331.

In-Vitro Metabolites of Coumestrol

Sabine Kulling and Manfred Metzler

Summary

Coumestrol (COUM) is a phytoestrogen found in many plants. However, very little is known about its metabolism and toxicity. We now report that COUM is a substrate for the cytochrome P-450-dependent monooxigenase (?) system. When COUM was incubated with liver microsomes from differently pretreated rats, at least five metabolites were formed. The major COUM metabolite (1) has been characterized by GC/MS. The mass spectrum of the trimethylsilyl derivative indicated a monohydroxylated COUM. Incubation of 1 with catechol-O-methyltransferase/S-adenosyl-L-methionine, followed by GC/MS analysis showed that 1 had a catechol structure. The effects of the different pretreatments on the pattern of oxidative COUM metabolites were studied. 3-Methylcholanthrene (MC)-induced microsomes yielded three times more 1 than control microsomes, whereas no significant differences in the pattern of COUM metabolites were noted among phenobarbital-, isosafrol-, and non-induced microsomes. This suggests that cytochrome P-450 isoenzymes inducible by MC, probably P-450 1A1 and/or 2A1, are involved in the formation of 1.

Introduction

Coumestrol (7,12-dihydroxycoumestan, COUM) is a phytoestrogen. The highest levels in edible plants were found in sprouts of soy beans (70 mg/g dw)(1). The weak estrogenic activity of COUM is well known, and COUM has been suggested as an anti-estrogen and chemopreventive agent (2). Although its chemical structure closely resembles that of diethylstilbestrol, little is known about the metabolism and toxicology of COUM. It is known that DES and 17β-estradiol undergo biotransformation to catechol estrogens (3). Quinones, semiquinones, and activated oxygen species generated from catechol estrogens have been implicated in estrogen-mediated carcinogenesis (4). Studies by Li, et al. (5), provide indirect evidence for the formation of a COUM catechol in hamster liver and kidney microsomal preparations. We have now obtained chromatographic and mass spectrographic evidence for the *in-vitro* formation of catechol metabolites of COUM.

Results

Spectral Interaction of COUM with Cytochrome P-450

Addition of COUM to liver microsomes from male rats pretreated with 3-methylcholanthrene (MC) induced a reverse type I spectrum with an absorption maximum at 418 nm and a minimum at 381 nm (Figure 1). It is known in the field of drug metabolism that compounds leading to reverse type I spectra are well metabolized (e.g., alcohols).

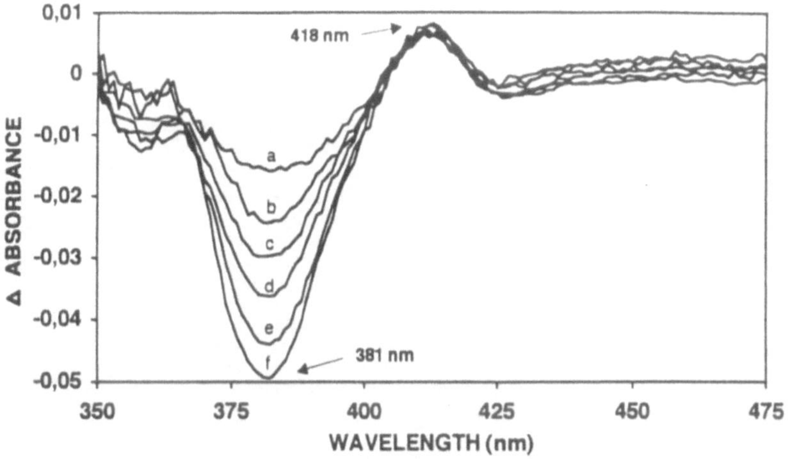

Figure 1. Reverse type spectral changes evokedby addition of different amounts of coumestrol: (a) 3.1 μM, (b) 6.2 μM,(c) 9.4 μM, (d) 12.6 μM, (e) 15.7 μM, (f) 18.8 μM

Metabolism of Coumestrol by Rat Liver Microsomes

COUM was incubated with hepatic microsomes from male rats pretreated with MC, phenobarbital (PB), and isosafrol (IS). The extracted metabolites were separated by HPLC and further characterized by UV and GC/MS after trimethylsilylation. At least five metabolites of COUM were clearly detectable by HPLC (Figure 2). The mass spectrum of the major metabolite 1 indicated the structure of a monohydroxylated COUM (Figure 2). When the HPLC fraction containing metabolite 1 was incubated with catechol-O-methyltransferase/ S-adenosyl-L-methionine, and subsequently trimethylsilylated and analyzed by GC/MS, two products were found which had mass spectra indicative of methoxy-COUM.

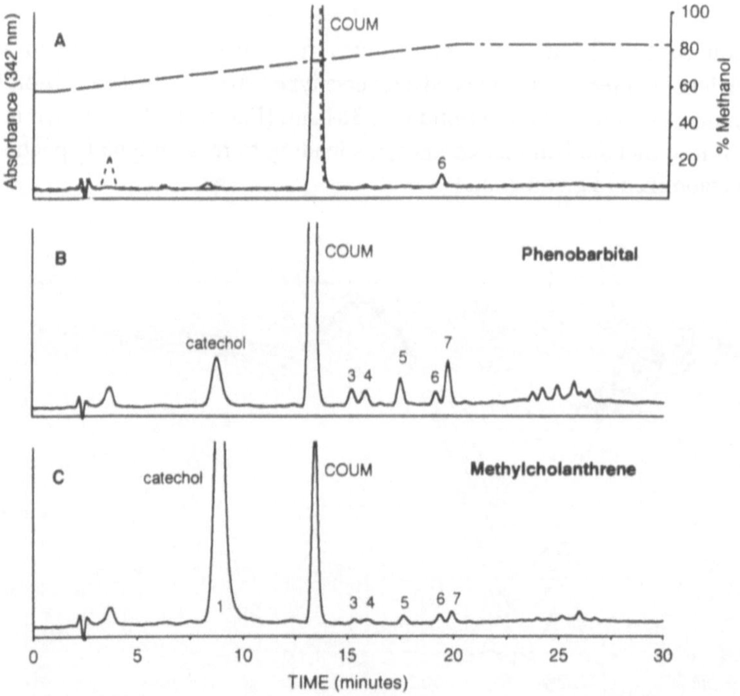

Figure 2. HPLC chromatograms of COUM metabolites formed in incubations with differently-induced rat liver microsomes. (A) Control incubations with boiled microsomes (broken lines) and intact microsomes omitting NADPH (unbroken lines), (B) Incubation with PB-induced microsomes, (C) Incubation with 3-MC-induced microsomes.

This provided strong evidence for a catechol structure of COUM metabolite 1. These two methoxy-derivatives may arise from one catechol metabolite of COUM by methylation of either hydroxy group, or from different catechols. Because microsomes from MC-pretreated rats yield three times more metabolite

1 than the other microsomal preparations (Figure 3), it is proposed that catechol formation of COUM is mediated by the MC-inducible isoenzymes P-450 1A1 and/or P-450 2A1. The structures of metabolites 3, 4, 5, and 7 are as yet unknown. However, the amounts of metabolites 5 and 7 increase about 2.0-fold in microsomal incubations when cumene hydroperoxide is used instead of NADPH, suggesting that these metabolites arise through peroxidative metabolism of COUM.

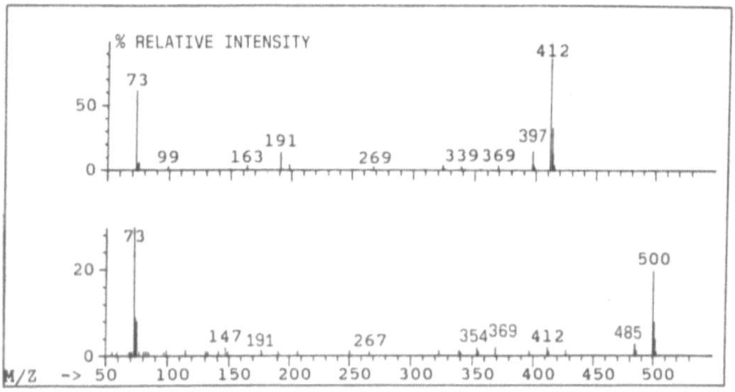

Figure 3. Electron impact mass spectra (70 eV) of the trimethylsilylated COUM and COUM metabolite 1 (bottom)

Conclusion

This study provided direct chromatographic and mass spectrometric evidence that a catechol of COUM is the major metabolite formed in rat liver microsomal preparations. As COUM has recently been shown to induce micronuclei in cultured cells (5), it may be important to study the role of this catechol and other metabolites to understand the mechanism of COUM toxicity.

Acknowledgments

Supported by the Deutsche Forschungsgemeinschaft (Grant Me 574/9-1). Coumestrol was kindly provided by Dr. Risto Santti, University of Turku, Finland.

References

1. Knuckles BE, de Fremery D, Kohler GO (1976) Coumestrol content of fractions obtained from wet-processing of alfalfa. J Agricul Food Chem 24:1177-1180.

3. Metzler M (1984) Metabolism of stilbene estrogens and steroidal estrogens in relation to carcinogenicity. Arch Toxicol 55:104-109.
4. Li JJ, Li SA, Klicka JK, Heller JA (1985) Some biological and toxicological studies of various estrogen mycotoxins and phytoestrogens. In McLachlan JA (ed): Estrogens in the Environment. II. Influence on Development. New York: Elsevier, pp 168-181.
5. Liehr JG (1990) Genotoxic effects of estrogens. Mutation Res 238:269-276.
6. Schuler M, Huber K, Zankl H, Metzler M (1995) Induction of micronucleation,spindle disturbances and mitotic arrest in human chorionic villi cells by 17β-estradiol, diethylstilbestrol and coumestrol. In Li JJ, Li SA, Nandi S, Gustafsson J-Å, Sekely L. (eds): Hormonal Carcinogenesis: Proceedings of the Second International Symposium. New York: Springer-Verlag, pp 458-462 (this volume).

Index